Chemical Pathology and the Sick Child

Chemical Pathology and the Sick Child

EDITED BY

BARBARA E. CLAYTON

CBE, PhD, MD, FRCPath, FRCP
Professor of Chemical Pathology and Human Metabolism
University of Southampton

AND

JOAN M. ROUND

PhD, FIMLS
Principal Biochemist
The Charles Dent Metabolic Unit
University College Hospital, London

Blackwell Scientific Publications

OXFORD LONDON
EDINBURGH BOSTON MELBOURNE

Editorial offices:
Osney Mead, Oxford OX2 0EL
8 John Street, London WC1N 2ES
9 Forrest Road, Edinburgh EH1 2QH
52 Beacon Street, Boston
Massachusetts 02108, USA
99 Barry Street, Carlton
Victoria 3053, Australia

First published 1984

Printed and bound in Great Britain at the University Press, Cambridge

DISTRIBUTORS

USA
Blackwell Mosby Book Distributors
11830 Westline Industrial Drive
St Louis, Missouri 63141

Canada
Blackwell Mosby Book Distributors
120 Melford Drive, Scarborough
Ontario, M1B 2X4

Australia
Blackwell Scientific Book Distributors
31 Advantage Road, Highett
Victoria 3190

First published 1984

British Library
Cataloguing in Publication Data

Clayton, Barbara E.
Chemical pathology and the sick child.
1. Children–Diseases–Diagnosis
2. Chemistry, Clinical
I. Title II. Round, Joan M.
618.9'200756 RJ50

ISBN 0-632-00889-X

Contents

Contributors

T. MARTIN BARRATT FRCP
Professor of Paediatric Nephrology, Institute of Child Health, Guilford Street, London WC1N 1EH.

NEVILLE R. BELTON BSc, PhD, CChem, MRSC
Senior Lecturer, Department of Child Life and Health, University of Edinburgh; and Honorary Biochemist Royal Hospital for Sick Children, Department of Child Life and Health, Royal Hospital for Sick Children, Edinburgh EH9 1LF.

BARBARA E. CLAYTON CBE, PhD, MD, FRCPath, FRCP
Professor of Chemical Pathology and Human Metabolism, University of Southampton, Southampton General Hospital, Tremona Road, Southampton SO9 4XY.

ALAN W. CRAFT MD, MRCP
Consultant Paediatrician, The Royal Victoria Infirmary, Queen Victoria Road, Newcastle-Upon-Tyne NE1 4LP.

VICTOR DUBOWITZ BSc, PhD, MD, FRCP, DCH
Professor of Paediatrics, Institute of Child Health, Hammersmith Hospital, Ducane Road, London W12 0HS.

JAMES W. FARQUHAR MD, FRCPE
Professor of Child Life and Health, University of Edinburgh; Consultant Paediatrician to the Simpson Memorial Maternity Pavilion, Royal Infirmary, Edinburgh and to the Department of Child Life and Health, The Royal Hospital for Sick Children, Edinburgh EH9 1LF.

JOSEPH M. GERTNER MA, MB, BChir, MRCP
Assistant Professor of Pediatrics, Yale University School of Medicine, Section of Pediatric Endocrinology, Yale University School of Medicine, 333 Cedar Street, New Haven, CT 06510, USA.

DAVID B. GRANT MD, FRCP, DCH
Consultant Paediatric Endocrinologist, The Hospital for Sick Children, Great Ormond Street, London WC1N 3JH.

JOSEPHINE E. HAMMOND MBBS, FRCP (Ed), DCH, DObst, RCOG
Consultant Paediatrician, Queen Mary's Hospital for Children, Carshalton, Surrey SM5 4NR.

JOHN Z. HECKMATT MB, ChB, MRCP
Lecturer in Paediatric Neurology, Department of Paediatrics and Neonatal Medicine, Institute of Child Health, Hammersmith Hospital, Ducane Road, London W12 0HS.

JOHN R. HOBBS BSc, MD, FRCP, FRCPath, DObst, RCOG
Professor of Chemical Pathology, Westminster Hospital Medical School, 17 Horseferry Road, London SW1 P2AR.

JAMES V. LEONARD PhD, FRCP
Senior Lecturer in Child Health, The Institute of Child Health and Honorary Consultant Physician, The Hospital for Sick Children, Great Ormond Street, London WC1N 3JH.

JUNE K. LLOYD MD, FRCP
Professor of Child Health, St Georges Hospital Medical School, Cranmer Terrace, London SW17 0RE.

DAVID P. R. MULLER PhD
Senior Lecturer in Biochemistry, The Institute of Child Health, Guilford Street, London WC1N 1EH.

JON PRITCHARD MRCP
Senior Lecturer in Paediatric Oncology, The Institute of Child Health, Guilford Street, London WC1N 1EH.

ALEXANDER T. PROUDFOOT BSc, MB, ChB, FRCP
Consultant Physician, Regional Poisons Treatment Centre, and Director of the Scottish Poisons Information Bureau, Regional Poisoning Treatment Centre, The Royal Infirmary, Edinburgh EH3 9YW.

PAMELA G. RICHES PhD, MRCPath
Senior Lecturer in Immunochemistry, Westminster Hospital Medical School, 17 Horseferry Road, London SW1 P2AR.

JOAN M. ROUND BSc, PhD, FIMLS
Principal Biochemist, Charles Dent Metabolic Unit, University College Hospital, Gower Street, London WC1N 6AU.

JAN STERN PhD, MRCPath
Clinical Biochemist, Queen Marys Hospital for Children, Carshalton, Surrey SM5 4NR.

PAUL H. SCOTT BSc
Senior Biochemist, Chemical Pathology Department, Selly Oak Hospital, Birmingham B29 6JD.

JAMES M. TANNER MD, DSc, FRCP, FRCPsych
Professor of Child Health and Growth, Institute of Child Health, 30 Guilford Street, London WC1N 1EH.

MICHAEL J. TARLOW *MB, BS, MSc, FRCP*
Senior Lecturer in Paediatrics University of Birmingham, East Birmingham Hospital, Bordesley Green, Birmingham B9 5ST.

JOHN H. TRIPP *BSc, MD, MRCP*
Consultant Paediatrician, Royal Devon and Exeter Hospital, Wonford, Barrack Road, Exeter EX2 5DW.

HANK K. A. VISSER *MD*
Professor of Paediatrics and Chairman, Erasmus University and University Hospital and Sophia Childrens Hospital, Gordelweg 160, 3038 Rotterdam, The Netherlands.

BRIAN A. WHARTON *MD, FRCP, FRCPE, DCH*
Consultant Paediatrician, Birmingham Area Health Authority, and Honorary Research Fellow in Child Nutrition, Institute of Child Health, Birmingham; and Special Care Baby Unit, Sorrento Maternity Hospital, Wake Green Road, Birmingham B13 9HE.

PAUL D. WHITEMAN *MD, BS, MSc*
Clinical Pharmacologist, Wellcome Research Laboratories, Beckenham, Kent, and Visiting Lecturer, Chemical Pathology and Human Metabolism, University of Southampton, Southampton General Hospital, Tremona Road, Southampton SO9 4XY.

Preface

We have felt for some time that there was a need for a chemical pathology book which was concerned entirely with the problems of infants and children. Although we realise that there are specialist text books devoted to disorders of particular systems, they may not readily be available to staff who are working outside major centres. It is particularly for such staff that this book has been written. We hope that it will be of help to general paeditricians and all those who work in chemical pathology departments. We have endevoured through our contributors to discuss the more common disorders which occur in infancy and childhood but have been deliberately selective regarding conditions which are rarely seen.

We have been very fortunate in our contributors, all of whom are experts in their own fields. To them we extend our grateful thanks. We should also like to thank the many colleagues who have helped us in numerous ways in completing this volume, and particularly Mrs B. Lloyd, Mrs V. Patel, Miss C. Sanders, Miss L. Snow and Miss J. Welch. We also thank Mrs J. Ross who so expertly compiled our index.

B. E. Clayton
J. M. Round

Chapter 1
Introduction

JOAN M. ROUND

In a childrens hospital the chemical pathology service, which is essential in the diagnosis and treatment of the sick child, is geared to the use of micro blood samples. For the sick infant and neonate the provision of a service based on micro quantities of blood is essential, since the removal of several macro samples may considerably reduce the blood volume in a small baby. In the past the use of micro blood samples has presented problems to chemical pathology departments in general hospitals which are serving a large adult population and have only one or two paediatric wards. With the advent of modern multichannel analyser systems and radioimmunoassay techniques it is now possible to make a wide range of biochemical investigations using very small quantities of plasma. These factors have considerably improved the service given to paediatric departments in most general hospitals.

Factors affecting the quality of biochemical results

The collection of adequate samples of capillary blood from a sick infant requires practice and should only be undertaken by staff specially trained in heel or finger prick techniques, who should preferably be medical laboratory scientific officers, biochemists or nurses based in the laboratory. The site of puncture should be warmed before sampling in order to increase the arterial blood flow and the efficiency of a heel prick can be greatly increased by using an 'Autolet'* pricking device. There is disagreement over the suitability of capillary blood for a few biochemical tests, in particular Po_2 for which many people consider an arterial sample should be collected.

The two major sources of error attributable to sample collection by heel prick are haemolysis and contamination by tissue fluid. Although these are minimised by good technique some specimens will inevitably be affected. The effect of haemolysis on biochemical measurements is inconsistent and although some elevation in plasma potassium, magnesium and phosphate is predictable its magnitude may vary from one analyte to another. An effective dilution of other plasma constituents also accompanies haemolysis, so that depressed levels of sodium and chloride result. This error is compounded by any contamination with tissue fluid as this has compara-

* Autolet, from Mumford Medical Division, Brook Hill, Woodstock, Oxon.

tively low concentrations of potassium, calcium and protein and so will also tend to lower plasma values.

The erythrocyte is a rich source of amino acids and enzymes. Haemolysis may therefore affect either the measured plasma enzyme activity, as in the case of lactate dehydrogenase, or a variety of other constituents which are substrates for the liberated enzymes. Examples include serum insulin, which is cleaved by an enzyme located in the red cell membrane, and carbonic anhydrase which will affect the HCO_3^-/H_2CO_3 equilibrium and so alter the pH of capillary blood. Further details regarding the collection of capillary blood can be found in *Paediatric Chemical Pathology* (Clayton *et al.* 1980). For older children venepuncture may be the preferred technique and if a toddler requires a whole range of investigations on a single occasion, venepuncture by an experienced venesector may be the least traumatic method of obtaining the blood.

Whenever possible plasma should be collected from children, as after centrifuging it will yield a larger specimen for analysis than the same volume of clotted blood, although the latter must be collected for some tests, for example protein electrophoresis. Plasma is also less likely to be haemolysed during separation but it is important to select the correct anticoagulant as an incorrect anticoagulant can interfere with, or even invalidate, the results of some tests. For example fluoride is an enzyme poison and anticoagulants containing sodium or potassium are unsuitable for electrolyte measurements. Containers provided by the laboratory for various biochemical tests will contain suitable anticoagulants, preservatives, etc. and if doubt exists as to the correct anticoagulant or test procedure a telephone call to the laboratory will usually resolve the problem and may save much time and energy. Laboratory staff usually know which wards are occupied by children and take extra care with specimens from these locations. The identification of blood specimens from children is aided by the use of paediatric specimen vials which contain the correct amount of anticoagulant for small volumes of blood.

Good liaison between paediatric wards and the laboratory is essential if the best use is to be made of the chemical pathology service. If a venous specimen is collected by a member of the medical staff and has to be transported to the biochemistry department the correct conditions for transport must be observed. Certain investigations must be made as soon as possible after the specimen has been collected, or the specimen may have to be processed quickly to a stage after which it can be stored safely. Heparinised specimens from newborn babies should always be separated with the minimum of delay as their erythrocytes have a greater mechanical fragility than those of older children and adults so that even a badly balanced centrifuge may cause some haemolysis. Specimens should never be left lying about on the ward after collection as this can lead to the passage of certain ions, for example potassium, across the red cell membrane and thus to erroneous plasma chemistry results. Blood or plasma for bilirubin determination should be protected from direct sunlight as this denatures the bilirubin and gives incorrectly low results. Blood pH is stable for about 2 hours if the specimen is immersed in ice water (NOT frozen!). Specimens for PO_2 and PCO_2 will be satisfactory for up to an hour after collection if kept on ice, but should be analysed more quickly if possible.

If a child is to receive a transfusion of either whole blood, lipid or intravenous dextran a pretransfusion blood specimen should always be collected, and all recent transfusions should be reported on the chemical pathology request forms. Tests on red blood cells such as galactose-1-phosphate uridyl transferase activity will be inaccurate until up to 6 weeks after blood transfusion and the lipaemia resulting from a lipid infusion can block automated analyser lines and may also cause falsely low values for non-lipid-soluble plasma constituents. The turbidity caused by lipid and sometimes by dextran can lead to incorrect bilirubin results and in the case of dextran this interference may last for several days.

Urine specimens

There are several types of container available for the collection of urine specimens from infants and young children; The Great Ormond Street pattern bag* which is available in 3 sizes, and the Hollister U-bag† with attached tubing can both be recommended.

It is inadvisable to collect urine for biochemical studies until at least 3 days after an intravenous pyelogram because contrast medium can cause errors, particularly if nephelometry is to be used.

Faecal specimens

For the collection of faecal specimens nappies may be lined with fine polythene sheeting; other materials have been used but polythene is the least likely to produce soreness if repeated collections are necessary.

DRUG INTERFERENCE

Any drugs which are being administered to a child should be recorded on all biochemistry request forms, as drugs can interfere with biochemical results both *in vivo* and *in vitro*. An index on the effects of drugs on biochemical analyses has been published by Young, Pestaner and Gibberman (1975). An up-to-date list of drug interferences entitled *Drug*

---

* The Great Ormond Street bag is obtainable from Down Surgical Ltd., Church Path, Mitcham, Surrey.

† Hollister U-bag. Abbot Laboratories, Ltd., Queenborough, Kent.

Effects in Clinical Chemistry is available on application from Uppsala University, Sweden (see Tryding & Lindblad 1981).

Some drugs are excreted as conjugates with glucuronic acid and may produce positive reducing tests in the urine and in some instances the metabolism of a drug may not be fully characterised and unknown metabolites may be troublesome on urine chromatograms. For example, paracetamol appears on chromatograms for phenolic acids and ampicillin can give a urinary amino acid pattern which simulates that seen in maple syrup urine disease (Smith & Seakins 1976). In breast-fed infants metabolites from drugs administered to a mother may appear in her baby's urine (O'Brien 1974). The excretion of drugs in human breast milk has been considered by Savage (1977), Wilson *et al.* (1980), Welch and Findlay (1981) and in a recent issue of *Drug and Therapeutics Bulletin* (1983).

THE MORIBUND INFANT

When an infant presents with symptoms which suggest that an inborn error of metabolism may be present, it is important to try and make a specific diagnosis as the condition may be treatable and the family will require a prognosis and genetic counselling.

Definitive diagnosis is likely to require the biochemical expertise available only at a specialised centre but to aid diagnosis, urine and plasma should always be collected from the sick infant, preferably during the acute phase. Preliminary biochemical tests should include plasma electrolytes, acid base studies, calcium, bilirubin and glucose. Further plasma and urine specimens should be stored frozen. If death appears imminent skin for fibroblast culture should be removed. The nearest specialist centre will give advice on culture medium for fibroblasts and on the correct methods of specimen storage. Reduced or absent activity of a specific enzyme in the cultured fibroblasts from a deceased infant can be a vital diagnostic finding which will allow subsequent genetic counselling to the parents and perhaps the possibility of prenatal diagnosis in a subsequent pregnancy. The diagnosis of inherited metabolic diseases in severely ill children has been reviewed in detail by Holton (1982) on behalf of the Association of Clinical Biochemists working party for Clinical Laboratory Investigation.

BIOCHEMICAL TESTS OUTSIDE THE MAIN LABORATORY

During the last few years there has been a considerable increase in the use of small instruments designed for the rapid monitoring of certain biochemical parameters at the patients' bedside or in a ward side-room. While these instruments may be a valuable aid to the often sorely taxed emergency pathology services, they must be used with discretion and ideally in conjunction with the staff of the main chemical pathology department, who should cooperate with the ward staff over quality assessment and instrument maintainance. This subject was reviewed as a contemporary theme for the *British Medical Journal* by Watson (1980) and more recently guidelines on the performance of chemical pathology tests outside the laboratory have been drawn up by McMillan and Lowe for the Royal College of Pathologists, Association of Clinical Pathologists and the Association of Clinical Biochemists (Anderson *et al.* 1981).

Recommended procedures are given which should help to reduce discrepancies in the performance of such tests, while assuring the reliability and quality of assays performed away from the routine service laboratory. The necessity for full cooperation between heads of routine laboratories and those responsible for ward assays regarding the selection and purchase of equipment used in wards and side-rooms is emphasised.

Each institution must develop a system for the supervision of both quality assessment and the health and safety aspects of 'bed-side' testing in which the Quality Assessment and Health and Safety officers from main pathology departments are fully involved. To paraphrase Drs McMillan and Lowe, 'If kits and instruments for performing biochemical tests away from the laboratory are to give results of the high quality necessary for good patient care, then it is clearly necessary for there to be not only full cooperation, but also a continual flow of information between clinical staff and the staff of main chemical pathology departments'. This is surely true not only for ward testing but also for paediatricians and other clinicians in order that they should all make the best use of the chemical pathology services provided by their particular hospital or institution.

REFERENCES

ANDERSON J. R., LINSELL W. D. & MITCHELL F. M. (1981) Guidelines on the performance of chemical pathology assays outside the laboratory. (letter) *Br. Med. J.* **282**, 743.

CLAYTON B. E., JENKINS P. & ROUND J. R. (1980) *Paediatric Chemical Pathology: Clinical Tests and Reference Ranges.* Blackwell Scientific Publications, Oxford.

Drug and therapeutics bulletin (1983) Drugs which can be given to nursing mothers. **21** (2), 5.

HOLTON J. B. (1982) Diagnosis of inherited metabolic diseases in severely ill children. *Ann. Clin. Biochem.* **19**, 389–95.

O'BRIEN T. E. (1974) The excretion of drugs in human milk. *Am. J. Hosp. Pharm.* **32**, 14.

SAVAGE R. L. (1977) Drugs and breast milk. *J. Hum. Nutr.* **31**, 459–66.

SMITH I. & SEAKINS J. W. T. (Eds.) (1976) *Chromatographic and electrophoretic techniques.* Heinemann, London.

TRYDING N. & LINDBLAD C. G. (Eds.) (1981) *Drug Effects in Clinical Chemistry*, 2nd edn. Apoteksbolaget, Uppsala.

WATSON D. (1980) Analytical investigations closer to the patient. *Br. Med. J.* **281**, 31–5.

WELCH R. M. & FINDLAY J. W. A. (1981) Excretion of drugs in human breast milk. *Drug Metab. Rev.* **12**, 261–77.

WILSON J. T., BROWN R. D., CHEREK D. R., DAILEY J. W., HILMAN B., JOBE P. C., MANNO B. R., MANNO J. E., REDETZKI H. M. & STEWART J. J. (1980) Drug excretion in human breast milk: principles, pharmacokinetics and projected consequences. *Clin. Pharmacokinet.* **5**, 1–66.

YOUNG D. S., PESTANER L. C. & GIBBERMAN V. (1975) Effects of drugs on clinical laboratory tests. *Clin. Chem.* **21**, 1D–432D.

Chapter 2
The Newborn

BRIAN A. WHARTON & PAUL H. SCOTT

This chapter describes the biochemical implications of routine neonatal care and presents a range of biochemical investigations in healthy newborn babies. It then approaches biochemical problems of the newborn in two ways: first, the investigation of those clinical presentations which may be a symptom of, or are accompanied by, biochemical upset, e.g. convulsions, diarrhoea; second the management of, and further investigation of, abnormal biochemical signs, e.g. hyponatraemia, hypoglycaemia. Wherever possible we have tried to use a problem-orientated, rather than an aetiological approach, and to describe our own personal practice. As in many other branches of clinical care, patients often require treatment while investigation is proceeding and, therefore, investigation plus treatment plus further evaluation, i.e. 'management', are often presented together.

We have not discussed in any detail individual biochemical mechanisms, particularly those which occur at all ages. Disorders which may present in the newborn, but are not specifically neonatal, are discussed only in so far as they affect differential diagnosis and emergency management.

CLINICAL PRINCIPLES

Major principles of neonatal care

The main points of neonatal care are to keep the baby warm, provide adequate nutrition, prevent infection, and ensure adequate oxygenation: they all have biochemical implications.

Warmth and prevention of hypothermia

Unfortunately 'primary' hypothermia, i.e. due to an inadequate thermal environment, still occurs in the newborn. Occasionally it results in hypoglycaemia (p. 32) which should be managed in the normal way. If the baby is warmed beneath an overhead radiant heater then attention to water balance is necessary to avoid hypernatraemia (p. 57) and dehydration.

Hypothermia often presents as a symptom of any illness, metabolic or otherwise, in the newborn, but has been a particular feature of babies who are later diagnosed as having 'kinky hair disease', which is an inborn error of copper transport (Menkes *et al.* 1962; Danks *et al.* 1972, 1973). Rarely hypothermia is a presenting feature of congenital hypothyroidism. Severe hypothermia in very low birth-weight babies, perhaps because it is often associated with hypoxia and/or hypotension, is sometimes followed by acute renal failure (p. 47) and disseminated intravascular coagulation.

Provision of adequate nutrition

Breast fed babies should generally be fed shortly after birth and frequently thereafter. If bottle fed, the normal healthy baby may be fed *ad libitum* and there is no need to prescribe specific amounts.

Low birth-weight babies are usually fed according to a schedule during the first 3–4 weeks of life. Our practice is to give babies of appropriate weight for gestational age 40 ml per kg bodyweight of either breast milk or an infant formula during the first 24 hours, increasing by 20 ml per kg each day, to reach 200 ml per kg by day 9. Light-for-gestational-age babies receive 60 ml per kg during the first 24 hours, increasing by 30 ml per kg each day, to reach 200 ml per kg by day 6. A full volume of 200 ml of a standard formula provides 3 g protein and 134 kcal per kg body weight (i.e. 2.25 g protein per 100 kcal). The composition of expressed breast milk is variable but on average 200 ml provides 2.4 g protein (nitrogen × 6.4) and 140 kcal. Many of the normal biochemical results given in Tables 2.3–2.13 are from babies on this regimen. Table 2.1 shows the nutrients present in breast milk and a typical modern demineralised whey infant formula.

Unadapted cows' milk is unsuitable for any child below the age of 6 months. Many 'normal' ranges established some years ago were from babies receiving cows milk and so are unrepresentative. Some modern infant formulae contain more protein than shown in Table 2.1 and contain unmodified cows' milk protein, e.g. up to 1.8 g per 100 ml, 80% as casein. These formulae will result in higher concentrations of urea and amino acids than shown in Tables 2.3–2.13.

We do not use either nasojejunal or parenteral nutrition extensively, but may do so when respiratory support is necessary or gastric stasis occurs. Nasojejunal feeding is sometimes associated with a degree of malabsorption, and continuous pump feeding often results in energy-rich fat being left behind in the syringe. (Dryburgh (1982) has reviewed the latter method of feeding.) There are various well-described regimen for parenteral nutrition; currently we use that of Hughes (1982). Parenteral nutrition is not infrequently associated with glycosuria (insulin may be indicated), hypophosphataemia, and moderate hyperammonaemia. The only permissable carbohydrate is glucose, there should not be a great excess of any one amino acid, and the solution should contain *all* amino acids. 'Older' style parenteral feed regimens (some of which are still used in adults), e.g.

Table 2.1 Typical chemical analyses of mature human milk, a modern demineralised whey infant formula, and cows' milk.

Concentration per litre	Mature human breast milk		Modern demineralised whey formula		Cows' milk	
Energy	700 (Kcal	2900 (kJ)	650 (kcal)	2700 (kJ)	670 (kcal)	2800 (kJ)
Protein	13 (g total)	11 (g amino)	15 (g total)	13 (g amino)	34 (g total)	30 (g amino)
Fat	42 (g)		36 (g)		39 (g)	
Lactose	74 (g)		72 (g)		48 (g)	
Cholesterol	160 (mg)	0.42 (mmol)	—	—	140 (mg)	0.36 (mmol)
	mg	mmol	mg	mmol	mg	mmol
Sodium	150	6.5	150	7	500	22
Potassium	600	15.4	560	14	1500	39
Chloride	430	12.1	370	10	950	27
Calcium	350	8.8	440	11	1200	30
Magnesium	28	1.2	55	2.4	120	5
Phosphorus	150	4.8	330	11	950	31
	µg	µmol	µg	µmol	µg	µmol
Iron	760	13.6	10000	18	500	9
Copper	390	6.1	400	6.2	200	3
Zinc	2950	45.3	3200	49	3500	54
Selenium	14	0.2			Variable	
Iodine	70	0.6	70	0.6	260	Variable
Fluoride	77	4.1			Variable	
	mg	mmol	mg	mmol	mg	mmol
Isoleucine	530	4	840	6	1960	15
Leucine	950	7	1440	11	3110	24
Lysine	700	5	1220	8	2460	17
Methionine	150	1	360	2.4	750	5
Cysteine	200	0.8	230	0.9	250	1
Phenylalanine	370	2.2	620	3.8	1610	10
Tyrosine	300	1.7	460	2.5	1610	9
Threonine	450	3.8	860	7.2	1450	12
Tryptophan	230	1.1	280	1.4	450	2.2
Valine	690	6	940	8	2210	19
	mg	mmol	mg	mmol	mg	mmol
Arginine	380	2.2	460	2.6	1150	6.6
Histidine	240	1.5	350	2.3	850	5.4
Alanine	410	4.6	610	6.8	1150	13
Aspartic acid	880	6.6	1350	10	2410	18
Glutamic acid	1760	12	2870	20	7020	48
Glycine	250	3.3	290	3.9	600	8
Proline	950	3.3	110	6.8	3060	26
Serine	420	4	780	7.4	1810	17
Taurine	48	0.4	—	—	—	—
	µg	IU	µg	IU	µg	IU
Vitamin D	0.1	4	10	400	0.3	12

those including fructose, sorbitol, ethanol, or very high glycine concentrations, are not appropriate in the newborn baby (nor in childhood) as they may lead to a variety of biochemical disorders, e.g. hyperammonaemia, lacticacidosis, hyperuricacidaemia.

Prevention and early diagnosis of infection

An antigenic stimulus *in utero*, e.g. from an intrauterine infection, results in fetal production of IgM. In the cord blood of normal babies IgM is less than 200 mg/l, but frank intrauterine infection is usually associated with very high concentrations, e.g. 700–1500 mg/l. It is doubtful whether cord blood IgM is a useful screen for intrauterine infection. If the diagnosis is suspected clinically, e.g. severe symmetrical growth retardation, cataracts, hepatosplenomegaly and bone lesions, it is probably better to proceed to the specific serological investigations and isolation of the microorganism.

Findings on cerebrospinal fluid are more difficult to interpret in the newborn. Mild bleeding is common during birth and so xanthochromia with protein concentrations as high as 2.4 g/l may be found in normal babies (Naidoo 1968). Its determination is of limited value. Unsaturated transferrin has an antibacterial effect. This mechanism is less effective in newborn, and particularly pre-term babies, because they have a transferrin concentration below the adult normal (mean concentration at birth in term babies 250 mg per 100 ml), which is more saturated with iron. Measures which lead to even greater saturation, e.g. parenteral iron, are therefore best avoided (Scott *et al.* 1975).

A rare biochemical aid is the very high plasma activity of 'cardiac' enzymes, e.g. AST, lactate dehydrogenase, found in viral myocarditis, which is a rare cause of heart failure.

Apart from these specific considerations however, severe infection may result in hypothermia, haemolysis, liver cell dysfunction, bacteraemic shock with hypotension, sick cell syndrome, peripheral circulatory failure, renal failure, and thence a variety of biochemical abnormalities. Treatment with antibiotics may interfere with the Guthrie screen and may produce 'strange' unidentified compounds on urinary chromatograms. Neonatal infection is very common, primary metabolic disorders are not!

Management of hypoxia and cyanosis

Apart from changes in arterial oxygen tension hypoxia results in a plethora of biochemical abnormalities, many of which persist for some time after successful treatment with oxygen and ventilatory support as necessary.

There are three major clinical presentations: ante and intrapartum hypoxia, postnatal hypoxia and suspected heart disease.

Ante and intrapartum hypoxia. The administration of oxygen through an adequate airway with ventilation as necessary, is the keystone of treatment. If, after 5 minutes of adequate therapy, there is no attempt at spontaneous respiration, intravenous sodium bicarbonate about 3 mmol per kg body weight and glucose about 250 mg per kg (e.g. 2.5 ml of 10% dextrose) may be given over 3–5 minutes and ventilation continued. There is rarely time for biochemical determinations at this acute stage.

Evidence of mild-to-moderate renal failure (p. 47) is commonly seen following severe intrapartum hypoxia and a number of babies develop hyperammonaemia (p. 42, and chapter 4) due probably to hepatic ischaemia.

Postnatal hypoxia. This is most commonly due to the respiratory distress syndrome (hyaline membrane disease). In untreated children there is hypoxaemia, hypercapnia and a mixed respiratory and metabolic acidosis. Oxygen therapy and mechanical respiratory support (continuous positive airways pressure or ventilation) again are the keystones of treatment. The aim is to maintain a $P\text{O}_2$ of 8–10 kPa; if this cannot be achieved with an inspired oxygen concentration of less than 60% then mechanical support is indicated. Ideally $P\text{O}_2$ is monitored continuously, either by an intravascular or transcutaneous electrode so that blood analysis is performed only to calibrate the monitor and to determine $P\text{CO}_2$ and acid–base status. Other biochemical measurements, particularly lactic acid, have been explored as a guide to therapy and prognosis, but they have little to offer over the routine measurement of arterial blood gases and acid–base. If there is marked hypercapnia as well, then mechanical treatment for the hypoxaemia often corrects the hypercapnia too, as alveolar ventilation improves. The determination of acid–base status is used more as an extra indicator of adequate respiratory support rather than an indicator for alkali therapy. If severe metabolic acidosis is present, together with hypoxaemia, the *main* treatment of the acidosis is adequate oxygenation. If the metabolic acidosis is not accompanied by hypoxaemia it may be a 'left over' metabolic complication of previously existing hypoxaemia; continued adequate oxygenation should dissipate it but alkali may also be given. In these circumstances, however, the possibility of the acidosis being due to other causes must be considered, e.g. hypothermia, infection.

Babies severely ill with respiratory distress syndrome (RDS) develop hyperkalaemia and ureamia, probably reflecting tissue breakdown and renal failure. Initial management is aimed at the underlying hypoxaemia, not the secondary biochemical signs.

Plasma protein concentrations bear some relationship to prognosis. Low total serum protein in cord blood (below 46 g/l) and a low concentration of plasma α1-antitrypsin (below 1.5 g/l) are associated with a poorer prognosis, but it is not clear to what extent this is merely reflecting immaturity.

Lecithin/sphyngomyelin ratios have been determined in saliva and tracheal aspirates, but more commonly this test is reserved for amniotic fluid. A ratio below 1.5 is associated with an incidence of RDS of 78%; 1.5–2.0 with 35%; > 2 with 1.5% (Harvey & Parkinson 1981).

Suspected heart disease. Despite the marked hypoxaemia (e.g. below 5 kPa) seen in babies with *cyanotic congenital heart disease*, severe metabolic disturbance, such as acidosis, does not occur in all of them, presumably because oxygen saturation is still adequate for tissue respiration and the situation is stable (e.g. in Fallot's tetralogy). Even in the deteriorating neonatal lesions which untreated will go on to death within a few weeks, a palliative procedure to improve oxygenation (e.g. a Rashkind atrial septostomy for transposition of the great arteries, or prostaglandin maintenance of a patent ductus arteriosis for pulmonary atresia) will usually have been performed before severe metabolic deterioration occurs. Metabolic acidosis in a child with cyanotic congenital heart disease generally indicates the need for expert cardiological assessment rather than alkali therapy. In a few, however, metabolic acidosis requiring alkali therapy may develop while awaiting a palliative procedure. If palliative or definitive cardiac procedures have nothing to offer, then, as the baby dies, there will naturally be profound disturbance, but biochemical investigation and management have no place at this stage.

The *hypoplastic left heart syndrome* may present as a shocked looking baby with acidosis and hypoglycaemia.

A blue baby who does not have hypoxaemia may have *methaemoglobinaemia*. This is most commonly secondary to toxins such as certain local anaesthetic agents given to the mother, but may be due to a rare inborn error of metabolism or haemoglobin M. A few drops of blood placed on a glass slide or filter paper look brown and do not become pink when oxygen is blown gently onto them. Precise concentrations of methaemoglobin are determined chemically and in blue infants vary from 10 to 70% (Oski & Naiman 1967).

Clinical presentation of metabolic disorders

Although some presentations will specifically suggest a metabolic disorder (e.g. coma occurring after a few days of life, an unusual fetor) signs are more often non-specific, e.g. poor feeding or an abnormal weight chart. An unexplained previous neonatal death or consanguinity in the parents will raise clinical suspicion. Metabolic disorder must, therefore, be considered in any clinical presentation where the cause is not absolutely clear or when a symptom is unusually prolonged, e.g. hypothermia. In practice if the investigations shown in Table 2.2 performed while the child is *on a normal diet* are normal, it is unlikely that the symptoms are due to primary metabolic disorder.

These investigations are all relatively quick, and straightforward, and

Table 2.2 Preliminary 'screening tests' if a metabolic disorder is suspected.

Clinical
History of previous neonatal deaths
Parental consanguinity
Biochemical
Plasma or blood*
Glucose, calcium, magnesium,
sodium, potassium, urea, acid base measurements,
conjugated bilirubin,
amino acid chromatography
Urine
Reducing substances, glucose,
ketones, bilirubin

* Plasma ammonia if symptoms are suggestive (see p. 42).

just as any ill baby might be quickly screened for infection it is also worthwhile quickly screening for metabolic disorders using these simple investigations.

Screening for hyperaminoacidaemia by chromotography will detect non-specific biochemical signs of many inborn errors (e.g. hyperglycinaemia in the organic acidaemias). Although plasma glutamine may be raised, the hyperammonaemias are often not accompanied by simply determined biochemical signs, so that if suspected (e.g. in a comatose child) plasma ammonia, by no means a routine test, must be determined. If symptoms of unknown cause persist it is worthwhile repeating the simple biochemical screen even if it was originally normal, since some primary metabolic disorders cause clinical symptoms before the secondary biochemical signs are apparent.

Clinical interpretation of biochemical signs

Clearly interpretation is impossible unless the specimen has been taken adequately, stored in a suitable container and transported to the laboratory properly, e.g. specimens of blood taken via intravascular catheters may easily be contaminated (usually diluted) by the solution infused to keep the catheter open. Not all biochemical disturbance indicates a primary metabolic disorder; metabolic acidosis is more commonly due to hypoxia or infection than (say) a renal tubular abnormality; hypoglycaemia may merely reflect inadequate feeding in a susceptible child.

When considering the aetiology of any abnormal biochemical sign it is wise to consider whether it may be explained by either an unsatisfactory specimen or one of the common causes of secondary biochemical disorders such as hypoxia, infection, hypothermia, unsuitable or inadequate enteral or parenteral feeding, or some other iatrogenic cause, e.g. maternal drugs. If not, then a primary metabolic disturbance can be considered.

Reference Ranges

Tables 2.3–2.13 give a variety of reference values found in the neonatal period. Unless otherwise stated they were obtained at the Biochemistry Department, Selly Oak Hospital, Birmingham, from babies in the Sorrento Maternity Hospital, Birmingham. Babies were either of low birth-weight, 1.7–2.2 kg and fed on Gold Cap SMA-S26 (a demineralised whey formula; Wyeth Laboratories, Berkshire) or were babies weighing over 2.2 kg at birth and fed either breast milk or a low solute formula. Some babies exhibited varying degrees of physiological jaundice but were otherwise well.

To facilitate reference, each measurement has a location number which can be obtained from the alphabetical list at the beginning of the Tables.

Tables 2.3–2.13 Reference ranges during the neonatal period. Alphabetical list of measurements with location numbers and the Tables in which they appear.

Measurement	Location no.	Table
Acid–base		
pH	9	2.3
base excess	10	2.3
$P\text{CO}_2$	11	2.3
Alanine aminotransferase	29	2.6
Albumin	14	2.5
Alkaline phosphatase	22	2.6
Alkaline ribonuclease	30	2.6
Amino acids (plasma & urine)	69	2.13
α-Amino nitrogen	55	2.9
Ammonia	39	2.7
α1-Antitrypsin	18	2.5
Aspartate aminotransferase	28	2.6
Base excess	10	2.3
Calcium		
plasma	3	2.3
urine	46	2.8
Carbon dioxide ($P\text{CO}_2$)	11	2.3
Ceruloplasmin	21	2.5
Chloride		
plasma	5	2.3
urine	47	2.8
Cholesterol	34	2.7
Copper	42	2.7
Cortisol	68	2.12
Creatine kinase	25	2.6
Creatinine		
plasma	8	2.3
urine	50	2.8
Creatinine nitrogen	52	2.9
Creatinine clearance	61	2.10
Ferritin	20	2.5
Free fatty acid	37	2.7
Galactose-1-phospate uridyl-transferase	31	2.6
Glucose	12	2.4
Glucose-6-phosphate dehydrogenase	32	2.6
γ-Glutamyl transferase	23	2.6
Growth hormone	66	2.12
3-Hydroxybutyrate	36	2.7
α-Hydroxybutyrate dehydrogenase	27	2.6
Hydroxyproline		
plasma & urine	69	2.13
total urine	58	2.9
free urine	59	2.9
peptide urine	60	2.9

Tables 2.3–2.13 (*Cont.*)

Measurement	Location no.	Table
Immunoglobulins		
IgG	15	2.5
IgA	16	2.5
IgM	17	2.5
Insulin	67	2.12
Iron	40	2.7
Lactate	38	2.7
Lactate dehydrogenase	26	2.6
Magnesium	4	2.3
Nitrogen (urine)		
total	51	2.9
urea	52	2.9
creatinine	53	2.9
uric acid	54	2.9
-amino	55	2.9
residual	56	2.9
5′-nucleotidase	24	2.6
Osmolality	43	2.7
pH (blood)	9	2.3
Phosphate		
plasma	6	2.3
urine	48	2.8
Potassium		
plasma	2	2.3
urine	45	2.8
Sodium		
plasma	1	2.3
urine	44	2.8
Sodium fractional excretion	62	2.10
Sulphate	57	2.9
Thyroid stimulating hormone	65	2.11
Thyroxine	63	2.11
Total protein	13	2.5
Transferrin	19	2.5
Triglyceride	33	2.7
Triiodothyronine	65	2.11
Urea		
plasma	7	2.3
urine	49	2.8
Urea nitrogen	52	2.9
Uric acid plasma	35	2.7
Uric acid nitrogen	54	2.9
Zinc	41	2.7

Table 2.3 Reference ranges for plasma electrolytes, urea and creatinine, and for blood acid–base measurements.

Location no.	Measurement (units)	Number of infants	Age	Mean	Range
1	Sodium (mmol/l)	102	1–3 weeks	138	130–145
2	Potassium (mmol/l)	98	1–3 weeks	4.8	3.6–5.8
3	Calcium (mmol/l)	28	1–3 weeks	2.41	1.90–2.85
4	Magnesium	28	1 week	0.80	0.59–1.05
	(mmol/l)	79[1]	1 month	0.83	0.51–1.15
5	Chloride (mmol/l)	83	1–3 weeks	101	92–109
6	Phosphate	28	1 day	1.84	1.00–2.60
	(mmol/l)	28	2–3 weeks	2.44	1.80–3.20
7	Urea (mmol/l)	28	1–3 weeks	3.0	1.0–5.0
8	Creatinine[2]	12	1 day	124	106–138
	(μmol/l)	12	1 week	82	62–106
9	pH	11	2 days	7.44	7.32–7.54
		12	2–3 weeks	7.41	7.33–7.47
10	Base excess	11	2 days	−0.1	base excess greater than −8 mmol/l indicative of significant acidosis
	(mmol/l)	12	1–2 weeks	−2.7	
		12	2–3 weeks	+0.2	
11	P_{CO_2} (kPa)	11	2 days	4.32	2.0–6.5
		12	1–2 weeks	4.60	2.5–6.0

1 Adapted from Fomon *et al.* (1970)
2 Adapted from Sertel and Scopes (1973)

Table 2.4 Blood glucose levels (mmol/l) for normal full term (n = 10–35) and low birth-weight (n = 26–48) infants after 4.5 and 3.5 hours fasting respectively (location number 12).

Age (days)	Full term		Low birthweight	
	Mean	Range	Mean	Range
1	3.17	1.7–4.7	2.44	1.0–4.0
2	3.17	2.5–4.3	2.17	0.8–4.0
3	3.89	2.7–4.9	2.22	1.1–3.5
4	3.83	3.2–4.5	2.33	1.2–4.4
5	3.78	2.0–5.0	2.39	1.0–4.3
6	3.89	3.0–4.5	2.39	1.2–4.6
7–13	3.94	3.2–4.9	2.50	1.5–3.4
14–20	3.83	2.5–5.8	3.11	1.3–5.4
21–27	4.00	3.4–4.5	2.89	1.0–4.3

Adapted from Baens *et al.* (1963) and Cornblath and Reisner (1965).

Table 2.5 Serum protein concentrations in neonates.

Location no.	Measurement (units)	Number of infants	Age	Mean	Range
13	Total protein (g/l)	20	first and second weeks	56	43–76
14	Albumin (g/l)	28	1–3 weeks	39	28–49
15	IgG (g/l)	10 (preterm)	birth	9.1	6.3–13.0
		8 (term LGA)	birth	12.2	8.4–17.6
		28	1–2 weeks	8.0	4.8–13.0
		28	2–3 weeks	6.3	3.5–11.0
16	IgA	28	1–2 weeks	16.9	0–72.2
	(mg/l)	28	1 week	44.0	15.6–124.0
17	IgM (mg/l)	28	birth	82	27–200
		28	1–2 weeks	291	140–603
		28	2–3 weeks	451	253–802
18	α-1-Antitrypsin[1] (g/l)	18	birth	2.26	a value less than 1.0 indicates possible deficiency of the PiZ type
		167	3 days	2.94	
		167	30 days	1.99	
19	Transferrin (g/l)	40 (preterm)	birth	1.60	0.90–2.60
		66 (term LGA)	birth	2.09	1.30–2.75
		13	1–3 weeks	1.71	1.05–2.65
20	Ferritin[2] (μg/l)	98	birth	160	65–395
		46	2 weeks	238	90–628
		46	4 weeks	240	144–399
21	Ceruloplasmin[3] (mg/l)	79	birth	135	20–380

1 Adapted from Kueppers and Offord (1979)
2 Adapted from Saarinen and Siimes (1978)
3 Adapted from Freer *et al.* (1979)

Table 2.6 Normal reference ranges for serum or plasma enzyme activities in infants, activities in U/l (* denotes value calculated from another temperature), and red cell enzyme activities.

Location no.	Measurement	Age	Number of infants	Mean	Range	
22	Alkaline phosphatase	3 weeks	49	349	124–574	at 37 °C
				(248	88–407	at 30 °C*)
23	γ-Glutamyl-transferase[1]	2 weeks	82	45	4–261	at 25 °C
24	5′-Nucleo-tidase[2]	4 weeks	24	2.8	up to 10	at 37 °C
25	Creatine kinase[3]	1–3 days	17	564	63–1137	at 37 °C*
				(349	40–747	at 30 °C*)
		4–10 days	21	222	63–1413	at 37 °C*
				(142	40–903	at 30 °C*)
		11–31 days	25	107	32–288	at 37 °C*
				(69	20–184	at 30 °C*)
26	Lactate dehydrogenase[4]	birth	49	874	470–1824	at 37 °C*
				(605	326–1263	at 30 °C*)
		to 3 weeks	24	672	326–1507	at 37 °C*
				(465	226–1044	at 30 °C*)
27	α-Hydroxy-butyrate dehydrogenase[4]	birth	49	377	172–735	at 37 °C*
				(342	156–667	at 30 °C*)
		to 3 weeks	24	273	127–669	at 37 °C*
				(248	116–608	at 30 °C*)
28	Aspartate[5] aminotransferase	3 weeks	100	37.8	22.9–72.8	at 37 °C*
				(24.9	15.1–47.9	at 30 °C*)
29	Alanine[5] aminotransferase	3 weeks	100	27.1	9.1–43.7	at 37 °C*
				(19.7	6.6–31.7	at 30 °C*)
30	Alkaline ribonuclease	birth	83	687	320–918	at 37 °C
		2 weeks	40	726	473–1003	at 37 °C
		3 weeks	20	645	439–918	at 37 °C
31	Galactose-1-phosphate uridyltransferase[6] (μmol/hr/gHb)	birth to 1 month	75	18.5	12.3–24.6	at 37 °C
32	Glucose-6-phosphate dehydrogenase[7] (μmol/min/gHb)	2 days	120	11.4	5.9–21.3	
		later	60	7.7	5.5–9.3	

1 Adapted from Shore *et al.* (1975)
2 Adapted from Belfield and Goldberg (1971)
3 Adapted from Kupke *et al.* (1980)
4 Adapted from Veit *et al.* (1975).
5 Adapted from Sitzmann *et al.* (1974)
6 Adapted from Ellis and Goldberg (1969)
7 Adapted from Herz *et al.* (1973)

Table 2.7 Other miscellaneous plasma or serum reference values, and blood lactate concentrations during the neonatal period.

Location no.	Measurement	Number of infants	Age	Mean	Range
33	Triglyceride (mmol/l)	37 (AGA)	birth	0.32	0.06–0.60
		53 (LGA)	birth	0.47	0.06–1.06
		28	2–3 weeks	0.86	0.24–1.58
34	Cholesterol	102	birth	1.81	0.55–3.15
	(mmol/l)	34	first to third weeks	2.93	1.93–3.93
35	Uric acid[1]	45	birth	0.42	0.09–0.58
	(mmol/l)	53	to 1 week	0.20	0.14–0.34
36	3-Hydroxy-butyrate[2] (mmol/l) fasting	7	to 1 week	2.45	Value less than 1.1, suggestive of hyperinsulin-aemia
37	Free fatty acid[2] (mmol/l) fasting	7	to 1 week	1.53	Value less than 0.46, suggestive of hyperinsulin-aemia
38	Lactate[3]		birth		1.5–4.5
	(mmol/l)		5 hours		0.9–2.0
			1 day		0.8–1.2
			to 1 month		0.5–1.4
39	Ammonia[4] (μmol/l)	80	to 2 weeks	89	32–192
40	Iron	106	birth	26	4–52
	(μmol/l)	41	2–3 weeks[5]	18	14–25
41	Zinc	15	birth[6]	12.7	10.7–15.6
	(μmol/l)	4	5 days[7]	13.5	11.6–15.6
		14	1 month[7]	10.1	7.9–12.5
42	Copper	15	birth[6]	4.5	2.7–7.8
	(μmol/l)	4	5 days[7]	7.4	6.0–9.3
		14	1 month[7]	9.9	7.8–16.3
43	Osmolality[8]	44	birth	286	275–300
	(mmol/kg)	17	day 7	291	276–305
		17	day 28	290	274–305

1 After Wharton *et al.* (1971)
2 Stanley and Baker (1976)
3 Koch and Wendel (1968)
4 After Beddis *et al.* (1980)
5 After Brozovic *et al.* (1974)
6 Henkin *et al.* (1971)
7 Ohtake (1977)
8 Davies (1973)

Table 2.8 Concentrations of electrolytes and other components in random urines from low birth-weight babies.

Location no.	Measurement	Number of infants	Age	Mean	Range
44	Sodium (mmol/l)	29	to 1 month	6.0	1–15
45	Potassium (mmol/l)	34	to 1 month	12.6	2–28
46	Calcium (mmol/l)	18	to 1 week	0.16	0.05–0.38
		7	second week	0.39	0.23–0.73
		8	third week	0.49	0.25–1.16
		11	at 1 month	1.26	0.50–1.65
47	Chloride (mmol/l)	35	to 1 month	13.6	5–30
48	Phosphate (mmol/l)	26	weeks 1 and 2	3.6	< 0.5–9.9
		19	weeks 3 and 4	7.5	0.5–12.0
49	Urea (mmol/l)	18	to 1 week	34	5–108
		20	weeks 2 and 3	24	4–59
		13	at 1 month	15	9–34
50	Creatinine (mmol/l)	17	to 1 week	1.61	0.5–3.5
		20	weeks 2 and 3	1.06	0.2–2.4
		13	at 1 month	1.09	0.5–1.8

Table 2.9 Urine nitrogen partition, urinary sulphate, and free and peptide hydroxyproline excretion in a group of eight low birth-weight infants.

Location no.	Measurement	Age (days)	Mean	Range
51	Total nitrogen	3	5.9	4.5–8.2
	(mmol/kg per	10	6.1	5.3–6.6
	24 h)	20	10.1	7.5–14.8
52	Urea nitrogen	3	3.5	2.7–4.9
	(mmol/kg per	10	3.7	3.0–4.2
	24 h)	20	7.5	5.5–11.7
53	Creatinine	3	0.27	0.23–0.36
	nitrogen	10	0.26	0.22–0.32
	(mmol/kg per	20	0.34	0.20–0.48
	24 h)			
54	Uric acid	3	0.30	0.14–0.58
	nitrogen	10	0.18	0.06–0.41
	(mmol/kg per	20	0.20	0.05–0.48
	24 h)			
55	α-Amino	3	0.40	0.28–0.66
	nitrogen	10	0.62	0.41–1.26
	(mmol/kg per	20	0.60	0.51–0.73
	24 h)			
56	Residual	3	1.2	0.7–2.2
	nitrogen	10	1.3	0.9–1.8
	(mmol/kg per	20	1.5	1.0–2.0
	24 h)			
57	Sulphate	3	0.14	0.09–0.18
	(mmol/kg per	10	0.16	0.13–0.20
	24 h)	20	0.34	0.20–0.51
58	Total	3	91	51–124
	hydroxyproline	10	170	113–220
	(μmol/24 h)	20	284	217–519
59	Free	3	38	13–66
	hydroxyproline	10	82	58–107
	(μmol/24 h)	20	129	96–242
60	Peptide	3	54	37–69
	hydroxyproline	10	88	55–115
	(μmol/24 h)	20	155	114–277

Table 2.10 Creatinine clearance and fractional sodium excretion in pre-term and full-term infants during the first weeks of life.

Location no.	Measurement (units)	Number of infants	Age	Mean
61	Creatinine clearance (ml/min per 1.73 m²)	5 (pre-term)	1–2 days	15.9
		7 (term)	1–2 days	20.8
		8 (pre-term)	4–6 days	24.1
		6 (term)	4–6 days	46.6
		6 (pre-term)	3–5 weeks	37.0
		5 (term)	3–5 weeks	60.1
62	Fractional sodium excretion (C_{Na}/C_{Cr} ratio as a percentage)	5 (pre-term)	1–2 days	2.11
		6 (term)	1–2 days	0.20
		8 (pre-term)	4–6 days	1.28
		6 (term)	4–6 days	0.30
		6 (pre-term)	3–5 weeks	0.09
		5 (term)	3–5 weeks	0.12

Adapted from Aperia *et al.* (1980)

Table 2.11 Thyroid hormone levels in cord blood of 27 full-term infants, and during the first week in venous blood from 94 full-term and 39 pre-term infants; also during the second, third and fourth weeks in 97 full-term and 66 pre-term infants. Hormone levels in pre-term infants extend over a greater range than those in full-term so only such differences are given.

Location no.		Age	Full-term range	Pre-term range
63	Thyroxine (T_4) (nmol/l)	at birth	103–206	down to 87
		0–24 hours	178–330	down to 118
		24–48 hours	196–327	down to 85
		48–72 hours	136–430	down to 99
		72–96 hours	139–437	down to 54
		96–120 hours	157–314	down to 142
		120–144 hours	154–265	as for full-term
		7–13 days	95–283	as for full-term
		14–20 days	85–260	as for full-term
64	Triiodothyronine (T_3) (nmol/l)	at birth	0.1–1.0	
		0–24 hours	3.0–6.3	0.8–1.6
		24–48 hours	1.4–5.3	0.9–2.1
		48–72 hours	0.9–5.3	0.8–1.3
		72–144 hours	1.6–4.5	0.8–1.7
		7–13 days	1.2–3.5	0.8–2.8
		14–29 days	1.7–5.0	0.7–3.0
65	Thyroid stimulating hormone (TSH) (mU/l)	at birth	less than 1–14	
		0–24 hours	3–120	down to 1
		24–48 hours	3–30	down to 1
		48–72 hours	less than 1–8	up to 13
		72–144 hours	less than 1–3	up to 5
		7–29 days	up to 5	as for full-term

Adapted from Jacobsen *et al.* (1977) and Jacobsen and Hummer (1979).

Table 2.12 Growth hormone levels in venous plasma from full-term and low birth-weight infants after a 3.5 hour fast (blood collected 9–11 am), serum insulin levels in full-term infants after an 8 hour fast and in pre-term infants after a 3 hour fast, and plasma cortisol levels in full-term infants (blood collected at 9 am).

Location no.	Measurement	Age	Full-term mean	Full-term range	Low birth-weight mean	Low birth-weight range
66	Growth hormone[1] (mU/l)	at birth	66	9–320		
		0–24 hours	52	9–167	59	38–83
		24–48 hours	72	28–221	76	17–150
		2–6 days	32	0–69	51	20–133
		7–13 days	20	1–51	30	5–51
		14–20 days	20	5–47	41	16–150
		21–27 days	16	4–26	47	17–125
			Full-term		Pre-term	
67	Insulin[2] (mU/l)		2	0–4	20	0–32
68	Cortisol[3] (nmol/l)	at birth	800	330–1700		
		first week	276	0–773		

1 Adapted from Cornblath *et al.* (1965)
2 Grant (1967)
3 Adapted from Stevens (1970)

Table 2.13 Plasma amino acid concentrations (μmol/l) from 17 pre-term and 11 term light-for-gestational-age (LGA) infants, and urinary excretion (μmol/24 h) of a combined group of 8 low birth-weight infants fed on a modern demineralised whey formula (Gold Cap SMA-S26; Wyeth Laboratories, Berkshire).

		Plasma					
Location no. 69	Age (days)	Pre-term mean	Pre-term range	Term LGA mean	Term LGA range	Urine mean	Urine range
Aspartic acid	11	55	23–114	52	38–66	11.4	7.5–19.7
	21	58	23–82	44	26–73	11.0	4.8–22.6
Hydroxyproline	11	65	50–109	52	33–81	77.8	48.9–107.1
	21	67	42–113	62	45–99	95.2	51.1–151.2
Threonine	11	256	137–391	165	114–250	54.3	20.8–101.9
	21	182	101–271	136	74–206	39.6	15.8–95.6
Serine	11	224	150–450	168	146–450	68.6	40.4–104.1
	21	183	113–299	144	111–208	54.2	24.5–123.2
Asparagine	11	34	9–57	30	21–48		
	21	33	13–49	29	11–53		
Glutamic acid + glutamine	11	552	327–772	528	438–651	56.2	31.9–101.9
	21	574	438–794	589	463–900	50.7	24.2–94.3
Proline	11	165	91–282	156	100–194	45.2	13.2–82.8
	21	136	100–210	137	110–187	56.6	31.8–102.3
Glycine	11	289	220–403	225	175–310	186	46–360
	21	277	193–457	228	182–321	263	139–443
Alanine	11	254	145–466	241	152–338	57.3	33.9–95.7
	21	241	162–492	263	187–371	46.5	20.9–87.3
Valine	11	155	114–265	157	104–252	3.5	2.2–5.2
	21	142	85–230	160	125–202	3.4	2.0–5.0
Cystine	11	44	31–62	41	27–77	8.4	3.8–12.5
	21	45	30–64	39	20–44	6.4	3.6–8.4
Methionine	11	27	15–39	25	14–41	2.2	1.5–2.6
	21	27	19–46	27	20–44	2.0	0.9–2.6
Cystathionine	11					2.1	1.1–3.1
	21					2.5	0.8–6.5
Isoleucine	11	46	28–74	45	34–80	0.7	0.4–1.0
	21	42	24–69	45	32–65	1.7	0.7–2.9
Leucine	11	97	65–168	92	61–153	2.2	1.2–3.0
	21	83	51–127	91	72–123	2.3	1.1–4.1
Tyrosine	11	103	61–159	78	53–125	10.5	7.9–13.3
	21	64	50–108	88	53–131	8.9	4.9–12.9
Phenylalanine	11	63	45–90	68	37–93	3.2	2.0–4.9
	21	52	39–78	63	46–94	3.3	1.6–5.3
Lysine	11	197	122–306	163	135–225	51.9	34.0–78.4
	21	163	119–272	149	90–233	27.7	13.0–36.8
Histidine	11	118	71–297	95	58–139	39.2	7.0–86.9
	21	102	71–123	95	77–111	57.6	24.6–98.4
Arginine	11	78	43–171	59	38–121	3.2	1.5–4.7
	21	83	31–153	63	50–109	3.4	2.0–4.9

Note Infants receiving a predominantly curd feeding formula will tend to have lower plasma concentrations of threonine, and higher plasma concentrations of branched-chain amino acids, tyrosine and phenylalanine.

JAUNDICE

Jaundice is the most common cause of biochemical investigation in the newborn, although in the majority it amounts to no more than a determination of plasma bilirubin. A recent short monograph specifically related to the newborn is that of Odell (1980).

Pathophysiology

So-called 'physiological jaundice' has usually been attributed to normal postnatal haemolysis and immaturity of the glucuronyl transferase enzyme system. Other factors must play a role since in pre-term babies the enzyme does not approach full functional capacity until the end of the first month, by which time the jaundice has long since gone. One such factor is the enterohepatic circulation of bilirubin. Any condition causing this to be exaggerated, e.g. inadequate feeding, impaired movement down the gut with delayed passage of meconium, intestinal atresia or Hirschsprung's disease, is often accompanied by jaundice.

Neonatal jaundice is common but always requires careful, albeit usually rapid, consideration because of the small but definite risk of kernicterus, i.e. encephalopathy due to unconjugated bilirubin. It is too simple to assess the risk of kernicterus merely in terms of the plasma concentration of unconjugated bilirubin. Other neurological insults, such as hypoxia, anaemia and acidosis are common in the neonatal period and may render the blood–brain barrier more permeable to bilirubin and clearance mechanisms are impaired. Conditions which lead to an increased proportion of free bilirubin, i.e. not bound to albumin, are also important. These include a low plasma albumin concentration (e.g. in the pre-term), dissociation of bilirubin from its high affinity binding site on albumin (e.g. acidosis, high concentrations of free fatty acids) and competition from other substances for the binding sites (e.g. sulphonamides and possibly cloxacillin).

Indications for an immediate reduction in plasma bilirubin by exchange transfusion, therefore, vary with the clinical circumstances: low gestational age, a short time since birth, the presence of anaemia or acidosis will all suggest the need for exchange transfusion at plasma bilirubin concentrations which in the term baby would be tolerated. In practice a concentration of less than 250 μmol/l, even in a very ill immature baby, is rarely cause for concern, whereas most paediatricians would proceed to exchange transfusion if the unconjugated bilirubin concentration were greater than 400 μmol/l, even in a fit one-week-old normal-sized baby. Between these limits individual consideration is necessary. Various objective measurements are available, e.g. free bilirubin, reserve albumin binding capacity, albumin–bilirubin affinity. They are not used widely in Britain but a detailed review of their clinical use is given by Lee and Gartner (1978).

A rule of thumb suggested by Odell (1980) is that exchange transfusion is indicated if molar ratio serum bilirubin: serum albumin is > 0.7, i.e. bilirubin in μmol/l: albumin in g/l is > 10 or bilirubin in mg/100 ml: albumin in g/100 ml is > 6.

'Routine' management of jaundice

Fig. 2.1 shows a problem-orientated case record and management regimen which we use (Matthew & Wharton 1981). It leads to investigation of a raised bilirubin concentration which is unusually *early*, *high* or *prolonged* and to investigation of conjugated hyperbilirubinaemia whenever it occurs. The action lines for extra fluid, phototherapy and exchange transfusion are horizontal. They are easily remembered by staff even when the protocol is not in front of them. We considered that this simplicity overrode the pathophysiological arguments for a series of sloping action lines. There is good evidence that higher volume *milk* feeds are associated with lower plasma bilirubin levels. Recent work has questioned whether water alone achieves the same result and after further evaluation it may be that our extra water therapy will need revision. We consider anyway that extra water is required during phototherapy.

The protocol is designed to detect in *unusually early jaundice*, abnormal haemolysis (Rhesus, ABO, G6PD deficiency); in *unusually severe jaundice*, abnormal haemolysis, infection, any cause of conjugated hyperbilirubinaemia (infection, metabolic); in *unusually prolonged jaundice*, all the preceding conditions and hypothyroidism.

Prolonged jaundice

In most instances the bilirubin is all unconjugated, there is no suggestion of haemolysis, anaemia or infection, and after demonstrating that TSH and T_4 are normal it is reasonable not to investigate further for some weeks, so long as the child continues to thrive. This course of events is commonly seen in mildly pre-term babies who are breast fed. Familial non-haemolytic jaundice will also present in this way and is currently classified into two types: the Type I disorder (Crigler–Najjar), is very rare, is an autosomal recessive condition and kernicterus is almost inevitable; Type II is an autosomal dominant and kernicterus does not occur, presumably because there is some residual glucuronyl transferase activity. Some instances of Type II may be in effect double heterozygote forms of Gilbert's disease. Gilbert's disease is similar to Type II but enzyme activity is about 50% of normal and *when bilirubin production rates are normal* the plasma bilirubin remains below 100 μmol/l; it is, therefore, rarely diagnosed in the newborn.

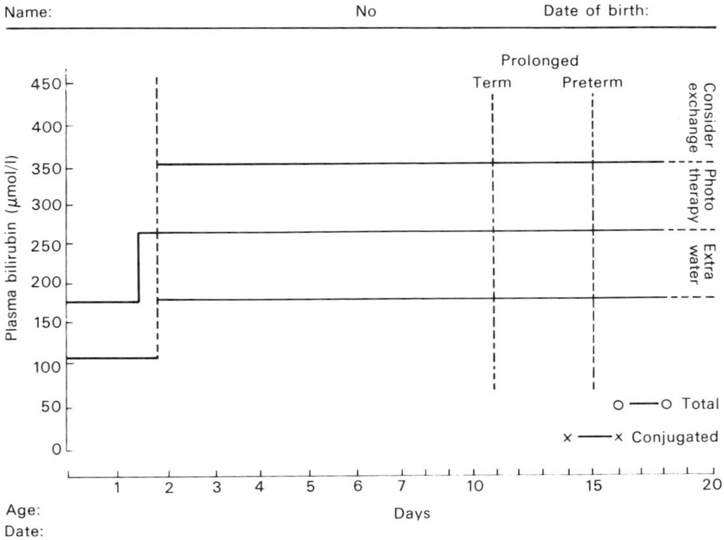

Plot plasma bilirubin concentration on chart above. Investigate and treat according to protocol below.

Put in date when investigation was sent off and record results in Table.

Zone	Suggested investigation	Management	Date Began/ended
Early jaundice	(H)	Watch carefully. Consider phototherapy if > 175 μmol/l. If Coombs's positive (i.e. blood group incompatibility) individual management required	
Extra water	Nil	Extra water (30% of normal feed volume) for one course of 4 days	
Phototherapy	(H) (U) (C)	Extra water as above. Phototherapy for 24-hour periods until end of 24-hour period in which bilirubin falls below 220 μmol/l	
Prolonged	(H) (U) (T) (C)	Initially, investigate only	

Haematology (H)	Urine (U)	Conjugated bilirubin (C); estimate and plot; if greater than 50 μmol/l and bile present in urine proceed to investigations below	
Mother	Clinitest:	Virus	IgM
Group	Clinistix:	Throat swab	Alpha-1-antitrypsin
Antibodies	Ictotest:	Urine	Toxoplasmosis
	Bacteriology		Cytomegalovirus
Baby			Rubella
Group		Other LFT	Wassermann reaction
Coombs	Thyroid function (T)	For example,	Australia antigen
Cord bilirubin		AST	
Hb	T4	ALT	
G6PD (in male immigrant)	T3 uptake	5 nucleotidase	
	TSH		
Other			

Fig. 2.1 Neonatal jaundice.

Conjugated hyperbilirubinaemia

A marked increase in conjugated bilirubin is most commonly a complication of either Rhesus haemolytic disease (the inspissated bile syndrome) or parenteral nutrition. If these have been excluded, infective and metabolic causes of hepatocellular dysfunction should be sought (serological evidence of intrauterine infections, bacterial sepsis, particularly Gram-negative infections of the urinary tract, galactosaemia, alpha-1-antitrypsin deficiency, cystic fibrosis).

The most common disorder presenting in this way is alpha-1-antitrypsin deficiency. Interpretation of plasma concentrations of this protein in the neonatal period is not straightforward. In normal babies (phenotype PiM or PiMZ) the protein concentration increases with gestational age, reaches a peak at one week postnatally, and then declines to reach adult levels by four weeks (see line 18, Table 2.5). Plasma levels in the neonatal period do not provide a definitive method of identifying deficient babies with the potentially pathological PiZZ or Pinull phenotype. A concentration of 1.0 g/l or less during the first month of life indicates possible deficiency of the pathological type and phenotyping should be performed (Morse 1978a, b). The Dubin–Johnson and Rotor syndromes may present this way and in these conditions plasma concentrations of other substances excreted by the liver, e.g. alkaline phosphatase and 5-nucleotidase, are often not increased.

The presence of a heart murmur or other abnormalities raises the possibility of trisomy 13, 18, the Alagille syndrome (odd face, systolic murmur) or arteriohepatic dysplasia (hypoplasia or stenosis of pulmonary arteries – main or peripheral, some with odd facies).

Commonly all aetiological and clinical clues and all investigations are negative and the diagnosis is either 'cryptogenic hepatitis' or biliary atresia. Whilst the infant is being investigated moderately raised concentrations of conjugated bilirubin often fall to normal and so it is reasonable to observe the infant for at least a month. At the end of this time if there is no evidence of biochemical improvement more definitive investigations should be considered, e.g. determination of serum bile acids, liver biopsy, percutaneous cholangiography, nuclear medicine scanning, mini laparotomy and cholangiography (Mowat 1979).

The severely ill child with jaundice

The jaundice may be a 'red herring', and represent no more than the 'physiological' causes of jaundice in a baby who is ill from some other disorder.

If the bilirubin is almost wholly unconjugated Rhesus incompatability may cause severe illness, mainly because of the anaemia. If conjugated bilirubin is present as well, Rhesus incompatability is still a possible cause, but in addition, infection (particularly with Gram-negative bacteria) and metabolic abnormalities such as galactosaemia should be considered. The occurrence of a biochemical abnormality does not necessarily indicate a metabolic aetiology; septicaemia commonly leads to acidosis, a severe hepatitis of infective aetiology may lead to relative galactose intolerance with galactosuria (although not sufficient to give 2% reducing substances) and tyrosinaemia, usually with accompanying raised plasma concentrations of phenylalanine and methione.

When the diagnosis is in doubt, specimens should be obtained for microbiological and biochemical studies. The child should be treated with antibiotics, in case there is bacterial sepsis, and with 'clear fluids', as a means of limiting substrate lest he has a metabolic disease.

Treatment of unconjugated hyperbilirubinaemia

Treatment of unconjugated hyperbilirubinaemia with phototherapy and exchange transfusion has a number of biochemical implications.

Phototherapy

Possible mechanisms whereby phototherapy reduces plasma concentrations of unconjugated hyperbilirubinaemia include direct photo-oxygenation of the bilirubin and subsequent hydrolysis into more water-soluble shorter length pyrroles, and the formation of 'photobilirubin' in the skin which more easily dissociates from skin binding sites and can be excreted *directly* into bile and its unconjugated form. Extra-sensory water losses from the skin and in the stool are increased and we provide additional water.

Exchange transfusion

Exchange transfusion may be indicated immediately after birth in Rhesus incompatibility, particularly if the cord haemoglobin concentration is below 12 g/100 ml and/or the cord bilirubin is above 85 μmol/l. In these circumstances we usually perform a single volume (80 ml/bodyweight) exchange with the normal red cell packs containing acid phosphate dextrose; Rhesus-negative cells, ABO compatible with mothers blood, to treat the anaemia followed immediately by a further single volume exchange with *whole* blood to maintain the haemoglobin and wash out bilirubin and antibody. In rare instances of cross-matching difficulty, cells frozen in liquid nitrogen may be resuspended in fresh frozen plasma; although resuspension in normal saline is acceptable in an adult this is not suitable for neonatal exchange transfusion. We do not find it necessary to give calcium routinely during the exchange, and if severe acidosis is present attention to its cause, e.g. hypoxia or hypothermia, is preferable to boluses of sodium bicarbonate, since they add to the hyperosmolality of the blood.

With the anaemia corrected, the indication for subsequent exchanges is usually the plasma bilirubin concentration which is interpreted in the light of the general condition of the baby, e.g. maturity, other biochemical abnormalities, etc. (see section on Pathophysiology, p. 27). The efficiency of an exchange transfusion when performed solely to lower the plasma bilirubin may be increased by priming with albumin. Salt-poor albumin, 1 g/kg bodyweight, is given intravenously one hour before the exchange. We have little experience of this technique; care is needed because of the resulting sudden increase in plasma volume.

Other treatments

Phenobarbitone (8 mg/kg bodyweight per day) which induces glucuronyl transferase, and oral agar (1 g/kg per day), which blocks the enterohepatic recirculation of bilirubin, may be used to reduce plasma bilirubin concentration. Their action is generally too slow to be of much value in acute neonatal management, but they have been used in the treatment of the inborn errors of unconjugated bilirubin metabolism (Arrowsmith *et al.* 1975; Poland *et al.* 1972).

HYPOGLYCAEMIA

Hypoglycaemia was recognised as a common problem in the newborn about 20 years ago, but changes in feeding regimen since then have reduced its prevalence considerably. A major review is given by Cornblath and Schwartz (1976) and a recent brief one by Rayner (1982); reference should also be made to Chapter 3.

Definition and recognition

Cornblath and Schwartz (1976) defined significant hypoglycaemia as a blood glucose < 1.2 mmol/l (< 20 mg/100 ml) or plasma glucose < 1.5 mmol/l (< 25 mg/100 ml) in a low birth weight infant, < 1.7 mmol/l (< 30 mg/100 ml) in a full-sized infant 0–72 hours and < 2.2 mmol/l (< 40 mg/100 ml) thereafter. These limits are clinically useful in that symptoms truly due to the hypoglycaemia *per se* are most unlikely to occur at blood glucose concentrations above this level, but in practice an adequate feeding regimen almost always ensures a blood glucose greater than 2.5 mmol/l and we view with suspicion even a small baby whose blood glucose remains below 2.5 mmol/l for more than 6 hours.

Blood glucose should be determined every 4 hours during the first 2 days of life in all babies weighing less than 2.25 kg, infants of known diabetic mothers, any baby sufficiently ill, whatever the cause, to warrant admission to a special or intensive care nursery and babies who although not of low birth weight are, nevertheless, light-for-gestational-age. Babies in the last category may easily slip through the net, e.g. the 2.3 kg baby born at 41 weeks who is breast fed. A stick test is usually performed at the cot-side and if the result is low a chemical determination of glucose should be made.

The rate of glycolysis exhibited by erythrocytes from pre-term and term infants is higher than that of the adult (Meites & Saniel-Banrey 1979). This affects the stability of blood glucose in samples of blood collected from venous or capillary sites. The use of fluoride as preservative may be inadequate and recent work has shown that even in combination with sodium iodoacetate (fluoride 17.7 g/l, iodoacetate 5.0 g/l) there was a 4% loss of glucose in one hour, rising to 10% in five hours. It is, therefore, important to minimise the time taken for such specimens to reach the laboratory and to be analysed.

Though most laboratories use the preferred glucose oxidase-based methods, chemical methods may be encountered. Care should be taken with results from some of these as, due to their non-specificity, the glucose level, particularly in the hypoglycaemic range, may be over-estimated. It has been suggested that only the hexokinase/glucose-6-phosphate dehydrogenase method is sufficiently accurate for this range in newborn babies. Glucose distributes itself throughout the aqueous phase of whole blood, so plasma glucose values may be up to 20% higher than the equivalent whole blood values when the haematocrit is high.

Hypoglycaemia in the small baby

The most probable diagnosis is that the light-for-gestational-age baby has poor glycogen stores, but if an adequate feeding regimen has been used the diagnosis should not be accepted too readily and the possibility of infection, hypothermia, perinatal hypoxia or a high insulin state, e.g. the infant of a missed diabetic mother, should be considered. The combination of hypoglycaemia with metabolic acidosis should raise suspicion of infection, perinatal hypoxia or occasionally an inborn error of carbohydrate or amino acid metabolism.

Management of symptomatic hypoglycaemia

If at any time symptoms such as collapse, irritability, apnoea or convulsions occur with a blood glucose < 1.2 mmol/l, dextrose should be given intravenously. A suitable procedure would be as follows:

1 If possible, collect blood for later assay (if necessary) of insulin and other metabolites, and arterial or arterialised blood for determination of acid–base status; collect urine to test for reducing substances and glucose.

2 Administer 50% glucose in a dosage of 2 ml (i.e. 1 g glucose) per kg body weight over 1–2 minutes into a peripheral vein, followed by a 10%

glucose drip hourly in order to maintain the blood glucose at 2.5–5.5 mmol/l.

3 Note whether symptoms were promptly abated as blood glucose rises; the occurrence of symptoms due to hypoglycaemia *per se* is unusual on modern feeding regimens; more commonly hypoglycaemia and symptoms occur as associated effects of some other cause, e.g. intracranial problems, in which case the symptoms are not alleviated as the blood glucose rises. Nevertheless, treatment to maintain a normal blood glucose level should be continued.

4 Once the immediate emergency is over, determine acid–base status, if this has not already been done, and consider whether the child could have an infection; if there are congenital abnormalities, particularly those affecting the mid-line or genitalia, obtain blood for later assay of growth hormone, cortisol, T_4 and TSH, since panhypopituitarism and cortisol deficiency are rare causes of neonatal hypoglycaemia.

5 If large volumes of 10% dextrose are necessary to maintain the blood glucose > 2.5 mmol/l or if after 12 hours it is proving difficult to wean the child from intravenous dextrose to enteral feeds, add oral prednisone and consider the possibility of a high insulin state.

6 Wean slowly, replacing intravenous dextrose by equal volumes of enteral feed over 2–3 days.

Management of asymptomatic hypoglycaemia

A suitable regime would be:

During the first 6 hours of life. Blood glucose < 2.5 mmol/l (< 45 mg/100 ml)—quite common, feed hourly or continuously.

After the first 6 hours. Blood glucose persistently 1.5–2.5 mmol/l (25–45 mg/100 ml)—not necessarily abnormal but unusual; continue frequent feeds and consider alternative diagnoses as above.

Blood glucose persistently < 1.5 mmol/l (25 mg/100 ml)—consider alternative diagnoses, begin intravenous treatment and proceed as for management of symptomatic hypoglycaemia.

Other causes of a reduced hepatic glucose release

The most common causes are perinatal hypoxia and infection, and these diagnoses should be carefully considered. A hypoplastic left heart may present as hypoglycaemia with acidosis.

Less common, but important causes, are inborn errors of metabolism, e.g. *glycogen storage disease type 1* may present at this age with hypoglycaemia and metabolic acidosis; after a few days, however, these biochemical symptoms may disappear as the child takes frequent feeds and so the diagnosis is delayed (Hufton & Wharton 1982). Because we have missed the diagnosis previously, we perform a glucagon test (30 μg/kg IM) in any child who has unexplained hypoglycaemia, particularly if metabolic acidosis is present. The baby is fasted for 4 hours only, since a longer period is potentially dangerous. If there is an inadequate response to glucagon (a rise in plasma glucose less than 1.5 mmol/l), then further differentiation of the metabolic error can be pursued. Since even a short fast is potentially dangerous for these babies, Fernandes *et al.* (1969) have suggested that changes in plasma lactate following a glucose load may be a safer but as effective method of preliminary diagnosis.

If the child is receiving a formula containing fructose or sucrose (a few available in continental Europe do so) then *fructose intolerance* or fructose-1,6-diphosphatase deficiency may present similarly. *Galactosaemia* also presents this way but in our limited experience of this condition in the neonatal period, evidence that the child is ill (e.g. vomiting, jaundice) precedes the hypoglycaemia. Urine should be tested for reducing substances and glucose in any episode of hypoglycaemia which is not easily explained and care must be taken when interpreting the result, that the child has received reasonably normal amounts of lactose in the previous 24 hours. If there is doubt about the diagnosis and the child is ill a 'clear fluid' dextrose and electrolyte regime should be given and blood taken for determination of galactose-1-phosphate uridyl transferase activity in the red cells. It may be dangerous to reintroduce lactose in an attempt to obtain meaningful urine results. An ill child in whom galactosaemia is suspected should not be transferred rapidly to a galactose-free formula, since these formulas have not been designed for acutely *ill* newborns and may precipitate further acidosis. Some have a very high casein content and others contain hydrolysed protein and/or medium chain triglycerides. It is wiser to introduce them gradually.

Inborn errors of metabolism affecting the branch chain amino acids, such as maple syrup urine disease and propionic acidaemia, often cause hypoglycaemia, but never as an isolated finding. Nevertheless, if hypoglycaemia is not readily explained the plasma amino acid pattern should be screened.

Endocrine deficiencies such as panhypopituitarism or adrenocortical insufficiency may occur. They are sometimes associated with 'mid-line' or genital abnormalities, e.g. cleft palate and lip, small penis, and are sometimes familial (Moncrieff *et al.* 1972). An antenatal clue of adrenal hypoplasia is the *very* low urinary oestriol excretion by the mother. Plasma growth hormone concentrations are normally much higher than in adults, throughout the neonatal period (see Table 2.12, line 66).

High insulin states

The most common cause is the infant of the inadequately controlled diabetic mother. As antenatal control has improved these babies present fewer problems concerned with blood glucose and management is often more directed towards respiratory difficulties, jaundice, occasionally hypocalcaemia and the increased incidence of congenital malformations. The babies should, if possible, be fed as if they are light-for-gestational-age, even though they may be heavy. Blood glucose must be monitored and hypoglycaemia, whether asymptomatic or symptomatic, managed as for the 'small baby' above. Occasionally a diabetic mother produces a poorly-grown baby and these often have persisting hypoglycaemia requiring intravenous therapy and corticosteroids. High insulin output also occurs in severe Rhesus incompatability and may lead to marked rebound hypoglycaemia about an hour after an exchange transfusion with acid citrate dextrose blood.

Less common, but very serious, causes of hyperinsulinism presenting in the newborn and persisting thereafter are β cell tumours, i.e. nesidioblastosis and islet cell adenoma. An early clue is the amount of intravenous glucose necessary to maintain a normal blood glucose, e.g. as much as 10 ml of 10% dextrose per kg bodyweight per hour, providing 16 mg per kg per minute. A better indication (so long as the precaution of taking blood for determination of plasma insulin before emergency intravenous therapy has been observed) is a plasma insulin of 10 IU per ml in the presence of a blood glucose below 2.3 mmol/l. This may not always be achieved and it has been suggested that blood 3-hydroxybutyrate and serum free fatty acid are more satisfactory indicators. If the glucose level is below 2.2 mmol/l then a 3-hydroxybutyrate level below 1.1 mmol/l or a free fatty acid level below 0.46 mmol/l suggest hyperinsulinaemia.

Hyperinsulinism may also occur in Beckwith's syndrome, i.e. large tongue, visceromegaly and omphalocele, and the infant giant syndrome.

Reducing substances present in urine

Glucose is the most common reducing substance in urine. It is most commonly iatrogenic and due to excessive amounts of intravenous dextrose. It occurs in ill small babies with excessive plasma glucose concentrations, presumably because their insulin output is limited. Renal glycosuria may be an indication of generalised renal tubular disease.

A non-glucose reducing substance in the urine will usually be galactose since newborn babies are no longer fed sucrose and they should not receive fructose or sorbitol intravenously. Galactosuria occurs in galactosaemia. If at the time the urine is tested the baby has received a reasonable amount of galactose in the previous 24 hours (e.g. 5 g per kg bodyweight, i.e. 10 g of lactose or 150 ml per kg of breast milk or most formulae); 2%

galactosuria would be expected. However, the child has commonly been ill and has been taken off milk feeds by the time the urine is tested, so interpretation is difficult. Galactosuria may also occur whenever there is severe liver cell dysfunction from whatever cause, but usually it will not be more than 1%. These children are also very ill and so galactosaemia is a possibility. A 'false positive' urinary test for galactosaemia is when profuse diarrhoea fluid contaminates a urine sample and the lactose present in the stool is detected as a urinary, non-glucose, reducing substance.

HYPOCALCAEMIA AND HYPERCALCAEMIA

Symptoms due to hypocalcaemia, such as irritability, twitching and convulsions, rarely occur if the plasma total calcium is more than 1.8 mmol/l (7.2 mg/100 ml). When ionised calcium measurements are available it seems that symptoms rarely occur if the concentration is above 0.8 mmol/l (3.2 mg/100 ml). This estimation is available in only a few centres, however, and specimens which must be taken anaerobically cannot be sent away, as they deteriorate rapidly. Corrections for plasma albumin are not usually made in the neonatal period but may require consideration in some babies with severe hypoproteinaemia. Substantial reviews of calcium metabolism in childhood are given by Harrison and Harrison (1979) and Tsang *et al.* (1979), and reference should also be made to Chapters 14 and 15.

Early onset hypocalcaemia (birth–72 hours)

This usually occurs in very small or ill babies, e.g. following hypoxia or in the infant of a diabetic mother. The aetiology is uncertain but is thought to represent an inbalance of calcitonin and parathormone. Prevention is important. Very small babies weighing < 1500 g, rapidly develop hypocalcaemia if they are not given calcium. If they are not receiving normal volumes of milk or formula, administered clear fluids, whether intravenous or enteral, should provide 1 mmol (40 mg) calcium and 0.5 mmol (12 mg) magnesium per kg body weight daily.

It is difficult to know whether symptoms are due to the hypocalcaemia *per se* or to other associated problems. If, in the presence of continued convulsions, the plasma calcium is below 1.8 mmol, it is useful firstly to take blood for magnesium and phosphate determinations, and secondly to set up a *slow* intravenous drip of calcium, preferably with heart rate monitoring so that if bradycardia occurs the infusion can be slowed; 1 ml per kg per hour of 10% calcium gluconate (0.22 mmol (9 mg) calcium) is a reasonable starting rate. Rapid intravenous injections of calcium salts are dangerous and the aim is to reduce the amount of calcium given intravenously as soon as possible. The plasma calcium should then be maintained at 2–2.5 mmol/l (8–10 mg/100 ml) by adding calcium and magnesium to clear fluid given intravenously or to an enteral feed, starting with the amounts described above under prevention. Generally a convulsion at this age is more likely to be due to another cause. In view of the potential danger of the procedure, therefore, rapid intravenous calcium is rarely indicated, if ever, without first establishing that there is indeed hypocalcaemia.

The plasma phosphate would usually be normal or low. It should be noted that normal concentrations are higher than in adults (Table 2.6, line 3) and haemolysis may lead to a spuriously raised value. A raised level suggests an unusual cause, e.g. accidental overload with phosphate during intravenous nutrition or hypoparathyroidism, although this rarely causes *early* hypocalcaemia.

Late onset hypocalcaemia (5–10 days or sometimes later)

Some years ago convulsions around the 7th day of life were most commonly due to hypocalcaemia precipitated by the high phosphate intake of a child receiving unmodified cows' milk. Now that more babies are breast fed and infant formulae contain quite low concentrations of phosphate, this cause has almost disappeared. Now, rather than indicating an unsuitable postnatal diet, late onset hypocalcaemia more probably indicates a poor antenatal vitamin D status, particularly if the mother is Asian. Less commonly it is due to one of the varieties of congenital hypoparathyroidism.

The following scheme of management is suggested. In most instances the hypocalcaemia is only a transient problem and there is no need to progress beyond step 4.

1 Confirm that the baby's plasma phosphate is raised. If it is not, the hypocalcaemia is not that of the typical late variety, but more probably a non-specific sign of infective or metabolic disease. Also determine plasma magnesium, since this is frequently below normal as well. Exclude uraemia as a cause.

2 Treat with (a) intravenous calcium gluconate to control the fits (see above) if absolutely necessary; it rarely is. (b) Give a low phosphate diet, i.e. breast milk or a demineralised whey formula. This is an essential step and attempts to raise the plasma calcium with calcium supplements whilst the plasma phosphate remains raised are often unsuccessful. (c) Give oral supplements of calcium 3 mmol/kg body weight in 24 hours (calcium gluconate 1.4 g) and magnesium 0.5 mmol/kg (magnesium chloride 100 mg). The main effect of the calcium is to precipitate the dietary phosphate as an insoluble salt in the intestine. (d) If the child is Asian we usually give vitamin D immediately (calciferol 25 μg (1000 IU) daily) on

the assumption that in the majority of instances the hypocalcaemia reflects a poor maternal and fetal vitamin D status. It is wise to save some of the baby's serum so that if the hypocalcaemia is prolonged the serum level of 25-hydroxycholecalciferol (25-hydroxyvitamin D) prior to treatment can be checked. Serum must be stored at −20 °C. The plasma calcium usually rises within 72 hours, but on occasion may take up to 10 days. Treatment is slowly withdrawn when the plasma calcium is normal and the usual prophylactic dose of vitamin D (7 μg) is continued.

3 Seek evidence of frank rickets in the baby; X-ray the skull and a knee. If rickets is present 25-hydroxycholecalciferol should be continued for about a month and then reduced to 7 μg daily. Lesions of the ends of long bones (admittedly somewhat different from rickets) also occur in syphilis which, like any severe neonatal illness, may be associated with the early onset of hypocalcaemia.

4 Seek circumstantial evidence of *maternal osteomalacia*, i.e. plasma calcium level low or normal, phosphate low or normal, alkaline phosphatase raised. If these investigations are delayed until one week post-partum, placental alkaline phosphatase has largely disappeared from the maternal circulation and the activity should be within the normal adult range. If this preliminary screening suggests osteomalacia, decisions must then be made whether to confirm the diagnosis in the mother, e.g. x-rays of bones, bone biopsy, determination of alkaline phosphatase isoenzymes, to accept a simple deficiency of diet and sunlight as the cause or whether to investigate further for evidence of malabsorption and renal tubular disease. In practice, in an Asian mother, it is probably reasonable to accept the 'simple' diagnosis but to investigate a European mother further. It is wise to save some of the mother's plasma whatever the result, so that concentrations of 25-hydroxycholecalciferol and parathyroid hormone may be determined should the hypocalcaemia not respond to the simple measures described in (2). If the neonatal hypocalcaemia is a reflection of maternal vitamin D deficiency the serum 25-hydroxycholecalciferol level will be below 10 ng/ml in both mother and baby, but normal concentrations have been found in adults with frank osteomalacia. This investigation may, very occasionally, suggest maternal hyperparathyroidism.

5 If the baby's plasma calcium level is not corrected within 10 days following the treatment described in (2), the possibility of *congenital hypoparathyroidism* requires careful consideration. During further investigation it may be possible to achieve a normal plasma calcium by increasing the dosage of 25-hydroxycholecalciferol. Occasionally doses of up to 50 μg/kg per day are necessary. These higher doses induce the formation of the active 1,25-dihydroxycholecalciferol even in the absence of parathyroid hormone, but care is necessary to avoid hypercalcaemia and thence nephrocalcinosis. Experience with more active metabolites is limited but 1,25-dihydroxycholecalciferol in doses of 1–8 μg per day have been used in older children.

6 Seek evidence of the *Di George syndrome* (i.e. absent thymus, absent parathyroids). If there is no thymic shadow on the chest X-ray, tests of cellular immunity should be pursued and fresh blood transfusion avoided if possible in view of the small, but definite, risk of a graft-versus-host reaction. Abnormalities of the heart, particularly of the great vessels, are often associated with abnormal development of the parathyroid glands. Plasma parathyroid hormone levels in these patients are low.

7 Obtain blood for a determination of parathyroid hormone. Most of the available assays are used mainly for the diagnosis of hyperparathyroidism and are much less reliable in the lower range. Hypoplasia of the parathyroids, whether isolated or part of the Di George syndrome, is associated with a low concentration of the hormone, despite the hypocalcaemia (estimated on the same sample as that used for the hormone assay), but other causes of hypocalcaemia lead to hormone concentrations in the upper normal or high range. Ideally the child should show a normal response to exogenous parathyroid hormone.

8 *Pseudohypoparathyroidism* should be considered, i.e. biochemical signs of hypoparathyroidism but with normal parathyroid glands producing large amounts of hormone. Children with the type 1 disorder are short and stocky with a round face and short 4th and 5th metacarpals, so that they cannot make a complete set of knuckles. There is often a family history of the disorder indicating 'autosomal dominant with variable penetrance'. Parathyroid hormone (PTH) levels are high normal or high but there is no metabolic response to administered PTH. PTH normally increases renal tubular cyclic AMP and inhibits tubular phosphate reabsorption so that urinary phosphate and cyclic AMP increase. It also promotes a natriuresis and decreases reabsorption of bicarbonate, which can create a mild hyperchloremic acidosis. The kidney of the newborn is initially unresponsive to PTH, but towards the end of the first week of life urinary cyclic AMP and phosphate excretion increase in response to PTH. In the type 2 disorder there are no anatomical abnormalities; there is normal increase in urinary cyclic AMP following exogenous PTH but *no* increase in urinary phosphate. Investigations using exogenous PTH require detailed consultation with a specialised laboratory. Hypoparathyroidism and pseudohypoparathyroidism are discussed further on p. 416 and in Chapter 14.

Other causes of hypocalcaemia

Hypocalcaemia may occur during and following exchange transfusion with acid citrate dextrose blood and may also be a non-specific sign of metabolic disturbance, e.g. in hypernatraemia and the organic acidaemias.

Hypercalcaemia

Hypercalcaemia is unusual in the newborn. In retrospect the infant with the severe variety of infant idiopathic hypercalcaemia (see Chapter 14) may have had symptoms which go back to the neonatal period, but the diagnosis is rarely made then. Plasma cholesterol is usually raised as well.

Four other diagnoses should be considered; (a) the phosphate depletion syndrome, particularly in very low birth-weight babies receiving a parenteral nutrition regimen which does not contain 'added' phosphorus; (b) disseminated or occasionally lone malignancy, e.g. neuroblastoma or hepatoblastoma due either to the osteolytic effect of secondaries or tumour production of PTH, prostaglandin PGE_2 or osteoclast stimulating factor (Tsang *et al.* 1979); (c) primary hereditary hyperparathyroidism (Spiegel *et al.* 1977); (d) transient secondary hyperparathyroidism in which hypercalcaemia is variable (Glass & Barr 1981).

NEUROLOGICAL SYMPTOMS

Convulsions and other abnormal movements

The more common metabolic causes of convulsions include hypoglycaemia, hypocalcaemia, hypomagnesaemia and occasionally hypernatraemia or hyponatraemia (see other sections of this chapter).

Inborn errors of metabolism affecting mainly organic substances usually cause other prominent symptoms as well, such as vomiting or coma. There are exceptions, however, e.g. B6-dependent convulsions (it is worth investigating for this condition only if hitherto intractable convulsions are controlled by pyridoxol (pyridoxine) up to 300 mg/24 hours); sulphite oxidase and/or xanthine oxidase deficiency (clues are a very low plasma concentration or uric acid and the sulphur amino acids).

Abnormal 'extrapyramidal' movements may occur following neonatal kernicterus, but in the neonatal period irritability and floppiness rather than abnormal movements are prominent. The Lesch–Nyhan syndrome has a raised plasma uric acid from birth but abnormal movements are not apparent till later. Plasma uric acid is well above adult levels in normal babies in the first week, falling to adult concentration within 2 weeks (Wharton *et al.* 1971). Many metabolic errors affecting the central nervous system lead to dystonic movements but are often associated with an abnormal conscious state (see 'coma' below).

Coma with and without seizures and/or vomiting

Intracranial haemorrhage and/or hypoxic brain damage are much the commonest causes, and usually cause symptoms from birth. Although creatine kinase is often raised following head injuries in adults it does not help in the diagnosis of intracranial injury in the perinatal period (Wharton *et al.* 1971). Metabolic causes should be suspected in babies who seem well initially and then deteriorate after a few days as their protein intake increases, but the occasional metabolic disorder may cause coma within the first 24 hours of life, e.g. non-ketotic hyperglycinaemia. Suspicion of a metabolic cause is strengthened if (a) there is nothing in the perinatal history to suggest hypoxia/haemorrhage (e.g. Apgar score is normal, endotrachael intubation is unnecessary), (b) lumbar puncture and ultrasound scan of the head are normal, (c) there has been a previous unexplained neonatal death, (d) there is parental consanguinity, or (e) if one of the simple metabolic screens discussed previously is positive. Surprisingly, some metabolic errors may present with neurological symptoms before the peripheral evidence of profound biochemical disorder is apparent, e.g. we have seen one baby with propionicacidaemia who was already comatose and on a ventilator for 24 hours before metabolic acidosis occurred. When symptoms persist it is worth repeating all the simple screening tests.

'Blind' treatment may have to be instituted before a definite diagnosis is made (see below). Hahn and Danks (1981) have described a clinical and biochemical approach to babies with suspected metabolic disease.

Coma with acidosis

An organic acidaemia or one of the pyruvate/lactate acidoses (see acidosis) is possible, but intraventricular haemorrhage and meningitis are more common.

Coma with a haematological disorder

Coma associated with abnormalities of blood such as thrombocytopenia or neutropenia is most probably due to overwhelming infection, but organic acidaemias with secondary hyperglycinaemia may present in this way. The combination of anaemia, acidosis, and coma is most commonly due to haemorrhage with hypoxia but may be a presentation of 5-oxoprolinuria.

Hyperammonaemia

It occurs most commonly following perinatal hypoxia probably due to hepatic ischaemia, when there is an excessive protein intake particularly during intravenous feeding, or as a transient phenomenon in pre-term babies. It may occur variably as a non-specific biochemical symptom in various metabolic disorders, e.g. organic acidaemias, periodic hyperlysinaemia. Urinary tract infection with stasis is an occasional cause. Reye's syndrome rarely occurs in the newborn. If these causes are excluded and the ammonia concentration is very high (more than three times normal) then one of the inborn errors of the urea cycle is likely. These are discussed further in Chapter 4.

Coma without any other metabolic symptom, such as hypoglycaemia, acidosis, hyperammonaemia, may occur in hypervalinaemia.

Management of a suspected metabolic disorder

If a metabolic error is suspected the screening tests described already should be performed and since a lumbar puncture will usually be performed an aliquot of cerebrospinal fluid (CSF) should be deep frozen for further examination if indicated, e.g. if non-ketotic hyperglycinaemia is a possibility following screening for plasma amino acids. In this condition plasma glycine concentration is between two and ten times normal, CSF glycine more than ten times normal, and the plasma/CSF glycine ratio, which is usually at least 30, is much reduced (blood and CSF should be collected at the same time).

Whilst waiting for results most babies will receive antibiotics, since severe infection cannot usually be excluded. If the child is very ill, he should be given dextrose-electrolytes only, but before doing so plasma and urine should be collected for more detailed biochemical analysis, e.g. quantitative amino acids, organic acids, short chain fatty acids. Although a high carbohydrate intake is the ultimate aim, it should be introduced gradually over 48–72 hours as large amounts introduced suddenly will often result in glycosuria and lead to secondary metabolic problems. Ventilate as necessary.

If the child is in extremis the aim of treatment is to (a) remove the potentially toxic metabolite, (b) prevent further production of the metabolite by excluding possible substrates from the diet and reducing breakdown of body proteins, (c) induce imperfect enzyme activity by supplying large amounts of coenzyme. The toxic metabolite may be removed by peritoneal dialysis or exchange transfusion. The first sample of blood removed should be saved so that analyses of enzymes in red cells or white cells may be performed later if indicated.

Dietary limitation of a suspected substrate is achieved by giving a reasonable energy, protein free (initially) diet by intravenous or enteral route. If the latter is possible glucose polymers may be helpful as a source of carbohydrate. An empirical vitamin cocktail may be given in an attempt to give a coenzyme: thiamine 10 mg, riboflavin 10 mg, biotin 10 mg, nicotinamide 100 mg, pyridoxine 10 mg, folic acid 5 mg, ascorbic acid 300 mg—all 8 hourly, plus B_{12} 1 mg daily.

If death is inevitable and no diagnosis has been reached, it is reasonable to introduce normal feeds, stop the vitamin therapy for 24 hours and then collect plasma, urine and possibly cerebrospinal fluid, for freezing. It may be helpful to make a preparation of leucocytes.

Ideally a skin biopsy for fibroblast culture should be taken and sometimes a liver biopsy may be helpful. Liver biopsies taken sometime after death are not generally helpful. Perry (1981) has described the method of autopsy in suspected disorders of amino acid metabolism.

Hypotonia

There are a number of reviews concerning the differential diagnosis of the 'floppy baby' (Dubowitz 1969). Many metabolic disturbances result in marked hypotonia and a similar biochemical approach to that described in the section on Coma (p. 41) might be used, even if the baby is fully conscious.

Conditions not mentioned in that section and which have floppiness as a symptom are hypercalcaemia, renal tubular acidosis and hyperthroidism. The neurolipidoses (e.g. Tay–Sachs), mucopolysaccharidoses (e.g. Hunter–Hurler) and the mucolipidoses, although causing hypotonia usually present later in infancy. Mucolipidosis type II (I-cell disease) may cause a floppy baby with decreased intrauterine movements, but the abnormal appearance and bones are more obvious features. The hypotonia of Pompe's disease (glycogenosis type II) may be apparent before heart failure occurs. The enzyme defect can be demonstrated in leucocytes.

Plasma creatine kinase activity might be determined but in practice its value at this age in the differential diagnosis of hypotonia is limited. In normal babies the activity is 'high' with a peak at about 24 hours, thereafter falling rapidly to reach 'adult' levels within 10 days; levels in low birth weight babies are slightly lower (Wharton *et al.* 1971). Boys with Duchenne muscular distrophy have markedly raised plasma concentrations of the enzyme in the neonatal period but the disease does not cause hypotonia at this time. Rarer autosomal recessive varieties of muscular dystrophy may present with floppiness in the early months of life (Wharton 1965), and have a moderately raised enzyme activity. A slightly raised plasma enzyme activity, even with profound hypotonia, does not necessarily indicate a myopathy since this may occur with neural causes, e.g. Werding–Hoffman disease (see Chapters 15 and 17).

DIARRHOEA AND DEHYDRATION

The main biochemical implication of diarrhoea is the assessment and management of dehydration and the associated electrolyte disorders, but there are a few aetiological points.

Aetiology

Infection is the most common aetiology and the episode is usually short lived.

If diarrhoea persists for more than a week the most likely cause is *secondary* food intolerance, e.g. secondary to gastroenteritis and/or cow's milk protein intolerance. Simple exclusion of lactose is rarely sufficient and modification of dietary protein and fat is often necessary, e.g. using a hydrolysed protein, soya-based or individually prescribed formula. Chronic intractable diarrhoea of undetermined cause presenting in the newborn, particularly in Asian babies or where it is familial, is often very difficult to manage (Larcher *et al.* 1977; Candy *et al.* 1981).

Surgical conditions may present with or be complicated by diarrhoea. Malabsorption of monosaccharides and disaccharides, secondary to malrotation, Hirschsprung's disease and following surgery of the gut in the newborn, for whatever cause, is well described (Anderson & Burke 1975) and a secondary chloride-losing diarrhoea following surgery may occur (Aaronson 1971).

Continuing diarrhoea may rarely be due to an inborn error of intestinal function. Chloride-losing diarrhoea is relatively common in Finland but does occur elsewhere (Launiala *et al.* 1968). It should be considered as a diagnosis if there has been maternal polyhydramnios which may be due to intrauterine diarrhoea, and there is alkalosis in the baby, since most diarrhoea states leading to dehydration are accompanied by acidosis. Investigation shows that the concentration of chloride in the stool fluid is greater than the sum of the concentrations of sodium and potassium in the fluid. Sucrase–isomaltase deficiency is unlikely to cause problems or be diagnosed in the newborn since sucrose is not usually included in the diet. Truly congenital lactase deficiency and glucose–galactose malabsorption are very rare but obviously do cause severe neonatal diarrhoea (Levin *et al.* 1970; Harries 1977). Certain inborn errors of intermediary metabolism have diarrhoea as a feature, e.g. lysinuric protein intolerance, 3-hydroxy-3-methylglutaryl aciduria and dicarboxylic aciduria. We have no experience of them, they seem to present later in infancy and diarrhoea is often accompanied by other serious signs such as hypoglycaemia, acidosis and coma.

Management of dehydration

Management of most instances of dehydration is possible without detailed biochemical help but it may be required when pre-renal failure (hypovolaemia) is changing to established renal failure (e.g. from acute tubular necrosis). A didactic regimen in 4 steps is described showing the place of biochemical determinations. The regimen is adequate for most causes of dehydration, e.g. due to gastroenteritis or the adrenogenital syndrome.

Usually it can be used irrespective of biochemical findings, but modifications may be based on a knowledge of electrolytes and acid–base results.

Step 1. Initially infuse 50 ml per kg body weight of N/2 saline (75 mmol/l) in dextrose over 4 hours. If the child is collapsed, about 10 ml per kg of this initial infusion may be given rapidly (over 30 minutes) as plasma.
Step 2. After 4 hours: begin to infuse 200 ml per kg body weight of N/5 saline (30 mmol/l) in dextrose over 24 hours. If by 4 hours the child has already passed at least 7 ml per kg body weight of urine, potassium may be added to the infusion (20 mmol/l), step 3 can be omitted and step 4 commenced. If metabolic acidosis was present initially it should now be improving. If it is not, reconsider the diagnosis and the cause of acidosis.
Step 3. After 6 hours: if by this time the child has passed a urine volume of at least 7 ml per kg (since 0 hours), continue as in step 2, adding potassium up to 20 mmol/l. If this urine volume has not been achieved (a) continue infusion as in 2 and give frusemide (5 mg per kg) intramuscularly; (b) if during the next 4 hours (6–10 hours) the urine excretion exceeds 7 ml per kg body weight continue as in step 2, but be prepared to repeat steps 3 a, b and c if the urine flow of 7 ml per kg every 6 hours is not maintained; (c) if a urine volume of 7 ml per kg is still not achieved following frusemide administration, consider administering intravenous mannitol 0.75 g per kg body weight and/or the regimen for acute renal failure. If at any time frusemide or mannitol is given, it is helpful to measure both urine and plasma osmolality, sodium, urea and creatinine, after the commencement of treatment. A urine to plasma ratio for osmolality of less than 1 and urine to plasma ratios for urea and creatinine of less than 5 suggest established renal failure, as does a fractional sodium excretion greater than 2.5% (see section on Acute Renal Failure, p. 47). When calculating a urine to plasma creatinine ratio care should be taken that both quantities are expressed in mmol.
Step 4. From 28–30 hours onwards: gradually change from intravenous N/5 saline (30 mmol/l) in dextrose with potassium, to oral glucose electrolyte solution and gradually reintroduce milk or formula, so that in 3–5 days a full strength full volume diet is achieved. Reducing substances in a watery stool are easily detected with a 'Clinitest' tablet (Ames Division, Miles Laboratories, England); care must be taken when collecting the fluid, for example onto a polythene sheet. Breast-fed term babies receiving 11 g lactose per kg daily (i.e. 150 ml per kg day of milk containing 7.2 g lactose per 100 ml) commonly excrete $\frac{1}{4}$% reducing substances (i.e. a green 'Clinitest') and concentrations below $\frac{1}{2}$% may be ignored. If the diarrhoea is thought to be due to gastroenteritis it is also possible to ignore stool reducing substances whatever their concentration during the first 48 hours of therapy and to press on with the management regimen described above.

If stool reducing substances in significant amounts do persist (i.e. > $\frac{1}{2}$%

for > 48 hours) then secondary food intolerance or one of the rarer causes of continuing diarrhoea should be considered. There is little point in attempting carbohydrate tolerance tests (rise in plasma glucose following oral loads of particular carbohydrates) at this stage. They may precipitate profuse diarrhoea and even if abnormal do not determine a primary cause or a secondary effect.

ACUTE RENAL FAILURE

The ability of the neonatal kidney to concentrate urine is limited, the maximum osmolality which can be achieved being 600–700 mmol/kg. The glomerular filtration rate is low, often only 30–40 ml/min per 1.73 m^2 SA. This limits the excretion of sodium surplus to requirements, though sodium retention may be further affected by the high circulating levels of renin and aldosterone found in the neonate. The capacity for excretion of an acid load is reduced, partly by the comparatively low availability of phosphate and partly by the reduced level of ammonia production by the cells of the renal tubule. Linked with this is a lowered renal bicarbonate threshold of 16–18 mmol/l, limiting bicarbonate conservation. The immaturity of the neonatal kidney is shown by the wide range of maturity of the nephrons such that tubular reabsorption can easily be overwhelmed, accounting for glycosuria and aminoaciduria. Renal function is discussed further in Chapter 5.

Acute renal failure results in azotaemia and oliguria (arbitrarily taken as less than 24 ml per kg in 24 hours), but non-oliguric renal failure may also occur. Now that babies receive much lower protein intakes than previously, a raised plasma urea concentration (> 5 μmol/l) indicates some degree of renal failure of whatever variety. Plasma creatinine concentration rises immediately after birth, reaching a peak as high as 115 μmol/l in pre-term babies during the first 48 hours of life, but thereafter falling to less than 60 μmol/l by the end of the neonatal period. A creatinine greater than 100 μmol/l after the first 48 hours of life is abnormally high.

Aetiology

Table 2.14 shows some causes of acute renal failure. There is a recent monograph on the subject by Lewy (1981). Most causes of neonatal renal failure are pre-natal and even in established intrinsic renal failure hypoxia, shock and hypotension are the major causes (Anand *et al.* 1978; Norman & Asadi 1979).

Perinatal hypoxia with or without hypotension is the most common cause of acute renal failure in the newborn (Dauber *et al.* 1976), but since most babies are only moderately affected and recover, this cause rarely

Table 2.14 Causes of acute renal failure in the newborn.

Pre-renal
Hypoxia ± shock
Dehydration of whatever cause

Renal
Renal vein thrombosis (often secondary to dehydration, shock, hypoxia, complication of uncontrolled maternal diabetes)
Renal arterial occlusion (complication of umbilical artery catheterisation)
Acute tubular necrosis (often secondary to a pre-renal cause)
Renal dysplasia and hypoplasia
Varieties of polycystic disease

Post-renal
Urethral values or stricture
Prune belly syndrome (absent abdominal musculature)

Although renal agenesis is an obvious cause of renal failure, these babies usually die because of pulmonary hypoplasia, well before significant azotaemia has occurred.

appears on a list prepared by a nephrology service where developmental problems such as renal dysplasia, varieties of polycystic disease and urinary obstruction predominate. A period of azotaemia (and sometimes hyponatraemia) together with proteinuria and microscopic haematuria is common in babies following a perinatal insult. Dehydration due to whatever cause is an important treatable cause.

Management

Whenever oliguria and azotaemia are noted the possibility of dehydration should be considered carefully. Signs of dehydration in the newborn are difficult to assess, particularly in the light-for-gestational-age marasmic baby, but serial weights are often available; only 3% of normal healthy term babies lose more than 8% of their body weight. If urine and blood are available for analysis then the urinary (U):plasma (P) ratios may be determined. Thus ratios for osmolality > 2, urea or creatinine > 10, fractional sodium excretion (U/P ratio for sodium ÷ U/P ratio for creatinine expressed as a percentage) $< 2.5\%$, favour a 'prerenal' cause for the renal failure and indicate the need for restoration of the circulation, whereas an osmolality ratio < 1, urea and creatinine ratios < 5, fractional sodium excretion $> 2.5\%$, suggest already established true renal failure (Matthew *et al.* 1980). When urine is not available, a molar plasma urea: creatinine ratio greater than 80 suggests pre-renal failure so long as urea is greater than 10 mmol/l and creatinine is greater than 0.2 mmol/l.

Urinary obstruction in the newborn usually results in a large bladder.

It is stressed that pre-renal failure is the most common variety in the newborn. Therefore, if this is a possibility, even though doubt exists, it is reasonable to follow the regimen given for dehydration, with the use of frusemide as indicated. This will help differentiate a pre-renal or intrinsic renal cause (see section on Diarrhoea and Dehydration, p. 44).

If true acute renal failure is established, fluid intake is reduced to 2 ml/kg bodyweight per hour. This may be given as a standard formula to which extra fat and carbohydrate have been added to boost the energy intake but in practice, this stage is rarely reached in the newborn and if it is progression is so rapid that dialysis is usually indicated.

HYPONATRAEMIA

We regard significant hyponatraemia as a plasma sodium below 130 mmol/l. It presents in four major ways, although there is considerable overlap.

The ill baby requiring intensive care

These babies often have indwelling cannulae or catheters and may be receiving nutrition parenterally. Contamination of blood samples taken via indwelling cannulae or catheters may cause spuriously low concentrations due to dilution, or occasionally spuriously high ones due to contamination with sodium heparin. The regimen for parenteral nutrition should incorporate a fat-free period of 4 hours so that blood for electrolytes may be taken at the end of this period, otherwise hyperlipaemia may lead to spuriously low sodium concentrations. Plasma osmolality should be normal, however, as should serum sodium measured with an ion selective electrode.

If these causes of 'pseudo hyponatraemia' are excluded, then inappropriate antidiuretic hormone (ADH) production, which would in infants be primarily hypothalamic in origin should be considered. Another cause is the 'sick cell' concept which involves increased cell permiability allowing efflux of, usually, non-diffusable solute. Flear *et al.* (1981) have suggested that these causes may co-exist, and certainly in clinical circumstances they should be considered together and often as one.

The possibility of sodium deficiency and the consequent need for replacement requires urgent consideration but the ill baby requiring intensive care is a prime candidate for hyponatraemia due to causes other than sodium deficiency:

1 Excessive amounts of water (as 5% dextrose) may have been given to the mother during labour.
2 Diminished venous return may arise either from raised intrathoracic pressure due to inappropriately high continuous positive airways pressure (CPAP), or from positive and expiratory pressure (PEEP) during ventilation leading to appropriate ADH secretion.
3 Intracranial injury (haemorrhage or oedema following hypoxia) involving the hypothalamus, may lead to truly inappropriate secretion of ADH or sick cells.
4 Intravenous nutrition may result in hyperglycaemia with a secondary fall in plasma sodium to maintain normal osmolality (it may also cause glycosuria resulting in polyuria, dehydration and sodium deficiency—see below).
5 Hypoxia, intracranial injury and severe infection (often secondary to the various foreign bodies in the baby such as an endotracheal tube, arterial monitoring cannulae or catheters—particularly 'main line' central venous catheters) may lead to the sick cell syndrome. In this condition the sick cells leak organic solutes and the extracellular sodium is diluted in an attempt to maintain osmolality.

It is, nevertheless, often impossible to confidently exclude true sodium depletion as the cause of the hyponatraemia. A moderately raised plasma urea (5–9 mmol/l) favours sodium *deficiency*. In practice the measurement of urinary sodium concentration is not always useful. It is very low in healthy newborn babies anyway and so cannot be taken as certain evidence of deficiency. Although a high urinary concentration in the presence of hyponatraemia (e.g. urine sodium > 10 mmol/l when plasma sodium < 130 mmol/l) occurs in renal or endocrine causes of sodium deficiency, it may also occur in inappropriate ADH secretion. Accurate daily weighing is invaluable. Generally hyponatraemia with loss of weight indicates sodium and water depletion, while hyponatraemia with weight gain indicates inappropriate ADH/sick cell syndrome but the extraneous paraphenalia of intensive care often make accurate weighing impossible. When in doubt it is reasonable to proceed as follows:
1 If the blood urea is very raised (> 10 mmol/l) proceed as for acute renal failure (p. 47) with a view to differentiating between pre-renal or renal causes. If the blood urea is only moderately raised (5–9 mmol/l) and the child seems to be passing urine reasonably, i.e. at least 4 wet nappies a day, regard the sodium deficiency as mild and give additional sodium and water, e.g. 150–200 ml per kg daily of N/5 saline (30 mmol/l) in 5% dextrose (i.e. 4.5–6 mmol Na per kg).
2 If the blood urea and plasma creatinine levels are normal, review all the possible factors which may result in hyponatraemia without sodium deficiency and proceed as necessary, i.e. has there been excessive water administration, are CPAP or PEEP pressures unnecessarily high, are

hyperglycaemia and glycosuria occurring, is there undiagnosed hypoxia or untreated infection? This review is the most important aspect of management. If one of these causes is very likely, first deal with the underlying cause and then restrict fluid, e.g. 80 ml per kg daily, which if given as N/5 saline (30 mmol/l) in 5% dextrose will provide 2.4 mmol sodium/kg body weight.

3 If there is uncertainty it may be necessary to 'sit on the fence'. Restrict the fluid intake to 120 ml per kg body weight of N/5 saline (30 mmol/l) in 5% dextrose; this regimen provides moderate fluid restriction in case of inappropriate ADH secretion while providing a sodium intake (3.6 mmol per kg day) in excess of normal requirements. Since this regime does 'sit on the fence' it should not be suitable for any cause, but in practice it seems to be useful! If this regimen is followed by a rise in blood urea, then the probable diagnosis is sodium deficiency which has not been corrected and more fluid and sodium are required. If the blood urea remains the same or falls and the plasma sodium falls, then inappropriate ADH/sick cell syndrome is more likely. Further restriction of both fluid and sodium is necessary and treatment should be aimed at the underlying cause.

4 If hyperglycaemia or glycosuria is present, the amount of glucose in the intravenous regimen should be reduced or insulin should be given (0.5–1.0 unit per kg 12 hourly subcutaneously). Although a glucose insulin regimen has been used in the management of sick cell syndrome in older children and adults, we are unaware of its use in the newborn except *en passant* during the management of hyperglycaemia secondary to parenteral nutrition.

Sodium deficiency due to excessive losses

This may occur at any time. It is usually associated with dehydration and is due to excessive loss from either the gastrointestinal or renal tracts.

Gastrointestinal losses

These are most commonly due to infective gastroenteritis but occasionally to inborn errors of absorption of chloride or sugar (see section on Diarrhoea, p. 44). Dangerous losses due to vomiting are uncommon in the newborn but pyloric stenosis can occur in the first week of life and intermittent obstruction of the upper gastrointestinal tract presents as vomiting, e.g. malrotation of the intestine with Ladd's bands causing intermittent duodenal obstruction or an ectopic pancreas in the wall of the duodenum. Generally the gastrointestinal disorders which lead to hyponatraemia will have clear additional presenting symptoms, e.g. vomiting, diarrhoea, abdominal distentions, but very severe water diarrhoea can be missed because it is mistaken for urine.

Gastrointestinal losses of sodium are often associated with hypokalaemia since there is a concomitant loss of potassium in the stool fluid, a response by the renal tubules to the alkalosis of vomiting or chloride losing diarrhoea, and an effect of hyperaldosteronism secondary to the sodium loss. Renin and aldosterone are, therefore, both raised and the urinary sodium : potassium ratio is below 1. Management is described in the section on Diarrhoea and Dehydration, p. 44.

Adrenocortical insufficiency

Disorders which lead to sodium loss from the tubules are described further in Chapter 12. Symptoms due to severe sodium deficiency occur at 7–14 days of age, perhaps because the effect of maternal salt-retaining hormones is disappearing. An early clue before this is the finding of ambiguous genitalia (p. 54). This may indicate masculinisation of a genotypic female due to 21-hydroxylase deficiency (the most common). Alternatively it may indicate the rarer feminisation of a genotypic male due to 20,22-desmolase or 3β-hydroxysteroid dehydrogenase deficiency. About a half of children with the 21-hydroxylase defect and all children with the other defects have life-threatening salt loss. Investigation of a baby with ambiguous genitalia before salt loss has presented is described later (p. 54).

Deficiencies of 21-hydroxylase defect in a male, 20-22 desmolase and 3β-hydroxysteroid dehydrogenase in the female, and the specific defects of aldosterone biosynthesis, i.e. 18-hydroxylase or 18-hydroxysteroid dehydrogenase deficiency do not result in ambiguous genitalia however, and so the diagnosis may not be considered until the baby becomes very ill with hyponatraemia dehydration.

A clue at the time of the acute illness is the presence of hyperkalaemia, reflecting an aldosterone deficiency. More sophisticated investigation shows a raised renin (normal mean 1.18 pmol A1/ml per hour at 6 days) and a *low* aldosterone (normal mean 1248 pmol/l on day 6) (Dillon *et al.* 1976).

If the child is ill and the diagnosis of adrenocortical insufficiency is suspected routine treatment for severe dehydration should be instituted. Biochemical measurements required for diagnostic purposes are described in Chapter 12.

Except in the classical 21-hydroxylase deficiency it may be impossible to reach an exact biochemical diagnosis during the clinical emergency. If in doubt it is better to risk treating the baby with steroids (which may prove to have been unnecessary) and possibly confuse the diagnosis than for the child to die: the diagnosis can always be sorted out later.

Adrenocortical insufficiency may rarely be due to adrenal hypoplasia. This may be isolated, associated with pituitary hypoplasia or associated with xanthomatosis (Wolmans disease). Oestrogen excretion by the mother is very low and may be an antenatal clue.

Renal losses

Pseudohypoaldosteronism, probably due to autosomal recessive inheritance (Cheek & Perry 1958), presents like the salt-losing crisis of the adrenogenital syndrome. Hyperkalaemia is present but the plasma and urinary glucocorticoid and androgenic steroids are normal, whilst aldosterone secretion is excessive. Treatment is with large amounts of sodium, e.g. 10–20 mmol per kg day; salt retaining corticosteroids have no effect.

In some cases of glomerular failure due to whatever cause, salt loss is sufficiently excessive to cause hyponatraemia. In older children this is often referred to as salt-losing nephritis and is associated for example with polycystic disease and dysplasia, or pyelonephritis. Some babies who have had an hypoxic insult pass through a brief azotaemic stage with mild proteinuria and microscopic haematuria and then develop hyponatraemia: we suspect this is the same phenomenon.

Excessive sodium loss can occur as one facet of a multiple disorder of tubular function (Fanconi syndrome) and in any renal tubular acidosis. Bartters syndrome may be associated with hyponatraemia, together with the more constant biochemical features of the syndrome, i.e. alkalosis and hypokalaemia. This triad of biochemical signs may raise the question of whether an upper gastrointestinal abnormality has caused vomiting, but vomiting is not a particular feature of Bartters syndrome. When the cause of sodium deficiency is a renal disorder, sodium loss may be relatively slow and so it is possible to demonstrate unequivocal evidence of an inappropriate loss of sodium, i.e. a urinary excretion greater than the dietary intake in the presence of hyponatraemia. This is often impossible in the renal losses secondary to adrenocortical insufficiency because the baby is seriously ill, intravenous therapy with sodium is imperative and thereafter urinary sodium concentrations are difficult to interpret.

Heart failure and hyponatraemia

Hyponatraemia in babies with heart failure is most commonly the result of over-vigorous diuretic therapy, and is usually associated with a raised blood urea. If potassium supplements are not given there may also be hypokalaemia and alkalosis. Although potassium supplements may not be necessary for many adults receiving diuretics, they are essential in infancy. Hyponatraemia may however be the result of the heart failure itself leading to a sick cell syndrome; this is a biochemical sign of severe failure and in these circumstances the blood urea is often moderately raised. Finally, many babies in heart failure will be ill ones requiring intensive care (p. 49) so that many of the factors discussed above will be acting. It may, therefore, be difficult to decide whether the hyponatraemia is due to the heart failure, the treatment, or some associated problem, and whether to relax diuretic therapy. Biochemical investigation rarely helps solve this dilemma.

If hyponatraemia and ureamia occur when the child is clearly not in heart failure it is reasonable to tentatively assume there is sodium depletion, restrict the fluid intake (120 ml per kg), stop the diuretics and give additional sodium, e.g. 4 mmol per kg bodyweight. The sodium should rise, and when (if) diuretics again become necessary a milder regimen should be introduced. If the hyponatraemic baby is in definite heart failure and this is due to a patent ductus arteriosus, pharmacological closure by indomethacin may be considered. If it is inappropriate to give indomethacin or if it fails then empirically we have used the following regimen: (a) reduce fluid intake to 120 ml per kg, (b) increase the strength of the feeds by one-third (104 kcal and 1 mmol sodium per kg body weight daily) and (c) substitute a 'milder' diuretic, e.g. chlorthiazide 50 mg instead of frusemide. This regimen has often been satisfactory and has prevented the need for surgical closure of a duct. Presumably it is effective because a reasonable intake of energy and sodium for growth is maintained despite the reduction in fluid intake, but the reduced intake enables a less potent diuretic to control the failure without severe effects on electrolyte metabolism.

We have little experience of spironolactone but recent reports suggest it may be helpful in neonatal heart failure and we suspect particularly so in this situation (Hobbins *et al.* 1981).

Late hyponatraemia in low birth-weight babies

This occurs usually in otherwise well, small babies around the third week of life. They are often growing quite rapidly and the plasma sodium concentration may be very low (e.g. < 120 mmol/l) in the face of apparently good health. There are probably two mechanisms involved. First, the rapid growth velocity outstrips the dietary supply of sodium. Calculations (using the factorial method) of sodium requirement by the rapidly growing babies are higher than can be obtained either from breast milk or most modern infant formulae (see section on Very Low Birth Weight Baby, p. 65). Secondly, renal tubular conservation of sodium in the presence of the hyponatraemia is relatively immature. This is probably due to an immaturity of mechanisms within the renal tubule itself, since aldosterone secretion is high (Honour *et al.* 1974).

Extra dietary sodium may be indicated for very low birth-weight babies since it prevents this syndrome and supports more rapid growth (Roy *et al.* 1976), but it is probably unnecessary in babies of birth weight > 1.80 kg (Berger *et al.* 1980).

AMBIGUOUS GENITALIA

The causes of ambiguous genitalia and their diagnosis are discussed in Chapter 12. The essential feature of early management is to determine whether the ambiguous genitalia are due to congenital adrenal hyperplasia, in which case a proportion of the children are in grave danger from a salt-losing crisis (see preceding section on Hyponatraemia—Adrenocortical insufficiency, p. 52).

Management in the first week

Determination of genotypic sex (by nuclear sexing, fluorescent Y chromosome staining, or full karyotype) should be arranged urgently. It is often possible to make a reasonable clinical 'estimate' of genotypic sex. Most babies born with ambiguous genitalia, but without a palpable gonad, are virilised XX babies. A palpable gonad is almost always a testicle and so the baby with a gonad and ambiguous genitalia is an incompletely masculinised XY baby. The possibility of virilising drugs having been given to the mother is explored. An early genitogram and ultrasound search for Müllerian structures (Fallopian tubes, uterus, upper vagina) is valuable but is not widely available and so is often delayed.

Meanwhile, preparation for, and early detection of potential salt loss proceeds. The child is weighed daily. There is little point doing serial plasma sodium determinations since the concentration remains normal until immediately prior to the crisis. Biochemical determinations are discussed in Chapter 12.

Later management

If salt loss does not occur or has been adequately treated, then a more leisurely anatomical and biochemical approach proceeds (see Chapter 12).

Although assignment of gender may be psychosocially urgent, life threatening events are not imminent. In many instances of ambiguous genitalia, however, there are no obvious endocrine or other causes, urgent diagnosis is not necessary and future management is mainly via plastic surgery and exclusion of possible associated urinary tract abnormalities. Useful reviews are given by Brook (1978) and Lippe (1979).

OTHER PLASMA ELECTROLYTE ABNORMALITIES

Hypokalaemia

As an acute neonatal problem, the most common causes are excessive losses from either the intestine in gastroenteritis or from the kidney as a complication of diuretic therapy, the cause usually being clear. The mild renal failure commonly seen with perinatal hypoxia may be followed by hypokalaemia, presumably because there is a diuresis. Potassium loss from the tubules may occur in any state leading to metabolic alkalosis (p. 61) e.g. vomiting, or hyperaldosteronism secondary to a deficiency of total body sodium. Finally, if potassium is inadvertently omitted from a total intravenous regimen hypokalaemia develops within 24–48 hours in very low birth weight babies (less than 1.5 kg). Treatment for all of these causes with oral potassium (e.g. 5 mmol per kg body weight daily) is simple. Whether extra water and sodium are required also must be considered carefully. The differential diagnosis of chronic hypokalaemia is little different from that in older children, i.e. mainly losses from the gut or kidney. Disorders affecting tubular function which lead to a potassium leak together with acidosis include the Fanconi syndrome, e.g. due to cystinosis (presentation in the neonatal period is unusual) and any distal renal tubular acidosis. Potassium loss and alkalosis occur in Liddle's syndrome together with hypertension (not reported before 10 months age), and in Bartters syndrome in which there is no hypertension.

Generally hypokalaemia with metabolic alkalosis is unusual in newborn babies. If chronic hypercapnia, diuretic therapy without potassium supplements, marked vomiting and chloride losing diarrhoea can be excluded, then a renal tubular disorder is likely (see Chapter 5; Schwartz & Spitzer 1978). Primary hyperaldosteronism is rare in childhood.

Hyperkalaemia

Hyperkalaemia is commonly spurious as a result of haemolysis of a heel prick sample of capillary blood.

True hyperkalaemia is frequently associated with uraemia. This may be a feature of renal failure (p. 47) or occur as a result of adrenal insufficiency (p. 52) which in itself causes sodium loss, dehydration and uraemia. Adrenal insufficiency should be considered particularly when the baby has become ill during the second week of life or when there are ambiguous genitalia. If there is difficulty in deciding the aetiology of hyperkalaemia and uraemia in an ill child, it is reasonable to treat with a regimen for dehydration (p. 45) while investigating further. This is suitable initial treatment for adrenal insufficiency, while if the cause is renal the treatment will help differentiate the pre- and intrinsic renal varieties.

A baby severely ill with hypoxia (p. 11) or infection may have hyperkalaemia, with only mild elevations of blood urea, probably reflecting tissue injury and thence protein catabolism. A blood transfusion if given too rapidly through very small needles may be followed by haemolysis and hyperkalaemia.

If the *true* plasma potassium (a venous sample will probably be necessary to be certain) is more than 8 mmol/l, then lowering of the concentration as an emergency is achieved by giving 1 unit soluble insulin and 2 g dextrose intravenously.

Hypomagnesaemia

Although the mean minus two standard deviations for plasma magnesium concentration in the newborn is 0.5 mmol/l, concentrations below this may not be associated with any symptoms (Harvey *et al.* 1970).

'Isolated' hypomagnesaemia, i.e. without any other biochemical abnormality, may present with convulsions but is rare (Dooling & Stern 1967).

It frequently accompanies late onset hypocalcaemia and treatment with magnesium as well as calcium may be necessary to achieve a normal calcium level (see section on Hypocalcaemia, p. 38).

It may also accompany hypokalaemia due to excessive losses of both metals from the gut in gastroenteritis, particularly in a malnourished marasmic child, or from the renal tubules in a rare hereditary tubular disorder (Gitelman *et al.* 1969).

Hypernatraemia

Hypernatraemia almost always occurs with fluid depletion. It is, therefore, associated with excessive weight loss or inadequate gain and most commonly due either to use of an overhead radiant heater, or to gastroenteritis. Urinary sodium and osmolality will initially be high but as acute renal failure develops urine may become more dilute. Some instances of isonatraemic dehydration with acidosis may be converted into hypernatraemic dehydration if attempts are made to correct the acidosis with high concentrations of sodium bicarbonate without correcting the dehydration (see section on Management of dehydration, p. 45).

Hypernatraemia also occurs in fluid depletion due to diabetes insipidus, but urinary sodium osmolality and specific gravity remain low. In practice, we have never diagnosed pituitary diabetes insipidus in the newborn and hereditary (sex linked) nephrogenic diabetes insipidus presents later in infancy, although theoretically there seems no reason why it should not be present in the newborn.

Hypernatraemia without primary fluid depletion can occur in feeding accidents, when excessive amounts of sodium are added to feeds but even then urinary losses of water, in an attempt to excrete the unnecessary sodium, will be excessive and so lead to secondary dehydration.

An immediate accurate diagnosis is not essential and in all instances if the child is ill it is reasonable to treat as for 'dehydration'. Even in iatrogenic hypernatraemia the extra 'free water' within the rehydration regimen will improve the situation.

AN ACID–BASE ABNORMALITY

The technology of blood acid–base determination has advanced greatly and today results can be obtained with relative ease and speed. As a result the importance of a good quality specimen may be overlooked. If the service is provided by the laboratory, then the specimens collected will be capillary and care must be taken in the training of staff. Unless a good flow of blood is achieved from a well-perfused limb the results will be worthless. If the blood-gas machine is on a special care unit and used by medical staff many specimens will be arterial. Here again, training is essential though the margin for error is much reduced with machines which are self-calibrating and have diagnostic fault-finding capability. It is important though to stress the need for an ongoing quality assessment programme.

It is very easy to introduce air bubbles into the blood sample on collection or on introduction to the machine. This often gives a spurious result and the various components of the acid–base profile lack internal consistency.

Metabolic acidosis

A base excess of below −8 mmol/l is abnormal. Moderate degrees of metabolic acidosis—'late metabolic acidosis'—were very common in pre-term babies fed on formulae containing 2 g or more of unmodified cows' milk protein per 100 ml, largely because of the endogenous acid load (i.e. acid formed from metabolism of protein, particularly the sulphur amino acids), but also reflecting some renal limitation in acid excretion (see Kildeburg & Winters 1972; Berger *et al.* 1980 for reviews). The use of demineralised whey formulae has reduced the prevalence of this phenomenon considerably, but it should still be considered amongst the diagnoses. Unsuitable intravenous feeding regimen, particularly ones containing fructose or sorbitol will cause acidosis, but these carbohydrates should not be used in the newborn. The diagnosis may be approached in three ways (see Table 2.15).

Preliminary assessment

The most common cause by far of metabolic acidosis in the newborn is hypoxia, and after this, infection and dehydration should be considered carefully before going on to consider intrinsically metabolic causes. Intrapartum asphyxia or postnatal hypoxia (p. 10), e.g. due to the respiratory distress syndrome, will usually be accompanied by a respiratory acidosis too, but this and the hypoxaemia may quickly disappear when adequate ventilation is achieved, leaving only the metabolic acidosis. Therefore, in assessing the possibility of hypoxia as a cause of metabolic

Table 2.15 Metabolic acidosis—causes and other features.

Preliminary assessment
Hypoxia
  arterial and tissue
  often hypercapnia too
Dehydration
Azotaemia
Infection

Overproduction
Inborn errors of branch-chain amino acid metabolism
  ketosis, abnormal amino acid pattern
Inborn errors of carbohydrate metabolism
  lactate raised, hypoglycaemia
  urinary reducing substances may be present
Lactate/Pyruvate acidoses
  normal amino acid pattern or hyperalaninaemia;
  lactate raised

Under excretion
Glomerular disease
  azotaemic renal failure
Tubular disease (hyperchloraemia present)
  Proximal
    variable urine pH
    may be part of general tubular diseases with
    leakage of other substances
  Distal
    urine pH always > 6
    transient and permanent varieties

acidosis the history may be more important than the contemporary blood gases. Infection may not be very severe and the child not yet ill, yet mild acidosis develops; certainly we have diagnosed urinary tract infections following further investigations of a mild persistant acidosis. Tissue hypoxia, despite normal arterial blood gases, may occur in hypothermia and other causes of a poor peripheral circulation, e.g. following an intraventricular haemorrhage or with the hypoplastic left heart syndrome. These diagnoses will usually be suspected on other grounds too, but may cause some confusion.

If the common neonatal causes (hypoxia, dehydration, infection) are excluded intrinsic metabolic causes may be suspected. The standard metabolic screen shown in Table 2.2 may be performed together with a plasma chloride determination. Generally if the plasma chloride is raised, the amino acid pattern is normal, and there is no ketonuria, renal tubular disease should be considered. If, however, the plasma chloride is low or normal, over-production of an 'undetermined anion' is probably responsible for the acidosis and this should be sought.

Over-production of acid

Most of the inborn errors of amino acid metabolism which result in over-production of acid have other serious symptoms, including coma, apnoea and convulsions, and these may occur before marked acidosis is apparent, e.g. maple syrup urine disease, organic acidaemias. There is usually ketonuria and the plasma amino acid screen is abnormal. Alternatively some other biochemical sign will give a diagnostic clue, e.g. a reducing substance in the urine in galactosaemia. The combination of metabolic acidosis with anaemia occurs in 5-oxoprolinuria and this may at first be ascribed to perinatal shock and haemorrhage.

If none of these symptoms are present and the preliminary metabolic screen is negative, lacticacidosis should be considered. A raised fasting plasma lactate (1.4 mmol/l) occurs in hypoxia, tissue hypoxia, dehydration and infection and so these diagnoses should again be considered.

Glycogen storage disease Type 1 results in lactic acidosis and the diagnosis may be easily missed (see Chapter 3). The congenital lactate/pyruvate acidoses are usually accompanied by hyperalaninaemia. In this group of disorders neurological symptoms may be intermittent (e.g. pyruvate carboxylase deficiency, pyruvate dehydrogenase deficiency) but others are associated with severe progressive neurological deterioration, e.g. Leigh's encephalopathy. Scharer (1972) has reviewed the congenital lactic acidoses.

Reduced excretion of acid

The acidosis may be due to renal failure leading to azotaemia and often mild hyponatraemia, secondary to reduction in the number of nephrons.

If the plasma urea and creatinine are normal and hyperchloraemia is present, renal tubular acidosis is probable. At this stage it is useful to collect three random urines for pH, protein, sugar, and sodium and potassium concentrations and as each urine sample is passed to take blood for the determination of acid–base status, sodium, potassium and chloride levels. In practice we sometimes arrange for the urines to be collected in the morning so that the blood sample can also be used for a lactate determination.

Distal renal tubular acidosis (Classical or Type 1) is associated with hyperchloraemia, an alkaline urine (pH > 6) irrespective of the blood standard bicarbonate concentration (but only when the standard bicarbonate is below 18 mmol/l should any significance be attached to a urine pH of > 6), a low urinary sodium to potassium ratio (probably reflecting mild secondary hyperaldosteronism) and often mild hypokalaemia. Correction of the acidosis is easily achieved, usually requiring much less than 5 mmol/kg daily of bicarbonate. Permanent primary distal renal tubular acidosis (Butler Allbright syndrome) may present in the neonatal period and one variant is associated with nerve deafness. Commonly, however, the problem is a transient one in pre-term babies and the disorder often seems to have disappeared by the time investigations are complete. It was more common some years ago and probably due to intoxication with vitamin D or mercury.

Proximal renal tubular acidosis (Type II) is also associated with hyperchloraemia but the urine pH may fall to 4 or 5 as the blood standard bicarbonate falls below 18 mmol/l; usually the urinary sodium potassium ratio is normal. Correction requires more than 5 mmol/kg daily of bicarbonate. The lesion may be a primary single hereditary one but more usually is a part of multiple tubular dysfunction due to, for example, cystinosis, galactosaemia, fructosaemia, cyanotic congenital heart disease or renal infarcts following withdrawal of umbilical artery catheters. In practice proximal renal tubular acidosis, whether primary or secondary, is rarely diagnosed in the newborn. The diagnosis and management of the renal tubular acidoses is discussed in detail by Rodriguez-Soriano (1978).

Alkalosis

An increased standard bicarbonate or base excess occurs most commonly as metabolic compensation for a raised $P\text{CO}_2$. Apart from this, alkalosis is rare in the newborn and, if found, the accuracy of the result should be questioned and an iatrogenic cause considered, e.g. administration of alkali.

If such causes are excluded, conditions leading to excessive loss of anion should be considered; they are often accompanied by hyponatraemia, hypochloraemia and hypokalaemia, e.g. excessive vomiting, chloride-losing diarrhoea (see section on Hyponatraemia, p. 49) and Bartter's syndrome.

The association of alkalosis with certain clinical presentations may point to the diagnosis, e.g. with vomiting, consider upper gastrointestinal obstruction or occasionally Bartter's syndrome; with diarrhoea consider chloride-losing diarrhoea; with drowsiness consider hyperammonaemia of whatever cause. Very prolonged gentamicin therapy has resulted in alkalosis in older children but is not described in the newborn.

OEDEMA AND HYDROPS

Hydrops fetalis

Table 2.16 shows the causes of hydrops fetalis. In reaching a diagnosis the following questions should be asked: (a) is the baby anaemic? (b) could he be suffering from an intrauterine infection? (c) is there evidence of disease of the heart, kidney or liver? In most instances the cause is severe

Table 2.16 Causes of hydrops fetalis.

Severe chronic anaemia in utero
Rhesus incompatability
Homozygous α-thalassaemia (usually pre-term stillbirths)
ABO incompatability (unusual)
Chronic bleeding into mother or other twin
Intrauterine infections
Syphilis, cytomegalovirus, toxoplasmosis
Cardiovascular disorders
Any variety of congenital heart disease
Angiomatous malformations in placenta or fetus
Renal tract disorders
Congenital nephrotic syndrome
Urinary ascites due to lower urinary tract obstruction
Polycystic disease
Liver disease
Congenital cirrhosis
Congenital Gaucher's disease
Others
Pulmonary lymphangieotasia (early death due to respiratory failure)
Cystic adenomatoid malformation of lung
Achondroplasia
Neuroblastoma
Congenital Gaucher's
Chylous ascites (no peripheral oedema)

Rhesus haemolytic disease but if this is excluded the most important chemical investigations are the plasma albumin and urinary protein.

Heavy proteinuria and hypoalbuminaemia indicate the nephrotic syndrome. This may be of the idiopathic variety (rare outside Finland) due to a specific nephropathy (diffuse mesangial sclerosis) or secondary to intrauterine infection, including treatable syphilis (of which the only other manifestation might be osteitis), renal vein thrombosis, and very rarely a nephroblastoma.

In the Finnish variety of congenital nephrotic syndrome the placenta is more than 25% of the fetal weight and the proteinuria is initially highly selective. Hyperlipidaemia is present, so spuriously low plasma concentrations of sodium may be obtained. Death occurs almost always within a year or so mainly due to infection; uraemia does not occur. Bilateral nephrectomy, haemodialysis and transplantation if still alive after the first birthday, may have a role so early diagnosis and treatment may be worthwhile (Hallman & Rapola 1978; Hoyer & Anderson 1981).

Postnatal oedema

Mild peripheral oedema is common in pre-term babies. 'Biochemical' causes of pitting oedema should be considered only after common perinatal complications have been excluded as a cause, e.g. hypothermia, intrapartum hypoxia (possibly due to a shift of fluid from mother to baby, and also increased capillary permeability), cardiac failure, severe chronic anaemia *in utero*, and iatrogenic overload.

Iatrogenic overload with sodium readily causes oedema in the newborn. It occurs in various circumstances: as an accident; because hyponatraemia (p. 49) due to inappropriate ADH secretion/sick cell syndrome has incorrectly been ascribed to sodium deficiency and excessive sodium supplements are given; when a diagnosis of oliguric renal failure has been delayed; when signs of marasmus are interpreted as signs of profound dehydration (the most experienced of us may do this); when large amounts of sodium bicarbonate are given to correct a metabolic acidosis without correcting the underlying cause such as dehydration.

All the causes of hydrops fetalis shown in Table 2.16 should be considered. Evidence of hypoproteinaemia and proteinuria are again sought (p. 62). Hypoproteinaemia occurs in the nephrotic syndrome and also occurs in the rare forms of congenital cirrhosis (Thaler 1964) and the rare neonatal, possibly familial, variety of protein-losing enteropathy (Cottom *et al.* 1961). We have seen plasma albumin concentrations as low as 25 g per litre in oedematous pre-term babies in whom the most probable cause of the oedema was iatrogenic sodium overload. It seems, therefore, that little diagnostic significance should be attached to a 'low' plasma albumin unless it is below 20 g/l.

Hypocalcaemia, due to various causes, may be associated with neonatal oedema. The pathogenesis is unclear (Benson & Parsons 1964; Chiswick 1971). Late oedema, sometimes in association with a haemolytic anaemia and thrombocytosis, is described in vitamin E deficiency (Hassan *et al.* 1966). The ratio in the diet of vitamin E to polyunsaturated fatty acids is adequate in modern infant feeding formulae, so that this syndrome should now be rare.

HYPERTENSION

Hypertension diagnosed in the newborn has usually presented as heart failure and is most commonly secondary to coarctation of the aorta. Renal vascular or cystic disease are the next most common, and iatrogenic hypertension occurs in fluid and sodium overload during acute renal failure (p. 47).

Metabolic causes are rare. They include over-production of catecholamines by neuroadrenal tissue in a phaeochromacytoma or neuroblastoma or overproduction of aldosterone or other substances with salt-retaining properties, e.g. 11β-hydroxylase or 17α-hydroxylase deficiency of the adrenal cortex. In practice 17α-hydroxylase deficiency does not present until there is delayed puberty because of deficient oestrogen production. The 11β-hydroxylase deficiency results in early hypertension but is more usually diagnosed in the newborn only during the investigation of ambiguous genitalia of an XX baby. Liddle's syndrome (pseudohyperaldosteronism) results in hypertension and hypokalaemic alkalosis but has not been diagnosed before the age of 10 months (see section on Hypokalaemia, p. 55).

BONE LESIONS

These conditions are discussed further in Chapter 14.

Frank *rickets* may occasionally be present *at birth* as a result of severe maternal osteomalacia (see section on Hypocalcaemia, p. 39). It also occurs some weeks after birth in very low birth-weight babies who are beginning to show rapid catch-up growth, and in a few the ribs may be so soft that there is respiratory difficulty (Editorial 1977). It is not clear whether this represents an inadequate intake of vitamin D (particularly by babies receiving breast milk), an inadequate conversion in the liver and/or kidney to form the active 1,25-dihydroxycholecalciferol, or a dietary deficiency of phosphate and/or calcium. A raised alkaline phosphatase without bone lesions in a rapidly growing low birth-weight baby may be an indication for giving extra calciferol (e.g. 25 μg daily) on the basis that the raised activity represents 'subclinical' rickets, but such a policy has not been assessed adequately. Some paediatricians prescribe this amount routinely to *all* very low birth-weight babies. The situation is complex since under normal circumstances rapid growth is physiologically associated with higher concentrations of alkaline phosphatase. Similarly, the place of 1α-or 1,25-hydroxycalciferol and phosphate supplements is not yet clear (see review by Tsang *et al.* 1981).

Bone X-rays with some features resembling florid rickets occur in *congenital hypophosphatasia*. Symptoms more commonly occur in mid-infancy but the full severe syndrome may be apparent in the neonatal period: hypotonia, very soft skull, demineralisation of bones with flaring and irregularities of the long ends of the bones, and shortening and bowing of the bones. Alkaline phosphatase activity is extremely low, calcium and phosphorus normal or high and there is an excess of phosphoethanolamine in the urine. Few babies survive (Goyer 1963).

Primary hyperparathyroidism in the newborn is rare, *transient neonatal hyperparathyroidism* secondary to maternal hypoparathyroidism is a little more common. Both present as bone deformities, demineralisation and a periosteal reaction. The plasma calcium is not necessarily raised in the secondary variety and serum alkaline phosphatase activity may initially be normal. Recorded plasma parathyroid hormone concentrations in both varieties were increased by four times (Spiegel *et al.* 1977; Glass & Barr 1981).

Periosteal lesions, together with anaemia, neutropenia and occasionally oedema due to *copper deficiency*, have been described in a few very low birth-weight babies and in one infant a fractured femur occurred (Blumenthal *et al.* 1980). Reported cases have had a very low plasma copper, e.g. 0.7, 2.5 μmol/l (stated normal 11–25) and sometimes a raised alkaline phosphatase and mild hypocalcaemia. There may be some overlap with rickets but the situation has not been fully elucidated (Sann *et al.* 1978; Yuen *et al.* 1979).

Neuroblastoma and other malignancies may present in the newborn, sometimes resulting in hypercalcaemia (p. 41) proceeding to uraemia due to nephrocalcinosis. *Mucolipidosis type II* (I cell disease) may be apparent in the newborn with hepatosplenomegaly, a coarse 'funny' face, and 'dysostosis multiplex'. There is no excess mucopolysacchariduria, but arylsulphatase A, hexoseaminidase, α- and β-galactosidase, α-mannosidase and α-fucosidase activities are markedly raised. Plasma acid phosphatase is normal.

In other bone lesions apparent in the newborn, e.g. osteogenesis imperfecta, thanatophoric dwarfism and occasionally achondroplasia, the biochemical defect is unclear and none of the presently available biochemical determinations are abnormal.

VERY LOW BIRTH-WEIGHT BABIES

Babies weighing less than 1500 g at birth are immature biochemically, often ill from such complications as hypoxia, hypothermia, and infection and even when 'healthy' their rapid catch-up growth imposes various metabolic strains.

Early critical days

Problems at this time have already been dealt with in other parts of this chapter, see sections on Early hypocalcaemia, Hypoglycaemia, Hyperammonaemia, Metabolic acidosis, etc.

Thyroid function

Screening tests for hypothyroidism may be more difficult to interpret than in the normal-sized baby. Serum concentrations of thyroxine and triiodothyronine are lower, and even more reduced in the very ill baby (Uhrmann *et al.* 1981) due partly to reduced concentrations of thyroid binding globulin. Mostly these babies will have normal thyroid stimulating hormone levels but instances of low thyroxine (T_4) levels with high TSH levels with high TSH subsequently becoming normal, are well-documented in ill pre-term babies (De Lange *et al.* 1978). Whether they require temporary thyroid supplements is not clear.

Hyperaminoacidaemia and hyperaminoaciduria

Eight factors should be considered when interpreting plasma and urinary amino acid concentrations. The first five are particularly important in very small babies but apply to all newborns.

1 *Adequacy of specimen.* Difficulty in collecting blood may distort the result because of haemolysis. Plasma protein should be precipitated within an hour or even earlier if concentrations of the sulphur amino acids are required. Loss of amino acids can ensue from storage of samples from which all protein has not been removed.

2 *Reference range of normal.* This should be adequately defined, e.g. age, maturity, diet, etc., since these factors affect the result (see below).

3 *Maturity.* The more immature have higher plasma and thence higher urinary concentrations of many amino acids, particularly phenylalanine, tyrosine and methionine. In addition, the short proximal tubule reduces reabsorption so that physiological aminoaciduria occurs in all newborn babies.

4 *Diet.* Higher protein intakes, particularly of unmodified cows' milk protein (mainly casein), cause higher plasma concentrations of phenylalanine, tyrosine and methionine (Valman *et al.* 1971) and an increased urinary excretion of cystathionine. Intravenous amino acid solutions obviously directly affect the result.

5 *Growth velocity.* Rapid growth rates (e.g. 30 g per day in babies below 2 kg birth weight) result in lower plasma concentrations of many amino acids, particularly the branched chain ones.

6 *The possibility of an inborn error of amino acid metabolism.* A specific pattern involving one or a characteristic group of amino acids is seen. This usually stands out on semi-quantitative screening without the need for exact quantitative determination.

7 *The possibility of other metabolic errors causing a secondary change in amino acid concentrations.* This includes hyperalaninaemia in pyruvate/lactate acidoses, hyperglycinaemia in the organic acidaemias, hyperglutamicacidaemia in the hyperammonaemias and a mixed aminoaciduria due to generalised tubular dysfunction, e.g. from cystinosis, galactosaemia, or vitamin D deficiency.

8 *The possibility of liver disease.* Plasma phenylalanine, methionine and particularly tyrosine are raised in hepatocellular disease irrespective of its aetiology.

Population screening tests for hyperaminoacidaemia were commonly positive in pre-term babies receiving a very high protein intake from unmodified cows' milk (e.g. 7 g protein per kg, 5.3 g per 100 kcal). This is seen much less commonly, indeed it occurs much less commonly in term babies also, following the substantial changes in infant feeding practice which occurred in the 1970s (Walker *et al.* 1981).

Catch-up growth

The factorial method of estimating nutritional requirements suggests that breast milk contains inadequate amounts of protein, calcium, and sodium to sustain a reasonably normal body composition. Infant formulae specially designed for the low birth-weight baby are available, but experience with them is still limited. In any case the factorial calculations depend on our limited knowledge of the body composition of the fetus, with little allowance for individual variation. In practice, therefore, management is aimed at the recognition of (a) deficiencies because the diet is inadequate for rapid catch-up growth, (b) excesses secondary to the infant's immature biochemical systems. The following points might be useful. If the baby is receiving breast milk only for more than about 2 weeks, it is useful to regularly check the levels of plasma sodium, phosphate, alkaline phosphatase and, if possible, copper. In babies receiving a formula these investigations may also be useful, but in addition, if the protein intake is greater than that given to term babies (e.g. more than 3 g per kg body weight or greater than 2.25 g per 100 kcal), screening for metabolic acidosis, hyperaminoacidaemia, and if possible hyperammonaemia, may be indicated.

These suggestions are made more with the aim of increasing our biochemical awareness of problems in very low birth-weight babies rather than as essential investigations, although we suspect they will be regarded as essential before this decade closes.

REFERENCES

AARONSON I. (1971) Secondary chloride-losing diarrhoea. Observations on stool electrolytes in infants after bowel surgery. *Arch. Dis. Child.* **46**, 479–82.

ANAND S. K., NORTHWAY J. D. & CRUSSI F. G. (1978) Acute renal failure in newborn infants. *J. Pediatr.* **92**, 985–8.

ANDERSON C. M. & BURKE V. (1975) *Paediatric Gastroenterology*, pp. 199–218. Blackwell Scientific Publications, Oxford.

APERIA A., BROBERGER O., ELINDER G., HERIN P. & ZETTERSTROM R. (1980) Postnatal development of renal function in preterm and full term infants. *Acta Paed. Scand.* **70**, 183–7.

ARROWSMITH W. A., PAYNE R. B. & LITTLEWOOD J. M. (1975) Comparison of treatments for congenital nonobstructive non haemolytic hyperbiliribinaemia. *Arch. Dis. Child.* **50**, 197–9.

Baens G. S., Lundeen E., Cornblath M., Morrison M. & Harr C. (1963) Studies of carbohydrate metabolism in the newborn infant. VI levels of glucose in blood in premature infants. *Pediatrics* **31**, 580–9.

Beddis I. R., Hughes E. A., Rosser E. & Fenton J. C. B. (1980) Plasma ammonia levels in newborn infants admitted to an intensive care baby unit. *Arch. Dis. Child.* **55**, 516–20.

Belfield A. & Goldberg D. M. (1971) Normal Ranges and Diagnostic Value of Serum 5-Nucleotidase and Alkaline Phosphatase activities in infancy. *Arch. Dis. Child.* **46**, 842–6.

Benson P. F. & Parsons V. (1964) Hereditary hyperparathyroidism presenting with oedema in the neonatal period. *Quart. J. Med.* **33**, 197–208.

Berger H. M., Scott P. H. & Wharton B. A. (1980) Metabolic acidosis, diet and growth. In Wharton B. A. (ed.) *Topics in Perinatal Medicine* I, pp. 88–97. Pitman Medical, London.

Blumenthal I., Lealman G. T. & Franklyn P. P. (1980) Fracture of the femur, fish odour, and copper deficiency in a preterm infant. *Arch. Dis. Child.* **55**, 229–31.

Brook C. G. (1978) *Practical Paediatric Endocrinology*, pp. 10–24. Academic Press, New York.

Brozovic B., Burland W. L., Simpson K. & Lord J. (1974) Iron status of preterm low birth weight infants and their response to oral iron. *Arch. Dis. Child.* **49**, 386–9.

Candy D. C. A., Larcher V. F., Cameron D. J. S., Norman A. P., Tripp J. H., Milla P. J., Pincott J. R. & Harries J. T. (1981) Lethal familial protracted diarrhoea. *Arch. Dis. Child.* **56**, 15–23.

Cheek D. B. & Perry J. W. (1958) A salt-wasting syndrome in infancy. *Arch. Dis. Child.* **33**, 252–5.

Chiswick M. L. (1971) Association of oedema and hypomagnesaemia with hypocalcaemic tetany in the newborn. *Br. med. J.* **3**, 15–18.

Cornblath M., Parker M. L., Reisner S. H., Forbes A. E. & Daughaday W. H. (1965) Secretion and Metabolism of growth hormone in premature and full term infants. *J. Clin. Endocrin.* **25**, 209–18.

Cornblath M. & Reisner S. H. (1965) Blood glucose in the neonate and its significance. *New Eng. J. Med.* **273**, 278–81.

Cornblath M. & Schwartz R. (1976) *Disorders of carbohydrate metabolism in infancy*, 2nd ed., pp. 155–205. W. R. Saunders, Philadelphia.

Cottom D. G., London D. R. & Wilson B. D. (1961) Neonatal oedema due to exudative enteropathy. *Lancet* **ii**, 1009–12.

Danks D. M., Campbell P. E. & Stevens B. J. (1972) Menkes kinky hair syndrome. An inherited defect in copper absorption with widespread effects. *Pediatrics* **50**, 188–201.

Danks D. M., Cartwright E., Stevens B. J. & Townley R. R. W. (1973) Menkes kinky hair disease: further definition of the defect of copper transport. *Science* **179**, 1140–2.

Dauber I. M., Krauss A. M., Symchych P. S. & Auld P. A. M. (1976) Renal failure following perinatal anoxia. *J. Pediat.* **88**, 851–55.

Davies D. P. (1973) Plasma osmolality and protein intake in preterm infants. *Arch. Dis. Child.* **48**, 575–9.

DeLange F., Dodion J., Wolter R., Bourdoux P., Dalhem A., Glinoer D. & Ermans A. M. (1978) Transient hypothyroidism in the newborn infant. *J. Pediat.* **92**, 974–6.

Dillon M. J., Gillin M. E. A., Ryness J. M. & Swiet M. (1976) Plasma renin activity and aldosterone concentration in the human newborn. *Arch. Dis. Child.* **51**, 537–40.

DOOLING E. C. & STERN L. (1967) Hypomagnesaemia with convulsions in a newborn infant. Report of a case associated with maternal hypophosphatemia. *Can. Med. Assoc. J.* **97**, 827–8.

DRYBURGH E. (1982) Transpyloric feeding. In Wharton B. A. (ed.) *Topics in Perinatal Medicine* II, pp. 75–86. Pitman Medical, London.

DUBOWITZ V. (1969) *The floppy infant: Clinics in Developmental Medicine No. 31*—passim. Spastics International and William Heinemann, London.

EDITORIAL (1977) Rickets, jaundice and late onset respiratory distress in premature babies. *Br. Med. J.* **2**, 78–9.

ELLIS G. & GOLDBERG D. M. (1969) The enzymological diagnosis of galactosaemia. *Ann. Clin. Biochem.* **6**, 70–3.

FERNANDES J., HUIJING F. & VAN DE KAMER J. H. (1969) A screening method for liver glycogen diseases. *Arch. Dis. Child.* **44**, 311–17.

FLEAR C. T. G., GILL G. V. & BURN J. (1981) Hyponatraemia: mechanisms and management. *Lancet* **2**, 26–31.

FOMON S. J., FILER L. J., THOMAS L. N. & ROGERS R. R. (1970) Growth and serum chemical values of normal breast fed infants. *Acta Paed. Scand. Suppl.* **202**.

FREER D. E., STATLAND B. E., JOHNSON M. & FELTON H. (1979) Reference values for selected enzyme activities and protein concentrations in serum and plasma derived for cord-blood specimens. *Clin. Chem.* **25**, 565–9.

GITELMAN H. J., GRAHAM J. B. & WELT L. G. (1969) A new familial disorder characterised by hypokalaemia and hypomagnesaemia. *Ann. N.Y. Acad. Sci.* **162**, 856.

GLASS E. J. & BARR D. G. D. (1981) Transient hyperparathyroidism secondary to maternal pseudohypoparathyroidism. *Arch. Dis. Child.* **56**, 565–7.

GOYER R. A. (1963) Ethanolamine phosphate excretion in a family with hypophosphatasia. *Arch. Dis. Child.* **38**, 205–8.

GRANT D. B. (1967) Fasting serum insulin levels in childhood. *Arch. Dis. Child.* **42**, 375–8.

HAHN E. A. & DANKS D. M. (1981) Clinical investigation of suspected metabolic disease. In Barson A. J. (ed.) *Laboratory Investigation of Fetal Disease*, pp. 410–28. John Wright, Bristol.

HALLMAN N. & RAPOLA J. (1978) Congenital nephrotic syndrome. In Edelmann C. M. (ed.) *Pediatric Kidney Disease*, pp. 711–17. Little Brown, Boston.

HARRIES J. T. (1977) Selective inborn errors of absorption. In Harries J. T. (ed.) *Essentials of Paediatric Gastroenterology*, pp. 199–209. Churchill Livingstone, Edinburgh.

HARRISON H. E. & HARRISON H. C. (1979) Disorders of calcium and phosphate metabolism in childhood and adolescence. *Major Problems in Clinical Pediatrics* Vol. 19. W. B. Saunders, Philadelphia.

HARVEY D. R., COOPER L. V. & STEVENS J. F. (1970) Plasma calcium and magnesium in newborn babies. *Arch. Dis. Child.* **45**, 506–9.

HARVEY D. R. & PARKINSON C. E. (1981) Prediction of the respiratory distress syndrome. In Barson A. J. (ed.) *Laboratory Investigation of Fetal Disease*, pp. 2ll67–98. John Wright, Bristol.

HASSAN H., HASHIM S. A., VAN ITALLIE T. B. & SEBRELL W. H. (1966) Syndrome in premature infants associated with low plasma Vitamin E levels and high polyunsaturated fatty acid diet. *Am. J. Clin. Nutr.* **19**, 147–57.

HENKIN R. I., MARSHALL J. R. & MERET S. (1971) Maternal–fetal metabolism of copper and zinc at term. *Am. J. Obstet. Gynecol.* **110**, 131–4.

HERZ F., KAPLAN E. & SCHEYE E. S. (1973) Erythrocyte acetylcholinesterase and glucose-6-phosphate dehydrogenase in newborn infants of low birth weight. *Clin. Chim. Acta* **46**, 147–52.

Hobbins S. M., Fowler R. S., Rowe R. D. & Korey A. G. (1981) Spironolactone therapy in infants with congestive heart failure secondary to congenital heart disease. *Arch. Dis. Child.* **56**, 934–8.

Honour J. W., Shackleton C. H. L. & Valman H. B. (1974) Sodium homeostasis in preterm infants. *Lancet* **2**, 1147.

Hoyer J. R. & Anderson C. E. (1981) Congenital nephrotic syndrome. *Clin. Perinatol.* **8**, 333–46.

Hufton B. & Wharton B. A. (1982) Type 1 glycogen storage disease presenting in the neonatal period. *Arch. Dis. Child.* **4**, 309–11.

Hughes C. A. (1982) Parenteral nutrition in preterm infants—personal practice. In Wharton B. A. (ed.) *Topics in Perinatal Medicine* II, pp. 87–102. Pitman Medical, London.

Jacobsen B. B., Andersen H. J., Peitersen B., Dige-Petersen H. & Hummer L. (1977) Serum levels of thyrotropin, thyroxine and triiodothyronine in full term, small for gestational age and preterm newborn babies. *Acta Paed. Scand.* **66**, 681–7.

Jacobsen B. B. & Hummer L. (1979) Changes in serum concentrations of thyroid hormones and thyroid hormone-binding proteins during early infancy. *Acta Paed. Scand.* **68**, 411–18.

Kildeburg P. & Winters R. W. (1972) Infant feeding and blood acid base status. *Pediatrics* **49**, 801–2.

Koch G. & Wendel H. (1968) Adjustment of arterial blood gases and acid base balance in the normal newborn infant during the first week of life. *Biol. Neonat.* **12**, 136–61.

Kueppers F. & Offord K. P. (1979) Alpha-1 antitrypsin elevation in healthy neonates. *J. Lab. Clin. Med.* **94**, 475–80.

Kupke I. R., Tritschler W., Kather B. & Bablok W. (1980) Creatinkinase "NAC-aktiviert": Referenzwerte bei Kindern. *Klin. Padiat.* **192**, 348–50.

Larcher V. F., Shepherd R., Francis D. E. M. & Harries J. T. (1977) Protracted diarrhoea in infancy. Analysis of 82 cases with particular reference to diagnosis and management. *Arch. Dis. Child.* **52**, 597–605.

Launiala K., Perheentupa J., Pasternack A. & Hallman N. (1968) Familial chloride diarrhoea—chloride malabsorption. *Bibl. Paediat.* **87**, 137–49.

Lee K. S. & Gartner L. M. (1978) Bilirubin binding by plasma proteins: a critical evaluation of methods and clinical implications. *Rev. Perinat. Med.* **2**, 319–43.

Levin B., Abraham J. M., Burgess E. A. Wallis P. G. (1970) Congenital lactose malabsorption. *Arch. Dis. Child.* **45**, 173–77.

Lewy J. E. (1981) Symposium on Perinatal Nephrology. *Clinic Perinatol.* **8**, 213–391.

Lippe B. M. (1979) Ambiguous genitalia and pseudo-hermaphroditism. *Paediat. Clin. N. Am.* **26**, 91–106.

Mathew O. P., Jones A. S., James E. *et al.* (1980) Neonatal renal failure: Usefulness of diagnostic indices. *Pediatrics* **65**, 57–60.

Matthew P. M. & Wharton B. A. (1981) Investigation and management of neonatal jaundice: a problem orientated case record. *Arch. Dis. Child.* **56**, 949–53.

Meites S. & Saniel-Banrey, K. (1979) Preservation, distribution, and assay of glucose in blood with special reference to the newborn. *Clin. Chem.* **25**, 531–4.

Menkes J. H., Alter M., Steigleder G. K., Weakley D. R. & Sung J. H. (1962) A sex-linked recessive disorder with retardation of growth, peculiar hair, and focal cerebral and cerebellar degeneration. *Pediatrics* **29**, 764–79.

Moncrieff M. W., Hill D. S., Archer J. & Arthur L. J. H. (1972) Congenital absence of pituitary gland and adrenal hypoplasia. *Arch. Dis. Child.* **47**, 136–7.

Morse J. O. (1978a) Alpha-1-Antitrypsin deficiency. *New Eng. J. Med.* **299**, 1045–8.

Morse J. O. (1978b) Alpha-1-Antitrypsin deficiency. *New Eng. J. Med.* **299**, 1099–105.

MOWAT A. P. (1979) *Liver Disorders in Childhood*, pp. 42–77. Butterworths, Sevenoaks.

NAIDOO B. T. (1968) The cerebrospinal fluid in the healthy newborn infant. *S. Afr. Med. J.* **42**, 933–5.

NORMAN M. E. & ASADI F. K. (1979) A prospective study of acute renal failure in the newborn infant. *Pediatrics* **63**, 475–9.

ODELL G. B. (1980) *Neonatal hyperbilirubinaemia*—passim. Grune and Stratton, New York.

OHTAKE M. TAMURA (1977) Serum zinc and copper levels in healthy Japanese infants. *Tohoku J. Exp. Med.* **120**, 99–103.

OSKI J. A. & NAIMAN J. L. (1967) *Hematologic problems in the newborn*, pp. 129–35. W. B. Saunders, Philadelphia.

PERRY T. L. (1981) Autopsy investigation of disorders of amino acid metabolism. In Barson A. J. (ed.) *Laboratory Investigation of Fetal Disease*, pp. 429–51. J. Wright, Bristol.

POLAND R. C., AVERY G. B., GOETCHERIAN E. & ODELL G. B. (1972) Treatment of Crigler–Najjar syndrome with agar. *Pediat. Res.* **6**, 377A.

RAYNER P. H. (1982) Endocrine emergencies in the early weeks. In Wharton B. A. (ed.) *Topics in Perinatal Medicine II*, pp. 114–25. Pitman Medical, London.

RODRIGUEZ-SORIANO J. (1978) Renal tubular acidosis. In Edelman C. M. (ed.) *Pediatric Kidney Disease*, pp. 995–1012. Little Brown, Boston.

ROY R. N., CHANCE G. W., RADDE I. C., HILL D. E., WILLIS D. M. & SHEEPERS J. (1976) Late hyponatraemia in very low birth weight infants (< 1.3 kilograms). *Pediatr. Res.* **10**, 526–31.

SAARINEN U. M. & SIIMES M. A. (1978) Serum ferritin in assessment of iron nutrition in healthy infants. *Acta Paed. Scand.* **67**, 745–51.

SANN L., DAVID L., GALY G. & ROMAND-MONIER M. (1978) Copper deficiency and hypocalcaemic rickets in a small for dates infant. *Acta Paed. Scand.* **67**, 303–7.

SCHARER K. (1972) Congenital Lactic Acidosis. In Stern J. and Toothill C. (eds.) *Organic Acidurias*, pp. 46–52. Churchill Livingstone, Edinburgh.

SCHWARTZ G. J. & SPITZER A. (1978) Disorders of renal transport of sodium potassium and magnesium. In Edelmann C. M. (ed.) *Pediatric Kidney Disease*, pp. 1079–93. Little Brown, Boston.

SCOTT P. H., BERGER H. M., KENWARD C., SCOTT P. & WHARTON B. A. (1975) Effect of gestational age and intrauterine nutrition on plasma transferrin and iron in the newborn. *Arch. Dis. Child.* **50**, 796–8.

SERTEL H. & SCOPES J. (1973) Rates of creatinine clearance in babies less than one week of age. *Arch. Dis. Child.* **48**, 717–20.

SHORE G. M., HOBERMAN L., DOWDEY A. B. C. & COMBES B. (1975) Serum Gamma-Glutamyl Transpeptidase activity in normal children. *Am. J. Clin. Pathol.* **63**, 245–50.

SITZMANN F. C., LOW C., KALOUD H. & PRESTELE H. (1974) Normalwerte der Serum-Transaminasen GOT und GPT mit neuen substratoptimierten standard-methoden. *Klin. Pediat.* **186**, 346–52.

SPIEGEL A. M., HARRISON H. E. & MARX S. J. (1977) Neonatal primary hyperpara-thyroidism with autosomal dominant inheritance. *Pediatrics* **90**, 269–72.

STANLEY C. A. & BAKER L. (1976) Hyperinsulinism in infants and children—diagnosis and therapy. *Adv. Pediat.* **23**, 315–55.

STEVENS J. F. (1970) Plasma cortisol levels in the neonatal period. *Arch. Dis. Child.* **45**, 592–4.

THALER M. M. (1964) Fatal neonatal cirrhosis: entity or end result? A comparative study of 24 cases. *Pediatrics* **33**, 721–34.

TSANG R. C., GREER F. & STEICHEN J. J. (1981) Perinatal metabolism of Vitamin D. *Clin. Perinatol.* **8**, 287–306.

Tsang R. C., Noguchi A. & Steichen J. J. (1979) Pediatric parathyroid disorders. *Pediat. Clin. N. Am.* **26**, 223–49.

Uhrmann S., Marks K. H., Maisels M. J., Kulin H. E., Kaplan M. & Utiger R. (1981) Frequency of transient hypothyroxinaemia in low birth weight infants. *Arch Dis. Child.* **56**, 214–17.

Valman H. B., Brown R. J. K., Palmer T., Oberholzer V. G. & Levin B. (1971) Protein intake and plasma amino acids of infants of low birth weight. *Br. med. J.* **4**, 789–91.

Veit S., Sitzmann F. & Prestele H. (1975) Normalwerte für Lactat—und Glutamatdehydrogenase, Sowie Leucinarylamidase, erstellt mit optimierten standardansätzen. *Klin. Pediat.* **187**, 244–51.

Walker V., Clayton B. E., Ersser R. S., Francis D. E. M., Lilly P., Seakins J. W. T., Smith I. & Whiteman P. D. (1981) Hyperphenylalaninaemia of various types among three quarters of a million neonates tested in a screening programme. *Arch. Dis. Child.* **56**, 759–64.

Wharton B. A. (1965) An unusual variety of muscular dystrophy. *Lancet* **i**, 248–9.

Wharton B. A., Bassi U., Gough G. & Williams A. (1971) Clinical value of plasma creatine kinase and uric acid levels during first week of life. *Arch. Dis. Child.* **46**, 356–62.

Yuen P., Lin H. J. & Hutchison J. H. (1979) Copper deficiency in a low birth weight infant. *Arch. Dis. Child.* **54**, 553–5.

Chapter 3
Recurrent Post-neonatal Hypoglycaemia

JAMES V. LEONARD

Recurrent hypoglycaemia is an important and potentially preventable cause of serious morbidity. It is essential to recognise the symptoms, identify the cause and start treatment as early as possible to avoid permanent neurological sequelae.

The purpose of this chapter is to assist with the diagnosis of children who present with recurrent hypoglycaemia after the neonatal period. Many reviews have been published to which the reader should refer for further details (Zuppinger 1975; Cornblath & Schwartz 1976; Marks & Rose 1981).

THE REGULATION OF BLOOD GLUCOSE

Healthy adults maintain their blood glucose concentrations within a narrow range (2.5–8.0 mmol/l) and can continue to do so even during prolonged starvation (Owen *et al.* 1967). However, children tolerate fasting much less well and may become frankly hypoglycaemic during short fasts, such as those before anaesthesia (Morrice *et al.* 1974; Kelnar 1976). After a 24-hour fast the mean blood glucose concentration of 56 children aged from 18 months to 9 years was 2.9 mmol/l, with a range of 1.0–4.7 mmol/l (Chaussain 1973). Broadly similar results have been observed in other studies (Kaye *et al.* 1961; Zuppinger 1975).

The measured blood glucose concentration represents the net balance of the glucose entering the blood from the liver and the gut, and the removal of glucose by the liver and the peripheral tissues. The regulation of these fluxes is complex, as they are modulated by changes in concentration of metabolites and hormones and by the autonomic nervous system. The liver plays a key role in maintaining blood glucose concentrations since under different conditions it may take up glucose or become a net producer.

Hepatic glucose metabolism

The pathway of glucose metabolism within the hepatocyte is shown in Fig. 3.1 (Huijing 1975). Glucose is stored in the liver as glycogen. Glucose enters the hepatocyte by a carrier-mediated process and is phosphorylated to form glucose-6-phosphate (G6P). Under conditions that favour glycogen

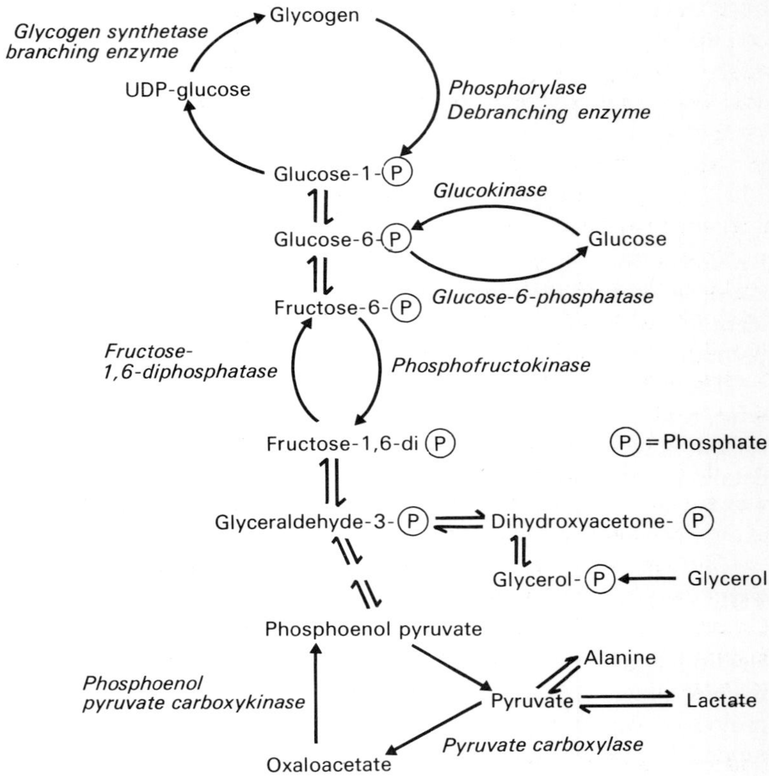

Fig. 3.1 Pathways of glucose metabolism in the hepatocyte.

synthesis G6P is converted to uridine diphosphoglucose (UDPG), the substrate for glycogen synthetase. This enzyme attaches the glucose residues by an α(1–4) linkage to the glycogen primer. At intervals the brancher enzyme transfers a number of glucose residues from the α(1–4) chain and re-attaches them through an α(1–6) bond to form a branch point, thereby producing the characteristic branched structure of glycogen. Glycogen is broken down by phosphorylase which hydrolyses the α(1–4) bonds. The α(1–6) ones are cleaved by the debrancher.

Both glycogen synthetase and phosphorylase exist in two forms, one of which is active and the other inactive. The interconversions are regulated by glucose, hormones and other effectors (Hers 1980). Glucagon and β-adrenergic agonists activate a series of enzymes that lead to the conversion of phosphorylase to the active form and glycogen synthetase to the inactive one. The net result is the breakdown of glycogen. In the presence of raised concentrations of glucose and insulin, glycogen synthetase becomes active and phosphorylase inactive so that there is formation of new glycogen.

Glucose is also synthesised in the liver by reversing glycolysis (gluconeogenesis), utilising several substrates including lactate, pyruvate, and certain amino acids especially alanine and glycerol (Fig. 3.1). Three additional enzymes are necessary to reverse thermodynamically unfavourable steps, fructose-1,6-diphosphatase, phosphoenolpyruvate carboxykinase and pyruvate carboxylase. The metabolic regulation of gluconeogenesis is complex and depends both on substrate supply and hormones (Exton 1973). Gluconeogenesis is stimulated by glucagon and catecholamines and is inhibited by insulin (Exton 1979).

Free glucose is released from the liver by the hydrolysis of glucose-6-phosphate by glucose-6-phosphatase. The glucose production rate has been estimated by the responses to stepwise increases of infused glucose and from studies using stable isotopes (Adam *et al.* 1968; King *et al.* 1969; Kalhan *et al.* 1976). From the neonatal period up to the age of 6 years the glucose production rate is 4–8 mg/kg/min, and after that age the rate decreases to the adult value of 2 mg/kg/min (Kalhan *et al.* 1976; Bier *et al.* 1977). Children have higher glucose turnover rates because the brain, which uses about 70% of the glucose output, forms a higher proportion of the total body mass (Bier *et al.* 1977).

Glycogen stores in the liver are small and it can be calculated that they would only be sufficient to maintain blood glucose concentrations for short periods (about 12 hours) (Cahill 1970). Thus, metabolic adaptations to fasting are essential to conserve glucose and maintain the glucose supply to the brain.

Endocrine and metabolic changes after food and during fasting

In the fed state, the secretion of insulin is stimulated by several mechanisms, including the rise in concentration of glucose and amino acids, gastro-intestinal hormones, particularly gastric inhibitory polypeptides (GIP) (Dupre *et al.* 1973) and by the autonomic nervous system (Unger 1981). At relatively low concentrations insulin stops the production of glucose by the liver (Brown *et al.* 1978) and also inhibits lipolysis and ketogenesis (Fain *et al.* 1966). Only at higher insulin concentrations is the uptake of glucose stimulated, with the formation of glycogen in muscle and liver (Newsholme & Start 1973).

As the period of fasting extends, the concentration of glucose and amino acids falls so that the rate of insulin secretion also falls, and that of glucagon rises (Aguilar-Parada *et al.* 1969). As a result of change in glucagon:insulin ratio glycogenolysis and gluconeogenesis are activated and there is also mobilisation of alternative fuels (Unger & Orci 1981). Free fatty acids (FFA) are released from adipose tissue and converted to ketone bodies in the liver (McGarry & Foster 1980). These substrates are metabolised preferentially by the tissues (Randle *et al.* 1966) and in particular ketone bodies are metabolised by the brain (Owen *et al.* 1967). Altogether they have an important glucose sparing effect (Newsholme & Start 1973; Randle 1980).

Catecholamines stimulate lipolysis, ketogenesis and both glycogenolysis and gluconeogenesis (Johnston *et al.* 1982). They may have a particular role in the response to exercise and stress, which is mediated via the hypothalamus (Unger 1981). Growth hormone has a hyperglycaemic effect (Johnston *et al.* 1982) but its precise role in the regulation of glucose metabolism is unclear. Cortisol has an important permissive effect on glycogenolysis and gluconeogenesis, making the processes more sensitive to the effects of glucagon and catecholamines (Exton 1979).

CLINICAL PRESENTATION

Hypoglycaemia is defined as blood glucose concentration of 2 mmol/l or less. It may cause a wide variety of symptoms including convulsions, coma, pallor, sweating, hunger, confusion, irritability, bizarre behaviour, weakness, headaches and abdominal pain (Cornblath & Schwartz 1976). However, in early childhood the most common presentation is convulsions and it is essential that all children with convulsions, whether febrile or not, are screened for hypoglycaemia. In older children more vague and less specific symptoms are more frequent.

If hypoglycaemia is suspected, the blood glucose level should always be measured since screening tests using glucose oxidase impregnated strips are not satisfactory for diagnostic purposes. Glucose should then be given either orally or intravenously and if symptoms improve rapidly, this helps to confirm that hypoglycaemia was the cause of the symptoms. However, if the hypoglycaemia has been severe and prolonged or is complicated by other problems recovery may be delayed.

Blood glucose values > 2.5 mmol/l at the time of symptoms exclude hypoglycaemia, but borderline values will need careful assessment. Although it is clearly important not to miss hypoglycaemia it should be remembered that even healthy children may develop low blood glucose without symptoms during fasting (Chaussain 1973).

Some patients who have chronic hypoglycaemia may have few signs despite almost unrecordable blood glucose concentrations. This is particularly common in severe glucose-6-phosphatase deficiency (glycogen storage disease type I) (Cornblath & Schwartz 1976).

INITIAL INVESTIGATIONS

History and examination

Once hypoglycaemia has been confirmed it is often possible to make a provisional diagnosis on the basis of the history and examination (Cornblath & Schwartz 1976). It is important to determine whether any factors appear to precipitate hypoglycaemia such as certain foods, intercurrent infections or fasting.

All children should, of course, be examined thoroughly and measured. Short stature may be a useful clue to the diagnosis of endocrine disease. In children with hypoglycaemia the presence of hepatomegaly is particularly important; sometimes the liver is so large that the symptom is missed. Excessive pigmentation or ambiguous genitalia are important signs of adrenal disease.

Biochemical investigations

Much useful information can be obtained by collecting appropriate blood specimens for metabolites (blood glucose, lactate, β-hydroxybutyrate, amino acids particularly alanine and branched-chain amino acids) and the hormones (plasma insulin, cortisol, growth hormone) at the time of hypoglycaemia. All these investigations can be undertaken on small quantities of blood, and from the results of the clinical examination and the investigations it will usually be possible to make a tentative diagnosis which can then be confirmed with the minimum of further investigation.

If hypoglycaemia is suspected but cannot be demonstrated then it is often helpful to try to provoke it. The most simple test is to fast the patient; the duration of the fast will depend on the patient's apparent tolerance to fasting, gauged from the history. Whilst fasting patients, an intravenous cannula should be *in situ* so that blood specimens for metabolites and hormones can be collected easily and frequently and so that intravenous glucose can be given if necessary. Fasts should not be continued for longer than 20 hours, since even normal children may develop asymptomatic hypoglycaemia after this length of fast (Chaussain 1973). Other tests will be needed to confirm certain diagnoses and these will be discussed in later sections.

CAUSES OF HYPOGLYCAEMIA

There are numerous causes of hypoglycaemia and several classifications have been proposed, but none is entirely satisfactory (Marks & Rose 1981). The classification used in this chapter is a practical one based on the biochemical findings (Table 3.1). There are 4 main categories:

1 Hyperinsulinism.

2 Hypoglycaemia with ketosis. Despite normal ketone body production the supply of glucose is inadequate to meet demands. This category includes disorders of carbohydrate metabolism in the liver and endocrine disorders.

3 Hypoglycaemia without ketosis but appropriate insulin concentrations.

4 Hepatocellular failure. In infants conditions that cause liver cell

Table 3.1 Causes of recurrent post-neonatal hypoglycaemia.

Elevated plasma insulin
Nesidioblastosis (includes leucine-sensitive hypoglycaemia)
Islet cell adenoma
Wiedemann–Beckwith syndrome
Malicious administration of insulin or hypoglycaemic drugs

Hypoglycaemia with ketosis

Fasting plasma lactate concentration raised
Glycogen storage disease type I (glucose-6-phosphatase deficiency)
Fructose-1,6-diphosphatase deficiency
Pyruvate carboxylase deficiency
Phosphoenol pyruvate carboxylase deficiency
Organic acidaemias

Fasting plasma lactate normal
Glycogen storage disease type III (debrancher deficiency)
Glycogen storage disease type VI (disorders of the phosphorylase system)
Glycogen synthetase deficiency
Endocrine disorders
- Hypopituitarism
- Isolated growth hormone deficiency
- Isolated ACTH deficiency
- Adrenal insufficiency
- Glucagon deficiency

Emotional deprivation
Idiopathic ketotic hypoglycaemia
Organic acidaemias

Hypoglycaemia without ketosis (normal insulin levels)
Inborn errors of metabolism
- 3-hydroxy-3-methyl glutaryl CoA lyase deficiency
- Multiple acyl CoA dehydrogenase deficiency (glutaric aciduria type II)
- Dicarboxylic acidurias
- Systemic carnitine deficiency

Hypoketotic–hypoglycaemic syndrome

Table 3.1 (*Cont.*)

Hepatocellular failure
Neonatal hepatitis
Hereditary fructose intolerance
Galactosaemia
Tyrosinosis (hepatorenal tyrosinaemia)

damage may impair the metabolic functions of the liver and lead to hypoglycaemia.

Hyperinsulinism

Hyperinsulinism is an inappropriate elevation of plasma insulin concentrations in relation to the blood glucose concentration. It is most readily evident when a patient is hypoglycaemic since the secretion of insulin should be fully suppressed. Thus, with the blood glucose below 2 mmol/l the plasma insulin should be < 5 mU/l. During hypoglycaemia with raised insulin concentrations, the concentrations of β-hydroxybutyrate and branched chain amino acids are also low.

Recurrent hypoglycaemia due to hyperinsulinism may be caused by nesidioblastosis, islet cell adenomata and the Wiedemann–Beckwith syndrome. The majority of patients with nesidioblastosis present with severe hypoglycaemia during the first months of life and almost all have developed symptoms within the first six months (Aynsley-Green 1981a). Those most severely affected present shortly after birth with intractable hypoglycaemia. Most, if not all patients with leucine-sensitive hypoglycaemia probably also had nesidioblastosis (Marks & Rose 1981). Islet cell adenomata may present at any age but they are more common during the first month of life or after the age of 4 years (Cornblath & Schwartz 1976). The distinction between these two conditions may not be as clear as has been thought since nesidioblastosis and islet cell adenomata may co-exist (Heitz *et al.* 1977).

Nesidioblastosis

Nesidioblastosis is the result of abnormal differentiation of the pancreatic islet tissue (Heitz *et al.* 1977). There is an increase in the total area occupied by the endocrine cells with an abnormal anatomical relationship to the ductular epithelial cells. There is also a decrease in the number and size of somatostatin cells (Aynsley-Green 1981a). The diagnosis can only be made conclusively by specific immunocytohistochemical techniques which identify the endocrine cells. Many terms have been used to describe the histological appearances of the pancreas in children with hyperinsulinism in infancy (diffuse β-cell hyperplasia, microadenomatosis, focal islet cell adenomatosis, etc.) and it now seems likely that they all refer to nesidioblastosis.

The aim of treatment is to maintain blood glucose within the normal range. However, this can be difficult and glucose infusion rates in excess of 15 mg/kg per min may be needed. Once the diagnosis of hyperinsulinism is proved, diazoxide (in combination with a thiazide diuretic) should be given in doses up to 20 mg/kg daily. Diazoxide inhibits insulin secretion, an effect that is potentiated by the thiazide diuretic, which also prevents sodium retention (Smith 1968). Mild cases often respond but those presenting early often do not (Stanley & Baker 1977), and for these subtotal, or even total, pancreatectomy may be needed.

Somatostatin is also effective but has not been used for prolonged periods. Glucagon transiently raises blood glucose and is useful in an emergency (Aynsley-Green *et al.* 1981).

Because of the high incidence of permanent neurological sequelae very intensive treatment is indicated, particularly for those who present in the neonatal period (Aynsley-Green *et al.* 1981).

Islet-cell adenoma

Hypoglycaemia due to an islet-cell adenoma in the neonatal period is usually severe and is difficult to distinguish from that due to nesidioblastosis. However, in older children the hypoglycaemia may be much milder and as a result it may be more difficult to demonstrate inappropriate insulin secretion. This may be provoked by fasting or exercise but none of the provocation tests are reliable. Measurements of C-peptide or pro-insulin can be useful (Turner & Heding 1977; Service *et al.* 1977).

In the young child, unless a tumour can be demonstrated by imaging techniques the management is similar to that of nesidioblastosis. In the older child, since a discrete lesion is more likely, surgical exploration should be undertaken at an earlier stage, and may be curative (Stanley & Baker 1977).

Wiedemann–Beckwith syndrome

The Wiedemann–Beckwith syndrome is characterised by exomphalos, macroglossia, visceromegaly and gigantism (Ichiba & Gardner 1975). Symptomatic hypoglycaemia due to hyperinsulinism occurs in 30% in the neonatal period (Sotelo-Avila *et al.* 1980) and is usually transient, however, it can be prolonged and severe (Roe *et al.* 1973). The management is similar to that of nesidioblastosis.

Malicious administration of insulin

With the increasing recognition of unusual causes of child abuse, the possibility of malicious administration of insulin or sulphonyl urea drugs to children with hyperinsulinaemic hypoglycaemia that does not conform to the usual pattern needs to be considered carefully (Dershewitz *et al.* 1976; Scarlett *et al.* 1977).

Hypoglycaemia with ketosis

Ketotic hypoglycaemia is not a diagnosis in itself and careful investigation is necessary to identify the cause.

Hypoglycaemia with elevated plasma lactate concentrations

If plasma lactate concentrations are raised during hypoglycaemia an inborn error of metabolism is likely but careful assessment of the result will be necessary if the child has had a convulsion or was struggling at the time of the venepuncture. Both may cause elevation of plasma lactate concentrations.

Glucose-6-phosphatase deficiency (Glycogen storage disease (GSD) type 1). Glucose-6-phosphatase deficiency characteristically presents with a combination of hepatomegaly, hypoglycaemia and lactic acidosis, although the severity of the hypoglycaemia and the lactic acidosis is very variable (Cornblath & Schwartz 1976). The degree of ketosis may be slight and much less than expected for the hypoglycaemia. Symptoms may start in the neonatal period (Hufton & Wharton 1982) but the majority present later during the first year of life. Although convulsions do occur, symptoms and signs of hypoglycaemia may be surprisingly slight and other common presenting symptoms are failure to thrive, diarrhoea, abdominal distension and excessive bleeding. The liver is grossly enlarged although this may not be striking in the neonatal period.

If the characteristic features of hypoglycaemia (lactic acidosis, hepatomegaly and hypertriglyceridaemia) are present, a liver biopsy should be performed to confirm the diagnosis. Other tests are not necessary and since most involve fasting these are potentially dangerous. In those in whom the findings are less clearcut a fasting glucagon test should be done, measuring both the blood glucose and lactate response (Dunger & Leonard 1982). If lactate concentrations are raised during this test a liver biopsy should be undertaken and both the glycogen content and the glucose-6-phosphatase activity measured. The latter should be done in both fresh liver and after freezing. In type I GSD, glucose-6-phosphatase activity is reduced but some patients who have all the clinical and biochemical features of type I GSD have normal glucose-6-phosphatase values in frozen liver (Senior & Loridan 1968). However, if the liver is assayed fresh the enzyme activity is reduced (Igarashi *et al.* 1979). This disorder, often called type Ib GSD, is probably one of membrane transport of glucose-6-phosphate.

In both GSD types I and Ib the liver is unable to liberate glucose and the aim of treatment is to maintain the blood glucose concentrations between 4–6 mmol/l, day and night (Leonard *et al.* 1979) by giving glucose. This is most easily done by giving frequent oral drinks of glucose (as glucose polymer) by day and a nasogastric infusion by night (Greene *et al.* 1976). Meticulous attention to detail is needed and the 24 hour profile of blood glucose values should be checked to ensure that the blood glucose levels are satisfactory. Excellent results can be achieved with this form of treatment (Fernandes *et al.* 1979).

Fructose 1,6-diphosphatase and other disorders of gluconeogenesis. Fructose 1,6-diphosphatase deficiency may present in a similar manner to GSD type I with hypoglycaemia and lactic acidosis (Eagle *et al.* 1974) but there is considerable variation in the severity (Baerlocher *et al.* 1980). Half of the published cases have presented in the neonatal period and 75% by 6 months of age, with hyperventilation being the most common presenting symptom. Hepatomegaly is not always found initially although it will develop in almost all patients. Severe hypoglycaemia and metabolic acidosis may be precipitated by fasting, infections or other metabolic stress and these can be fatal if not treated adequately. Fructose, glycerol and dihydroxyacetone may all also provoke hypoglycaemia and lactic acidosis (Baerlocher *et al.* 1980). The diagnosis should be confirmed by measuring the enzyme activity in leucocytes or liver (Melancon *et al.* 1973).

Children should be treated with regular meals and with moderate restriction of fructose. If early morning hypoglycaemia or acidosis is a problem a feed should be given during the night. To prevent complications during intercurrent infections, regular glucose supplements should be given and in severe cases a regimen similar to that used in GSD type I may be necessary (Stacey *et al.* 1980).

The other disorders of gluconeogenesis appear to be less common and in both pyruvate carboxylase deficiency and phosphoenol pyruvate carboxykinase (PEPCK) deficiency the phenotypic expression is varied. The mode of presentation and the biochemical findings have been reviewed recently (Leonard 1982).

Rather surprisingly, hypoglycaemia is not a constant feature of pyruvate carboxylase deficiency (Hommes *et al.* 1980) and, when present, has been relatively mild (van Biervliet *et al.* 1977; Atkin *et al.* 1979).

Few patients with PEPCK deficiency have been described in detail though all have had hypoglycaemia. However, the lactic acidosis has been variable and there is some doubt in some of the patients that the observed enzyme deficiency has been the primary disorder (Leonard 1982). Intermittent fasting hypoglycaemia with lactic acidosis has also been described in combined deficiency of pyruvate and 2-oxoglutarate dehydrogenases (Haworth *et al.* 1976).

Organic acidaemias. Children with organic acidaemias present in various ways, most commonly with a toxic encephalopathy and keto-acidosis although this may not be marked. Rarely these patients present with recurrent hypoglycaemia (Haymond *et al.* 1973; Garnica *et al.* 1976). Plasma amino acids and urine organic acids should be examined in order to establish the diagnosis (Chalmers & Lawson 1982).

Hypoglycaemia with normal lactate concentrations

Glycogen storage disease type III. In type III GSD fasting hypoglycaemia is not uncommon in infants and young children, but is unusual in older children. Fasting blood lactate concentrations are normal but there is a rise in blood lactate following a meal or a carbohydrate load (Fernandes *et al.* 1969). Ketonaemia and ketonuria may be marked (Fernandes & Pikaar 1969) and hepatomegaly is often massive. The rise in blood glucose in the fasting glucagon test is less than 2 mmol/l. Although there may be a greater rise in blood glucose if the test is repeated 2 hours after a carbohydrate meal (Hug *et al.* 1963) the response is not specific (Dunger & Leonard 1982). The diagnosis should be confirmed by measuring red cell glycogen and the amylo-1,6-glucosidase in leucocytes (Sidbury *et al.* 1961; Spencer Peet *et al.* 1971). Liver biopsy is not normally necessary.

Disorders of the phosphorylase system. Although hypoglycaemia can occur in infants with disorders of the phosphorylase system and the activating enzymes, it is uncommon and the glucose response to glucagon is usually normal (Dunger & Leonard 1982). Hepatomegaly is often marked but becomes less obvious at puberty. The diagnosis is established by measuring the components of the phosphorylase cascade in red and white blood cells (Lederer *et al.* 1975). Only in those patients in whom the diagnosis remains uncertain should it be necessary to do a liver biopsy.

Intensive treatment is rarely needed for these forms of glycogen storage disease, but regular meals with a high protein content should be given (Fernandes & van de Kamer 1968). A snack, late at night, will normally prevent any early morning hypoglycaemia. Both symptoms and growth improve with age.

Glycogen synthetase deficiency. Glycogen synthetase deficiency appears to be extremely rare (Aynsley-Green 1981b) but the diagnosis could easily be missed as the liver is not enlarged and the disorder is not easily distinguished on clinical grounds from other causes of ketotic hypoglycaemia. However, there is no rise in blood glucose during the fasting glycogen test, and after meals there is a sharp rise in blood glucose with glycosuria because of the failure of the liver to utilise glucose for glycogen synthesis (Aynsley-Green *et al.* 1977a).

The diagnosis should be confirmed by liver biopsy and enzyme assay. Treatment with a high protein diet prevents hypoglycaemia (Aynsley-Green *et al.* 1977b).

Endocrine disorders. Hypoglycaemia may complicate many endocrine conditions including adrenal insufficiency of any cause, growth hormone deficiency, ACTH deficiency and combined deficiencies of the pituitary hormones. Disorders of glycagon and catecholamine secretion are less well defined.

Since intermediary metabolism is normal in these patients there is an appropriate rise in ketone bodies. During hypoglycaemia specimens taken for cortisol and growth hormone at the time of hypoglycaemia can give useful clues to the diagnosis, but not infrequently the results are equivocal. Thus all patients with ketotic hypoglycaemia, particularly those who are short or growing slowly, should be assessed carefully for endocrine disease. This is discussed in detail in Chapter 12 and only a brief outline will be given in this chapter.

Hypopituitarism. Up to 70% of children with congenital hypopituitarism may develop hypoglycaemia in the neonatal period, which may be very severe; in one series 60% of the patients died (Cornblath & Schwartz 1976). Many of the patients had midline anomalies or microphallus which provided important clues towards recognising the underlying cause of the illness (Moncrieff *et al.* 1972) since treatment with intravenous glucose and corticosteroids is urgent.

In isolated growth hormone deficiency hypoglycaemia is rare in the neonatal period, but approximately 15% of infants, particularly those who are thin, may have spontaneous hypoglycaemia with a higher incidence during starvation (Hopwood *et al.* 1975).

Isolated ACTH deficiency is very rare in childhood but those patients who have been described have presented with recurrent hypoglycaemia (Aynsley-Green *et al.* 1978; Hung & Migeon 1968). One patient was shown to have hyperketonaemia and hypoalaninaemia and in addition to ACTH and cortisol deficiency, the catecholamine response to hypoglycaemia is reduced. Hypoglycaemia is common in patients with multiple deficiencies of pituitary hormones especially if ACTH is missing (Goodman *et al.* 1968).

Adrenocortical insufficiency. Although hypoglycaemia is a rare presenting feature of adrenal insufficiency (D'Albora & Martin 1966) it is an important complication of these conditions regardless of the cause of the adrenal disease. It is most likely to occur if glucocorticoid replacement therapy is not increased during infections, starvation or with a stress (such as surgery).

Familial glucocorticoid deficiency requires special mention since this condition characteristically presents with recurrent episodes of hypoglycaemia and collapse which often start in the second or third year of life (Migeon *et al.* 1968). Generalised pigmentation is usually present particularly on the exposed parts. Plasma cortisol concentrations are low and those of ACTH high, but salt-wasting is not a feature and aldosterone concentrations are normal. This condition is treated by appropriate glucocorticoid replacement (Forest 1981).

It has also been suggested that an impaired cortisol response could contribute to the hypoglycaemia caused by adrenomedullary hyporesponsiveness (Tietze *et al.* 1972). This is discussed further below.

Adrenomedullary hyporesponsiveness (AMH). Broberger *et al.* (1959) described a group of children with spontaneous hypoglycaemia in whom there were none of the characteristic symptoms of the adrenergic response (such as sweating, pallor, tachycardia) and reduced urinary excretion of catecholamines in response to hypoglycaemia. They suggested that this was responsible for the hypoglycaemia but it could be secondary to it. Reduced clinical signs of the adrenergic response to hypoglycaemia have been noted in other conditions including diabetics who have had recurrent hypoglycaemia, therapeutically induced insulin hypoglycaemia, and several other disorders (Goldfien *et al.* 1961; Sussman *et al.* 1963; Marks & Rose 1981). Furthermore, patients who have had a total adrenalectomy do not have recurrent hypoglycaemia provided they have adequate corticosteroid replacement (Senior 1973).

It has been suggested that this condition is caused by a primary disorder of the hypothalamus which affects the neural regulation of ACTH and catecholamines (Brunjes *et al.* 1963) since some patients, but not all, are cortisol-deficient (Tietze *et al.* 1972). However, since catechol-n-methyl transferase, an enzyme necessary for the synthesis of adrenaline is cortisol-dependent and inducible (Pohorecky & Wurtman 1971), AMH could be secondary to ACTH deficiency. Patients with isolated ACTH have been shown to have reduced catecholamine secretion (Aynsley-Green *et al.* 1978; Hung & Migeon 1968).

At present there remains considerable doubt about the significance and importance of the apparent failure of catecholamine secretion. Ephedrine in small doses is said to prevent the hypoglycaemia (Broberger *et al.* 1959), but its true value has not been proved.

Glucagon deficiency. McQuarrie *et al.* (1950) suggested that glucagon deficiency was the cause of hypoglycaemia in two patients who had reduced numbers of A cells in the pancreatic islets. However, this finding has been questioned because the histological techniques were not sufficiently specific. Several other patients have been described in whom the diagnosis has been suspected, but only recently have patients been reported in whom plasma glucagon concentrations have been measured. Two patients, both of whom presented with severe hypoglycaemia in the neonatal period, had very low plasma glucagon concentrations with normal insulin levels (Vidnes 1976; Kollee *et al.* 1978). Zinc protamine glucagon successfully corrected the hypoglycaemia.

Emotional deprivation. Recurrent hypoglycaemia is an unusual feature of emotional deprivation. When first admitted to hospital these children may show evidence of growth hormone and ACTH deficiency (Powell *et al.* 1967; Schutt-Aine *et al.* 1973).

Idiopathic ketotic hypoglycaemia. It is apparent from the preceding sections of this chapter that ketotic hypoglycaemia occurs in several conditions and is not a diagnosis in itself. In any child presenting with ketotic hypoglycaemia meticulous investigation is indicated to identify the cause of the hypoglycaemia. However, there remains a group of children with a characteristic history in whom no cause for hypoglycaemia associated with marked ketosis can be identified (Colle & Ulstrom 1964). Typically a child aged between 18 months and 3 years presents with symptomatic hypoglycaemia during minor infections. There is often a history of anorexia with a missed meal and then the following morning the child is found pale, floppy and unresponsive or even convulsing. The parents may notice the smell of ketone in the child's breath. Recovery is normally rapid once glucose is given. Children may have recurrent attacks but these cease by the age of 8 years.

The cause of this condition remains unclear despite extensive investigation. No consistent abnormality of the endocrine responses have been demonstrated (Haymond *et al.* 1974; Dahlquist *et al.* 1979) although adrenomedullary hyporesponsiveness has been described in some patients (Zuppinger 1975). The metabolic responses to infusions of ketones, alanine, glycerol and fructose are also normal (Cornblath & Schwartz 1976). The response to glucagon after overnight fasting is normal but the rise in glucose is reduced or absent at the time of hypoglycaemia implying glycogen depletion (Colle & Ulstrom 1964). It has been suggested that the supply of gluconeogenetic substrate is reduced (Haymond *et al.* 1974) but glucose production is normal and the low plasma alanine levels are probably secondary rather than the primary abnormality (Wolfsdorf *et al.* 1982).

During starvation the respiratory quotient is raised, implying that glucose is being utilised (Kerr *et al.* 1975). Dahlquist *et al.* (1979) studied glucose disappearance rates and also found that affected children did not restrict glucose utilisation during fasting. Since it is known that the blood glucose of healthy normal children may fall into the hypoglycaemic range after prolonged fasting this condition may represent no more than the extreme spectrum of the response of normal children to calorie deprivation and stress (Senior 1973).

Hypoglycaemia may be provoked by giving a ketogenic diet (Cornblath & Schwartz 1976). After a 12-hour fast a diet is started in which 67% of the calorie requirements are given as fat and the rest divided equally between protein and carbohydrate. An intravenous cannula should be inserted so that the endocrine and metabolic responses can be monitored carefully. Affected children become ketotic within 8–12 hours and hypoglycaemic within 16–24 hours. The test is terminated with glucose but since one wishes to avoid severe symptomatic hypoglycaemia it may be surprisingly difficult to decide when to stop the test, so that in the author's experience equivocal results are not uncommon. The test is not specific as it may provoke hypoglycaemia, particularly in disorders of intermediary metabolism (Greene *et al.* 1972) and so, in addition to measuring plasma glucose, ketone bodies and the counter-regulatory hormones, it is the author's practice to measure plasma lactate and to examine the urine organic acids at the end of the test.

Treatment is usually straightforward. Patients should eat meals regularly and avoid very fatty foods. When the child is unwell or anorexic, carbohydrate supplements of sucrose or glucose-containing drinks should be given and urine should be tested for ketones. If ketonuria develops the child should be given extra sugary drinks both by day and during the night as well.

Hypoglycaemia without ketosis (and appropriate insulin concentrations)

If, for any reason, there is a failure to form ketone bodies during starvation or illness severe hypoglycaemia may develop. The most common cause for hypoglycaemia without ketosis is inappropriate elevation of insulin levels and this must always be excluded. However, if insulin levels are normal, a disorder of the oxidation of fat or of the formation of ketone bodies is likely. Examination of the urine organic acids during hypoglycaemia will often give useful clues towards establishing the diagnosis, which includes 3-hydroxy-3-methylglutaryl CoA lyase deficiency (Chalmers & Lawson 1982); multiple acyl CoA dehydrogenase deficiency (Dusheiko *et al.* 1979); systemic carnitine deficiency (Ware *et al.* 1978) and other dicarboxylic acidurias (Gregersen *et al.* 1976). However, detailed metabolic studies may be needed in some cases (Bougneres *et al.* 1981).

Idiopathic hypoketotic hypoglycaemia

Some patients with symptomatic hypoglycaemia have been described who were hypoketotic compared to normal controls (Teijema *et al.* 1980). No cogent explanation for the findings was given and the exact significance and importance of this condition remains to be determined.

Liver disease

Since the liver is the key organ in glucose homoeostasis it is not surprising that hypoglycaemia may complicate liver failure regardless of its cause.

Acute liver failure in childhood is commonly complicated by hypoglycaemia (Psacharopoulos *et al.* 1980) and in young infants it may be recurrent in chronic hepatocellular liver disease. The hypoglycaemia probably has several causes and as a result the metabolic responses are rather variable. Hepatocellular failure should be assessed in all children with hepatomegaly. Tests should include conventional liver function tests, clotting studies and plasma amino acids.

Neonatal hepatitis

Hypoglycaemia is an unusual complication in neonatal hepatitis regardless of its cause (Leblanc *et al.* 1981). This may be sufficiently severe to cause convulsions but commonly it is asymptomatic. Leblanc *et al.* (1981) have suggested that it may be secondary to cortisol deficiency.

Galactosaemia

Galactosaemia most commonly presents in the neonatal period with vomiting, failure to thrive and evidence of liver failure. Cataracts may be present from an early age. Hypoglycaemia is not a common finding and does not recur once the patient is on a galactose-free diet (Zuppinger 1975). If galactosaemia is suspected, intake of galactose should be stopped and the activity of the galactose-1-phosphate uridyl transferase in red blood cells assayed. Testing the urine for reducing substances is not a reliable screening test (Mowat 1982).

Fructosaemia

Like galactosaemia, symptoms in fructosaemia only occur when offending carbohydrate is eaten. In most children, the introduction of fructose in the diet is delayed for a few weeks or months and when fructose is given symptoms are often dramatic with vomiting and other acute symptoms, but some children present with chronic illness and failure to thrive. Hepatomegaly is usually present with evidence of hepatocellular dysfunction and clotting disorders. Hypoglycaemia is present in about one-third of the patients (Odievre *et al.* 1978; Baerlocher *et al.* 1980). Fructosaemia should be suspected from the history, and the diagnosis is confirmed by measuring the fructose-1-aldolase activity in liver.

Tyrosinosis (hepatorenal tyrosinaemia, tyrosinosis type 1)

Tyrosinosis may present in several ways but most commonly with failure to thrive with rickets. Hepatosplenomegaly is usually present with evidence of severe hepatic dysfunction. Hypoglycaemia may be marked but it is usually asymptomatic, and may be due to hyperinsulinism with islet cell hyperplasia (La Du & Gjessing 1978), but in addition there may be little or no rise in blood glucose after glucagon (J. V. Leonard, unpubl. obs.). After treatment with a low tyrosine and methionine diet and vitamin supplements the liver failure and the hypoglycaemia improve but the treatment does not prevent the eventual development of cirrhosis and hepatoma.

This condition is probably caused by an inborn error of fumaryl acetoacetase (Lindblad *et al.* 1977; Berger *et al.* 1981) and as a result succinyl-acetone and succinyl-acetoacetate accumulates. The compounds can be detected in the urine even whilst the patient is receiving strict dietary therapy.

TREATMENT

Symptomatic hypoglycaemia should always be treated immediately because of the risk of permanent neurological sequelae. After appropriate blood specimens have been taken, glucose up to 1 g/kg is given orally or intravenously depending on the clinical state. In small infants or in those with hyperinsulinism there is a risk of rebound hypoglycaemia so care should be taken to ensure that the supply of glucose is continued.

In those conditions in which normal glycogen stores are present and can be utilised, parenteral glucagon (20 μg/kg) can be used to correct hypoglycaemia if the patient is uncooperative or there is difficulty giving the necessary glucose.

ACKNOWLEDGEMENTS

I should like to thank Dr David Grant for his suggestions and constructive criticism of the manuscript.

REFERENCES

Adam P. A. J., King K. & Schwartz R. (1968) Model for the investigation of intractable hypoglycaemia: insulin–glucose interrelations during steady state infusions. *Pediatrics* **41**, 91–105.

Aguilar-Parada E., Eisentraut A. M. & Unger R. (1969) Effects of starvation on plasma pancreatic glucagon in normal man. *Diabetes* **18**, 717–23.

ATKIN B. M., BUIST N. R. M., UTTER M. F., LEITER A. B. & BANKER B. Q. (1979) Pyruvate carboxylase deficiency and lactic acidosis in a retarded child without Leigh's disease. *Pediatr. Res.* **13**, 109–16.

AYNSLEY-GREEN A. (1981a) Nesidioblastosis of the pancreas in infancy. In Randle P. J., Steiner D. F., Whelan W. J. (eds.) *Carbohydrate Metabolism and its Disorders*. Vol. 3, pp. 181–204. Academic Press, New York.

AYNSLEY-GREEN A. (1981b) Hepatic glycogen synthetase deficiency. In Randle P. J., Steiner D. F., Whelan W. J. (eds.) *Carbohydrate Metabolism and its Disorders*, pp. 139–56. Academic Press, New York.

AYNSLEY-GREEN A., MONCRIEFF M. W., RATTER S., BENEDICT C. R., STORRS C. N. & WILKINSON R. H. (1978) Isolated ACTH deficiency. *Arch. Dis. Child.* **53**, 499–502.

AYNSLEY-GREEN A., POLAK J. M., BLOOM S. R., GOUGH M. H., KEELING J., ASHCROFT S. J. H., TURNER R. C. & BAUM J. D. (1981) Nesidioblastosis of the pancreas: definition of the syndrome and the management of the severe neonatal hyperinsulinaemic hypoglycaemia. *Arch. Dis. Child.* **56**, 496–508.

AYNSLEY-GREEN A., WILLIAMSON D. H. & GITZELMANN R. (1977a) Hepatic glycogen synthetase deficiency. *Arch. Dis. Child.* **52**, 573–9.

AYNSLEY-GREEN A., WILLIAMSON D. H. & GITZELMANN R. (1977b) The dietary treatment of hepatic glycogen synthetase deficiency. *Helv. Paediatr. Acta* **32**, 71–5.

BAERLOCHER K., GITZELMANN R. & STEINMANN B. (1980) Clinical and genetic studies of disorders in fructose metabolism. In Burman D., Holton J. B. & Pennock C. A. (eds.) *Inherited Disorders of Carbohydrate Metabolism*, pp. 163–90. MTP Press, Lancaster.

BERGER R., SMIT G. P. A., STOKER DE VRIES S. A., DURAN M., KETTING D. & WADMAN S. K. (1981) Deficiency of fumaryl acetoacetase in a patient with hereditary tyrosinaemia. *Clin. Chim. Acta* **114**, 37–44.

BIER D. M., LEAKE R. D., HAYMOND M. W., ARNOLD K. J., GRUENKE L. D., SPERLING M. A. & KIPNIS D. M. (1977) Measurement of 'true' glucose production rates in infants and childhood with 6,6-dideuteroglucose. *Diabetes* **26**, 1016–23.

BOUGNERES P. F., SAUDUBRAY J.-M., MARSAC C., BERNARD O., ODIEVRE M. & GIRARD J. (1981) Fasting hypoglycaemia resulting from hepatic carnitine palmitoyl transferase deficiency. *J. Pediatr.* **98**, 742–6.

BROBERGER O., JUNGNER I. & ZETTERSTROM R. (1959) Studies in spontaneous hypoglycaemia in childhood: failure to increase the epinephrine secretion in insulin-induced hypoglycaemia. *J. Pediatr.* **55**, 713–19.

BROWN P. M., GUNDOGDU S., WILSON N. & SONKSEN P. H. (1978) Regulation of glucose production in vivo. *Biochem. Soc. Trans.* **6**, 1134–8.

BRUNJES S., HODGMAN J., HOVACK J. & JOHNS V. J. (1963) Adrenal medullary function in idiopathic spontaneous hypoglycaemia of infancy and childhood. *Am. J. Med.* **34**, 168–76.

CAHILL G. F. (1970) Starvation in man. *New Engl. J. Med.* **282**, 668–75.

CHALMERS R. A. & LAWSON A. M. (1982) *Organic acids in man.* Chapman & Hall, London.

CHAUSSAIN J. L. (1973) Glycemic response to 24-hour fast in normal children and children with ketotic hypoglycaemia. *J. Pediatr.* **82**, 438–43.

COLLE E. & ULSTROM R. A. (1964) Ketotic hypoglycaemia. *J. Pediatr.* **64**, 632–51.

CORNBLATH M. & SCHWARTZ R. (1976) *Disorders of Carbohydrate Metabolism in Infancy*, 2nd Edn. Saunders, Philadelphia.

DAHLQUIST G., GENTZ J., HAGENFELDT L., LARSSON A., LOW H., PERSSON B. & ZETTERSTROM R. (1979) Ketotic hypoglycaemia of childhood—a clinical trial of several unifying aetiological hypotheses. *Acta Paediat. Scand.* **68**, 649–56.

D'Albora J. B. & Martin M. M. (1966) Addison's disease in childhood. *Am. J. Dis. Child.* **111**, 208–14.

Dershewitz R., Vestal B., MacLaren N. K. & Cornblath M. (1976) Transient hepatomegaly and hypoglycaemia. A consequence of malicious insulin administration. *Am. J. Dis. Child.* **130**, 998–9.

Dunger D. B. & Leonard J. V. (1982) The value of the glucagon test in screening for hepatic glycogen storage disease. *Arch. Dis. Child.* **57**, 384–9.

Dupre J., Ross S. A., Watson D. & Brown J. C. (1973) Stimulation of insulin secretion by gastric inhibitory polypeptide in man. *J. Clin. Endocr.* **37**, 826–8.

Dusheiko G., Kew M. C., Joffe B. I., Lewin J. R., Mantagos S. & Tanaka K. (1979) Recurrent hypoglycaemia associated with glutaric aciduria type II in an adult. *New Engl. J. Med.* **301**, 1405–9.

Eagle R. B., MacNab A. J., Ryman B. E. & Strang L. B. (1974) Liver biopsy data on a child with fructose-1,6-diphosphatase deficiency that closely resembled many aspects of glucose-6-phosphatase deficiency. *Biochem. Soc. Trans.* **2**, 1118–21.

Exton J. H. (1973) Gluconeogenesis. *Metabolism* **21**, 945–90.

Exton J. H. (1979) Hormonal control of gluconeogenesis. *Adv. Exp. Med. Biol.* **111**, 125–67.

Fain J. N., Kovacev V. P. & Scow R. O. (1966) Antilipolytic effect of insulin in isolated fat cells of the rat. *Endocrinology* **78**, 773–8.

Fernandes J., Huijing E. & Van der Kamer J. H. (1969) A screening method for liver glycogen diseases. *Arch. Dis. Child.* **44**, 311–17.

Fernandes J., Jansen H. & Jansen T. C. (1979) Nocturnal gastric drip feeding in glucose-6-phosphate deficient children. *Pediatr. Res.* **13**, 225–9.

Fernandes J. & Pikaar N. A. (1969) Ketosis in hepatic glycogenoses. *Arch. Dis. Child.* **47**, 41–6.

Fernandes J. & van de Kamer J. H. (1968) Hexose and protein tolerance tests in children with liver glycogenosis caused by a deficiency of the debranching enzyme system. *Pediatrics* **41**, 935–44.

Forest M. G. (1981) Adrenal steroid deficiency states. In Brook C. G. H. (ed.) *Clinical Paediatric Endocrinology*, pp. 396–428. Blackwell Scientific Publications, Oxford.

Garnica A., Desfrosiers P. & Sargent T. (1976) Methylmalonic aciduria and hypoglycaemia. *New Engl. J. Med.* **295**, 1136–7.

Goldfien A., Moore R., Zileli S., Havens L. L., Boling L. & Thorn G. W. (1961) Plasma epinephrine and nor-epinephrine levels during insulin-induced hypoglycaemia in man. *J. Clin. Endocr.* **21**, 296–304.

Goodman H. G., Grumbach M. M. & Kaplan S. L. (1968) Growth and growth hormone II. A comparison of isolated growth-hormone deficiency and multiple pituitary-hormone deficiencies in 35 patients with idiopathic hypopituitary dwarfism. *New Engl. J. Med.* **278**, 57–68.

Greene H. L., Slonim A. E., O'Neill J. A. & Burr I. M. (1976) Continuous nocturnal intragastric feeding for management of type I glycogen storage disease. *New Engl. J. Med.* **294**, 423–5.

Greene H. L., Stifel F. B. & Herman R. H. (1972) Ketotic hypoglycaemia due to hepatic fructose-1,6-diphosphatase deficiency. Treatment with folic acid. *Am. J. Dis. Child.* **124**, 415–18.

Gregersen N., Lauritzen R. & Rasmussen K. (1976) Suberylglycine excretion in the urine of a patient with dicarboxylic aciduria. *Clin. Chim. Acta* **70**, 417–25.

Haworth J. C., Perry T. L., Blass J. P., Hansen G. & Urquhart N. (1976) Lactic acidosis in three sibs due to defects in both pyruvate dehydrogenase and α-ketoglutarate dehydrogenase complexes. *Pediatrics* **58**, 564–72.

HAYMOND M. W., KARL I. E., FEIGIN R. D., DEVIVO D. & PAGLIARA A. S. (1973) Hypoglycaemia and maple syrup urine disease: defective gluconeogenesis. *Pediatr. Res.* 7, 500–508.

HAYMOND M. W., KARL I. E. & PAGLIARA A. S. (1974) Ketotic hypoglycaemia: An amino acid substrate limited disorder. *J. Clin. Endocr. Metab.* **38**, 521–30.

HEITZ P. U., KLOPPELL G., HACKI W. H., POLACK J. M. & PEARSE A. G. E. (1977) Nesidioblastosis: the pathologic basis of persistent hyperinsulinaemic hypoglycaemia in infants. *Diabetes* **26**, 632–42.

HERS H. G. (1980) Carbohydrate metabolism and its regulation. In Burman D., Holton J. B., Pennock C. A. (eds.) *Inherited Disorders of Carbohydrate Metabolism*, pp. 3–18. MTP Press, Lancaster.

HOMMES F. A., SCHRIJER J. & DIAS TH. (1980) Pyruvate carboxylase deficiency, studies on patients and on an animal model system. In Burman D., Holton J. B., Pennock C. A. (eds.) *Inherited Disorders of Carbohydrate Metabolism*, pp. 269–86. MTP Press, Lancaster.

HOPWOOD N. J., FORSMAN P. J., KENNY F. M. & DRASH A. L. (1975) Hypoglycaemia in hypopituitary children. *Am. J. Dis. Child.* **129**, 918–26.

HUFTON B. R. & WHARTON B. A. (1982) Glycogen storage disease type I presenting in the neonatal period. *Arch. Dis. Child.* **57**, 309–11.

HUG G., KRILL C. E., PERRIN E. V. & GUEST G. M. (1963) Cori's disease (amylo-1,6-glucosidase deficiency). *New Engl. J. Med.* **268**, 113–20.

HUIJING F. (1975) Glycogen metabolism and glycogen storage diseases. *Physiol. Rev.* **55**, 609–58.

HUNG W. & MIGEON C. J. (1968) Hypoglycaemia in a two-year-old boy with adrenocorticotropic hormone (ACTH) deficiency (probably isolated) and adrenal medullary unresponsiveness to insulin induced hypoglycaemia. *J. Clin. Endocr.* **28**, 146–52.

ICHIBA Y. & GARDNER L. I. (1975) Metabolic and genetic syndromes of overgrowth. In Gardner L. I. (ed.) *Endocrine and Genetic Diseases of Childhood and Adolescence*, pp. 1314–38. W. B. Saunders, Philadelphia.

IGARASHI Y., OTOMO H., NARISAWA K. & TADA K. (1979) A new variant of glycogen storage disease type I probably due to a defect in the glucose-6-phosphate transport system. *J. Inher. Metab. Dis.* **2**, 45–9.

JOHNSTON D. G., PERNET A., MCCULLOCH A., BLESAMALPICA G., BURRIN J. & ALBERTI K. G. M. M. (1982) Some hormonal influences on glucose and ketone body metabolism in normal human subjects. Ciba Symposium No. 87, *Metabolic Acidosis*, pp. 168–91. Pitman, London.

KALHAN S. C., SAVIN S. M. & ADAM P. A. J. (1976) Measurement of glucose turnover in the human newborn with glucose-1-^{13}C. *J. Clin. Endocr.* **43**, 704–707.

KAYE R., DAVIDSON M. H., WILLIAMS M. L., KUMAGAI M. & PICOU D. M. (1961) The response of blood glucose, ketones and plasma nonesterified fatty acids to fasting and epinephrine injection in infants and children. *J. Pediatr.* **59**, 836–47.

KELNAR C. J. H. (1976) Hypoglycaemia in children undergoing adenotonisillectomy. *Br. Med. J.* **i**, 751–2.

KERR D. S., STEVENS M. C. G. & PICOU D. I. M. (1975) Estimation of fasting glucose flux in malnourished and hypoglycaemic children by constant infusion of U-^{13}C glucose. Second International Conference of Stable Isotopes, Argonne, Illinois, U.S.A.

KING K., ADAM P. A. J., CLEMENTS G. A. & SCHWARTZ R. (1969) Infants of diabetic mothers: Attenuated glucose uptake without hyperinsulinaemia during continuous glucose infusion. *Pediatrics* **44**, 381–92.

KOLLEE L. A., MONNENS L. A., CEJKA V. & WILMS R. H. (1978) Persistent neonatal hypoglycaemia due to glucagon deficiency. *Arch. Dis. Child.* **53**, 422–4.

LA DU B. N. & GJESSING L. R. (1978) Tyrosinosis and tyrosinemia. In Stanbury J. B., Wyngaarden J. B., Fredrickson D. S. (eds.) *The Metabolic Basis of Inherited Disease*, 4th Edition, Chapter 12, pp. 256–67. McGraw-Hill, New York.

LEBLANC A., ODIEVRE M., HADCHOUEL M., GENDREL D., CHAUSSAIN J.-L. & RAPPAPORT R. (1981) Neonatal cholestasis and hypoglycaemia: possible role of cortisol deficiency. *J. Pediatr.* **99**, 577–80.

LEDERER B., VAN HOOF F., VAN DER BERGHE G. & HERS H. G. (1975) Glycogen phosphorylase and its converter enzymes in haemolysates of normal human subjects and patients with type VI glycogen storage disease. *Biochem. J.* **147**, 23–35.

LEONARD J. V. (1982) Problems in the congenital lactic acidoses. Ciba Symposium No. 87, *Metabolic Acidosis*, pp. 168–91. Pitman, London.

LEONARD J. V., FRANCIS D. E. M. & DUNGER D. B. (1979) Dietary management of hepatic glycogen storage disease. *Proc. Nutr. Soc.* **38**, 321–4.

LEONARD J. V., SEAKINS J. W. T., BARTLETT K., HYDE J., WILSON J. & CLAYTON B. E. (1981) Inherited disorders of 3-methyl crotonyl CoA carboxylation. *Arch. Dis. Child.* **56**, 53–9.

LINDBLAD B., LINDSTEDT S. & STEEN G. (1977) On the enzymic defects in heriditary tyrosinaemia. *Proc. Nat. Acad. Sci. (USA)* **74**, 4641–5.

MCGARRY J. D. & FOSTER D. W. (1980) Regulation of fatty acid oxidation and ketone body production. *Ann. Rev. Biochem.* **49**, 395–420.

MCQUARRIE I., BELL E. T., ZIMMERMANN B. & WRIGHT W. S. (1950) Deficiency of alpha cells of pancreas as possible etiological factor in hypoglycaemosis (Abstract). *Fed. Proc.* **9**, 337.

MARKS V. & ROSE F. C. (1981) *Hypoglycaemia*, 2nd Edition. Blackwell Scientific Publications, Oxford.

MELANCON S. B., KHACHARDURIAN A. K., NADLER H. L. & BROWN B. I. (1973) Metabolic and biochemical studies in fructose-1,6-diphosphatase deficiency. *J. Pediatr.* **82**, 650–7.

MIGEON C. J., KENNY F. M., KOWARSKI A., SNIPES C. A., SPAULDING J. S., FINKELSTEIN J. W. & BLIZZARD R. M. (1968) The syndrome of congenital adrenocortical unresponsiveness to ACTH: Report of six cases. *Pediatr. Res.* **2**, 501–13.

MONCRIEFF M. W., HILL D. S., ARCHER J. & ARTHUR L. J. H. (1972) Congenital absence of pituitary gland and adrenal hypoplasia. *Arch. Dis. Child.* **47**, 136–7.

MORRICE J. J., TAYLOR K. M., BLAIR J. I. & YOUNG D. G. (1974) Pre-operative plasma glucose level. *Arch. Dis. Child.* **49**, 898–900.

MOWAT A. P. (1982) Hepatic disorders. *Clin. Gastroenterol.* **11**, 171–205.

NEWSHOLME E. A. & START C. (1973) *Regulation in metabolism*. John Wiley, London.

ODIEVRE M., GENTIL C., GAUTIER M. & ALAGILLE D. (1978) Hereditary fructose intolerance in childhood. *Am. J. Dis. Child.* **132**, 605–608.

OWEN O. E., MORGAN A. O., KEMP H. G., SULLIVAN J. M., HERRARA M. G. & CAHILL G. F. (1967) Brain metabolism during fasting. *J. Clin. Invest.* **46**, 1589–95.

POHORECKY L. A. & WURTMAN R. (1971) Adrenocortical control of epinephrine synthesis. *Pharmacol. Rev.* **23**, 1–35.

POWELL G. F., BRASEL J. A., RAITI S. & BLIZZARD R. M. (1967) Emotional deprivation and growth retardation simulating idiopathic hypopituitarism. II. Endocrinologic evaluation of the syndrome. *New Engl. J. Med.* **276**, 1279–83.

PSACHAROPOULOS H. T., MOWAT A. P., DAVIES M., PORTMANN B., SILK D. B. A. & WILLIAMS R. (1980) Fulminant hepatic failure in childhood. *Arch. Dis. Child.* **55**, 252–58.

RANDLE P. J. (1980) The biochemical basis of the relation between glucose and fatty acid metabolism. *Acta Chir. Scand.* Suppl. 498, 111–15.

RANDLE P. J., GARLAND P. B., HALES C. N., NEWSHOLME E. A., DENTON R. M. & POGSON C. I. (1966) Interactions of metabolism and physiological role of insulin. *Recent Prog. Horm. Res.* **22**, 1–44.

ROE T. F., KERSHNAR A. K., WEITZMAN J. J. & MADRIGAL L. S. (1973) Beckwith's syndrome with extreme organ hyperplasia. *Pediatrics* **52**, 372–81.

SCARLETT J. A., MAKO M. E., RUBENSTEIN A. H., BLIX P. M., GOLDMAN J., HORWITZ D. L. *et. al.* (1977). Factitious hypoglycaemia. Diagnosis by measurement of serum C-peptide immunoreactivity and insulin-binding antibodies. *New Engl. J. Med.* **297**, 1029–32.

SCHUTT-AINE J. C., DRASH A. L. & KENNY F. M. (1973) Possible relationship between spontaneous hypoglycaemia and the maternal deprivation syndrome. *J. Pediatr.* **82**, 809–13.

SENIOR B. (1973) Ketotic hypoglycaemia. *J. Pediatr.* **82**, 555–6.

SENIOR B. & LORIDAN L. (1968) Studies on liver glycogenosis with particular reference to the metabolism of intravenously administered glycerol. *New Engl. J. Med.* **279**, 958–70.

SERVICE F. J., HORWITZ D. L., RUBENSTEIN A. H., KUZUYA H., MAKO M. E., REYNOLDS C. & MOLNAR G. D. (1977) C-peptide suppression test for insulinoma. *J. Lab. Clin. Med.* **90**, 180–6.

SIDBURY J. R., CORNBLATH M., FISHER J. & HOUSE E. (1961) Glycogen in erythrocytes of patients with glycogen storage disease. *Pediatrics* **27**, 103–11.

SMITH H. M. (Ed.) (1968) Diazoxide and the treatment of hypoglycaemia. *Ann. N.Y. Acad. Sci.* **150**, 191–467.

SOTELO-AVILA C., GONZALEZ-CRUSSI F. & FOWLER J. W. (1980) Complete and incomplete forms of Beckwith–Wiedemann syndrome: their oncogenic potential. *J. Pediatr.* **96**, 47–50.

SPENCER-PEET J., NORMAN M. E., LAKE B. D., MCNAMARA J. & PATRICK A. D. (1971) Hepatic glycogen storage disease. Clinical and laboratory findings in 23 cases. *Quart. J. Med.* (New Series) **40**, 95–114.

STACEY T. E., STRANG L. B., FLEISCHMAN A., EAGLE R. & RYMAN B. E. (1980) Treatment of fructose-1,6-diphosphatase deficiency. Abstract: International Symposium on Inborn Errors of Metabolism in Humans, Interlaken.

STANLEY C. A. & BAKER L. (1977) Hyperinsulinism in infants and children: Diagnosis and therapy. *Adv. Pediatr.* **23**, 315–55.

SUSSMAN K. E., CROUT J. & ALEXANDER R. M. (1963) Failure of warning in insulin-induced hypoglycemic reactions. *Diabetes* **12**, 38–45.

TEIJEMA H. L., VAN GELDEREN H. H. & GIESBERTS M. A. H. (1980) Hypoketosis as a cause of symptoms in childhood hypoglycaemia. *Eur. J. Pediatr.* **134**, 51–5.

TIETZE H. U., ZURBRUGG R. P., ZUPPINGER K. A., JOSS E. E. & KASER H. (1972) Occurrence of impaired cortisol regulation in children with hypoglycaemia associated with adrenal medullary hyporesponsiveness. *J. Clin. Endocr.* **34**, 948–58.

TURNER R. C. & HEDING L. G. (1977) Plasma pro-insulin, C-peptide and insulin in diagnostic suppression tests for insulinomas. *Diabetologia* **13**, 571–7.

UNGER R. H. (1981) The milieu interieur and the islets of Langerhans. *Diabetologia* **20**, 1–11.

UNGER R. H. & ORCI L. (1981) Glucagon and the A-cell. *New Engl. J. Med.* **304**, 1518–24.

VAN BIERVLIET J. P. G. M., BRUINVIS L., VAN DER HEIDEN C., KETTING D., WADMAN S. K., WILLEMSE J. L. & MONNENS L. A. H. (1977) Report of a patient with severe chronic lactic acidaemia and pyruvate carboxylase deficiency. *Develop. Med. Child. Neurol.* **19**, 392–401.

VIDNES J. (1976) Persistent hereditary neonatal hypoglycaemia caused by glucagon deficiency. *Pediatr. Res.* **10**, 881 (Abst).

WARE A. J., BURTON W. C., MCGARRY J. D., MARKS J. F. & WEINBERG A. G. (1978) Systemic carnitine deficiency. *J. Pediatr.* **93**, 959–64.

WOLFSDORF J. I., SADEGHI-NEJAD A. & SENIOR B. (1982) Hypoalaninaemia and ketotic hypoglycaemia: Cause or consequence? *Eur. J. Pediat.* **138**, 28–31.

ZUPPINGER K. A. (1975) *Hypoglycaemia in childhood. Evaluation of diagnostic procedures.* S. Karger, Basel.

Chapter 4
Hyperammonaemia in Childhood

JAMES V. LEONARD

Hyperammonaemia is an important cause of mortality and morbidity in childhood and is still commonly misdiagnosed.

Ammonia is one of the major products of the catabolism of nitrogen compounds and it is rapidly metabolized by the liver, skeletal muscle and brain (Lockwood *et al.* 1979) so that normally the concentration in the blood remains low. At higher concentrations ammonia is highly toxic, particularly to the central nervous system.

THE UREA CYCLE

Depending on the protein intake, 50–90% of waste nitrogen is excreted as urea (Allison & Bird 1964; Barness *et al.* 1957), one of the nitrogen atoms of which is derived from ammonia and is mostly formed from glutamate (Krebs *et al.* 1978). The sequence of reactions in the synthesis of urea from ammonia is shown in Fig. 4.1 and within the cycle ornithine is catalytic since none is formed or removed.

The condensation of ammonia with bicarbonate and ATP is catalyzed by the intra-mitochondrial carbamoyl phosphate synthetase (CPS 1) which requires N-acetyl glutamate as an allosteric activator. Carbamoyl phosphate is also a precursor of pyrimidines and is synthesised in the cytosol by a distinct enzyme (CPS 2). Details of the biochemistry of the urea cycle, particularly the enzymology, have recently been reviewed (Snodgrass 1981).

The regulation of the cycle is complex since several mechanisms are involved (Evered 1981) and the subject will only be considered briefly. Over several days the total activity of the enzymes varies with the protein intake (Schimcke 1962), this induction being mediated at least in part by glucagon and corticosteroids (Snodgrass *et al.* 1978; Gebhardt & Mecke 1979). More rapid changes are mediated through N-acetyl glutamate, which is synthesized in the mitochondria by a specific synthetase (Shigesada *et al.* 1978). This enzyme is activated by arginine and no other amino acid. N-acetyl glutamate is broken down by an acylase so that a regulatory cycle is formed. Variation in ornithine concentrations may also have an important regulatory function (Evered 1981).

A metabolic block in the urea cycle leads to the accumulation of ammonia and other metabolites.

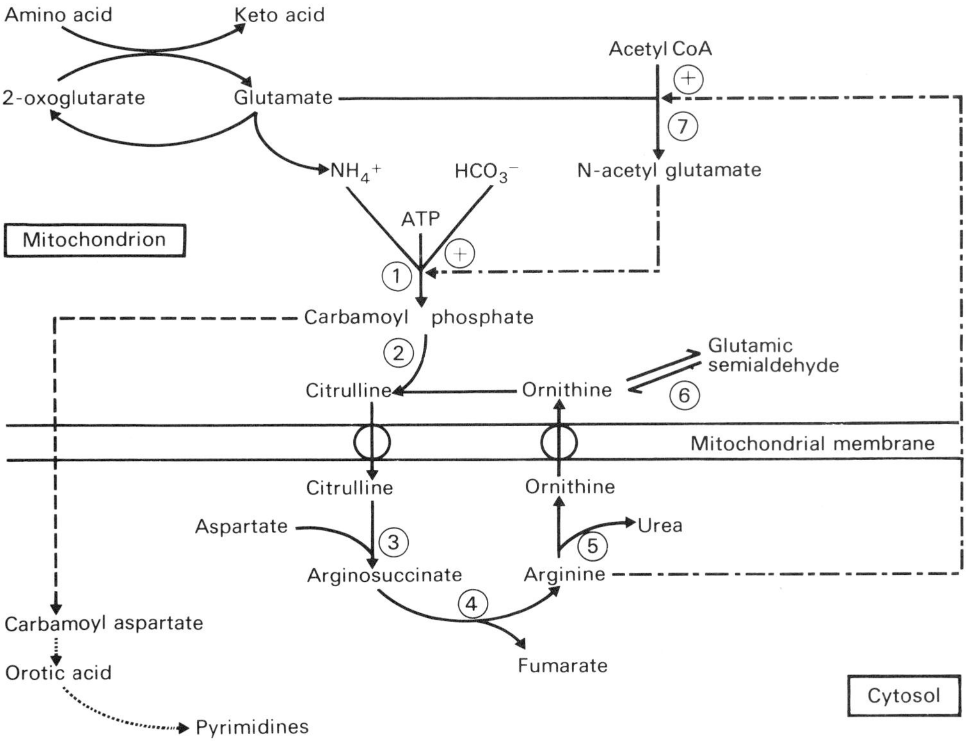

Fig. 4.1 The urea cycle and related metabolic pathways. Enzymes: 1 Carbamoyl phosphate synthetase; 2 Ornithine carbamoyl transferase; 3 Arginosuccinate synthetase; 4 Arginosuccinate lyase; 5 Arginase; 6 Ornithine keto-acid aminotransferase; 7 N-acetyl glutamate synthetase.

PLASMA AMMONIUM CONCENTRATIONS

When considering plasma ammonium concentrations in blood it is necessary to take several factors into account.

The early methods for measuring plasma ammonia used strong alkali to liberate the ammonia which was then assayed using a colour reaction (Seligson & Seligson 1951; Colombo 1971). However, under the conditions of the assay nitrogenous compounds such as glutamine decomposed, liberating ammonia, so that the values obtained using these methods are considerably higher than present methods, which do not use extreme conditions. Present methods include the ammonia-selective electrode (Park & Fenton 1973), ion-exchange chromatography using a small column of cation exchange resin (Oberholzer *et al.* 1976), and an enzymatic reaction (Jacobs & Althuis 1973). None of the methods is entirely satisfactory, especially at low ammonium concentrations, but they are adequate for clinical purposes.

The plasma ammonium concentrations in venous blood of healthy adults and children are usually < 40 μmol/l (Oberholzer *et al.* 1976) but values are dependent on protein and energy intake (Raiha *et al.* 1976) so that a dietary assessment of what the patient is taking is often helpful when interpreting plasma ammonium values. Plasma ammonium values of healthy full-term neonates are similar to those of older children (Batshaw & Brusilow 1978) although some studies have found higher values (Oberholzer *et al.* 1976). Pre-term and small-for-dates babies have higher plasma ammonium concentrations; the mean determined by Batshaw & Brusilow (1978) was 47 μmol/l (2 SD above the mean was 65 μmol/l). The ammonium concentration of arterial or free-flowing capillary blood, measured using a satisfactory analytical method, is approximately 8 μmol/l higher than that in venous blood (Oberholzer *et al.* 1976) because there is a net uptake of ammonia by muscle (Lockwood *et al.* 1979).

Blood specimens for the determination of ammonium need to be collected meticulously. Falsely elevated values are common in children in whom there have been difficulties with blood collection. The ammonium concentration within red blood cells is more than twice that of plasma so that haemolysis increases the apparent plasma concentration (Colombo 1971). Exercise also raises plasma levels (McCullough 1968). Plasma should be separated immediately and analysed without delay. If it is necessary to store the plasma for short periods it should be kept at −70 °C since breakdown of nitrogenous compounds proceeds at −20 °C (Oberholzer *et al.* 1976).

CAUSES OF HYPERAMMONAEMIA

Hyperammonaemia may be caused by inborn errors of the urea cycle and organic acid metabolism. Plasma ammonium levels are also elevated in advanced liver disease, Reye's syndrome and occasionally during therapy with sodium valproate. The causes in the neonatal period and those during infancy and childhood are listed in Table 4.1.

The presentation of hyperammonaemia is very variable, and may present with acute toxic encephalopathy or with more chronic symptoms such as vomiting, failure to thrive or growth retardation. The symptoms and signs of ammonia intoxication are most clearly demonstrated in the congenital disorders of the urea cycle.

Table 4.1 Causes of hyperammonaemia.

Neonatal period

- Inherited disorders of the urea cycle
  - N-acetyl glutamate synthetase deficiency
  - Carbamoyl phosphate synthetase deficiency
  - Ornithine carbamoyl transferase deficiency
  - Citrullinaemia
  - Arginosuccinic aciduria
- Organic acidaemias
  - Propionic acidaemia
  - Methylmalonic acidaemia
  - β-ketothiolase deficiency
  - Isovaleric acidaemia
- Transient neonatal hyperammonaemia
- Perinatal asphyxia
- Intravenous feeding
- Pre-term and sick neonates

Infancy and childhood

- Inherited disorders of the urea cycle
  - Carbamoyl phosphate synthetase deficiency
  - Ornithine carbamoyl transferase deficiency
  - Citrullinaemia
  - Arginosuccinicaciduria
  - Arginase deficiency
- Other inherited disorders
  - Lysinuric protein intolerance
  - Hyperornithinaemia, hyperammonaemia and homocitrullinuria
  - Hyperlysinaemia with hyperammonaemia
  - Organic acidaemias
- Liver disease
  - Advanced liver disease (acute and chronic)
  - Reye's syndrome
- Other causes
  - Sodium valproate therapy
  - Infection and stasis in the urinary tract
  - Systemic illness (shock, congestive cardiac failure)

Inborn errors of the urea cycle and related disorders

Clinical presentation

These conditions vary considerably in their severity and may present at almost any age during childhood (Hsia 1974; Shih 1978; Snyderman 1981).

Neonatal period. Those most severely affected present in the neonatal period. At birth they appear normal but within the first week there is a gradual onset of lethargy, poor feeding, vomiting, tachypnoea (often with grunting), seizures and alteration in tone. The encephalopathy progresses to loss of all reflex activity, apnoea and death. Liver failure may develop with a bleeding diathesis and a pulmonary or intracranial haemorrhage may be the terminal event.

Infancy and childhood. In patients with less severe disease symptoms develop during infancy and childhood. Persistent vomiting and failure to thrive with developmental retardation are common, and often it is possible to elicit a history that symptoms are exacerbated by protein. All patients, even those who are mildly affected, may have episodes of more acute illness during which neurological symptoms and signs such as headaches, irritability, ataxia, slurring of speech, confusion, hemiplegia, alteration in consciousness and seizures may predominate. Behavioural problems are also frequent. The pattern of the whole illness may have a rather characteristic episodic or fluctuating course with symptoms often being precipitated by intercurrent infection, or any stress that causes protein catabolism. During acute exacerbations, the patient's condition may deteriorate rapidly due to the development of acute cerebral oedema, and survival is infrequent once the patient is deeply comatose. Hyperammonaemia caused by a urea cycle disorder should always be considered in a child who is thought to have 'encephalitis' or undiagnosed encephalopathy, particularly if the child has had more than one attack.

Inborn errors of metabolism

N-acetyl glutamate synthetase deficiency (NAGSD)

This condition has recently been described in a child who developed symptoms in the neonatal period (Bachmann *et al.* 1981).

Carbamoyl phosphate synthetase deficiency (CPSD)

Although in early clinical descriptions the clinical and biochemical features of CPSD were rather varied (Shih 1978), most patients with this condition have developed characteristic symptoms of hyperammonaemia during the early weeks of life.

Ornithine carbamoyl transferase deficiency (OCTD)

Ornithine carbamoyl transferase deficiency is the most common of the urea cycle disorders and its mode of inheritance is X-linked. Most affected males have little residual enzyme activity and develop symptoms shortly after birth. Even with energetic treatment these boys will die in the early weeks or months of life (McReynolds *et al.* 1978; Donn *et al.* 1979; Wiegand *et al.* 1980). In contrast the symptoms of the female heterozygotes is very varied even within one kindred. Those who are most severely affected develop symptoms in early infancy whilst the most mildly affected remain asymptomatic, although they may have a slight reduction in IQ (Batshaw *et al.* 1980a). This variation is thought to arise because of the random inactivation of one X chromosome at an early stage of fetal development (Lyon hypothesis). As a result the liver contains two populations of cells, one with normal enzyme levels and the other with absent or reduced activity (Ricciuti *et al.* 1976). The relative proportions of the two populations will determine the total OCT activity and hence the individual's tolerance of protein and her symptoms. Males with milder symptoms similar to those of the females are not uncommon (Levin *et al.* 1969; Saudubray *et al.* 1975).

Citrullinaemia (arginosuccinate synthetase deficiency, ASSD)

This condition often presents with hyperammonaemia in the neonatal period or infancy but mild and atypical forms have been described (Shih 1978).

Arginosuccinic aciduria (arginosuccinate lyase deficiency, ASLD)

Many patients with this condition present with neurological problems including fits, intermittent ataxia, irritability and delayed development. About half of these patients have friable short hair with a characteristic microscopic appearance (trichorrhexis nodosa). There is however wide variation in the mode of presentation with some patients developing symptoms in the neonatal period or in early infancy whilst others remain asymptomatic (Shih 1978).

Arginase deficiency

This is a rare disorder characterised by recurrent vomiting, fits, mental retardation and spastic diplegia (Terheggen *et al.* 1969). One child treated from birth has developed normally (Snyderman *et al.* 1979).

Diagnosis

Physical examination

There are no diagnostic physical signs. Short friable hair in a mentally retarded child suggests the diagnosis of arginosuccinic aciduria but it is not specific. In all urea cycle disorders hepatomegaly may be present, and is more likely to be found during acute attacks.

Biochemical tests

Blood urea. The blood urea is normal or only slightly reduced in all but the most severe neonatal forms.

Plasma ammonium levels. The plasma ammonium concentrations of children with urea cycle disorders are very variable. In the neonatal forms plasma ammonium levels often exceed 500 µmol/l when the diagnosis is first made and may rise rapidly to 1000–5000 µmol/l. Patients with coma due to ammonia intoxication usually have values > 300 mmol/l but serious encephalopathy may occur with lower values. In contrast patients with milder disease who are well or on a low protein diet may have normal ammonium levels but in many symptomatic patients who are on treatment values up to 100 µmol/l are common.

Liver function tests. The aminotransferases (alanine and aspartate) are commonly raised during acute exacerbations but jaundice is unusual. Abnormalities of clotting may develop, particularly in the neonatal forms.

Plasma and urine amino acids. The results of plasma and urine amino acids are diagnostic in citrullinaemia, arginosuccinic aciduria and arginase deficiency since the metabolic block causes the accumulation of a characteristic compound that is easily detected (Table 4.2). In NAGSD, CPSD, and OCTD only glutamine and alanine accumulate. Arginine levels are often reduced, except in arginase deficiency, but ornithine levels are normal.

Orotic acid and pyrimidines. Carbamoyl phosphate (CP) is a precursor of pyrimidines as well as of urea. If reactions in the urea cycle distal to the formation of CP are blocked then CP is diverted to the synthesis of orotic acid and pyrimidines (Fig. 4.1). The excess orotic acid is excreted in the urine and can be estimated in a random urine specimen, the result being most conveniently expressed as an orotic acid/creatinine ratio. The ratio is elevated in all the urea cycle defects except NAGSD and CPSD (Table 4.1) (Bachmann & Colombo 1980). The measurement of orotic acid after a protein load is useful for the detection of OCTD carriers as it is more

Table 4.2 Biochemical findings in inherited disorders of the urea cycle. Abnormalities of plasma and urine amino acids, urine orotic acid excretion and the tissues in which enzyme defect is expressed.

Disorder	Plasma	Amono acids Urine	Urine orotic acid excretion	Tissues in which enzyme defect is expressed
N-acetyl glutamate synthetase deficiency	Glutamine + Alanine +	As plasma	Normal	Liver
Carbamoyl phosphate synthetase deficiency	Glutamine + Alanine +	As plasma	Normal	Liver Jejunal mucosa
Ornithine carbamoyl transferase deficiency	Glutamine + Alanine +	As plasma	+	Liver Jejunal mucosa
Citrullinaemia (arginosuccinate synthetase deficiency	Citrulline + +	Citrulline, homocitrulline and other metabolites +	+	Liver Skin fibroblasts
Arginosuccinic aciduria (argino-succinate lyase deficiency	Argino succinate +	Arginosuccinic acid and other metabolites + +	+ (during acute attacks)	Liver Red blood cells Skin fibroblasts
Arginase deficiency	Arginine + +	Arginine, ornithine, cystine, lysine +	+	Liver Red blood cells ? skin fibroblasts

+ = increased concentrations.
For diagnostic purposes the amino acid concentrations should be measured by column chromatography. The precise concentrations of amino acids in all these disorders are dependent on many factors such as the protein and energy intake.
Arginine concentrations are low or low normal in all the conditions except arginase deficiency. Details of the amino acid concentrations in plasma and urine of healthy children are given in Scriver and Rosenberg (1973) and Bremer *et al.* (1981).

sensitive for this purpose than other indices such as plasma ammonium (Haan *et al.* 1982).

Enzyme assay. The diagnosis of a urea cycle disorder should be confirmed by measurement of enzyme activity in suitable tissue (Table 4.2), although in difficult or atypical cases a needle biopsy of the liver should be obtained. However, the interpretation of enzyme assays can be difficult since many factors effect the enzyme activity and as a result there is a wide range of activity even in 'normal' controls (Snodgrass 1981).

Treatment

The mainstay of the treatment of the urea cycle disorders is the restriction of dietary protein to approximately 0.75–1.5 g/kg/day (depending on age and protein tolerance) with a generous energy intake. Many affected patients spontaneously select a low protein diet as they realize that their symptoms improve. The effect of the diet should be monitored by measurement of plasma ammonium and amino acids, and growth. However, the protein restriction necessary to control the plasma ammonia may be insufficient to support normal growth. By giving supplements of essential amino acids, the natural protein intake may be reduced and the requirements of essential amino acids for growth still be met (Snyderman *et al.* 1975). Alternatively, keto-analogues of the essential amino acids may be used. A small percentage of the administered dose of these compounds is transaminated to form the respective amino acid, thereby both providing additional amino acids for growth and reducing the waste nitrogen load (Brusilow *et al.* 1979a).

Arginine has an important place in the treatment of urea cycle disorders. Since this amino acid is normally synthesised in the urea cycle it is not an essential amino acid, but in urea cycle disorders which have the metabolic block, it may become semi-essential or essential. Arginine deficiency may be one factor responsible for the skin rashes, growth retardation and poor intellectual development of patients with these disorders (Danks *et al.* 1974; Kline *et al.* 1981). As it is difficult to estimate requirements we routinely give arginine supplements to patients with CPSD or OCTD, 3–12 mmol/day (0.5–2 g/d) and monitor the plasma level.

In ASSD and ASLD the removal of ammonia is limited by the supply of ornithine which may be replenished by giving supplements of arginine sufficient to control the ammonia levels (up to 5 mmol/kg/day) (Brusilow *et al.* 1979b). Although in ASSD plasma citrulline concentrations may rise to high levels this seems less harmful than ammonia accumulation. In ASLD, arginosuccinate is rapidly cleared by the kidneys.

If these measures fail to control hyperammonaemia other forms of treatment may be used. By promoting the excretion of nitrogenous compounds other than urea the load on the urea cycle can be reduced (Brusilow *et al.* 1979b). Sodium benzoate seems particularly promising. This compound is conjugated with glycine to form hippuric acid which has a renal clearance close to the renal plasma flow. When given to patients with urea cycle disorders urinary nitrogen increases and plasma ammonium falls. It may be used both in acute hyperammonaemia and in long-term management. However, its safety in the neonatal period remains uncertain because it may displace bilirubin off plasma albumin and could therefore potentially cause kernicterus. Oral citrate has been given to increase the supply of Krebs cycle intermediates and in some patients it reduces post-prandial elevation of ammonia (Levin *et al.* 1969; Sunshine *et al.* 1972).

In the majority of patients, ammonia levels can be satisfactorily controlled for much of the time but all patients are liable to develop episodes of severe hyperammonaemia. These may be precipitated by infection, a protein load or surgery but are often unexplained. In the acute episodes even if apparently mild, the protein intake should be stopped and a high energy intake, as fat and carbohydrate together with sodium benzoate, given orally or intravenously. Ammonia levels may still rise rapidly and if simple measures are not quickly effective peritoneal or haemodialysis should be performed (Herrin & McCredie 1969). Details of the acute and long-term management of these inborn errors have recently been published (Batshaw & Brusilow 1980; Batshaw *et al.* 1981).

Outcome

Even with the most vigorous treatment the outlook for the majority of babies who present in the neonatal period is not good. Many will die during the first year of life and the survivors are often mentally retarded (Snyderman *et al.* 1975; McReynolds *et al.* 1978; Brusilow *et al.* 1979a). Delayed development and failure to thrive are common problems in those who are less severely affected, but with early diagnosis and appropriate management normal growth and development are possible. Those with mild disease will do well with little treatment but acute hyperammonaemia may develop in all patients, including those with apparently mild disease and despite active treatment patients may sustain a severe cerebral insult. Plasma ammonium is not a good predictor of outcome since, although high levels carry a bad prognosis, severe encephalopathy may occur with moderate levels ($\sim$ 250 μmol/l).

Bessman and Bessman (1955) have suggested that depletion of Krebs cycle intermediates by the synthesis of glutamine from 2-oxoglutarate is responsible for hepatic encephalopathy. In two infants with OCTD hyperammonaemia coma was preceded by a fall in plasma 2-oxoglutarate and it was suggested that this might be of predictive value (Batshaw *et al.* 1980b).

Genetic considerations

All the conditions are inherited as autosomal recessives except OCTD which is X-linked. Prenatal diagnosis of ASLD and citrullinaemia has been achieved by measuring enzyme activity in cultured amniotic fluid cells (Goodman 1981) but the methods are not without difficulty (Cathelineau *et al.* 1981b; Jacoby *et al.* 1981). Recently prenatal diagnosis of OCTD by liver biopsy at 18 weeks' gestation through a fetoscope has been performed. OCT activity is present at this age and reliable prediction of the genotype of males is possible (Rodeck *et al.* 1982).

Detailed family studies of those with OCTD are indicated to detect the female carriers. Many, but not all, carriers can be identified using a standard protein load and measuring the orotic acid excretion in the urine. An alanine load may be more sensitive. When counselling parents it is important to note that mild disease in the mother does not imply mild disease in the offspring.

OTHER RELATED INBORN ERRORS OF METABOLISM

Two other inborn errors that cause hyperammonaemia are probably caused by disorders of membrane transport of dibasic amino acids.

Lysinuric protein intolerance

This condition usually presents with diarrhoea, vomiting and failure to thrive. Symptoms are exacerbated by the ingestion of protein. Other features include hepatosplenomegaly, osteoporosis, mental retardation and leucopenia (Simell *et al.* 1975). The plasma concentrations of lysine, ornithine and arginine are reduced and there is increased urinary excretion of these amino acids. Hyperammonaemia is rather variable, the plasma ammonium concentrations often being normal whilst fasting but elevated after protein ingestion. It is postulated that this disorder is due to a defect of transport of dibasic amino acids at the basolateral membrane of the cell, including those of the liver, gut and kidney (Rajantie *et al.* 1980a). This reduces the supply of ornithine and arginine and in the liver limits the activity of the urea cycle causing hyperammonaemia. This and the protein intolerance can be corrected by giving supplements of citrulline, since this amino acid is transported into cells and will replenish the urea cycle intermediates. With these supplements there may be catch-up growth but the osteoporosis and hepatosplenomegaly persist and are probably caused by lysine deficiency which cannot be corrected by oral supplements (Rajantie *et al.* 1980b).

Hyperornithinaemia, hyperammonaemia and homocitrullinuria

Several patients have now been described who have had a combination of hyperammonaemia, hyperornithinaemia and homocitrullinuria (Shih 1978). They have presented with fits, a variable degree of mental retardation and episodes of lethargy, ataxia and other symptoms associated with hyperammonaemia. On a low protein diet plasma ammonium levels fall to normal and plasma ornithine concentrations decrease with clinical improvement. No conclusive enzyme defect has been described although the activity of ornithine decarboxylase and CPS 1 have been found to be reduced in some patients (Shih & Mandell 1974; Gatfield *et al.* 1975). The metabolism of ornithine to glutamate and proline and the incorporation of radioactive label derived from ornithine into protein is reduced (Shih *et al.* 1982; Gray *et al.* 1982). In one kindred the ultrastructure of the mitochondria was abnormal (Gatfield *et al.* 1975). It seems probable that this disorder is caused by a defect of the transport of ornithine into mitochondria and decreased ornithine uptake has recently been demonstrated (Hommes *et al.* 1982).

HYPERAMMONAEMIA IN THE NEONATAL PERIOD

In addition to the disorders of the urea cycle several other conditions may cause hyperammonaemia during the neonatal period.

Organic acidaemias

Inborn errors of organic acid metabolism commonly present in the neonatal period and several disorders may be complicated by undetected hyperammonaemia. Those most commonly responsible are propionic acidaemia, methylmalonic acidaemia, and isovaleric acidaemia (Coude *et al.* 1982; Cathelineau *et al.* 1981a).

Clinical presentation (Rosenberg 1978; Dancis & Levitz 1978)

Neonatal period. The baby is initially well and after a variable interval becomes lethargic, tachypnoeic and feels poorly. Vomiting may also be prominent. Later the child develops an acute encephalopathy with fits, alterations in tone and loss of normal reflex activity that may be indistinguishable from that caused by urea cycle disorders (Harris *et al.* 1980). Later in childhood children may present with failure to thrive and metabolic acidosis, neurological problems and intermittent attacks of acidosis. Hyperammonaemia is not usually as marked as in the neonatal period.

Biochemical tests. In the organic acidaemias there is usually a marked metabolic acidosis with ketonuria which is unusual in urea cycle disorders. Plasma ammonium concentrations may range up to 1500 μmol/l (Cathelineau *et al.* 1981a). Plasma amino acids often show a raised glycine and the concentrations of several other amino acids may be elevated (Bremer *et al.* 1981). Since it may be difficult to distinguish urea cycle disorders and organic acidaemias, urine organic acids should always be examined in children with hyperammonaemia by gas-liquid chromatography (Goodman & Markey 1981).

Treatment. The aim of treatment is to correct the metabolic derangement as quickly as possible. If the child is still receiving protein this is stopped and replaced with generous calories from fat and glucose given intravenously. Unless the symptoms are particularly mild peritoneal dialysis should be started using a bicarbonate dialysate to remove organic acids, ammonia and other toxic metabolites and to correct acidosis. With this therapy glucose concentrations often rise and since insulin inhibits protein catabolism at low concentrations, a continuous infusion can be given. Any infection should always be treated vigorously.

The contribution of the raised ammonium to the symptoms is uncertain but the treatment of the organic acidaemia will also reduce the ammonium concentrations. Many suggestions have been made to explain the hyperammonaemia and these have been reviewed recently (Walser & Stewart 1981).

The long-term treatment of children with organic acidaemias is beyond the scope of this chapter.

Transient hyperammonaemia of the newborn

Pre-term babies may develop a condition that is indistinguishable from those with inborn errors of the urea cycle (Ballard *et al.* 1978). However, with vigorous treatment to reduce ammonium concentrations these babies may make a complete recovery without any persistent biochemical abnormalities. Plasma ammonium level may range up to 5000 μmol/l and plasma glutamine concentrations are often raised, but there is no raised urinary excretion of orotic acid. The explanation for this disorder remains unknown.

Perinatal asphyxia and other illness in sick neonates

Goldberg *et al.* (1979) have examined the plasma ammonium concentrations in severely asphyxiated infants and found markedly elevated values that decreased over several days. In another study (Beddis *et al.* 1980) plasma ammonium measured by a more specific method was also found to be raised in asphyxiated neonates but to a less marked degree. Plasma ammonium concentrations were mildly elevated in other sick neonates regardless of the cause.

The contribution of raised ammonium levels to the neurological problems of these neonates and whether any specific treatment should be given remains unclear.

Intravenous feeding

Parenteral feeding in neonates may cause hyperammonaemia sufficient to evoke symptoms of drowsiness and lethargy (Johnson *et al.* 1972; Heird *et al.* 1972). This may have been in part due to the high protein and ammonia load but an imbalance in the amino acid composition of the infusate which contained insufficient arginine also contributed. The addition of this amino acid abolished the hyperammonaemia in one study (Heird *et al.* 1972).

HYPERAMMONAEMIA DURING INFANCY AND CHILDHOOD

In addition to the inborn errors of the urea cycle and other inherited disorders that have already been discussed, acute encephalopathy with hyperammonaemia may occur in Reye's syndrome, severe liver disease and several other conditions (see Table 4.1).

Reye's syndrome (Pollack 1975; De Vivo & Keating 1976; Crocker 1979)

Hyperammonaemia is one of the biochemical manifestations of Reye's syndrome which is an important cause of acute encephalopathy in childhood. It commonly affects children between the ages of 2 and 7 years, but has been described at all ages.

Clinical presentation

The illness usually starts with an apparently mild infection but after a variable interval persistent vomiting develops with irritability and lethargy. This stage is often followed by a period of restlessness, confusion and hyperventilation but if the condition progresses rapidly this phase may be brief and pass unnoticed. The child then loses consciousness and develops neurological signs such as extensor plantar responses and ankle clonus. In the severely affected child the deterioration progresses relentlessly to deep coma with decerebrate posturing, opisthotonos and evidence of raised intracranial pressure. Finally, there is flaccidity, apnoea and fixed dilated pupils. To facilitate the assessment of children with Reye's syndrome classifications of the clinical stages have been proposed (Huttenlocher 1972; Lovejoy *et al.* 1974).

Many children who become deeply comatose will die and at autopsy there is cerebral oedema without any cellular reaction, and fatty infiltration of the liver and other organs including kidney and skeletal muscle.

Pathogenesis. It seems probable that Reye's syndrome is the clinical manifestation of hepatic mitochondrial insult (DeVivo 1978). A wide variety of different agents including viruses, toxins and drugs as well as inborn errors of metabolism have been implicated (Table 4.3) (DeVivo & Keating 1976; DeVivo 1978) and an interaction between several factors may be important (Thaler 1976). Outbreaks of Reye's syndrome may occur during epidemics of certain viruses, such as influenza B but it is unusual

Table 4.3 Causes of Reye's syndrome (De Vivo & Keating 1976, and other sources).

Viruses
Varicella
Influenza A and B
Parainfluenza
ECHO
Coxsackie A and B
Adenoviruses
Reoviruses, and many others
Toxins
Aflatoxins
Pesticides—DDT
Hypoglycins (Jamaican vomiting sickness)
Drugs
Aspirin
Sodium valproate
Inborn errors
Systemic carnitine deficiency
Ornithine carbamoyl transferase deficiency
Isovaleric acidaemia
3-hydroxy-3-methyl glutaryl CoA lyase deficiency

for a child to have more than one attack of Reye's syndrome. Investigations to identify the precipitating cause are indicated in all patients with this illness. An inborn error of metabolism should particularly be suspected in those who are very young or in those who have recurrent attacks. This condition appears to be considerably more common in the United States than in Europe, and the reason for this is not clear.

Many hypotheses have been suggested to account for the encephalopathy of Reye's syndrome and there remains much uncertainty (DeVivo & Keating 1976; DeLong & Glick 1982). Huttenlocher *et al.* (1969) proposed that ammonium was responsible but this hypothesis became discredited because of the apparent lack of correlation between the plasma ammonium concentrations and the clinical state (Haller 1975). In particular some deeply comatose patients had normal plasma ammonium concentrations (Glasgow *et al.* 1972). However with improvement in the understanding of ammonia metabolism and toxicity there has been renewed interest in the role of ammonia (DeLong & Glick 1982).

Laboratory findings. Many biochemical abnormalities have been recorded in Reye's syndrome and are caused by the acute and severe disruption of liver function (Romshe 1975). The transaminases are usually markedly elevated and clotting is severely disordered. Hypoglycaemia is more common in younger children and hyperammonaemia is both transient and variable in its severity. Concentrations may exceed 500 μmol/l but may be only mildly elevated (Glasgow *et al.* 1972). Plasma glutamine, alanine, lysine and amino-N-butyrate levels are raised (Romshe *et al.* 1981) but this pattern is not specific. Urine orotic acid levels are only slightly raised in the urine (Sinatra *et al.* 1975; Glasgow & Chase 1975) unless the illness is caused by an inborn error of the urea cycle (Yokoi *et al.* 1981). Lipid metabolism is disordered and free fatty acid concentrations in the plasma, especially short and medium chain length, are increased (Trauner *et al.* 1975; Mamunes *et al.* 1975). Urine organic acids are usually normal (Harrington *et al.* 1977). Examination of the cerebrospinal fluid shows no cellular reaction but the glucose concentration may be reduced and the pressure increased.

Although the biochemical findings may be highly suggestive they are rarely conclusive and it is necessary to establish the diagnosis by tissue biopsy. Histology of the liver shows marked fatty infiltration with small droplets of fat evenly distributed throughout the lobule without any cellular reaction or necrosis.

Electron microscopy of hepatocytes shows disorganised mitochondria which may be swollen and form amoeboid shapes with a decrease in matrix density (Partin *et al.* 1979). Enzyme studies of liver tissue demonstrate a transient loss of mitochondrial enzymes, including carbamoyl phosphate synthetase and ornithine carbamoyl transferase (Sinatra *et al.* 1975; Brown *et al.* 1976) but preservation of enzymes in the cytosol (Robinson *et al.* 1978). The hyperammonaemia is due to both increased protein catabolism and failure of the intramitochondrial reactions of the urea cycle (Snodgrass & DeLong 1976).

Treatment. The detailed management of children with Reye's syndrome is beyond the scope of this review. It is essential that patients are treated in units with all the necessary facilities for intensive care including monitoring intracranial pressure (Trauner 1980). Glucose is given to prevent hypoglycaemia and provide a substrate for glycolysis. It may also help to minimise cerebral oedema to keep the plasma concentration between 10–20 mmol/l. Hypoxia must be prevented and in all but the mildest cases, elective endotracheal intubation and assisted ventilation are indicated, since this can also be used to reduce intracranial pressure by hyperventilation. The patient's body temperature should be controlled to prevent hyperthermia. It seems probable however that the single most important factor is the control of the intracranial pressure (Shaywitz *et al.* 1980). This should be monitored with the simple devices that are now available so that any rise in pressure can be controlled with infusions of mannitol. The value of exchange transfusion, peritoneal dialysis and haemoperfusion in Reye's syndrome not complicated by renal failure remains unclear. Citrulline reduces the half-time of the decline in plasma ammonium (DeLong & Glick

1982), but there is no evidence that this alters the natural history of the disease.

Prognosis. The prognosis was very poor and the majority of children died (Reye *et al.* 1963; Huttenlocher *et al.* 1969). However, with meticulous intensive care the outcome has improved even for patients in deep coma and in one study 88% of severely affected children survived and only 10% of the survivors had neurological sequelae (Shaywitz *et al.* 1980).

ACUTE AND CHRONIC LIVER DISEASE

Plasma ammonium concentrations are raised in liver failure in both acute fulminating hepatic necrosis and in advanced chronic liver disease, regardless of the cause. These concentrations do not correlate closely with the clinical state and many other factors have been proposed to account for the encephalopathy (Schenker *et al.* 1974; James *et al.* 1979; Zieve 1979). Although the liver disease is usually clinically fully evident it is important that liver function is carefully assessed in patients with undiagnosed hyperammonaemia by both clinical examination and tests of liver function including conventional liver function tests, clotting studies and plasma amino acids.

Many of the measures that are used to treat hepatic encephalopathy such as low protein diet and measures to alter the gut flora (Mowat 1979; Hoyumpa *et al.* 1979) will reduce plasma ammonium levels.

OTHER CAUSES OF HYPERAMMONAEMIA

Sodium valproate therapy

Children who are treated with sodium valproate have higher plasma ammonium levels than controls receiving other anticonvulsants (Coulter & Allen 1981). A small minority of these patients develop more marked hyperammonaemia with plasma levels > 60 μmol/l. This usually presents as lethargy and drowsiness and the symptoms quickly disappear when the drug is discontinued. Rarely the illness progresses to Reye's syndrome (Gerber *et al.* 1979). In some children there has been evidence of hepatitis, pancreatitis and other complications.

Coude *et al.* (1981) have suggested the hyperammonaemia is due to inhibition of carbamoyl phosphate sythetase by the acyl CoA esters derived from the catabolism of sodium valproate.

Urinary tract infection

Patients with urinary tract abnormalities that predispose to stasis may develop an acute encephalopathy during urinary tract infections (Samtoy & De Beukelaer 1980; Drayna *et al.* 1981). If the infecting organisms are capable of splitting urea large quantities of ammonia may be liberated and absorbed from the urinary tract. This may overwhelm the capacity of the liver to detoxify ammonia and hence cause encephalopathy. This resolves once the urine is drained and the infection treated with antibiotics.

Patients with a ureterosigmoidostomy also become hyperammonaemic because of the breakdown of urea and subsequent reabsorption of ammonia from the bowel (Silberman 1958).

Fits and other general illness

Mild elevation of plasma ammonium concentrations is common in babies having frequent fits. The cause of this is uncertain. Shock and congestive cardiac failure may cause mildly raised values (Cowley *et al.* 1969; Valero *et al.* 1974).

CONCLUSIONS

Hyperammonaemia is under-diagnosed and plasma ammonium concentrations should be measured more frequently. The clinical conditions in which hyperammonaemia should particularly be suspected include:

1 Any unexplained acute encephalopathy or 'encephalitis'
2 Failure to thrive, with mental retardation
3 Any unusual recurrent or episodic illness

With early diagnosis the outlook for many conditions associated with hyperammonaemia is good, but if the diagnosis is not made the outcome is often disastrous.

ACKNOWLEDGEMENT

I would like to thank Mr V. G. Oberholzer for his helpful discussions in the preparation of this chapter.

REFERENCES

ALLISON J. B. & BIRD J. W. C. (1964) Elimination of nitrogen from the body. In Munro H. N. & Allison J. B. (eds.) *Mammalian Protein Metabolism*, Vol. I, pp. 483–512. Academic Press, New York.

Bachmann C. & Colombo J. P. (1980) Diagnostic value of orotic acid excretion in heritable disorders of the urea cycle and in hyperammonaemia due to organic acidurias. *Eur. J. Pediatr.* **134**, 109–13.

Bachmann C., Krahenbuhl S., Colombo J. P., Schubiger G., Jaggi K. H. & Tonz O. (1981) N-acetylglutamate synthetase deficiency: A disorder of ammonia detoxication. *New Engl. J. Med.* **304**, 543 (Letter).

Ballard R. A., Vinocur B., Reynolds J. W., Wennberg R. P., Merritt A., Sweetman L. & Nyhan W. L. (1978) Transient hyperammonaemia of the preterm infant. *New Engl. J. Med.* **299**, 920–5.

Barness L. A., Baker D., Guilbert P., Torres F. E. & Gyorgy P. (1957) Nitrogen metabolism of infants fed human and cow's milk. *J. Pediatr.* **51**, 29–39.

Batshaw M. L. & Brusilow S. W. (1978) Asymptomatic hyperammonaemia in low birthweight infants. *Pediatr. Res.* **12**, 221–4.

Batshaw M. L. & Brusilow S. W. (1980) Treatment of hyperammonaemic coma caused by inborn errors of urea synthesis. *J. Pediatr.* **97**, 893–900.

Batshaw M. L., Painter M. J., Sproul G. T., Shaffer I. A., Thomas G. H. & Brusilow S. (1981) Therapy of urea cycle enzymopathies: Three case studies. *Johns Hopkins Med. J.* **148**, 34–40.

Batshaw M. L., Roan Y., Jung A. L., Rosenberg L. A. & Brusilow S. W. (1980a) Cerebral dysfunction in asymptomatic carriers of ornithine transcarbamylase deficiency. *New Engl. J. Med.* **302**, 482–5.

Batshaw M. L., Walser M. & Brusilow S. W. (1980b) Plasma α-ketoglutarate in urea cycle enzymopathies and its role as a harbinger of hyperammonaemic coma. *Pediatr. Res.* **14**, 1316–19.

Beddis I. R., Hughes E. A., Rosser E. & Fenton J. C. B. (1980) Plasma ammonia levels in newborn infants admitted to an intensive care baby unit. *Arch. Dis. Child.* **55**, 516–20.

Bessman S. P. & Bessman A. N. (1955) The cerebral and peripheral uptake of ammonia in liver disease with an hypothesis for the mechanism of hepatic coma. *J. Clin. Invest.* **34**, 622–8.

Bremer H. J., Duran M., Kamerling J. P., Przyrembel H. & Wadman S. K. (1980) *Disturbances of Amino Acid Metabolism: Clinical Chemistry and Diagnosis.* Urban & Schwarzenberg, Munich.

Brown T., Hug G., Lansky L., Bove K., Scheve A., Ryan M., Brown H., Schubert W. K., Partin J. C. & Lloyd-Still J. (1976) Transiently reduced activity of carbamyl phosphate synthetase and ornithine transcarbamylase in liver of children with Reye's syndrome. *New Engl. J. Med.* **294**, 861–7.

Brusilow S., Batshaw M. & Walser M. (1979a) Use of keto-acids in inborn errors of urea synthesis. In Winick M. (ed.) *Nutritional Management of Genetic Disorders.* John Wiley, New York.

Brusilow S. W., Valle D. L. & Batshaw M. L. (1979b) New pathways of nitrogen excretion in inborn errors of urea synthesis. *Lancet* **ii**, 452–4.

Cathelineau L., Briand P., Ogier H., Charpentier C., Coude F. X. & Saudubray J. M. (1981a) Occurrence of hyperammonaemia in the course of 17 cases of methylmalonic acidaemia. *J. Pediatr.* **99**, 279–80.

Cathelineau L., Dinh D. P., Boue J., Saudubray J. M., Farriaux J. P. & Kamoun P. (1981b) Improved method for the antenatal diagnosis of citrullinaemia. *Clin. Chim. Acta* **116**, 111–15.

Colombo J. P. (1971) *Congenital Disorders of the Urea Cycle and Ammonia Detoxication.* S. Karger, Basel.

Coude F. X., Ogier H., Grimber G., Parvy Ph., Pham Dinh D., Charpentier C. & Saudubray J. M. (1982) Correlation between blood ammonia concentration and organic acid accumulation in isovaleric and propionic acidaemia. *Pediatrics* **69**, 115–17.

COUDE F. X., RABIER D., CATHELINEAU L., GRIMBER G., PARVY P. & KAMOUN P. P. (1981) A mechanism for valproate-induced hyperammonaemia. *Pediatr. Res.* **15**, 974–5.
COULTER D. L. & ALLEN R. J. (1981) Hyperammonaemia with valproic acid therapy. *J. Pediatr.* **99**, 317–19.
COWLEY R. A., ATTAR S., LA BROSSE E., MCLAUGHLIN J., SCANLAN E., WHEELER S., HANASHIRO P. *et al.* (1969) Some significant biochemical parameters found in 300 shock patients. *J. Trauma* **9**, 926–36.
CROCKER J. F. S. (ed.) (1979) *Reye's Syndrome.* Vol. II. Grune & Stratton, New York.
DANCIS J. & LEVITZ M. (1978) Abnormalities of branched chain amino acid metabolism. In Stanbury J. B., Wyngaarden J. B. & Fredrickson D. S. (eds.) *The Metabolic Basis of Inherited Disease*, 4th Edition, pp. 397–410. McGraw Hill, New York.
DANKS D. M., TIPPETT P. & ZENTNER G. (1974) Severe neonatal citrullinaemia. *Arch. Dis. Child.* **49**, 579–81.
DELONG G. R. & GLICK T. H. (1982) Encephalopathy of Reye's syndrome: A review of pathogenetic hypotheses. *Pediatrics* **69**, 53–63.
DEVIVO D. C. (1978) Reye's syndrome: A metabolic response to an acute mitochondrial insult? *Neurology* **28**, 105–108.
DEVIVO D. C. & KEATING J. P. (1976) Reye's syndrome. *Adv. Pediatr.* **22**, 175–229.
DONN S. M., SWARTZ R. D. & THOENE J. G. (1979) Comparison of exchange transfusion, peritoneal dialysis and haemodialysis for the treatment of hyperammonaemia in an anuric newborn infant. *J. Pediatr.* **95**, 67–70.
DRAYNA C. J., TITCOMB C. P., VARMA R. R. & SOERGEL K. H. (1981) Hyperammonaemic encephalopathy caused by infection in a neurogenic bladder. *New Engl. J. Med.* **304**, 766–8.
EVERED D. F. (1981) Advances in amino acid metabolism in mammals. *Biochem. Soc. Trans.* **9**, 159–69.
GATFIELD P. D., TALLER E., WOLFE D. M. & HAUST M. D. (1975) Hyperornithinaemia, hyperammonaemia, and homocitrullinuria associated with decreased carbamoyl phosphate synthetase 1 activity. *Pediatr. Res.* **9**, 488–97.
GEBHARDT R. & MECKE D. (1979) Permissive effect of dexamethasone or glucagon induction of urea cyclic enzymes in perifused primary monolayer cultures of rat hepatocytes. *Eur. J. Biochem.* **97**, 29–35.
GERBER N., DICKINSON R. G., HARLAND R. C., LYNN R. K., HOUGHTON D., ANTONIAS J. I. & SCHIMSCHOCK J. C. (1979) Reye-like syndrome associated with valproic acid therapy. *J. Pediatr.* **95**, 142–4.
GLASGOW A. M. & CHASE H. P. (1975) Reye's syndrome. *Lancet* **i**, 100–101.
GLASGOW A. M., COTTON R. B. & DHIENSIRI K. (1972) Reye's syndrome: Blood ammonia and consideration of the non-histologic diagnosis. *Am. J. Dis. Child.* **124**, 827–33.
GOLDBERG R. N., CABAL L. A., SINATRA F. R., PLAJSTEK C. E. & HODGMAN J. E. (1979) Hyperammonaemia associated with perinatal asphyxia. *Pediatrics* **64**, 336–41.
GOODMAN S. I. (1981) Antenatal diagnosis of defects of ureagenesis. *Pediatrics* **68**, 446–7.
GOODMAN S. I. & MARKEY S. P. (1981) *Diagnosis of Organic Acidaemias by Gas Chromatography–Mass Spectrometry.* Alan Liss, New York.
GRAY R. G. F., HILL S. E. & POLLITT R. J. (1982) Reduced ornithine catabolism in cultured fibroblasts and phytohaemagglutinin stimulated lymphocytes from a patient with hyperornithinaemia, hyperammonaemia and homocitrullinuria. *Clin. Chim. Acta* **118**, 141–8.
HAAN E. A., DANKS D. M., GRIMES A. & HOOGENRAAD N. J. (1982) Carrier detection in ornithine transcarbamylase deficiency. *J. Inherit. Metab. Dis.* **5**, 37–40.

HALLER J. S. (1975) Clinical experience with Reye's syndrome. In Pollack J. D. (ed.) *Reye's Syndrome*, pp. 3–14. Grune & Stratton, New York.

HARRINGTON W., LIU A.-S., LONSDALE D. & IGOU D. (1977) Urine organic acid profiles of Reye's syndrome patients. *Clin. Chim. Acta* **74**, 247–54.

HARRIS D. J., YANG B. I. Y., WOLF B. & SNODGRASS P. J. (1980) Dysautonomia in an infant with secondary hyperammonaemia due to propionyl CoA carboxylase deficiency. *Pediatrics* **65**, 107–10.

HEIRD W. C., NICHOLSON J. F., DRISCOLL J. M., SCHULLINGER J. N. & WINTERS R. W. (1972) Hyperammonaemia resulting from intravenous alimentation using a mixture of synthetic L-amino acids: A preliminary report. *J. Pediatr.* **81**, 162–5.

HERRIN J. T. & MCCREDIE D. A. (1969) Peritoneal dialysis in the reduction of blood ammonia levels in a case of hyperammonaemia. *Arch. Dis. Child.* **44**, 149–51.

HOMMES F. A., HO C. K., ROESEL R. A., CORYELL M. E. & GORDON B. A. (1982) Decreased transport of ornithine across the inner mitochondrial membrane as a cause of hyperornithinaemia. *J. Inherit. Metab. Dis.* **5**, 41–7.

HOYUMPA A. M., DESMOND P. V., AVANT G. R., ROBERTS R. K. & SCHENKER S. (1979) Hepatic encephalopathy. *Gastroenterology* **76**, 184–95.

HSIA Y. E. (1974) Inherited hyperammonaemia syndromes. *Gastroenterology* **67**, 347–74.

HUTTENLOCHER P. R. (1972) Reye's syndrome: relation of outcome to therapy. *J. Pediatr.* **80**, 845–50.

HUTTENLOCHER P. R., SCHWARTZ A. D. & KLATSKIN G. (1969) Reye's syndrome: Ammonia intoxication as a possible factor in encephalopathy. *Pediatrics* **43**, 443–54.

JACOBS H. A. M. & ALTHUIS F. M. F. G. (1973) A kinetic determination of ammonia in plasma. *Clin. Chim. Acta* **43**, 81–6.

JACOBY L. B., SHIH V. E., STRUCKMEYER C., NIERMEIJER M. F. & BOUE J. (1981) Variation in arginosuccinate synthetase activity in amniotic fluid cell cultures: implications for prenatal diagnosis of citrullinaemia. *Clin. Chim. Acta* **116**, 1–7.

JAMES J. H., ZIPARO V., JEPPSSON B. & FISCHER J. E. (1979) Hyperammonaemia, plasma amino acid imbalance and blood brain amino acid transport: A unified theory of portalsystemic encephalopathy. *Lancet* **ii**, 772–5.

JOHNSON J. D., ALBRITTON W. L. & SUNSHINE P. (1972) Hyperammonaemia accompanying parenteral nutrition in newborn infants. *J. Pediatr.* **81**, 154–61.

KLINE J. J., HUG G., SCHUBERT W. K. & BERRY H. (1981) Arginine deficiency syndrome. *Am. J. Dis. Child.* **135**, 437–42.

KREBS H. A., HEMS R., LUND P., HALLIDAY D. & READ W. W. C. (1978) Sources of ammonia for mammalian urea synthesis. *Biochem. J.* **176**, 733–7.

LEVIN B., DOBBS R. H., BURGESS E. A. & PALMER T. (1969) Hyperammonaemia. A variant type of deficiency of liver ornithine transcarbamylase. *Arch. Dis. Child.* **44**, 162–9.

LOCKWOOD A. H., MCDONALD J. M., REIMAN R. E., GELBARD A. S., LAUGHLIN J. S., DUFFY T. E. & PLUM F. (1979) The dynamics of ammonia metabolism in man. Effects of liver disease and hyperammonaemia. *J. Clin. Invest.* **63**, 449–60.

LOVEJOY F. H., SMITH A. L., BRESNAN M. J., WOOD J. N., VICTOR D. I. & ADAMS P. C. (1974) Clinical staging in Reye syndrome. *Am. J. Dis. Child.* **128**, 36–41.

MAMUNES P., DEVRIES G. H., MILLER C. D. & DAVID R. B. (1975) Fatty acid Quantitation in Reye's syndrome. In Pollack J. D. (ed.) *Reye's Syndrome*, pp. 245–54. Grune & Stratton, New York.

MCCULLOUGH H. (1968) A simple micro-technique for the determination of blood ammonia and a note on the effect of exercise. *Clin. Chim. Acta* **19**, 101–105.

MCREYNOLDS J. W., MANTAGOS S., BRUSILOW S. & ROSENBERG L. E. (1978) Treatment of complete ornithine transcarbamylase deficiency with nitrogen-free analogues of essential amino acids. *J. Pediatr.* **93**, 421–7.

MOWAT A. P. (1979) *Liver Disorders in Childhood.* Butterworths, London.

OBERHOLZER V. G., SCHWARZ K. B., SMITH C. H., DIETZLER D. N. & HANNA T. L. (1976) Microscale modification of a cation-exchange column procedure for plasma ammonia. *Clin. Chem.* **22**, 1976–81.

PARK N. J. & FENTON J. C. B. (1973) A simple method for the estimation of plasma ammonia using an ion specific electrode. *J. Clin. Path.* **26**, 802–804.

PARTIN J. C., BOVE K., PARTIN J. S. & SCHUBERT W. K. (1979) Liver and muscle ultrastructure in Reye's syndrome. In Crocker J. F. S. (ed.) *Reye's Syndrome*, Vol. II, pp. 215–32. Grune & Stratton, New York.

POLLACK J. D. (ed.) (1975) *Reye's Syndrome.* Grune & Stratton, New York.

RAIHA N. C. R., HEINONEN K., RASSIN D. K. & GAULL G. E. (1976) Milk protein quantity and quality in low birth weight infants: I. Metabolic responses and effects on growth. *Pediatrics* **57**, 659–74.

RAJANTIE J., SIMELL O. & PERHEENTUPA J. (1980a) Basolateral-membrane transport defect for lysine in lysinuric protein intolerance. *Lancet* **i**, 1219–21.

RAJANTIE J., SIMELL O., RAPOLA, J. & PERHEENTUPA J. (1980b) Lysinuric protein intolerance: A two-year trial of dietary supplementation therapy with citrulline and lysine. *J. Pediatr.* **97**, 927–32.

REYE R. D. K., MORGAN G. & BARAL J. (1963) Encephalopathy and fatty degeneration of the viscera. A disease entity in childhood. *Lancet* **ii**, 749–59.

RICCIUTI F. C., GELEHRTER T. D. & ROSENBERG L. E. (1976) X-chromosome inactivation of human liver: confirmation of x-linkage of ornithine transcarbamylase (OTC). *Am. J. Human Genet.* **28**, 332–8.

ROBINSON B. H., TAYLOR J., CUTZ E. & GALL D. G. (1978) Reye's syndrome: Preservation of mitochondrial enzymes in brain and muscle compared with liver. *Pediatr. Res.* **12**, 1045–47.

RODECK C. H., PATRICK A. D., PEMBERY M. E., TZANNATOS C. & WHITEFIELD A. E. (1982). Fetal liver biopsy for prenatal diagnosis of ornithine carbamyl transferase deficiency. *Lancet* **ii**, 297–9.

ROMSHE C. A. (1975). Laboratory diagnosis of Reye's syndrome. In Pollack J. D. (ed.) *Reye's Syndrome*, pp. 15–26. Grune & Stratton, New York.

ROMSHE C. A., HILTY M. D., MCCLUNG H. J., KERZNER B. & REINER C. B. (1981) Amino acid pattern in Reye's syndrome. Comparison with clinically similar entities. *J. Pediatr.* **98**, 788–90.

ROSENBERG L. E. (1978) Disorders of propionate, methylmalonate and cobalamin metabolism. In Stanbury J. B., Wyngaarden J. B., Frederickson D. S. (eds.) *The Metabolic Basis of Inherited Disease*, 4th Edition, pp. 411–29. McGraw Hill, New York.

SAMTOY B. & DEBEUKELAER M. M. (1980) Ammonia encephalopathy secondary to urinary tract infection with *Proteus mirabilis*. *Pediatrics* **65**, 294–97.

SAUDUBRAY J. M., CATHELINEAU L., LAUGIER J. M., CHARPENTIER C., LEJEUNE J. A. & MOZZICONACCI P. (1975) Hereditary ornithine transcarbamylase deficiency. Report of two male cases with residual enzymatic activity. *Acta Paed. Scand.* **64**, 464–72.

SCHENKER S., BREEN K. J. & HOYUMPA A. M. (1974) Hepatic encephalopathy: Current status. *Gastroenterology* **66**, 121–51.

SCHIMCKE R. T. (1962) Adaptive characteristics of the urea cycle enzymes in the rat. *J. Biol. Chem.* **237**, 459–68.

SCRIVER C. R. & ROSENBERG L. E. (1973) *Amino acid Metabolism and its Disorders.* W. B. Saunders, Philadelphia.

SELIGSON D. & SELIGSON H. A. (1951) A microdiffusion method for the determination of nitrogen liberated as ammonia. *J. Lab. Clin. Med.* **38**, 324–30.

SHAYWITZ B. A., ROTHSTEIN P. & VENES J. L. (1980) Monitoring and management of increased intracranial pressure in Reye syndrome: results in 29 children. *Pediatrics* **66**, 198–204.

SHIGESADA K., AOYAGI K. & TATIBANA M. (1978) Role of acetylglutamate in ureotelism. Variations in acetylglutamate level and its possible significance in control of urea synthesis in mammalian liver. *Eur. J. Biochem.* **85**, 385–91.

SHIH V. E. (1978) Urea cycle disorders and other congenital hyperammonaemic syndromes. In Stanbury J. B., Wyngaarden J. B. & Fredrickson D. S. (eds.) *The Metabolic Basis of Inherited Disease*, 4th Edition, pp. 362–86. McGraw Hill, New York.

SHIH V. E. & MANDELL R. (1974) Metabolic defect in hyperornithinaemia. *Lancet* **ii**, 1522–3.

SHIH V. E., MANDELL R. & HERZFELD A. (1982) Defective ornithine metabolism in cultured skin fibroblasts from patients with the syndrome of hyperornithinaemia, hyperammonaemia and homocitrullinuria. *Clin. Chim. Acta* **118**, 149–57.

SILBERMAN R. (1958) Ammonia intoxication following ureterosigmoidostomy in a patient with liver disease. *Lancet* **ii**, 937–9.

SIMELL O., PERHEENTUPA J., RAPOLA J., VISAKORPI J. K. & ESKELIN L.-E. (1975) Lysinuric protein intolerance. *Am. J. Med.* **59**, 229–40.

SINATRA F., YOSHIDA T., APPLEBAUM M., MASON W., HOOGENRAAD J. & SUNSHINE P. (1975) Abnormalities of carbamyl phosphate synthetase and ornithine transcarbamylase in liver of patients with Reye's syndrome. *Pediatr. Res.* **9**, 829–33.

SNODGRASS P. J. (1981) Biochemical aspects of urea cycle disorders. *Pediatrics* **68**, 273–83.

SNODGRASS P. J. & DELONG C. R. (1976) Urea cycle enzyme deficiencies and an increased nitrogen load producing hyperammonaemia in Reye's syndrome. *New Engl. J. Med.* **294**, 855–60.

SNODGRASS P. J., LIN R. C., MULLER W. A. & AOKI T. T. (1978) Induction of urea cycle enzymes of rat liver by glucagon. *J. Biol. Chem.* **253**, 2748–53.

SNYDERMAN S. E. (1981) Clinical aspects of disorders of the urea cycle. *Pediatrics* **68**, 284–9.

SNYDERMAN S. E., SANSARICQ C., NORTON P. M. & GOLDSTEIN F. (1979) Argininaemia treated from birth. *J. Pediatr.* **95**, 61–3.

SNYDERMAN S. E., SANSARICQ C., PHANSALKAR S. V., SCHACHT R. G. & NORTON P. M. (1975) The therapy of hyperammonaemia due to ornithine transcarbamylase deficiency in male neonate. *Pediatrics* **56**, 65–73.

SUNSHINE P., LINDENBAUM J. E., LEVY H. L. & FREEMAN J. M. (1972) Hyperammonaemia due to a defect in hepatic ornithine transcarbamylase. *Pediatrics* **50**, 100–11.

TERHEGGEN H. G., SCHWENK A., LOWENTHAL A., VAN SANDE M. & COLOMBO J. P. (1969) Argininaemia with arginase deficiency. *Lancet* **ii**, 748–9.

THALER M. M. (1976) Reye's syndrome: Cause and effect relationships. In Berenberg S. R. (ed.) *Liver Diseases in Childhood*, pp. 72–83. Martinus Nyhoff Medical Division, The Hague.

TRAUNER D. A. (1980) Treatment of Reye's syndrome. *Ann. Neurol.* **7**, 2–4.

TRAUNER D. A., NYHAN W. L. & SWEETMAN L. (1975) Short chain organic acidaemia and Reye's syndrome. *Neurology* **25**, 296–8.

VALERO A., ALROY G., EISENKRAFT B. & ITSKOVITCH J. (1974) Ammonia metabolism in uronic obstructive pulmonary disease with special reference to congestive right ventricular failure. *Thorax* **29**, 703–709.

WALSER M. & STEWART P. M. (1981) Organic acidaemia and hyperammonaemia: A review. *J. Inherit. Metab. Dis.* **4**, 177–82.

WIEGAND C., THOMPSON T., BOCK G. H., MATHIS R. K., KJELLSTRAND C. M. & MAUER S. M. (1980) The management of life-threatening hyperammonaemia: A comparison of several therapeutic modalities. *J. Pediatr.* **96**, 142–4.

YOKOI T., HONKE K., FUNABASHI T., HAYASHI R., SUZUKI Y., TANIGUCHI N., HOSOYA M. & SAHEKI T. (1981) Partial ornithine transcarbamylase deficiency simulating Reye syndrome. *J. Pediatr.* **99**, 929–31.

ZIEVE L. (1979) Hepatic encephalopathy: summary of present knowledge with elaboration on recent developments. In Popper H. & Schaffner F. (eds.) *Progress in liver diseases*, Vol. VI, pp. 327–41. Grune & Stratton, New York.

Chapter 5
Renal Disorders

T. MARTIN BARRATT

The Renal Unit is a major consumer of the services of the Department of Chemical Pathology, and it is therefore important that each discipline has a full and sympathetic understanding of the needs, limitations and resources of the other. A biochemical investigation may be conceived as part of a communication loop, initiated and terminating in the mind of the clinician, and can be no more effective than the weakest point of that loop, be it the insight of the clinician, the efficiency of the laboratory, or the interaction between the two.

The kidney is directly responsible for the maintenance of volume and composition of the extracellular fluid, and simple measurements of the concentrations of the major solutes in plasma therefore constitute the principal data-base of the clinical practice of nephrology: accuracy, specificity, reproducibility, rapidity of turn-round and efficiency of data handling in the routine output of the chemical pathology laboratory are of as much concern to the nephrologist as a sophisticated repertoire of tests of renal function. During the long chronic course that characterises most serious renal disease in childhood a very large volume of biochemical data may be accumulated: automation in the laboratory has unfortunately not been matched by an equivalent sophistication of data handling in the ward, and it must be a matter of distress for the pathologist to appreciate, on inspection of the clinical record, how much of his information has been squandered or irretrievably lost—but if so, he must ask himself if the format of his output is not one of the weaker links in the communication loop.

The progress in renal physiology during recent times has led to a sophisticated understanding of kidney function in health and disease. There has not, however, been a commensurate development of simple techniques of functional assessment that are easily and reliably applicable in clinical practice, particularly in the investigation of children with renal disease (Barratt 1974a). In this chapter, therefore, the emphasis will be on those methods of estimating renal function in children which are suitable for everyday use and are within the scope of non-specialized laboratories. Complex techniques which properly belong to the research laboratory will not be discussed.

PRACTICAL CONSIDERATIONS

Trauma

Tests of renal function should not inflict trauma, mental or physical, on the child. Investigations which require repeated venepuncture, intravenous infusion or bladder catheterization are for this reason of limited value in paediatric practice. The clearance techniques of classical renal physiology (Smith 1951) therefore require some modification. It must be further recognized that even the simplest tests of renal function make substantial demands on nursing time if they are to be undertaken in a kindly manner without error. Some of the most useful tests last a few hours and are thus not suitable for out-patient use, but do not require formal hospital admission; for these, day-care facilities are ideal.

Blood samples

Nearly all techniques of clinical chemistry can be scaled down to require only 100 μl or less of plasma, and thus be suitable for capillary blood samples; it is now unacceptable for laboratories to require venepuncture samples from infants for routine chemical analyses. Capillary blood sampling is, however, a more difficult technique than may appear at first sight, and some experience is necessary to avoid haemolysis (which particularly distorts potassium estimations) in young babies.

Urine collection: the creatinine correction

Many techniques necessitate an accurately timed urine collection. It is possible to collect urine from infants by the application of bags to the perineum, although close nursing supervision is essential in order that the bags are emptied immediately after micturition to prevent leakage. Short-term urine collections are obviously less liable to losses, but are more susceptible to error from incomplete bladder emptying and from diurnal variation, and therefore 24-hour collections are often preferred. The completeness of a 24-hour collection should be checked by estimation of the creatinine content, which should fall within the following limits in a well-nourished child (Ghazali & Barratt 1974):

$$U_cV \text{ (mmol/kg/24 h)} = 0.132 + 0.004 \times \text{age(y)} \pm 0.026\text{(SD)}$$

where U_c = urine creatinine concentration and V = urine flow rate. Indeed, it is often preferable to relate the excretion of a urinary constituent to creatinine and thus eliminate the necessity for timed urine collections and at the same time introduce a correction for body size; examples of this approach are given for albumin, phosphate and calcium below.

AGE AND KIDNEY FUNCTION

Anatomical development

The nephrogenic zone is in the outer cortex, and new glomeruli continue to be formed up to the 36th week of gestation; the full-term neonate thus enters extrauterine life with his definitive complement of one million nephrons in each kidney. At birth, the combined kidney weight is 25 g and DNA content, reflecting cell number, is 20% of the adult level; mature levels of DNA are attained by 6 months of age, at which time the combined kidney weight is 50 g, and subsequent growth of the kidneys to 300 g in maturity is achieved by increase in cell size rather than cell number. Microdissection studies show that mean glomerular diameter in the neonate is half that of the adult, whereas proximal tubular length is only 10% of the mature value: the anatomical data suggest glomerular preponderance in the delicate balance of filtration–reabsorption at this stage of development (McCrory 1972).

Physiological development

Renal blood flow

Blood flow to the infant kidney is low compared to the adult organ, with a preferential distribution to the juxtamedullary area (Spitzer 1978): during the first year of life there is a four-fold increase in the proportion of the cardiac output perfusing the kidneys. In the newborn puppy, the ratio of blood flow in the superficial cortical nephrons to that in the juxtamedullary nephrons is 2:1, rising to 10:1 in the mature animal (Olbing *et al.* 1973). Renal vascular resistance is high in the newborn and in the piglet falls ten-fold during the first months of life (Gruskin *et al.* 1970). Much of the rapid rise in blood flow in early life is due to the opening up of the vasculature in the superficial cortex.

Glomerular filtration

Even allowing for body size (see below), the glomerular filtration rate (GFR) is low in infancy (Table 5.1), principally due to the vascular factors described above. Permselectivity studies with polydisperse dextrans indicate restricted filtration of macromolecules by the neonatal glomerulus (Arturson *et al.* 1971) to an extent that the validity of inulin as a glomerular marker in premature infants has been questioned, though general consensus still hold that inulin clearance is the definitive estimate of GFR.

Table 5.1 Glomerular filtration rate.

Age	GFR mean	(ml/min per 1.73 m² SA) ±2 SD
Premature	47	29–65
2–8 days	38	26–60
4–28 days	48	28–68
35–95 days	58	30–86
1–5.9 months	77	41–103
6–11.9 months	103	49–157
12–19 months	127	63–191
2–12 years	127	89–165
Adult male	131	88–174
Adult female	117	87–147

GFR measured by inulin clearance: data for premature infants from Barratt *et al.* (1974), for children collected by Goldsmith (1978) and for adults from Smith (1951).

Proximal tubular function

The anatomical evidence of proximal tubular immaturity is reflected in a decreased tubular reabsorption of endogenous amino acids, a low bicarbonate threshold, and a low tubular maximum secretory capacity for para-aminohippurate (related to GFR); tubular maximum capacity for glucose reabsorption in relationship to GFR, however, is similar in neonates and adults.

Salt and water excretion

Full-term infants are able to maintain salt balance on intakes of sodium as low as 0.1 mmol/kg/24 h, though premature infants are less efficient in this respect. During hypotonic saline volume expansion the infant excretes a greater fraction of filtered water than the adult, suggesting that a larger proportion of the filtered sodium is escaping reabsorption in the proximal nephron (Rodriguez-Soriano *et al.* 1981); there is thus greater dependence on mineralocorticoid sensitive distal sodium reabsorption in the neonate, reflected perhaps in the much higher plasma renin activity and aldosterone concentration (Dillon & Ryness 1974), as well as in the increased severity of the clinical expression of defects of mineralocorticoid biosynthesis such as congenital adrenal hyperplasia.

The infant is less able to excrete a sodium load than the adult. For example, the fractional excretion of filtered sodium (sodium clearance/GFR) in the adult dog rises to 17% following an infusion of isotonic saline

100 ml/kg, but only reaches 5% in the puppy (Goldsmith *et al.* 1975). This phenomenon may be related to diminished perfusion of the superficial nephrons which, with their short loops of Henle, are attuned to the excretion of a sodium load.

In response to water deprivation or exogenous antidiuretic hormone, newborn infants only concentrate their urine to 600–700 mmol/l, whereas adults generally achieve 1200 mmol/l; the difference is largely due to low rates of urea excretion for urinary concentrations of non-urea solute are similar and the discrepancy is minimised by urea supplements or a high protein diet (Edelmann *et al.* 1960).

METABOLIC ASPECTS OF YOUNG CHILDREN

Growth

The role of the kidney should be seen as but one aspect of an integrated system of nutrition, growth and excretion. The growing infant gains nearly 1% of body weight every 24 hours; most of this is water, but about 1 g of protein per kg body weight is incorporated into growing tissues, along with potassium, phosphate and other minerals, thus sparing the load presented for renal excretion. McCance (1959) described growth as the 'third kidney' and reported the protection by anabolism (provided energy intake was adequate) of nephrectomised puppies from the effects of uraemia, hyperkalaemia and hyperphosphataemia. The equation might be written thus:

$$\text{Nutrition} - \text{Growth} = \text{Renal Excretion}$$

Nutrition

A consideration of major importance in the metabolism of young children and the interpretation of biochemical data from them is the wide range in protein and mineral composition of various diets (Table 5.1). A breast-fed baby will have a protein intake of about 2 g/kg per 24 h, whereas on unmodified cow's milk he would receive 5 g/kg per 24 h (Table 5.2). Surplus protein not incorporated into growing tissue is broken down to urea, each gram of protein producing 0.3 g (5 mmol) of urea. With net protein synthesis of 1 g/kg per 24 h, urea production on a protein intake of 2 g/kg per 24 h will be 5 mmol/kg per 24 h, whereas on a protein intake of 5 g/kg per 24 h it will be 20 mmol/kg per 24 h: the plasma urea concentration may therefore vary four-fold on conventional diets without implying any change in renal function. Although deeply embedded in the professional consciousness as a test of renal function, the plasma urea concentration is in fact determined in large measure by nutritional and metabolic factors, and is only loosely related to GFR.

Table 5.2 Representative dietary intakes.

Intake per kg/24 h	Cow's milk	Breast milk
Water (ml)	150	100
Calories	100	100
Protein (g)	5	2
Sodium (mmol)	4	1
Potassium (mmol)	5	2
Phosphorus (mg)	150	23
Renal solute load (mosmol)	33	13
Renal H^+ load (mEq)	6	1

Data from Barratt (1974b)

INTERPRETATION OF DATA

Normal Range

For many tests, comprehensive normal data are not available. In some circumstances, however, the change in a parameter in the same individual may be of more importance than its absolute value. In other situations it may not be clear which the correct control group should be: is it more appropriate, for example, to compare the calcium excretion of children with urolithiasis to that of healthy children or those with other urological disorders? The data should often be related to the physiological state of the child at the time of specimen collection: thus a urinary sodium excretion of 1 mmol/kg per 24 h may be observed in a healthy child, but is inappropriately high in a child who is salt depleted. Similarly, a given level of plasma renin activity may be interpreted as high, normal or low, depending on the concomitant salt status. Finally, for a measurement such as the selectivity of proteinuria, which may only be meaningfully applied to the pathological state of heavy protein excretion, the concept of a normal range is inappropriate.

Correction for body size

Estimates of renal function must inevitably be related to the size of the child: the problem exists for the physiological parameters of many other systems as well (Tanner 1949). The establishment of normal ranges at different ages does not resolve the problem, for there is no *a priori* reason why age should be preferred to some aspect of body size as the reference parameter. The best reference standard(s) would permit the most precise prediction of renal function in a healthy child, but the data are not available for this type of analysis. Another approach is to select a reference standard that minimises age-dependent variation; for this purpose, surface area (SA) is often chosen. No causal connection between the development of renal function and the increase in body surface area is implied, and from the physiological point of view a better case can be made for relating GFR, for example, to extracellular fluid volume (Friis-Hansen 1961) or total body water (McCance & Widdowson 1952), the domains over which the kidney exercises its regulatory functions. Nevertheless, from the age of 2 years, GFR is approximately proportional to body surface area and, as the turnovers of urea, salt and water are also roughly proportional to surface area, there is some merit in relating GFR to this standard, for similar reductions in GFR thus corrected in individuals of different size represent similar threats to homeostasis.

Body surface area is difficult enough to define, let alone to measure, and there is a substantial error in its estimation, of the order of 20% (Krovetz 1965). The most satisfactory nomogram is that of Sendroy and Cecchini (1954), but the formula of Dubois and Dubois (1916) is almost as good, though based on woefully inadequate data for younger children:

$$\text{SA(cm}^2) = \text{wt(kg)}^{0.425} \times \text{ht(cm)}^{0.725} \times 71.84$$

The logarithmic transformation is more convenient:

$$\log \text{SA(m}^2) = 0.425 \times \log \text{wt(kg)} + 0.725 \times \log \text{ht(cm)} - 2.144$$

A recent analysis of the estimation of surface area in infants is given by Haycock *et al.* (1978).

GLOMERULAR FILTRATION RATE

There is no single test that reflects overall kidney function or mass of renal parenchyma. GFR commands greatest attention, for it determines the load presented to the renal tubules and thus the flexibility of renal response to the changing demands of the body as a whole. Filtration equilibrium is achieved in the glomerulus, and thus there is a considerable reserve filtration area, so that substantial glomerular damage may be sustained before reduction of GFR occurs (Brenner *et al.* 1981). There is also feed-back control of GFR within the nephron, mediated by the sodium concentration of the tubular fluid perfusing the macula densa in the juxtaglomerular apparatus (Schnermann *et al.* 1970); tubular phenomena thus strongly influence GFR, and in a histological study GFR was found to correlate better with tubular-interstitial than with glomerular changes (Risdon *et al.* 1968).

The glomerulus is functionally an ultrafilter, which in disease may become leaky, blocked or both. Leak is reflected by the appearance in the urine of abnormal amounts of macromolecules which are effectively excluded from the glomerular filtrate in health. On the other hand, block (reduced GFR) is recognised by the accumulation in the plasma of solutes normally excreted by glomerular filtration. As it is the degree of block that usually determines the outcome of the illness, methods of estimation of GFR will be reviewed in some detail.

Inulin clearance

The definitive measure of GFR is the inulin clearance (Smith 1951) and normal ranges are given in Table 5.1. However, the technique requires constant intravenous infusion and timed urine samples; moreover, even modern automated methods of inulin analysis are technically demanding, and the method as a whole is ill-suited for routine clinical practice.

Creatinine

Metabolism

Creatinine is formed from creatine; most of the body creatine is in skeletal muscle, and 2% is converted non-enzymatically each day to creatinine and excreted in the urine. Creatinine excretion is thus a function of muscle mass, and is relatively constant in one individual from day to day; it is roughly proportional to body weight (see above) but is also affected by diet and subject to diurnal variation (Doolan *et al.* 1962).

Methodology

Creatinine estimation depends upon the orange colour produced with alkaline picrate—the Jaffé reaction. Unfortunately the reaction is not specific and other substances (non-creatinine chromogens) in plasma react with the reagent; specific estimation of creatinine necessitates its adsorption onto Fuller's earth or a cation exchange resin (Stoten 1968) and subsequent elution. The estimation of true plasma creatinine is tedious and demands considerable technical skill, particularly at low concentrations. The true plasma creatinine concentration is less in children than in adults (Table 5.3).

Non-creatinine chromogens are approximately equivalent to 18 μmol/l (Dodge *et al.* 1967), and their concentration is independent of age or renal function; thus they make a relatively greater contribution to the total Jaffé chromogen in children than in adults, and specific methods for plasma creatinine estimation are therefore essential in the young. Automated methods eliminate some of the non-creatinine chromogens, but still overestimate the true plasma creatinine concentration in children by an average of 13 μmol/l (Counahan *et al.* 1976). Greater specificity of the estimation is achieved by reaction rate analysis as, in general, colour generation is faster with creatinine than with the non-creatinine chromogens.

Table 5.3 Plasma creatinine concentration.

	Plasma creatinine (μmol/l)			
	Female		Male	
Age (years)	mean	±2 SD	mean	±2 SD
1	31	22–40	36	18–54
2	40	28–52	38	17–59
3	37	23–51	40	21–59
4	41	20–62	40	21–59
5	40	21–59	44	25–63
6	42	23–61	46	25–67
7	47	26–68	48	23–73
8	47	28–66	50	25–75
9	48	29–67	52	24–80
10	48	25–71	54	15–93
11	53	30–76	55	30–80
12	52	29–75	57	15–93
13	55	30–80	60	30–80
14	57	34–80	63	29–85
15	59	20–98	67	23–97
16	57	31–83	65	21–109
17	62	27–97	70	38–102
18–20	63	30–96	81	51–111

'True' plasma creatinine concentration measured by autoanalyser (Schwartz *et al.* 1976)

The measurement of plasma creatinine concentration presents a severe technical challenge to the paediatric clinical chemistry laboratory. As it is the single most valuable estimate of renal function, a rapid high volume service must be achieved without loss of accuracy or specificity. A coefficient of variation of replicate analyses of less than 5% is required by the clinician, with at least a daily service; after renal transplantation, twice-daily plasma creatinine determinations are usually required. The lower concentrations of creatinine in children than in adults require that the methodology be established to give the greatest accuracy in the 40–80 μmol/l range, a fact not always appreciated by laboratories geared to adult nephrological practice. Adequate investment in resources to achieve this quality of service should without doubt be a high priority of the biochemical/nephrological team.

Plasma concentration

The plasma creatinine concentration must be interpreted in relation to the size of the child. Glomerular filtration rate (related to surface area; GFR/SA) can be predicted from the plasma creatinine concentration by the following argument:

1 Creatinine clearance (C_c) is proportional to GFR:

$$GFR \propto C_c = U_cV/P_c$$

where C_c = creatinine clearance, U_c = urine creatinine concentration, P_c = plasma creatinine concentration, and V = urine flow rate.

2 Creatinine excretion (U_cV) is proportional to weight:

$$U_cV \propto Wt$$

3 Weight is proportional to the cube of body height, surface area to the square of body height:

$$Wt \propto Ht^3$$
$$SA \propto Ht^2$$

4 Substituting the proportionalities in 2 and 3 into 1:

$$GFR/SA \propto Ht/P_c$$

Counahan *et al.* (1974) found that in a population of children of widely varying size and renal function there was a linear relation between GFR/SA and Ht/P_c, and that GFR could be predicted with reasonable confidence from the following formula:

$$GFR(ml/min\ per\ 1.73\ m^2\ SA) = 40 \times Ht(cm)/P_c(\mu mol/l)$$

In malnutrition (common in children with renal disease) muscle bulk may be reduced, and GFR hence be lower than predicted from P_c.

A reasonable estimate of the rate of progression of renal failure in an individual patient can be derived from regression analysis of a plot of the reciprocal plasma creatinine concentration as a function of time: for this purpose it is usually helpful to gather several data points in children with chronic renal disease (Mitch *et al.* 1976).

Clearance

There is some tubular secretion of creatinine, hence its clearance exceeds GFR. In children aged 3–13 years the creatinine clearance is reported to be 94–142 ml/min per 1.73 m² SA (Winberg 1959). If the plasma creatinine is over-estimated by a non-specific method, the error will reduce the estimate of clearance, and may result in a figure that is closer to the GFR than the true creatinine clearance. However, the agreement is fortuitous, and does not hold in pathological circumstances. Even under research conditions, estimation of creatinine clearance is an imprecise technique, and has a coefficient of variation of replicate estimates in the same individual of about 12% (Chantler & Barratt 1972); hence two estimates in an individual must differ by more than 30% before it can be inferred with 95% confidence that a real change in clearance has in fact taken place, and in the hurly-burly of clinical practice it is always difficult to be confident about the accuracy of timed urine collections. As an estimate of GFR, the creatinine clearance does not have the convenience of the plasma creatinine concentration alone, nor the precision of the plasma clearance of ^{51}Cr-EDTA discussed below. The author's current practice, therefore, is to rely upon the plasma creatinine concentration for routine purposes and to use the plasma clearance of ^{51}Cr-EDTA if a more accurate estimate of GFR is required. The creatinine clearance remains, however, a useful internal standard to which other clearances such as that of albumin or phosphate may be related (see below).

GFR estimates without urine collection

There have been several modifications of standard clearance techniques which are more convenient because of the elimination of the need for timed urine collections. First, GFR can be estimated from the rate of fall of plasma concentration after intravenous injection of a substance excreted by glomerular filtration—the plasma clearance or single injection technique. Secondly, GFR may be estimated from the equilibrium plasma concentration attained during constant intravenous infusion. Finally, the rate of renal uptake of a radiolabelled marker may be measured with a gamma camera, enabling individual kidney GFR (IKGFR) to be measured. All methods require that the administered substance has a urinary clearance equal to the GFR with a negligible extrarenal clearance.

Single injection techniques

Following intravenous injection of a substance that is rapidly mixed in its volume of distribution and cleared solely by glomerular filtration, the plasma concentration falls in a monoexponential fashion which can be described by the y-intercept, i.e. the zero-time plasma concentration (P_0), and the slope, most conveniently represented by the half-time ($T\frac{1}{2}$), of the plot of the logarithm of the plasma concentration against time. Clearance (GFR) can be derived by single compartmental analysis of these parameters, but in practice mixing of a substance distributed in the extracellular fluid volume is not instantaneous: immediately after injection the concentration in plasma is higher than in the interstitial fluid (Chantler *et al.* 1969). As an approximation, the half-time can be derived from the later linear part of the curve and a theoretical value for P_0 obtained by extrapolation of this portion. The volume of distribution (V_D) can be calculated by dividing the amount injected by P_0, and the clearance calculated as follows:

$$GFR = V_D \times \log_e 2/T\tfrac{1}{2}$$

If plasma concentrations P_1 and P_2 are measured at times T_1 and T_2, the following relationships (helpful for programmable calculators) hold:

$$T\tfrac{1}{2} = \log 2(T_2 - T_1)/(\log P_1 - \log P_2)$$
$$P_0 = \text{antilog}\,((T_2(\log P_1 - \log P_2)/(T_2 - T_1)) + \log P_2)$$

More complex analyses based on a two-compartmental model (intravascular and interstitial volumes) are possible and give results that are closer to the true GFR, but are more troublesome in practice, requiring several blood samples to define the early part of the curve (Barratt & Chantler 1975).

Accurate measurement of low plasma concentrations is essential, and therefore the method is really only suitable for use with radioisotopically labelled substances. The two compounds most frequently used are technetium-99^m diethylenetriamine pentacetic acid (^{99m}Tc–DTPA) and chromium-51ethylenediaminotetra-acetic acid (^{51}Cr–EDTA): the latter is used in the author's laboratory, but as the former is in routine use for dynamic renal scanning the possibility of obtaining quantitative data during the same investigation is worthy of emphasis.

^{51}Cr–EDTA is diluted with physiological saline and a dose of approximately 1 μCi/kg body weight is given by intravenous injection; the exact dose administered is measured by weighing the syringe before and after injection. The precise time of injection is noted, and blood samples are taken at approximately two and four hours after injection, with the exact time of sampling being recorded. Five ml venepuncture samples are usually obtained, but if necessary 1 ml capillary samples will suffice. Chromium-51 is counted in the plasma samples and injected solution by standard methods and the clearance, calculated by the formula given above, is multiplied by an empirical factor of 0.87 to allow for the error inherent in the single exponential analysis due to delayed mixing (Chantler & Barratt 1972). Errors arise principally from faulty injection techniques with extravasation, or from the presence of oedema fluid which delays equilibration.

Using the single shot technique, GFR can be predicted with 95% confidence limits of ±17%. However, the greatest virtue of the method lies in its reproducibility on separate occasions in the same individual: the coefficient of variation is 3.9%, and thus two estimates of clearance which differ by 11% have a 95% probability of representing a real change in GFR. The radiation dosage is very small: the kidney is the critical organ and receives only 0.29 mrad, which is less than 0.1% of that administered during intravenous urography. The method is simple in clinical and laboratory execution and has much to recommend it for the assessment of GFR in children with renal disease.

Constant infusion techniques

Constant infusion techniques have recently attracted renewed interest, which is unfortunate, for they are inherently flawed. The principle is that a loading dose of a glomerular marker is administered, followed by a constant infusion so that, in the equilibrium state, the rate of infusion is equalled by the rate of excretion and can be substituted for the UV term in the clearance equation. At the beginning the plasma concentration is determined by the loading dose and its volume of distribution, at the end by the rate of infusion and GFR. The gap between these two concentrations will be traversed at a rate determined by the plasma half-life of the marker substance, and equilibration will not be 95% complete until at least 5 half lives have elapsed—several hours or more. The error will be small if the relationship between the loading dose and infusion rate has been nicely judged, but large if it has not.

Uptake analysis and individual kidney GFR

In many children, particularly those with urological disorders in whom the kidneys are unequally affected by disease, an estimate of IKGFR is of great value. The rate of uptake of a glomerular substance such as ^{99m}Tc–DTPA by a kidney is proportional to the plasma concentration at that moment in time and the IKGFR. These data can be derived from computer analysis of the gamma camera scan, the principal technical problem being the background correction for extrarenal counts. The technology has advanced to the point at which other methods of estimating IKGFR involving ureteric catheterisation are now obsolete (Piepsz *et al.* 1978).

PROTEINURIA

Protein is present in the urine of healthy individuals so that the distinction between normal and pathological urine in this respect is quantitative rather than qualitative: normal protein excretion rates are reported to be 80 ± 24(SD) mg/24 h in adults (Berggard 1970) and 100 ± 94 mg/24 h in children aged 5–14 years (Wagner *et al.* 1968); as a provisional standard, a protein excretion of 4 mg/m^2 SA per hour has been proposed as the upper limit of normal. However, data on young children are scanty, the appropriateness of a surface area correction in doubt, and technical methods unsatisfactory. In addition, the logic of grouping together the many heterogeneous urinary proteins with different physicochemical characteristics and physiological pathways of excretion as a single estimate of 'total protein' is questionable.

Excess protein in the urine may arise from an increased leak of plasma proteins through a damaged glomerulus ('glomerular'), failure of proximal tubular reabsorption of the low-molecular-weight proteins normally present in the glomerular filtrate ('tubular') or local secretion into the urinary tract (Hardwicke *et al.* 1970).

Detection of proteinuria

Proteinuria is most conveniently detected by Albustix: paper strips impregnated with tetrabromophenol blue buffered with citrate at pH 3.5. Protein binds the dye, and causes a colour change by displacement of the transformation range of the indicator. False positive results may be obtained in urine of high pH, otherwise there is a good correlation between the Albustix reading and urine albumin concentration (Barratt 1974a).

Occasionally a child will be found to have protein in urine formed in the upright posture, but not whilst recumbent. Such postural proteinuria may be detected by finding the overnight urine to be free of protein, and does not indicate renal disease. It is obviously important to exclude such a postural effect before investigating a child with proteinuria more extensively.

Glomerular proteinuria

Quantitation

Some consideration should be given to the most suitable parameter of glomerular leak. Urinary protein concentration is modified by variations in the urine flow rate, and it is therefore more appropriate to measure the protein excretion rate, perhaps corrected for body surface area. However, in view of the heterogeneous nature of the urinary proteins it is preferable to select a single molecular species, such as albumin, of a size and charge that in health is just effectively excluded from the glomerular filtrate. The albumin excretion rate should be factored both by the plasma albumin concentration and by an estimate of the GFR, such as the creatinine clearance, for reduction in both these parameters will result in a fall in the albumin excretion without implying any change in glomerular permeability. The albumin–creatinine clearance ratio thus approximates to the ideal parameter of glomerular permeability (Barratt *et al.* 1970). It has the advantage that only an untimed urine and coincident plasma sample are required, for the urine flow rate term does not appear in the final equation:

$$C_A = U_A V / P_A$$
$$C_C = U_C V / P_C$$
$$C_A / C_C = \frac{U_A}{P_A} \bigg/ \frac{U_C}{P_C}$$

where U_A, P_A, C_A = urine concentration, plasma concentration and clearance of albumin; U_C, P_C, C_C = urine concentration, plasma concentration and clearance of creatinine; and V = urine flow rate.

In healthy adults C_A/C_C is 3.3 (range 1.3–8.7) $\times 10^{-6}$, in neonates it is 11 (range 2.8–40) $\times 10^{-6}$.

In the derivation of the albumin–creatinine clearance ratio, the urinary concentration terms vary over a much wider range than the plasma concentrations; changes in C_A/C_C are largely due to U_A/U_C and the latter parameter can therefore be used with little loss of precision and considerable gain in convenience. The upper limit of normal U_A/U_C in neonates is 0.4 mg/mg (45 mg/mmol) and in children over 10 kg in weight is 0.1 mg/mg (11 mg/mmol). As a rough guide, a figure for U_A/U_C (mg/mg) of 0.1 can be considered as the upper limit of normal, 1.0 as significant albuminuria, 10 as a value usually associated with the nephrotic syndrome and 100 as an extremely high level occasionally observed in states of very heavy albuminuria such as the congenital nephrotic syndrome. Another way of viewing the data, recalling that urine creatinine excretion is approximately 15 mg/kg per 24 h, is to note that the value of U_A/U_C (mg/mg) is numerically similar to the albumin excretion rate expressed as G/24 h per 70 kg body weight—drawing attention to the compensation for body size inherent in the creatinine correction. Albumin is easily measured immunochemically (Mancini *et al.* 1965), and the method is very suitable for routine paediatric practice.

Selectivity

Even in severe pathological states there is usually some selectivity of proteinuria, that is larger proteins have lower clearances than smaller ones (Hardwicke *et al.* 1970). If several proteins are measured it is commonly found that there is an approximately linear inverse relationship between the logarithm of the clearance of a protein and the logarithm of its molecular size, and some authors have sought to summarise selectivity data as the angle of slope (θ) of the calculated regression line. A convenient simplification is to select two proteins only and measure the ratio of the clearance of the larger to the smaller one: either IgG and transferrin (Cameron & Blandford 1966) or IgG and albumin (Barratt 1974) will do. As a clearance ratio, selectivity (S) estimates are not dependent on timed urine collections:

$$S = \frac{U_G}{P_G} \Big/ \frac{U_A}{P_A}$$

where U_G and P_G = urine and plasma concentrations of IgG.

Estimates of selectivity are only appropriate in states of moderate or heavy proteinuria ($U_A/U_C > 1.0$ mg/mg). In the nephrotic syndrome of childhood, a value of S less than 0.1 suggests steroid responsiveness, but if S is greater than 0.2, steroid resistance is more probable. The investigation is of considerable help in the management of such cases particularly in the decision to undertake a renal biopsy or to embark on steroid therapy without confirmation of minimal-change histology. However, the logistic challenge is for the laboratory to produce the data fast enough to influence the clinical judgement: too often, the data are only available for a retrospective assessment of a decision already taken.

Tubular proteinuria

The proteinuria of patients with renal tubular diseases is qualitatively different to that resulting from glomerular disease. The major constituents are α_2- and β-globulins, which are normally filtered at the glomerulus and reabsorbed in the proximal tubule. Tubular proteinuria may be recognised by electrophoretic analysis, but a simpler method is the measurement of the urinary excretion of a representative protein of low molecular weight, such as lysozyme (Barratt & Crawford 1970) or beta-2-microglobulin. Lysozyme is a basic protein with a molecular weight of 15 000; it is synthesized in neutrophils, liberated into the extracellular fluid on their destruction, filtered at the glomerulus and reabsorbed and catabolised in the proximal tubule. The kidney is the principal site of lysozyme catabolism, and plasma concentrations therefore rise in chronic renal failure. Lysozyme excretion (expressed as the lysozyme–creatinine clearance ratio) is raised about 100-fold in children with the Fanconi syndrome. The upper limit of the urine lysozyme/creatinine concentration ratio (U_L/U_C) in neonates is approximately 20 μg/mg (2.2 μg/μmol), and in older children 10 μg/mg (1.1 μg/μmol). There are some limitations, however: as indicated above, plasma concentrations are elevated in chronic renal failure, and the filtered load in the residual functioning nephrons may overwhelm tubular reabsorptive mechanisms leading to lysozymuria without implying specific proximal tubular impairment. It should further be noted that neonates do not have increased lysozyme excretion, even in conditions such as asphyxia, which might be expected to lead to tubular damage. Clearance studies with dextrans of varying molecular size (Arturson *et al.* 1971) suggest that the neonatal glomerulus is less permeable than the adult glomerulus to molecules of the size of lysozyme, and therefore that the filtered load presented for reabsorption is less. Lysozyme excretion data should therefore be interpreted with caution in the first few months of life.

Local secretion

Tamm–Horsfall glycoprotein is prominent amongst the proteins locally secreted into the urinary tract, but the significance of variation in urinary secretion rates is not clear. Renal tubular damage, for example from an ischaemic insult or nephrotoxic agent, results in increased excretion of the tubular enzymes N-acetyl-glucosaminidase (NAG) (Wellwood *et al.* 1975) and gamma-glutamyl transpeptidase; both are quite sensitive tests. IgA is secreted into the urine, and can be demonstrated by immunofluorescent techniques bound to bacteria if infection involves the renal parenchyma (Thomas *et al.* 1974).

PROXIMAL TUBULAR FUNCTION

In the proximal tubule there is isosmotic reabsorption of about 80% of the glomerular filtrate; sodium is actively reabsorbed, drawing water and chloride along the osmotic and electrochemical gradients thus generated. There are also specific transport systems for the reabsorption of filtered glucose, amino acids and phosphate, and secretory mechanisms for hydrogen ion (mediating bicarbonate reabsorption), organic acids and organic bases. Proximal tubular dysfunction may arise from highly specific genetically-determined defects of individual transport systems, or be multifactorial (the Fanconi syndrome): the latter is characterised by renal glycosuria, generalised aminoaciduria, phosphaturia leading to hypophosphataemia and rickets, low-molecular-weight proteinuria, bicarbonate wastage with proximal (Type II) renal tubular acidosis, defective sodium conservation with secondary hyperreninaemia, potassium loss leading to hypokalaemia and hypouricaemia (Roth *et al.* 1981).

The commonest cause of the Fanconi syndrome in childhood is cystinosis, an autosomal recessive disorder characterised by widespread intracellular deposits of cystine, particularly in the reticuloendothelial system, renal tubular cells and cornea (Schneider *et al.* 1978). The primary defect is unknown, but the crystals are bounded by the lysosomal membrane so a disorder of lysosome function is suspected. The diagnosis can be confirmed by the demonstration of birefringent cystine crystals in bone-marrow specimens, or more precisely by the measurement of leucocyte cystine content.

Glycosuria

There are trace amounts of glucose in normal urine, up to 0.5 mmol/l; indeed the absence of urinary glucose is abnormal, and suggests significant bacteriuria. Measurement of urinary glucose concentration by copper reduction or glucose oxidase methods tend to underestimate due to inhibitory substances in urine; the most satisfactory technique is based on a coupled hexokinase–glucose-6-phosphatase reaction.

Glycosuria may be due to an increased filtered load of glucose (overflow) resulting from hyperglycaemia, or to incomplete reabsorption of the normal filtered load (renal glycosuria). The critical observation is the simultaneous determination of urinary glucose and plasma glucose concentrations throughout the period of formation of the urine sample;

is actively transported followed by sodium, but not by water, so that tubular fluid emerges into the distal nephron hypotonic to plasma, but medullary interstitial fluid becomes hypertonic. The third site of sodium reabsorption in the distal nephron effects an exchange of the sodium ion for potassium or hydrogen, and is under the control of aldosterone. Under conditions of water diuresis, antidiuretic hormone is suppressed and the distal nephron becomes relatively impermeable to water: the urine flow rate is then determined by the fraction of the glomerular filtrate that escapes proximal reabsorption, in turn dependent upon the proportion of the filtered sodium load reabsorbed in the proximal tubule. Thus clearance studies under conditions of hypotonic volume expansion give some idea of the relative magnitudes of sodium reabsorption at different sites within the nephron (Rodriguez-Soriano *et al.* 1981).

Salt-wasting states

Inappropriate urinary losses of sodium may be due to renal or adrenal disorders. The underlying renal disease may be a diffuse parenchymal abnormality or a more specific tubular abnormality such as Bartter's syndrome, distal renal tubular acidosis, the Fanconi syndrome or pseudohypoaldosteronism. Adrenal salt-wasting may be a feature of global adrenal failure or a consequence of a specific enzyme deficiency affecting aldosterone biosynthesis. The hallmark of adrenal salt-wasting is that the plasma aldosterone concentration is inappropriately low in relation to plasma renin activity.

Renin and hypertension

Renin is an enzyme secreted by the juxtaglomerular apparatus in response to diminished renal perfusion pressure; it acts on its substrate angiotensinogen to cleave the decapeptide angiotensin I which in its turn is processed to an octapeptide angiotensin II by a converting enzyme in the plasma and pulmonary vascular bed. Angiotensin II is a very potent pressor substance and also stimulates aldosterone production by the adrenal gland. Plasma renin activity (PRA) is estimated by the rate of generation of angiotensin I in the presence of inhibitors of converting enzyme. Peripheral venous PRA is influenced by many physiological variables: age (Table 5.4), posture, and dietary sodium intake, as well as by drugs particularly diuretics and antihypertensive agents: all these variables must be defined before a peripheral venous PRA value can be evaluated.

Renin-dependent hypertension may be due to diffuse renal parenchymal disease or to focal inappropriate renin production by tumour or more commonly by an area of renal tissue which is ischaemic due to a vascular anomaly or scarring. Cases amenable to cure by removal of a kidney or revascularisation can be identified by demonstrating elevated PRA in the

Table 5.4 Plasma renin activity (PRA) and aldosterone concentration.

	PRA (ng A1/l per h)		Aldosterone (pmol/l)	
Age (y)	mean	range	mean	range
< 1	1459	472–3130	788	164–2929
1–4	757	110–2610	294	69–946
5–9	417	131–834	147	28–616
10–15	321	55–899	211	72–577
Adult	85	22–311	230	39–422

Data from Dillon and Ryness (1975).

main renal vein or segmental vein draining the affected area (Dillon *et al.* 1978). In health, the PRA ratio between the two main renal veins is less than 1.5 (Gertds *et al.* 1979), and a value greater than this strongly suggests significant unilateral renal ischaemia, particularly if renin production by the contralateral kidney is suppressed, as indicated by the PRA ratio between its main vein and the low inferior vena cava being close to unity.

Concentration of the urine

Urine concentration is achieved by the reabsorption of water under the permissive influence of antidiuretic hormone (ADH) along the osmotic gradient established by the ascending limb of the loop of Henlé. Defects of urinary concentration arise from failure to produce ADH or to respond to it, either because of diffuse renal parenchymal disease or a specific tubular unresponsiveness—nephrogenic diabetes insipidus. It is the osmolality of the urine that is relevant, and its measurement with contemporary osmometers is an easy matter requiring only small volumes of urine. The specific gravity of the urine correlates poorly with the osmolality and its measurement should be abandoned. A more convenient quick estimate of concentration, which can be done on the ward with a drop of urine, is the refractive index.

Water deprivation

In response to water deprivation, normal children aged 2–16 years achieve a urine osmolality of 1127 ± 128(SD) mmol/l; there is a slight increase in maximum osmolality with age. Edelmann *et al.* (1967a) report a simplified test of renal concentrating capacity which is of considerable practical value: a normal noon meal is given on the day that the test is initiated following which, except for a dry supper, the child is thirsted and fasted totally until the test is complete. The bladder is emptied to start the collection between 7 and 9 pm and all urine passed between that time and a specimen collected between 7 and 8 am the next morning is pooled. In healthy children the osmolality of this specimen is 1089 ± 110 mmol/l and further water deprivation only results in an increment of osmolality of 10% or less. As an overnight specimen in healthy individuals is usually acid and free of protein its examination provides a useful screening test of renal function.

During water deprivation tests, the child should not be allowed to lose more than 3% of his body weight. Infants with water-losing disease may have urine flow rate of the order of 10 ml/kg per hour and therefore reach 3% weight loss in three hours. In these children, water deprivation tests are dangerous unless adequately supervised, and the protocol should be designed for each individual child.

ADH response

Failure to concentrate the urine in response to water deprivation may be due to lack of ADH or resistance to its action; the alternatives may be discriminated by the administration of intramuscular vasopressin tannate in oil (0.1 units/kg). The test is best carried out in the water-depleted state to avoid the danger of water intoxication in young infants. Winberg (1959b) reported the urinary osmolality after pitressin in children aged 3–12 years to be 1069 ± 128(SD)/mmol per l; under 3 years somewhat lower values were observed. The coefficient of variation of repeated tests in the same individual is 13%.

Vasopressin is now difficult to obtain, and as an alternative the synthetic analogue 1-deamino-8-D-arginine vasopressin (DDAVP) may be used in an intramuscular dosage of 0.05 μg/kg, or by nasal installation of 10 μg in infants or 20 μg in older children (Aronson & Svenningsen 1974): of the two routes, the intramuscular test is more secure as, in the event of an inadequate response, there is no question about the absorption of the drug.

Some consideration should be given to whether it is preferable to investigate the child with water deprivation or exogenous ADH first. It is logical to start with water deprivation and this is the safest test in infants who seem likely to have normal renal function. With gross polyuria, however, in children in whom ADH resistance seems probable, the safer investigation is the DDAVP test and if the response is poor, the need for a water deprivation test is averted.

HYDROGEN ION EXCRETION

There are two major threats to hydrogen ion homeostasis: the first is the potential loss of filtered bicarbonate, amounting to some 50 mmol/kg per 24 hour, and the second arises from the generation of hydrogen ion unaccompanied by bicarbonate during the metabolism of the diet: the metabolic hydrogn ion load varies from 1 mmol/kg per 24 hour in the breast-fed baby to 6 mmol/kg per 24 hour in babies on high protein intakes. The metabolic component of the acid/base status of the child is best assessed by the standard bicarbonate (bicarbonate concentration at $P\text{CO}_2$ of 40 mmHg) or base excess (amount of hydrogen ion required to bring the pH of the blood to 7.40 at a $P\text{CO}_2$ of 40 mmHg) (Sigaard-Andersen 1963).

Hydrogen ion is secreted by the proximal tubule and neutralises filtered bicarbonate; distal tubular fluid is essentially bicarbonate free and at this site hydrogen ion, secreted in exchange for sodium, is buffered by phosphate and ammonia. The titratable acidity is measured by the amount of alkali required to bring the pH of urine to that of plasma, and net hydrogen ion excretion is given by the sum of urinary titratable acidity, and ammonium less urinary bicarbonate. The urinary excretion of bicarbonate, titratable acidity and ammonium are all effectively determined by the urinary pH and, although of great importance to the study of hydrogen ion homeostasis, their measurement does not contribute to an understanding of the renal handling of hydrogen ion, any more than can be ascertained from the careful measurement of urine pH and its relationship to plasma bicarbonate concentration.

Assessment

Urine pH should be measured in a fresh sample, preferably collected under oil. A urinary infection, particularly with a urea-splitting organism such as *Proteus*, invalidates any estimate of hydrogen ion excretion, as the ammonia formed will elevate the pH. The hydrogen ion concentration should be accurately measured with a glass electrode; most indicator papers are insufficiently precise, although narrow-range indicator papers are available which are suitable for clinical use out of laboratory hours.

Infants and children usually pass acid urine during the night, and the observation of a pH of less than 5.3 in the overnight urine specimen excludes distal renal tubular acidosis (RTA). This simple screening test should always be done before more elaborate assessments of hydrogen ion secretion are undertaken.

The bicarbonate threshold is defined as the plasma bicarbonate concentration at which bicarbonate appears in the urine (in effect, when the urine pH exceeds 6.1; at this pH the bicarbonate concentration is approximately 1 mmol/l. In infants it is lower (tCO_2 = 22 mmol/l) than in older children (tCO_2 = 26 mmol/l) (Edelmann *et al.* 1967c). The threshold may be observed as the plasma bicarbonate concentration rises during the correction of a metabolic acidosis, or as it falls during hydrogen ion loading with ammonium chloride. Observation during the correction of an acidosis is often more informative, as it may be possible under such conditions to determine the renal response to a much more severe metabolic acidosis than can safely be induced subsequently with ammonium chloride, but it may necessitate the gathering of data from the first moment of an emergency admission.

The ingestion of ammonium chloride results in a metabolic acidosis, and can thus be used as a standardised challenge to renal hydrogen ion secretion (Wrong & Davies 1958). Two baseline urine specimens are collected and ammonium chloride, dissolved in fruit juice, is administered by mouth. Urine is collected in approximately hourly aliquots for the next six hours, and urine pH, titratable acidity and ammonium concentration are estimated. A blood sample should be taken three to four hours after the ammonium chloride ingestion to ensure that a metabolic acidosis has indeed been induced.

The amount of ammonium chloride used has varied in different studies. Edelmann *et al.* (1967b) used two dose levels of 75 mEq (4 g) and 150 mEq (8 g) per m^2 SA. The urine pH normally falls to 5.3 or less after the ammonium chloride, but the lower dose may sometimes be insufficient to achieve this. The actual dose of ammonium chloride is of less importance than the demonstration that a significant reduction of plasma tCO_2 has in fact occurred.

Renal acidosis

The acidosis of chronic renal failure results from low rates of excretion of hydrogen ion acceptors, principally ammonia. There is retention of phosphate, sulphate and other anions so that the chloride concentration is low and the anion gap large. In contrast, with renal tubular acidosis the anion gap is small and thus there is usually hyperchloraemia.

Three varieties of renal tubular acidosis (RTA) have been described (McSherry 1981). In Type I (distal, gradient) RTA, the urine pH remains above 6.5 whatever the plasma bicarbonate concentration: there is an inability to generate a significant tubular fluid/plasma hydrogen ion gradient. In Type II RTA, there is a quantitative restriction of the rate at which hydrogen ion can be secreted to neutralise filtered bicarbonate: the bicarbonate threshold is low, and the urine inappropriately alkaline during mild metabolic acidosis, but acidification occurs normally if the plasma bicarbonate, and thus the filtered bicarbonate load, is substantially reduced. Type IV RTA (Type III has been applied to cases with mixed features of Type I and Type II) results from a failure of the distal exchange between sodium and hydrogen ion/potassium. It is due to aldosterone deficiency or resistance, and, unlike other forms of RTA, is associated with hyperkalaemia.

Dissection of Types I and II of RTA can be made by bicarbonate loading. In Type II RTA the normal functioning of the distal hydrogen ion secretory mechanism is revealed during bicarbonate loading by the finding of a urine to plasma $P\text{CO}_2$ gradient greater than 25 mmHg, whereas in Type I RTA the defect of hydrogen ion secretion results in a lesser $P\text{CO}_2$ gradient (Halperin *et al.* 1974; Stinebaugh *et al.* 1981). Attempts to characterise the syndromes by measurement of tubular maximum reabsorption capacity for bicarbonate are frustrated by the overriding effects of volume expansion inherent in such studies.

CALCIUM AND PHOSPHORUS

Calcium excretion

Urinary excretion of calcium is 0.060 ± 0.017 (SD) and of magnesium is 0.122 ± 0.034 (SD) mmol/kg per 24 hour (Ghazali & Barratt 1974); the upper limit of calcium excretion can be conveniently taken as 0.1 mmol/kg per 24 hour. As a simple but less accurate measure, the urine calcium/creatinine ratio in the first urine sample passed after the overnight urine has been voided can be used, and should be less than 0.74 mmol/mmol—there is considerable diurnal variation in this ratio, it being higher in the afternoon and at night than in the morning. Calcium excretion is increased by bed rest, particularly postoperative convalescence—a point to be remembered when investigating children with urinary calculi.

Phosphate

The urinary excretion of phosphate, like that of other inorganic ions, is determined by the amount absorbed from the diet less that portion retained for growth—thus metabolic rather than renal factors determine urinary phosphate excretion, and the role of the kidney is reflected in the plasma phosphate or parathormone (PTH) concentration at which phosphate homeostasis is achieved. Parathormone is the principal factor controlling phosphate excretion, suppressing proximal tubular reabsorption.

Plasma phosphorus

Plasma inorganic phosphorus concentration should be interpreted in relation to the quite marked variations in the normal range at different ages (see Chapters 2 and 14). Hypophosphataemia arises from either excessive urinary loss or diminished gastrointestinal absorption. Hyperphosphataemia tends to occur in chronic renal failure due to diminished renal excretion.

Renal excretion of phosphate

Assessment of the role of the kidney in phosphate excretion and deciding whether the renal tubule is responding appropriately to the prevailing state of glomerular filtration rate, phosphate balance and parathyroid activity is not an easy matter. The phosphate/creatinine clearance ratio (C_p/C_c) and percentage tubular reabsorption of phosphate (%TRP) are useful parameters, and are, of course, mathematically related:

$$C_p/C_c = \frac{U_p V}{P_p} \bigg/ \frac{U_c V}{P_c} = \frac{U_p}{P_p} \bigg/ \frac{U_c}{P_c}$$

$$\%TRP = 100(F_p - U_p V)/F_p$$

$$= 100(1 - U_p V/F_p)$$

$$F_p = P_p(U_c V/P_c)$$

$$\therefore \quad \%TRP = 100(1 - U_p V/P_p(U_c V/P_c))$$

$$= 100(1 - C_p/C_c)$$

where U_p, P_p, C_p and F_p = urine concentration, plasma concentration, clearance and filtered load of phosphate respectively.

TRP in healthy children is greater than 80% (Kruse *et al.* 1982), but is influenced by the state of phosphate balance, and should with phosphate depletion, rise to 98% or more: in effect, if the plasma phosphate is more than 2 SD below the mean for age, there should be very little phosphate in the urine, and if this is not so, then the hypophosphataemia can be attributed to a renal tubular leak. The functional capacity of the renal tubule can only be fully assessed however by measurement of the tubular maximum capacity for phosphate reabsorption (TmP), but this is an elaborate technique requiring phosphate infusion. In adults TmP/GFR can be predicted with considerable confidence from C_p/C_c (Walton & Bijovet 1975) and similar data have been reported in children (Kruse *et al.* 1982) though the relationship between the two parameters has not been rigorously established in children.

Phosphate depletion states

Renal phosphate loss may be due to intrinsic tubular disease or a consequence of hyperparathyroidism; the latter may be primary or secondary to states of calcium depletion such as vitamin D deficiency rickets. A primary renal tubular phosphate leak may either be an isolated renal tubular abnormality, as in vitamin D-resistant rickets, or part of the spectrum of the Fanconi syndrome. Hypophosphataemia may also arise from inadequate dietary absorption of phosphate, either because the diet itself is inadequate (e.g. with parenteral feeding) or due to an excessive use of phosphate binding gels, particularly Aludrox.

Phosphate retention

The bone disease that accompanies chronic renal failure, termed uraemic osteodystrophy, has two elements: rickets and hyperparathyroidism. The functional abnormalities underlying these are a deficiency of 1,25-dihydroxycholecalciferol due to inadequate 1-hydroxylation (which only takes place in the kidney), and parathyroid overactivity linked to the tendency to phosphate retention. As GFR falls, phosphate is retained, there is reciprocal depression of the plasma calcium concentration, stimulation of parathormone and re-establishment of phosphate balance by depression of tubular reabsorption of phosphate: hyperparathyroid bone disease is the 'trade-off' for the compensatory responses necessary to restore phosphate homoeostasis (Bricker 1972). The sequence can be interrupted by reducing phosphate intake in proportion to reduction in GFR (Slatopolsky *et al.* 1971): in practical terms this means blocking phosphate absorption with Aludrox so that PTH is suppressed, plasma phosphate maintained around the mean normal for age, and TRP kept above 80%.

CONCLUSION

The material presented in this chapter reflects the dependence of the nephrologist on the services of the Chemical Pathology Department, and, it is hoped, go some way to explain the heavy demands that the Renal Unit make on the laboratory. The onus is now on the clinician to make best use of the results provided, and to develop methods of information handling so that trends and patterns in the data are clearly perceived.

ACKNOWLEDGEMENT

I would like to thank Professor Barbara Clayton for introducing me to the art and science of laboratory medicine, an apprenticeship that was the foundation of this chapter.

REFERENCES

Aronson A. S. & Svenningsen N. W. (1974) DDAVP test for estimation of renal concentrating capacity in infants and children. *Arch. Dis. Child.* **49**, 654–9.

Arturson G., Groth T. & Grotte G. (1971) Human glomerular membrane porosity and filtration pressure: dextran clearance analysed by theoretical models. *Clin. Sci.* **40**, 137–58.

Barnett H.L., Hare W., McNamara H. & Hare R. (1948) Measurement of glomerular filtration rate in premature infants. *J. Clin. Invest.* **27**, 691–9.

BARRATT T. M. (1974a) Assessment of renal function in children. In Apley J. (ed.) *Modern Trends in Paediatrics*, Vol. 4, pp. 181–215. Butterworths, London.

BARRATT T. M. (1974b) The nephrological background to urology. In Williams D. I. (ed.) *Encyclopaedia of Urology*, Vol. 15 (Supplement), pp. 1–12. Springer, Heidelberg.

BARRATT T. M. & CHANTLER C. (1975) Clinical assessment of renal function. In Rubin M. I. & Barratt T. M. (eds.) *Pediatric Nephrology*, pp. 55–83. Williams & Wilkins, Baltimore.

BARRATT T. M. & CRAWFORD R. (1970) Lysozyme excretion as a measure of renal tubular dysfunction in children. *Clin. Sci.* **39**, 457–65.

BARRATT T. M., MCLAINE P. N. & SOOTHILL J. F. (1970) Albumin excretion as a measure of glomerular dysfunction in children. *Arch. Dis. Childh.* **45**, 496–501.

BERGGARD I. (1970) Plasma proteins in normal human urine. In Manuel Y., Revillard J. P. & Betuel H. (eds) *Proteins in Normal and Pathological Urine*, pp. 7–23. S. Karger, Basel.

BIANCANI F., RAPPAPORT R., BALSAN S. & GUIVARCH J. (1972) L'épreuve de restriction sodée chez l'enfant normal. Étude de la conservation rénale du sodium et des variations de la tétrahydro-aldosterone urinaire. *Arch. Franc. Pédiat.* **29**, 609–19.

BRENNER B. M., ICHIKAWA I. & DEEN W. M. (1981) Glomerular filtration. In Brenner B. M. & Rector F. C. (eds.) *The Kidney*, 2nd edition, Vol. I, pp. 289–327. W. B. Saunders, Philadelphia.

BRICKER N. S. (1972) On the pathogenesis of the uremic state. An exposition of the 'trade-off' hypothesis. *New Engl. J. Med.* **286**, 1093–9.

BRODEHL J. (1978) Renal hyperaminoaciduria. In Edelmann C. M. (ed.) *Pediatric Kidney Disease*, pp. 1047–78. Little Brown, Boston.

CAMERON J. S. & BLANDFORD G. (1966) The simple assessment of permeability in heavy proteinuria. *Lancet* **ii**, 242–5.

CHANTLER C. & BARRATT T. M. (1972) Estimation of glomerular filtration rate from plasma clearance of 51-chromium edetic acid. *Arch. Dis. Childh.* **47**, 613–17.

CHANTLER C., PARSONS F. & VEALL N. (1969) Glomerular filtration rate with measurement in man by the single injection method using ^{51}Cr–EDTA. *Clin. Sci.* **37**, 169–80.

COUNAHAN R., CHANTLER C., GHAZALI S., KIRKWOOD B., ROSE F. & BARRATT T. M. (1976) Estimation of glomerular filtration rate from plasma creatinine concentration in children. *Arch. Dis. Childh.* **51**, 875–8.

DILLON M. J. & RYNESS J. M. (1975) Plasma renin activity and aldosterone concentration in children. *Br. Med. J.* **4**, 316–19.

DILLON M. J., SHAH V. & BARRATT T. M. (1978) Renal vein renin measurements in children with hypertension. *Br. Med. J.* **2**, 168–70.

DODGE W. F., TRAVIS L. B. & DAESCHNER C. W. (1967) Comparison of endogenous creatinine clearance with inulin clearance. *Am. J. Dis. Child.* **113**, 683–92.

DOOLAN P. D., ALPON E. L. & THEIL G. B. (1962) A clinical appraisal of the plasma concentration and endogenous clearance of creatinine. *Am. J. Med.* **32**, 65–79.

DUBOIS D. & DUBOIS E. F. (1916) Clinical calorimetry: a formula to estimate the approximate surface area if height and weight be known. *Arch. Int. Med.* **17**, 863–81.

EDELMANN C. M., BARNETT H. L. & TROUPKOU V. (1960) Renal concentrating mechanisms in newborn infants: effect of dietary protein and water content, role of urea and responsiveness to antidiuretic hormone. *J. Clin. Invest.* **36**, 1062–9.

EDELMANN C. M., BARNETT H. L. & STARK H. (1967a) A standardised test of renal concentrating capacity in infants and children. *Am. J. Dis. Child.* **114**, 639–44.

EDELMANN C. M., BOICHIS H., RODRIGUEZ-SORIANO J. & STARK H. (1967c) The renal response to acute ammonium chloride acidosis. *Pediat. Res.* **1**, 452–60.

EDELMANN C. M., RODRIGUEZ-SORIANO J., BOICHIS H., GRUSKIN A. B. & ACOSTA M. (1967c) Renal bicarbonate reabsorption and hydrogen ion excretion in infants. *J. Clin. Invest.* **46**, 1309–17.

FRIIS-HANSEN B. (1961) Body water compartments in children: changes during growth and related changes in body composition. *Pediatrics* **28**, 169–81.

GERDTS K. G., SHAH V., SAVAGE J. M. & DILLON M. J. (1979) Renal vein renin measurements in normotensive children. *J. Pediatr.* **95**, 953–8.

GHAZALI S. & BARRATT T. M. (1974) Urinary excretion of calcium and magnesium. *Arch. Dis. Child.* **49**, 97–101.

GOLDSMITH D. I. (1978) Clinical and laboratory evaluation of renal function. In Edelmann C. M. (ed.) *Pediatric Kidney Disease*, Vol. 1, pp. 213–24. Little Brown, Boston.

GOLDSMITH D. I., DRUKKER A., BLAUFOX M. D., SPITZER A. & EDELMANN C. M. (1975) Response of the neonatal canine kidney to acute saline expansion. In Zum Winkel K., Blaufox M. D. & Funck-Brentano J. L. (eds.) *Radionuclides in Nephrology*, pp. 45–54. G. Thieme, Stuttgart.

GRUSKIN A. B., EDELMANN C. M. & YUAN S. (1970) Maturational changes in renal blood flow in piglets. *Pediatr. Res.* **4**, 7–13.

HALPERIN M. L., GOLDSTEIN M. B., HAIG A., JOHNSON M. D. & STINEBAUGH B. J. (1974) Studies on the pathogenesis of Type I (distal) renal tubular acidosis as revealed by the urinary PCO_2 tensions. *J. Clin. Invest.* **53**, 669–77.

HARDWICKE J., CAMERON J. S., HARRISON J. F., HULME B. & SOOTHILL J. F. (1970) Proteinuria, studied by clearances of individual macromolecules. In Manuel Y., Revillard J. P. & Betuel H. (eds.) *Proteins in Normal and Pathological Urine*, pp. 111–52. S. Karger, Basel.

HAYCOCK G. B., SCHWARTZ G. J. & WISOTSKY D. H. (1978) Geometric method for measuring body surface area: a height–weight formula validated in infants, children and adults. *J. Pediatr.* **93**, 62–6.

KROVETZ L. J. (1965) The physiologic significance of body surface area. *J. Pediatr.* **67**, 841–62.

KRUSE K., KRACHT U. & GÖPFERT G. (1982) Renal threshold phosphate concentration. *Arch. Dis. Child.* **57**, 217–23.

MCCANCE R. A. (1959) The maintenance of stability in the newly born. I. Chemical exchange. *Arch. Dis. Child.* **34**, 361–70.

MCCANCE R. A. & WIDDOWSON E. M. (1952) The correct physiological basis on which to compare infant and adult renal function. *Lancet* **ii**, 860–2

MCCRORY W. W. (1972) *Developmental Nephrology*. Harvard University Press, Cambridge, Mass.

MCSHERRY E. (1981) Current issues in hydrogen ion transport. In Gruskin A. B. & Norman M. E. (eds.) *Pediatric Nephrology. Proceedings of the Fifth International Pediatric Nephrology Symposium, Philadelphia, October 1980*, pp. 403–15. Martinus Nijhoff, The Hague.

MANCINI G., CARBONARA A. O. & HEREMANS J. F. (1965) Immunochemical quantitation of antigens by single radial immunodiffusion. *Immunochemistry* **2**, 235–54.

MITCH W. E., WALSER M., BUFFINGTON G. A. & LEMANN J. (1976) A simple method of estimating progression of chronic renal failure. *Lancet* **ii**, 1326–8.

OLBING H., BLAUFOX M. D., ASCHINBERG L. C., SILKALNS G. I., BERNSTEIN J., SPITZER A. & EDELMANN C. M. (1973) Post-natal changes in blood flow distribution in puppies. *J. Clin. Invest.* **52**, 2885–95.

PIEPSZ A., DENIS R., HAM H. R., DOBBELEIR A., SCHULMAN C. & ERBSMANN F. (1978) A simple method for measuring separate glomerular filtration rate using a single injection of ^{99m}Tc–DTPA and the scintillation scanner. *J. Pediatr.* **93**, 769–74.

RISDON R. A., SLOPER J. C. & DE WARDENER H. E. (1968) Relationship between renal function and histological changes found in renal biopsy specimens from patients with persistent glomerular nephritis. *Lancet* **ii**, 363–6.

RODRIGUEZ-SORIANO J., VALLO A., CASTILLO G. & OLIVEROS R. (1981) Renal handling of water and sodium in infancy and childhood: a study using clearance methods during hypotonic saline diuresis. *Kidney Int.* **20**, 700–4.

ROTH K. S., FOREMAN J. W. & SEGAL S. (1981) The Fanconi syndrome and mechanisms of tubular transport dysfunction. *Kidney Int.* **20**, 705–16.

SCHNEIDER J. A., SCHULMAN J. D. & SEEGMILLER J. E. (1978) Cystinosis and the Fanconi syndrome. In Stanbury J. B., Wyngaarden J. B. & Fredrickson D. S. (eds.) *The Metabolic Basis of Inherited Disease*, 4th edition, pp. 1660–82. McGraw-Hill, New York.

SCHNERMANN J., WRIGHT F. S., DAVIS J. M., VON STACKELBERG W. & GRILL G. (1970) Regulation of superficial nephron filtration by tubuloglomerular feedback. *Pflügers Arch. ges. Physiol.* **318**, 147–75.

SCHWARTZ G. J., HAYCOCK G. B. & SPITZER A. (1976) Plasma creatinine and urea concentration in children. Normal values for age and sex. *J. Pediatr.* **88**, 828–30.

SENDROY J. & CECCHINI L. P. (1954) Determination of human body surface area from height and weight. *J. Appl. Physiol.* **7**, 1–12.

SIGGAARD-ANDERSEN O. (1963) The acid–base status of the blood. *Scand. J. Clin. Lab. Invest.* **15** (Suppl. 70), 1–134.

SLATOPOLSKY E., CAGLAR S., PENNELL J. P., TAGGART D. D., CANTERBURY J. M., REISS E. & BRICKER N. S. (1971) On the pathogenesis of hyperparathyroidism in chronic experimental renal insufficiency in the dog. *J. Clin. Invest.* **50**, 492–9.

SMITH H. W. (1951) *The Kidney: Structure and Function in Health and Disease*. Oxford University Press, Oxford.

SPITZER A. (1978) Renal physiology and functional development. In Edelmann C. M. (ed.) *Pediatric Kidney Disease*, pp. 25–128. Little Brown, Boston.

STINEBAUGH B. J., SCHLOEDER F. X., TAM S. C., GOLDSTEIN M. B. & HALPERIN M. L. (1981) Pathogenesis of distal renal tubular acidosis. *Kidney Int.* **19**, 1–17.

STOTEN A. (1968) A micromethod for creatinine using resin to remove interfering substances. *J. Med. Lab. Tech.* **25**, 240–7.

TALBOT N. B., RICHIE R. H. & CRAWFORD J. D. (1959) *Metabolic Homeostasis. A Syllabus for those concerned with the Care of Patients*. Harvard University Press, Cambridge, Mass.

TANNER J. M. (1949) The fallacy of per-weight and per-surface area standards and their relation to spurious correlation. *J. Appl. Physiol.* **2**, 1–15.

THOMAS V., SHELOKOV A. & FORLAND M. (1974) Antibody-coated bacteria in the urine and the site of urinary tract infection. *New Engl. J. Med.* **290**, 588–90.

WAGNER M. G., SMITH F. G., TINGLOF B. O. & CORNBERG E. (1968) Epidemiology of proteinuria: a study of 4807 school children. *J. Pediatr.* **73**, 825–32.

WALTON R. J. & BIJVOET O. L. M. (1975) Normogram for derivation of renal threshold phosphate concentration. *Lancet* **ii**, 309–10.

WELLWOOD J. F., ELLIS B. G., PRICE R. G., HAMMOND K., THOMPSON A. E. & JONES N. P. (1975) Urinary N-acetyl-β-D-glucosaminidase activities in patients with renal disease. *Br. Med. J.* **3**, 408–11.

WINBERG J. (1959a) The 24-hour true endogenous creatinine clearance in infants and children without renal disease. *Acta Paediat.* **48**, 443–52.

WINBERG J. (1959b) Determination of renal concentration capacity in infants and children without renal disease. *Acta Paediat.* **48**, 318–28.

WRONG O. M. & DAVIES H. E. F. (1958) The excretion of acid in renal disease. *Quart. J. Med.* **28**, 259–313.

Chapter 6

The Gastrointestinal Tract

JOHN H. TRIPP & DAVID P. R. MULLER

The request for investigations and chemical tests should be informed and selective, based on either a specific premeditated diagnosis or a detailed knowledge of the normal physiological processes involved. General observations such as the sequential documentation of height and weight and inspection of stools can, however, provide important information with regard to the presence of underlying organic disease before more specific and detailed investigations are undertaken.

This chapter will consider the more important biochemical investigations currently available for the assessment of gastrointestinal problems in childhood and their limitations. For convenience the chapter has been divided into two sections; first the investigation of the separate organs of the gastrointestinal tract and second, the investigation of defects in the handling of nutrients, water, electrolytes, etc. There will of necessity be overlap between these two sections and cross-references will be made where appropriate.

INVESTIGATION OF SPECIFIC ORGANS

Stomach

Gastric acid secretion

The secretion of gastric acid is controlled by a feedback mechanism that appears to be competent at birth. The gastric parietal cells are functional and secrete acid within a few hours of birth (Avery *et al.* 1966a) and a normal fasting pH is achieved rapidly. On a body weight or surface area basis the volumes and total acid secretion are fairly constant throughout childhood.

Measuring the fasting pH of gastric contents may be useful in situations where a near neutral pH is clinically desirable. For example, in the treatment of severe reflux oesophagitis or bleeding varices with antacids or H2 blocking agents, a fasting pH of > 4 compared to normal values of 1–2 indicates a satisfactory control of acid secretion.

Tests of acid output are well standardised in adults but have been relatively little used in children. The procedure is as follows: after an overnight fast a nasogastric tube is passed under fluoroscopic control and the resting juice is aspirated and discarded. Gastric juice is then collected for four 15-minute periods to establish the basal acid output (BAO) and for a further eight 15-minute periods after the intravenous injection of pentagastrin (6 μg/kg) to establish the maximal acid output (MAO) and the peak acid output (PAO). The MAO refers to the rate of acid secretion during the first hour after stimulation and the PAO to the sum of the two highest consecutive 15 minute secretory periods multiplied by 2. The titratable acidity is measured by titration to pH 7 with 0.1 M sodium hydroxide.

Intermittent blockage of the holes at the end of the nasogastric tube by the gastric mucosa can be a practical problem, and an appropriately designed low pressure suction pump which provides intermittent suction and has an automatic 'unblocking' facility, such as the automatic gastric suction pump supplied by United Analysts Limited, East Boldon, Tyne and Wear, UK, is a good investment.

Normal values of BAO and MAO in children are probably close to those of adults when expressed on a body weight basis (Baron 1970). As in adults the changes seen in children with peptic (duodenal) ulceration are not consistent although an increase in MAO and PAO with a normal BAO has been reported (Christie & Ament 1976). In contrast in the Zollinger–Ellison syndrome, which is very rare in children, the BAO is increased to near or even greater than normal levels of MAO with little increase after the administration of pentagastrin (Schwartz *et al.* 1974). The diagnosis is confirmed by the measurement of increased concentrations of plasma gastrin. An increased BAO may also occur after massive small bowel resection (Avery *et al.* 1966b) secondary to increased circulating gastrin levels (Strauss *et al.* 1974). A reduced acid output occurs in conditions such as malnutrition, gastric mucosal hypertrophy (Menetrier's disease; Frank & Kern 1967), the Verner–Morrison syndrome (Verner & Morrison 1958) and in classical pernicious anaemia which is, however, very rare in children.

Pepsin secretion

Pepsin secretion normally parallels acid secretion and has been little investigated in children, although methods are available (Agunod *et al.* 1969) and have been used in research studies (Cameron *et al.* 1979).

Gastric emptying

The simplest method to assess gastric emptying is to instil a test meal containing a non-absorbable marker such as polyethylene glycol into the stomach and to measure the volume recovered after defined periods. A description of the rate of emptying requires a continuous sampling technique and such an approach has been used in infants by Cavell (1979).

Exocrine pancreas

Prior to detailed and specific investigations, an examination of the faeces can provide useful information regarding pancreatic function. For example, the appearance of fat will often suggest the diagnosis of pancreatic insufficiency and in the rare condition of isolated pancreatic lipase deficiency the appearance of the faeces (a normal stool surrounded by yellow/orange oil) is characteristic.

Pancreatic stimulation and duodenal intubation. The assessment of pancreatic function involves the quantification of secreted pancreatic enzymes, volume and bicarbonate in duodenal juice following stimulation of the pancreas. Enzyme activities reflect the function of the acinar cells whereas volume and bicarbonate outputs are indicators of tubular function. The pancreas can be stimulated either by a test meal or by the intravenous administration of cholecystokinin and secretin (consecutively or combined). The former will provide information on the physiological response of the gland to intraluminal substrates, whereas the latter gives information on the maximal secretory capacity of the pancreas. The choice of stimulation may depend on the specific clinical case and the precise questions being asked, but it is important to stress that the pancreas must be stimulated as it is impossible to interpret results from fasting or unstimulated specimens.

Stimulation of the pancreas by a modified Lundh meal has been described and validated in infants and children by McCollum *et al.* (1977). The test meal consists of glucose, corn oil and comminuted chicken (each of 4 g/100 ml) which enables it to be given to patients with various food intolerances without precipitating gastrointestinal symptoms. Following, where possible, an overnight fast the patient is sedated one hour before the test (intramuscular chlorpromazine, 1 mg/kg) and a nasogastric tube is passed for administering the meal (30 ml/kg to a maximum of 240 ml). The collecting tube is then passed under fluoroscopic control to the fourth part of the duodenum and post-prandial juice is collected on ice for 2 hours. Assays are performed on aliquots of this 2-hour pooled specimen. Bicarbonate is neutralised by gastric acid and cannot be usefully assayed.

Stimulation by either consecutive or combined intravenous injections of cholecystokinin and secretin have been validated with both relatively complete collection systems when volume and quantitative data may be collected (Hadorn 1975a) and with incomplete collection systems when bicarbonate concentrations and enzyme activities per millilitre of juice are obtained. We have used an incomplete collection system and two different procedures for hormonal stimulation: either cholecystokinin (2 IU/kg) was given by a slow infusion and duodenal juice collected on ice for 20 minutes followed by infusion of secretin (2 IU/kg) and a further 30 minutes collection period or alternatively, secretin was infused immediately after the cholecystokinin and juice collected for a total of 30 minutes.

Table 6.1 Activities of pancreatic enzymes, bicarbonate and total bile salt concentrations in control[1] children following various methods of pancreatico–biliary stimulation.

	Trypsin[2] µeq/min/ml			Lipase[3] µeq/min/ml			Amylase[4] units/ml			Bicarbonate[5] mmol/l			Total bile salts[6] mmol/l		
	n	mean	range	n	mean	range	n	mean	range	n	mean	range	n	mean	range
Test meal (McCollum *et al.* 1977)	13	42.8	19.4–83.3	13	768	270–1920							9	6.8	2.7–16.0
Post-chole-cystokinin and secretin	21	55.8	29.4–105.8	21	1195	696–2664	20	185	30–401	6	59.1	36–70	13	17.3	6.2–32.0
Post-chole-cystokinin	9	71.4	48.1–108	10	1040	363–1930	8	282	119–525				9	37.1	17.7–54.7

1 Children (aged 9 months–16 years) with suspected malabsorption who were proven to be normal following intensive investigation.
2 Titrimetric assay using p-tosyl arginine methyl ester as substrate (Wiggins 1967).
3 Titrimetric assay (Erlanson & Borgstrom 1970).
4 Amylochrome kit supplied by Roche Products Limited.
5 Colorimetric assay using cresol red (Skeggs & Hochstrasser 1964).
6 Enzymatic method (Iwata & Yamasaki 1964).

Table 6.2 Volume, bicarbonate and pancreatic enzyme activities[1] in control children[2] following stimulation with cholecystokinin and secretin (Hadorn 1975a).

	Volume ml/kg per 50 min	Bicarbonate meq/kg per 50 min	Trypsin μg/kg per 50 min	Chymotrypsin μg/kg per 50 min	Lipase IU/kg per 50 min	Amylase IU/kg per 50 min
Mean	3.5	0.15	770	750	1420	354
Range	1.2–7.5	0.05–0.45	300–2000	350–2700	300–5000	90–1050
n	71	71	69	72	65	49

1 For the analytical methods used see Hadorn (1975b)
2 Children without pancreatic disease aged from 3 months–14 years

For routine tests of pancreatic function we generally use combined hormonal stimulation rather than a test meal as the investigation is easier to perform at the bedside and is relatively quick. The intubation can also be combined with a jejunal biopsy, with the latter being carried out at the end of the procedure.

Reference ranges (from our laboratory) for the different procedures are given in Table 6.1, and the total outputs following cholecystokinin/secretin reported by Hadorn (1975a) are shown in Table 6.2. It should be stressed that normal values will vary with different assay procedures and that each centre should obtain its own reference data. The collection and storage of the juice prior to assay is also important as pancreatic lipase and trypsin activities in hormonally stimulated juices are labile at −20 °C and either must be assayed immediately or diluted with an equal volume of 50% glycerol (Muller & Ghale 1982). In contrast the enzymatic activities in juices collected after a test meal are stable for many months when stored at −20 °C.

In addition to the assay of enzymatic activities, etc. post-stimulated juices may be used for other investigations, e.g. bile salt concentrations (see below), bacterial culture and a study of the products of digestion, if appropriate substrates are included in the test meal.

Indirect methods

Faecal proteolytic activity. The determination of faecal proteolytic activity has been used as a test of pancreatic function. Serial dilutions of stool are incubated with gelatin and the end point at which no digestion occurs is established. Normal values are said to be at dilutions of > 1:80. The quantitation of faecal chymotrypsin has also been used as a measure of pancreatic enzyme secretion (e.g. Hadorn 1975a). False positive and negative results occur, however, with both procedures and we consider their use should be discouraged in clinical practice.

Immunoreactive serum trypsin. The estimation of immunoreactive serum trypsin concentrations was suggested as a screening test for pancreatic disease by Lake-Bakaar *et al.* (1979). Durie *et al.* (1981) have reported it to be a sensitive indicator of exocrine pancreatic dysfunction in patients with proven cystic fibrosis and Shwachman's syndrome, and Heeley *et al.* (1982) have recently reported its use as a screening test for cystic fibrosis. Heeley *et al.* (1982) measured immunoreactive trypsin in dried blood specimens from a total of 20000 infants. They then performed a second test in infants with an increased trypsin concentration (> 80 ng/ml) and detected eight infants with cystic fibrosis. There was a single case of an increased trypsin concentration where a diagnosis of cystic fibrosis could not be established but no false negatives were identified.

Ruddell *et al.* (1981) however reported a study of almost 300 adults with and without chronic pancreatic disease. They concluded that serum concentrations of immunoreactive trypsin had a low specificity and sensitivity for pancreatic disease and did not reflect the degree of exocrine pancreatic insufficiency in patients with proven pancreatitis.

Hydrolysis of N-benzoyl-L-tyrosyl-p-aminobenzoic acid (BTP). The synthetic peptide BTP is hydrolysed by chymotrypsin to give p-aminobenzoic acid (PABA) which is then absorbed, conjugated in the liver and excreted in the urine. It has, therefore, been suggested that the estimation of PABA and its metabolites in urine following ingestion of BTP provides an index of exocrine pancreatic function (de Benneville *et al.* 1972). Some hydrolysis of BTP, however, also takes place at the brush border as a result of peptidase activity (Sterchi *et al.* 1981). Misleading results may occur in patients with malabsorption or liver disease but these can be minimised by either performing a second test with oral PABA (Mitchell *et al.* 1979) which is rather cumbersome for routine clinical use, or by adding a tracer dose of radiolabelled ^{14}C-PABA to the BTP (Mitchell *et al.* 1981). There are, however, ethical problems in giving radioactively-labelled substrates to children, especially when other more direct tests are available.

Liver

With respect to intestinal function the principal role of the liver is the secretion of bile, which is an isosmotic alkaline solution (pH 7.5–9.5) rich in organic solutes such as conjugated bile salts, bilirubin, cholesterol and lecithin. Adequate intraluminal concentrations of bile salts are important for a number of digestive events but their major function is the solubilisation of dietary lipids in the small intestine prior to their absorption. In health, bile salts are conserved by a highly efficient enterohepatic circulation (Hofmann 1977) and if this becomes deranged, pool size and luminal concentrations of the bile salts may become reduced. Methods of investigating the enterohepatic circulation of the bile salts will now be discussed.

Table 6.3 Total and individual bile salt concentrations in control[1] children following various methods of pancreaticobiliary stimulation.

	Total bile salts[2] mmol/l			Individual bile salts[3] Glycine: taurine ratio		Dihydroxy bile salts (% of total)	
	n	mean	range	mean	range	mean	range
Test meal (McCollum *et al.* 1977)	9	6.8	2.7–16.0	2.8	1.3–4.0	60	53–70
Post-cholecystokinin and secretin	13	17.3	6.2–32.0				
Post cholecystokinin	9	37.1	17.7–54.7				

1 Children (aged 16 months–16 years) with suspected malabsorption who were proven to be normal following intensive investigation
2 Enzymatic method of Iwata and Yamasaki (1964)
3 As described by McCollum *et al.* (1977)

Intraluminal bile salts

A qualitative and/or quantitative assessment of bile salts in the gut lumen can be made on duodenal juice after pancreaticobiliary stimulation, as discussed previously for pancreatic enzyme activities. Total bile salt concentrations can be conveniently estimated following their extraction on Amberlite XAD resin (Makino & Sjovall 1972) by an enzymatic procedure using 3,α-hydroxy steroid dehydrogenase (Iwata & Yamasaki 1964). The same method can also be used for the quantification of individual bile salts after their separation by thin-layer chromatography (McCollum *et al.* 1977). Reference ranges for total and individual bile salt concentrations are shown in Table 6.3.

Above a certain concentration, known as the critical micellar concentration, bile salt molecules aggregate to form micelles which are then able to solubilise lipids. Under normal physiological conditions, the critical micellar concentration is approximately 2 mmol/l and bile salt concentrations below this will result in impaired fat solubilisation and absorption. This can arise for a number of reasons: obstruction to bile flow, as in biliary atresia, choledochal cyst, or other anatomical anomalies of the biliary tree, hepatitis, cholestatic jaundice, cholangitis, or increased excretion as in ileal resection and pancreatic steatorrhoea.

Qualitative changes (an increased ratio of glycine to taurine conjugates and an increased proportion of dihydroxy bile salts) can also occur in situations of a broken enterohepatic circulation. Both these qualitative disturbances may cause impaired micellar solubilisation, as glycine conjugates have a higher pKa and are therefore less soluble than the taurine conjugates, and the dihydroxy bile salts have a higher critical micellar concentration than the trihydroxy species. The presence of significant quantities of deconjugated bile salts in the upper small intestinal contents implies bacterial degradation and abnormal colonisation of the small intestine.

Faecal bile salts

As mentioned previously the enterohepatic circulation is a highly efficient process in health, with greater than 95% of the bile salts being reabsorbed by the ileum and returning to the liver. Estimation of faecal bile salts will give a measure of the excretion and, therefore, of the efficiency of the reabsorptive process, but methods are in general cumbersome and laborious.

^{14}C-Glycocholate breath test

If increased amounts of glycine-1-^{14}C labelled glycocholate reach the colon, bacterial deconjugation will result in increased amounts of radiolabelled glycine being absorbed and metabolised to radiolabelled carbon dioxide. Thus, impaired reabsorption and increased excretion of bile salts can be assessed by the measurement of $^{14}CO_2$ in expired breath. The test can also be used as a measure of bacterial overgrowth of the small intestine, when a large peak of $^{14}CO_2$ is seen within 2–8 hours after ingestion, while in controls the excretion is over a period of a few days without a large peak (Fromm & Hofmann 1971). Generally this method has not been used in young children since it requires radiolabelled substrates and cooperation in the collection of breath.

Serum bile salts

The concentration of serum bile salts can also provide useful information on the enterohepatic circulation, but the timing of blood specimens in relation to meals is important. Ideally estimations should be carried out at half-hour intervals between one and three hours after a meal, when peak concentrations are reached. Total serum bile salts can be measured by the enzymatic assay as described above, but with a fluorimetric detection system. Radioimmunoassays have been devised for groups of the more common bile salts (e.g. Simmonds *et al.* 1973). These provide rapid and accurate quantitation of the particular bile salts for which antibodies are available and only require small volumes (0.5 ml) of blood. Gas–liquid chromatographic methods have also been described (e.g. Clayton & Muller 1980) but are normally used for research rather than routine applications. When combined with mass spectrometry they do, however, provide a complete qualitative and quantitative description of the circulating bile acids.

The small intestine

The small intestine is the major absorptive organ and specific investigations related to the absorption of fat, carbohydrate and protein etc. will be discussed in the sections dealing with the handling of these nutrients. Small intestinal overgrowth and the morphology of the small intestine will be discussed here.

Bacterial overgrowth

In health the upper small intestine is virtually sterile with total bacterial counts of $< 10^4$/ml, increasing slightly in the terminal ileum. Bacterial overgrowth of the small intestine in for example the blind loop syndrome, strictures and duplications can result in malabsorption and it is, therefore, important to make a correct diagnosis. As already discussed the presence of deconjugated bile salts in duodenal juice demonstrates the presence of bacteria. Duodenal, jejunal or ileal juice may be cultured but adequate anaerobic culture facilities are necessary. The combination of vitamin B_{12} malabsorption and increased folate absorption may suggest the diagnosis. The former results from micro-organisms binding the vitamin and intrinsic factor, and the increased folate results from synthesis by the bacteria.

The excretion of urinary indicans has also been suggested as a test for bacterial overgrowth, as concentrations tend to be increased following bacterial degradation of tryptophan (Greenberger *et al.* 1968). The test is, however, unreliable as not all gastrointestinal micro-organisms can metabolise tryptophan to indican and those that do require a luminal pH of < 6.5. Malabsorption of protein or tryptophan will also lead to increased indican production, and pancreatic supplements may break down unabsorbed protein to suitable substrates for bacterial metabolism.

Morphology

D-xylose, a pentose sugar, is poorly metabolised and absorbed passively by the proximal small intestine and excreted in the urine. Its absorption, measured either as a 5-hour collection of urine or in blood 1 hour after the oral load, has been used as a measure of the relative surface area and hence morphology of the jejunum, and has been suggested as a screening test for coeliac disease (Rolles *et al.* 1973). However, the value of this test is limited, as Lamabadusuriya *et al.* (1975) reported false positive and negative results even when both blood and urine were analysed. The xylose test is not a substitute for the definitive investigation which is an intestinal biopsy.

Large intestine and rectum

The main functions of the large intestine are the storage of intestinal contents prior to excretion, the absorption of water and electrolytes entering from the ileum and in particular the conservation of sodium ions and water. Methods which we have used and found helpful in assessing the function of the large intestine will now be discussed.

Potential difference

A potential difference of the order of 25 mV exists across the mucosal epithelium of the colon with the luminal side negative with respect to the serosa. This results largely from the active transport of sodium ions from the lumen to the blood and thus, the colonic potential difference gives a measure of the overall functional state of the colonic epithelium. The measurement of the transcolonic electrical potential difference has been described in detail by Edmonds and Godfrey (1970). An increase in the colonic potential difference has been found, for example, in congenital chloridorrhoea and a decrease in a patient with pseudohypoaldosteronism (Milla 1982).

Dialysis method

Edmonds (1971) described a dialysis method to study transport in the rectum of human subjects. A closed sac made from dialysis membrane and containing a solution of known composition and volume is left in contact with the mucosa of an empty segment of intestine for a timed period. The transport properties of the epithelium at that region can then be deduced from the changes in composition of the solution.

Stool electrolytes

Determination of stool electrolytes, in particular sodium, potassium and chloride ions, is a useful diagnostic test in patients with explosive watery stools. For example, stool electrolytes may be helpful in distinguishing an osmotic from a secretory diarrhoeal state (Krejs *et al.* 1980). In a secretory state, stool electrolytes (Na^+, K^+ and their respective anions) account almost completely for the total faecal osmolarity, whereas in osmotically induced diarrhoea there is a large deficit (solute gap) which results from the presence of metabolic products arising from bacterial action on malabsorbed carbohydrate. A stool chloride concentration which exceeds the sum of the sodium and potassium concentrations is pathognomonic for congenital chloridorrhoea (Launiala *et al.* 1968) which is a primary defect of electrolyte absorption.

DEFECTS IN ABSORPTION OF SPECIFIC NUTRIENTS

Fat

The diagnosis of steatorrhoea is generally a clinical one. When large pale fatty stools are associated with failure to thrive in a child, the quantitation of the steatorrhoea is of relatively little importance compared to the need to identify the cause. In our opinion, therefore, an estimate of fat absorption which is very time-consuming and expensive is only necessary when fat malabsorption is in doubt or when it is essential for the assessment of therapy.

Quantitation of faecal fat

A quantitative chemical determination of faecal fat over at least a 72-hour period is the most reliable measure of steatorrhoea. A number of conditions should be met in order to obtain an accurate and meaningful determination. Faecal excretion of fat is related to intake, so a dietary intake of fat sufficient to stress the absorptive capacity of the gastrointestinal tract is necessary. In infants this should be at least 5 g/kg per day, in children (1–5 years) 40–50 g/day and older children (6–12 years) 50–80 g/day. Regular bowel movement should be ensured and the stool collection must be complete. The stools are pooled and homogenised and the long-chain fatty acid content is then most commonly analysed by the method of Van de Kamer *et al.* (1949). The extraction procedure should be modified if medium chain fatty acids are a significant proportion of the dietary fat (Leyland *et al.* 1969). Results are expressed as either absolute values of stool fat per day with an upper limit of normal of 5 g/day or as a coefficient (percentage) of fat absorption, which is preferable as fat excretion varies with age. The following are normal limits for varying age groups:

Preterm infant	> 60%
Term infants	> 80%
1–6 months	> 85%
6 months–2 years	> 90%
> 2 years	> 95%

For metabolic balance studies some form of continuous marker such as chromium dioxide or radio-opaque markers is essential, but since a stabilisation period of 5–10 days may be required this is not practical for routine clinical practice. The use of carmine or edecol blue markers at the beginning and end of the period of study results in a more accurate collection than merely collecting over a fixed time but errors are still large compared with continuous marker techniques (Roce 1964).

Screening tests

A variety of screening tests have been advocated for the assessment of fat absorption. These include measurement of serum turbidity or triglyceride concentrations following an oral fat load, vitamin A absorption tests, the measurement of serum carotene and cholesterol concentrations and stool microscopy. None of these methods are reliable, although if a screening test is used, stool microscopy is simple and rapid in experienced hands.

Breath tests

Radioactively-labelled triglycerides and more recently triglycerides labelled with stable isotopes have been used to assess fat absorption in adults. Some of the absorbed triglyceride is metabolised to isotopically labelled carbon dioxide, which is then expired in breath, so giving an estimate of absorption. The disadvantages of this approach are that cooperation of the patient is essential and therefore it cannot be used in infants and young children; there are ethical problems in giving radioactively-labelled substrates to children; the costs of the instrumentation for the quantitative measurement of stable isotopes are high; and the method is dependent on normal gastric emptying and intermediary metabolism.

Fat-soluble vitamins

An important clinical result of fat malabsorption is the malabsorption of the fat-soluble vitamins A, D, E and K. Methods are available for the direct estimation of these vitamins in serum, but abnormal clotting tests (prothrombin and partial thromboplastin times) with a normal fibrinogen are more routinely used to indicate a deficiency of vitamin K.

Hyperoxaluria

A possible consequence of fat malabsorption is a tendency to develop hyperoxaluria and to form calcium oxalate stones (Ogilvie *et al.* 1976). The cause of this hyperoxaluria appears to be increased colonic absorption of oxalate (Chadwick *et al.* 1973; Dobbins & Binder 1977). This probably results from unabsorbed fatty acids and bile salts causing a non-specific increase in colonic permeability (Dobbins & Binder 1976) and/or an increase in oxalate solubility as a result of fatty acids forming calcium soaps and thus preventing the luminal formation of insoluble calcium oxalate. Ogilvie *et al.* (1976) have shown that treatment of the underlying condition is accompanied by a return of urinary oxalate concentrations to normal.

Causes of fat malabsorption

Once fat malabsorption has been suspected it is necessary to define the cause. Suspected coeliac disease is confirmed by peroral jejunal biopsy. Luminal events can be investigated by pancreaticobiliary stimulation and duodenal intubation, as described earlier. For normal fat absorption there must be adequate lipolytic activity and solubilisation of the lipolytic products and it is, therefore, necessary to assay the activity of pancreatic lipase, the concentration of bile salts and measure the luminal pH. A reduced pH is able to denature pancreatin and lipase and precipitate the bile salts. Co-lipase has recently been recognised as an essential cofactor for lipase activity (Borgstrom *et al.* 1979), and two siblings with suspected co-lipase deficiency have been described (Hildebrand *et al.* 1982).

The commonest cause of pancreatic insufficiency and a deficiency of pancreatic lipase in childhood is cystic fibrosis. Other causes of very low or absent pancreatic lipase activities are congenital pancreatic hypoplasia (Shwachman's disease) and the very rare condition of isolated lipase deficiency. The differential diagnosis of these three conditions can be made on clinical and biochemical grounds. Patients with cystic fibrosis have abnormal sweat electrolytes. The finding of a sweat sodium concentration of > 70 mmol/l on at least two separate collections of more than 100 mg of sweat is consistent with the diagnosis of cystic fibrosis. Borderline sweat electrolytes (sodium concentration of 50–70 mmol/l) present a difficult clinical problem, requiring a careful clinical and family history and detailed pulmonary and pancreatic function studies. Patients with congenital pancreatic hypoplasia have a number of non-pancreatic features (Aggett *et al.* 1980); those with an isolated deficiency of pancreatic lipase have a characteristic leakage of oil (Sheldon 1964) while the activities of the other pancreatic enzymes are either normal or only moderately reduced.

Reduced luminal concentrations of bile salts can arise for a number of reasons, as discussed earlier.

In abetalipoproteinaemia fat malabsorption results from a defect in the transport of lipid away from the mucosal cell (Herbert *et al.* 1978). The primary abnormality is a failure to synthesise apoprotein B and as a result chylomicrons are not formed and fat cannot be transported via the lymphatic route. The appearance of a small intestinal biopsy is characteristic with an accumulation of lipid in the mucosal cells which is particularly evident at the villus tip (Muller 1982). The diagnosis can be made by documenting the absence of low density lipoproteins, very low density lipoproteins and chylomicrons on lipoprotein electrophoresis, and by the presence of abnormal spiky red cells (acanthocytes) on a wet blood film.

Patients with intestinal lymphangiectasia have a mild fat malabsorption as a result of impaired lymphatic transport of lipid. The major problem is, however, an excessive loss of protein into the gut lumen and this will be discussed when considering protein-losing enteropathies.

Carbohydrate

Tests of carbohydrate absorption are indicated in children with chronic diarrhoea and in those who fail to thrive or have persistent diarrhoea following an episode of gastroenteritis. Although carbohydrate malabsorption can result from a derangement of any part of the digestive or absorptive process, in practice it is principally mucosal abnormalities of hydrolysis or transport that result in clinical disease. The disorders may be genetic or acquired, temporary or permanent (Table 6.4), and with the exception of late-onset lactose intolerance, which affects older children and adults, the majority of severe clinical problems arise in infancy.

Significant carbohydrate malabsorption in the small intestine results in an increased osmotic load and a movement of water and electrolytes into the lumen of the gut until an equilibrium is reached. A proportion of the carbohydrate may be excreted unchanged in the stools but the majority is partially metabolized by colonic bacteria to short-chain organic acids,

Table 6.4 Clinical disorders of carbohydrate absorption.

Primary disorders (see Harries & Muller 1980)
Lactase deficiency
Sucrase/isomaltase deficiency
Trehalase deficiency
Congenital glucose–galactose malabsorption
Acquired disorder
Late onset lactase deficiency
Secondary disorders (see McNeish & Harran 1980)
Disaccharides
Gastroenteritis
Cows' milk protein intolerance
Malnutrition
Coeliac disease
Tropical sprue
Surgery to gastrointestinal tract
Phototherapy
Immune deficiency syndromes
Inflammatory bowel disease
Giardiasis
Monosaccharides
Gastroenteritis
Malnutrition
Necrotising enterocolitis

hydrogen, carbon dioxide and other fermentation products. The organic acids lower the faecal pH, inhibit water absorption from the colon and cause osmotic diarrhoea (Cummins 1981). Tests are now described to document carbohydrate malabsorption and to establish its cause.

Examination of faeces

Stools of patients with carbohydrate malabsorption are characterised by being liquid, frothy and acidic and as a result the perianal region is typically excoriated. A simple pH determination has not been found to be a reliable screening test for sugar malabsorption (Soeparto *et al.* 1972) and its use should be discouraged.

The most useful screening procedure is to test the faeces for reducing substances, as described by Kerry and Anderson (1964). Five drops of a freshly collected aliquot of liquid stool are diluted with 10 drops of water and a Clinitest tablet added. The presence of more than 0.5% of reducing substances is abnormal although 1–2% would be a typical finding in situations of clinical malabsorption of carbohydrate. If sucrose malabsorption is suspected, the faeces should be boiled for 2 minutes with dilute hydrochloric acid before testing, in order to hydrolyse the sucrose, which is not itself a reducing sugar. The separation of the faecal carbohydrates by thin-layer chromatography will give a qualitative and semi-quantitative description of the carbohydrates present and can be helpful in reaching a specific diagnosis. For example, lactose, galactose and glucose may be demonstrated in lactase deficiency whereas only glucose and galactose are present in glucose–galactose malabsorption.

Carbohydrate loading tests

The observed response to an oral load of a mono- or disaccharide is a standard test of carbohydrate absorption. As in the investigation of fat malabsorption the ability of the intestine to absorb a given sugar should be examined after an adequate dietary intake. This can be done by gradually increasing the intake to 4–6% over a period of days or by a single load of the sugar under investigation. After a fast (the length being dependent on the age and clinical state of the patient) the sugar is given in a dose of 1–2 g/kg body weight (maximum of 50 g) as a 10% solution together with a carmine marker. Body weight, stool frequency and volume are carefully recorded prior to and throughout the test, and the stools containing the marker are tested for reducing substances and individual sugars. Urine may also be collected for 4 hours following the oral load and tested for sugars, as in disaccharide malabsorption small amounts are absorbed and excreted intact in the urine. Serial blood glucose concentrations can also be determined. These normally increase by at least 1.7 mmol/l whereas in sugar intolerance the increase is $<$ 1.1 mmol/l. The measurement of blood levels has, however, no obvious advantage over the testing of faeces and is also dependent on the rate of gastric emptying.

Assay of mucosal enzymes

The assay of disaccharidase activity in small intestinal mucosa obtained by biopsy is in general an unnecessary procedure for the diagnosis and clinical management of patients with sugar intolerance. The only firm indication for disaccharidase estimations is when a primary genetic defect is suspected. A careful clinical history will almost invariably differentiate such a patient from those with the much commoner secondary malabsorptive disorders. For all mucosal enzyme disorders it is essential that the biopsy specimen should be appropriately handled. For disaccharidase estimations, the biopsy should be snap frozen, stored at -20 °C prior to analysis, and the disaccharidase activity is then typically estimated by the enzymatic method of Dahlqvist (1968). As a control several disaccharidase activities should be determined so that results may be expressed as ratios of two activities.

Breath tests

Carbon dioxide. If an oral load of a sugar containing radiolabelled ^{14}C is administered and $^{14}CO_2$ is recovered in the breath, then a proportion of the sugar has been absorbed and metabolised. This principle has been used to assess the absorption of ^{14}C-lactose (Sasaki *et al.* 1970) and a good correlation has been obtained between breath ^{14}C and lactase activity. More recently ^{13}C-labelled (non-radioactive) isotopes have been used. These methods are, however, indirect and have the disadvantages inherent in all breath tests as discussed on p. 161.

Hydrogen. The amount of hydrogen excreted in breath correlates well with the amount that is produced in the intestine by fermentation of unabsorbed carbohydrate and has been used to quantitate carbohydrate absorption (Bond & Levitt 1972). It does not involve the use of radioactive substrates but has the other disadvantages inherent in breath tests, particularly in the paediatric age group.

Perfusion studies

Perfusion studies have been used to answer specific questions related to carbohydrate absorption (Milla *et al.* 1978) but this technique provides a powerful research tool rather than a routine diagnostic procedure.

Dietary exclusion

In practice many infants with carbohydrate intolerance may be too ill for detailed investigation or for formal oral challenges with carbohydrate. Thus an empirical diagnosis based on a detailed history, clinical examination and examination of faeces will lead to an exclusion diet. The diagnosis can then be confirmed at a later date after clinical recovery, although this may not be possible in many secondary disorders when a spontaneous recovery may occur during the initial period of empirical treatment.

Protein

Total faecal nitrogen may be used to give the coefficient of nitrogen absorption. This provides information on net absorption but does not discriminate between nitrogen loss into the lumen and malabsorption. In general, an estimation of faecal nitrogen is less useful clinically than faecal fat and is rarely measured. A few specific investigations relating to protein and amino acid absorption will now be considered.

Investigation of protein-losing enteropathy

In protein-losing enteropathies, the protein loss is non-selective so that the plasma protein pattern reflects the different synthetic rates of the proteins. Thus concentrations of albumin and immunoglobulins are reduced to the greatest extent, followed by transferrin and caeruloplasmin, while fibrinogen may be normal. Peripheral lymphocyte counts are generally reduced. A pattern reflecting these differences in the absence of severe liver disease or renal protein loss is strongly suggestive of the diagnosis.

A number of radioactive substances have been used to measure the magnitude of the gastrointestinal loss of protein in both normal and pathological states. Details of the various methods together with their relative merits have been described by Waldmann (1972). The most widely employed methods use either ^{51}Cr-labelled albumin or blood mixed directly with ^{51}Cr-chloride, as chromium is not reabsorbed or secreted by the gastrointestinal tract. Twenty-five μCi of ^{51}Cr-albumin is given intravenously or 25 μCi of ^{51}Cr chloride is mixed with 10 ml of the patient's blood and reinjected, and a 4-day stool collection made. Since most of the ^{51}Cr is excreted in the urine, particularly if the chromium chloride method is used, it is essential to be certain that the stool collection is entirely free of urine. If there are a reasonable number of stools in the collection a single contaminated specimen can be discarded as the results will usually remain unequivocal. In patients with abnormal protein loss 4–40% is excreted in a 4-day collection compared to 0.1–0.7% in normal controls.

More sophisticated studies of rates of albumin turnover and synthesis may be carried out by combining the test with a second label on the albumin such as ^{131}iodine (Walker *et al.* 1973).

Proteolytic enzymes

The digestion of protein is initiated in the stomach by pepsin and then continued in the small intestine by the proteolytic enzymes secreted by the pancreas. These pancreatic enzymes are stored and secreted as inactive pro-enzymes or zymogens and are only activated on arrival at the duodenum. The key enzyme in this activation is enterokinase which converts trypsinogen to trypsin, the trypsin in turn activating the other zymogens. Absence of trypsinogen (Townes *et al.* 1967) and enterokinase (Tarlow *et al.* 1970) has been reported but the relationship between the two disorders has not been fully elucidated. In the report of trypsinogen deficiency an absence of trypsin, chymotrypsin and carboxypeptidase activities in duodenal juice could be corrected by the addition of exogenous trypsin. In cases of enterokinase deficiency the enzyme is undetectable in both the mucosa and duodenal juice and the addition of exogenous enterokinase to the duodenal juice, or incubation of the juice with normal duodenal juice, or incubation of the juice with normal duodenal mucosa results in activation of all the proteolytic enzymes.

Amino acid absorption tests

The selective inborn errors of amino acid absorption are listed in Table 6.5. Deficient amino acid transport is usually expressed in both the gut and kidney, so that patients both malabsorb and leak the relevant amino acids. Gastrointestinal tolerance tests of specific amino acids have been used, but have limited clinical application, since measurement of plasma and urinary amino acids and calculation of renal clearance rates give more information. Patients with Hartnup disease also excrete indicans in the urine as a result of bacterial metabolism of unabsorbed tryptophan in the colon.

Table 6.5 Selective inborn errors of amino acid absorption (see Silk 1972).

Disease	Amino acids involved
Iminoglycinuria	Glycine, proline, hydroxyproline
Hartnup disease	Neutral amino acids
Cystinuria	Dibasic amino acids, cystine
Lysinuric protein intolerance	Lysine, arginine, ornithine
Blue diaper syndrome	Tryptophan
Oast house syndrome	Methionine
Lowe's syndrome	Arginine, lysine (partial)

Fluid and electrolytes

Abnormalities of fluid and electrolyte absorption in the small and/or large intestine result in malabsorption or secretion and therefore in diarrhoea. A simple measurement of stool volume and electrolytes can provide important information in assessing the course and response to therapy of children with severe and protracted diarrhoeal states. As discussed previously congenital chloridorrhoea is the only condition where the measurement of stool electrolytes is diagnostic.

In patients with severe protracted diarrhoea (Candy *et al.* 1981) a detailed investigation of water and electrolyte transport in the jejunum by an *in vivo* perfusion technique (Milla *et al.* 1978) or in the large intestine or rectum using dialysis sacs or measurement of the potential difference (as discussed previously) may help to establish the site of the transport abnormality and determine the approach to therapy.

Haematinics

Tests which assess the capacity of the small intestine to absorb haematinic substances such as iron, folic acid and vitamin B_{12} provide important information with regard to the presence and site of gastrointestinal disease. Defective iron and folate absorption occur in diseases of the proximal small intestine, whereas vitamin B_{12} malabsorption is associated with lesions of the terminal ileum. Malabsorption of vitamin B_{12} may, however, also occur in situations of bacterial overgrowth of the proximal small intestine, in pancreatic disease, and in conditions where there is impaired secretion of gastric intrinsic factor (see below).

Iron and folate

Specific tests for iron and folate malabsorption are not usually undertaken although the reticulocyte response to oral, as compared to systemic, treatment may be compared. In the absence of blood loss, malabsorption of iron is presumed to exist in the presence of a microcytic anaemia with a low serum iron and high total iron binding capacity. Similarly, folate malabsorption is presumed if serum and red cell folate levels are low in the absence of conditions causing folate deficiency such as malignancy or therapy with anticonvulsants or folic acid antagonists. A rare disorder of congenital folate malabsorption has been described (Lanzkowsky 1970) and was associated with mental retardation, ataxia and convulsions.

Vitamin B_{12}

For absorption to take place, vitamin B_{12} must first be attached to a specific carrier glycoprotein (gastric intrinsic factor) which is derived from the gastric parietal cell. This complex is recognised by a specific receptor in the ileum and once the B_{12} has entered the enterocyte it complexes with another carrier protein, transcobalmin II, which transports it via the systemic circulation to the liver and other tissues.

Normal pancreatic (Allen *et al.* 1978) and biliary function (Teo *et al.* 1980) are important for the efficient absorption of vitamin B_{12} and abnormalities of them must be excluded as causes for its malabsorption (as should bacterial overgrowth of the intestine) before specific defects of the intrinsic factor, of ileal uptake and transport are investigated. The Schilling test will demonstrate malabsorption of the vitamin and together with other information such as the presence of antibodies, measurement of gastric acid, etc. will identify the specific defect. The Schilling test is carried out using the 'Dicopac' dual isotope test kit supplied by Amersham International Limited. The radiolabelled cyanocobalamin capsules (^{58}Co–B_{12} and ^{57}Co–B_{12} bound to gastric intrinsic factor) are given orally after a 2-hour fast. After a further 2-hour fast, 1 mg of non-radioactive B_{12} is given intramuscularly to ensure saturation of body stores so that significant amounts of the labelled B_{12} are excreted in the urine following absorption. Normally $> 10\%$ of both labelled preparations are excreted in the urine during the 24-hour period following ingestion. If the percentage excretion of one or both of the isotopes is less than 10% the ratio of the percentage excretion of the two preparations should be computed. A ratio of ^{57}Co to ^{58}Co of > 2 suggests a deficiency of active intrinsic factor whereas a ratio near unity suggests other causes of malabsorption such as ileal disease, specific B_{12} malabsorption or a transport defect. A congenital deficiency of intrinsic factor (Arthur 1972) is differentiated from pernicious anaemia by demonstrating normal gastric acid secretion, normal gastric mucosal histology and absence of antibodies to parietal cells and intrinsic factor. A biologically inert but immunologically active intrinsic factor has been reported by Katz *et al.* (1972). This latter diagnosis may be made if the malabsorption is corrected by exogenous intrinsic factor even though the assay for intrinsic factor is normal. The diagnosis of transcobalamin II deficiency depends on the demonstration of the absent protein in serum by techniques such as gel filtration (Burmann *et al.* 1979).

Minerals

A number of conditions associated with malabsorption of minerals have been described.

Zinc. Moynahan and Barnes (1973) were the first to suggest that the condition of acrodermatitis enteropathica resulted from a deficiency of zinc. A number of subsequent studies (Atherton *et al.* 1979) have provided evidence that the deficiency results from an impairment of the normal absorptive process. The diagnosis should be considered on clinical grounds (severe skin lesions, alopecia and diarrhoea) and confirmed by a greatly reduced serum zinc concentration, and by reduced activities of alkaline

Table 6.6 Conditions which can result in increased plasma concentrations of gastrointestinal hormones and the clinical effects produced.

Hormone	Condition resulting in increased plasma concentration	Clinical effect
Gastrin	Gastrinoma—Zollinger–Ellison syndrome	↑ gastric acidity, duodenal ulcers
	Isolated retained antrum	↑ gastric acidity, duodenal ulcers
	Intestinal resection	Peptic ulcers, ↑ malabsorption
	Pernicious anaemia	None known
	Atrophic gastritis	None known
Vasoactive intestinal polypeptide (VIP)	Pancreatic tumours (VIP-omas)	Watery diarrhoea, hypokalaemia, ↓ gastric acidity (Verner–Morrison syndrome)
	Some ganglioneuromas and neuroblastomas	↓ gastric acidity (Verner–Morrison syndrome)
Pancreatic polypeptide	High proportion of pancreatic endocrine tumours	None known
	Renal failure	None known
Gastric inhibitory polypeptide	Uraemia, duodenal ulcer, diabetes, obesity	None known
Somatostatin	Somatostatinoma	Diabetes, gall stone, steatorrhoea, ↓ gastric acidity

phosphatase, which is a zinc-dependent enzyme. There is a substantial risk of sample contamination and great care must be taken when estimating serum zinc concentrations. Plastic-ware should be used in the collection and processing of samples, and rubber stoppers must be avoided. The use of zinc oxide ointment on the skin lesions is a potent potential source of contamination and zinc estimations must not be performed on haemolysed specimens. Once a deficiency of zinc has been established it is necessary to distinguish acrodermatitis enteropathica from acquired zinc deficiency states and a rare condition of defective zinc secretion in breast milk (Hambidge & Walravers 1982).

Magnesium. Primary hypomagnesaemia is a rare disorder which presents in early infancy with convulsions and tetany. The primary abnormality appears to be impaired absorption of magnesium (Milla *et al.* 1979). The diagnosis depends on the demonstration of severe hypomagnesaemia and hypocalcaemia in the absence of other causes for secondary magnesium deficiency and thereafter on the clinical and biochemical response to magnesium therapy.

Copper. Impaired absorption of copper has been demonstrated in Menkes' disease (Danks *et al.* 1972) and was further defined by this group who reported an increased concentration of copper in duodenal mucosal cells. This was suggestive of normal uptake, but impaired transport away from the mucosal cells (Danks *et al.* 1973). The diagnosis is based on the finding of low plasma copper and caeruloplasmin (copper containing glycoprotein) concentrations and the microscopic demonstration of pili torti (twisting of the hair shaft around its axis).
Phosphate. In familial hypophosphataemic rickets, absorption of phosphate and calcium is reduced (Rasmussen & Anast 1978), although the principal abnormality is impaired renal tubular reabsorption of phosphate. Hypophosphataemia in patients with rickets or osteomalacia is suggestive of the condition. Serum alkaline phosphatase activity is also increased.

Gastrointestinal hormones

This subject has been reviewed by Creutzfeldt (1980).

The estimation of an ever-increasing number of gastrointestinal hormones by radioimmunoassay has become important for the assessment of gastrointestinal function. Table 6.6 shows some of the clinical effects of increased concentrations of certain gastrointestinal hormones. An absence of these hormones has not yet been described. Blood samples are usually analysed from fasting patients and must be placed immediately in containers with a proteolytic inhibitor such as trasylol to prevent their degradation.

REFERENCES

Aggett P. J., Cavanagh N. P. C., Matthew D. J., Pincott J. R., Sutcliffe J. & Harries J. T. (1980) Shwachman's syndrome. A review of 21 cases. *Arch. Dis. Child.* **55**, 331–47.

Agunod M., Yamaguchi N., Lopez R., Lubey A. L. & Glass G. B. J. (1969) Correlative study of hydrochloric acid, pepsin and intrinsic factor secretion in newborns and infants. *Am. J. Digest. Dis.* **14**, 400–14.

Allen R. H., Seetharam B., Podell E. & Alpers D. H. (1978) Effect of proteolytic enzymes on the binding of cobalamin to R protein and intrinsic factor. *J. Clin. Invest.* **61**, 47–54.

Arthur L. J. H. (1972) Juvenile pernicious anaemia. *Proc. Roy. Soc. Med.* **65**, 728–9.

Atherton D. J., Muller D. P. R., Aggett P. J. & Harries J. T. (1979) A defect in zinc uptake by jejunal biopsies in acrodermatitis enteropathica. *Clin. Sci.* **56**, 505–7.

Avery G. B., Randolph J. G. & Weaver T. (1966a) Gastric acidity on the first day of life. *Pediatrics* **37**, 1005–7.

Avery G. B., Randolph J. G. & Weaver T. (1966b) Gastric response to specific disease in infants. *Pediatrics* **38**, 874–8.

Baron J. H. (1970) The clinical use of gastric function tests. *Scand. J. Gastroenterol.* **5** (Suppl 6), 9–46.

de Benneville P. L., Godfrey W. J., Sims H. J. & Imondi A. R. (1972) New substrates for a pancreatic exocrine function test. *J. Med. Chem.* **15**, 1098–1100.

BOND J. H. & LEVITT M. D. (1972) Use of pulmonary hydrogen (H_2) measurements to quantitate carbohydrate absorption. *J. Clin. Invest.* **51**, 1219–25.

BORGSTROM B., ERLANSON-ALBERTSSON C. & WIELOCH T. (1979) Pancreatic co-lipase: chemistry and physiology. *J. Lipid Res.* **20**, 805–16.

BURMANN J. F., MOLLIN D. L., SOURIAL N. A. & SLADDEN R. A. (1979) Inherited lack of transcobalamin II in serum megaloblastic anaemia: a further patient. *Br. J. Haematol.* **43**, 27–38.

CAMERON D. J. S., MULLER D. P. R. & HARRIES J. T. (1979) The validation and application of a technique for simultaneously determining gastric emptying and secretion after a test meal in children. *Acta Paediatr. Belgique* **32**, 156–7.

CANDY D. C. A., LARCHER V. F., CAMERON D. J. S., NORMAN A. P., TRIPP J. H., MILLA P. J., PINCOTT J. R. & HARRIES J. T. (1981) Lethal familial protracted diarrhoea. *Arch. Dis. Child.* **56**, 15–23.

CAVELL B. (1979) Gastric emptying in preterm infants. *Acta Paediatrica Scand.* **68**, 725–30.

CHADWICK V. S., MODHA K. & DOWLING H. (1973) Mechanism for hyperoxaluria in patients with ileal dysfunction. *New Engl. J. Med.* **289**, 172–6.

CHRISTIE D. L. & AMENT M. E. (1976) Gastric acid hypersecretion in children with duodenal ulcer. *Gastroenterology* **71**, 242–4.

CLAYTON P. T. & MULLER D. P. R. (1980) A simplified gas–liquid chromatographic method for the estimation of non-sulphated plasma bile acids. *Clin. Chim. Acta* **105**, 401–5.

CREUTZFELDT W. (1980) Gastrointestinal Hormones. *Clin. Gastroenterol.* **9**, No. 3.

CUMMINS J. H. (1981) Short chain fatty acids in the human colon. *Gut* **22**, 763–79.

DAHLQVIST A. (1968) Assay of intestinal disaccharidases. *Analyt. Biochem.* **22**, 99–107.

DANKS D. M., CAMPBELL P. E., WALKER-SMITH J., STEVENS B. J., GILLESPIE J. M., BLOMFIELD J. & TURNER B. (1972) Menkes' kinky-hair syndrome. *Lancet* **i**, 1100–3.

DANKS D. M., CARTWRIGHT E., STEVENS B. J. & TOWNLEY R. W. (1973) Menkes' kinky-hair disease: Further definition of the defect in copper transport. *Science* **179**, 1140–2.

DOBBINS J. W. & BINDER H. J. (1976) Effect of bile salts and fatty acids on the colonic absorption of oxalate. *Gastroenterology* **70**, 1096–1100.

DOBBINS J. W. & BINDER H. J. (1977) Importance of the colon in enteric hyperoxaluria. *New Engl. J. Med.* **296**, 298–301.

DURIE P. R., LARGMAN C., BRODRICK J. W., JOHNSON J. H., GASKIN K. J., FORSTNER G. G. & GEOKAS M. C. (1981) Plasma immunoreactive pancreatic cationic trypsinogen in cystic fibrosis: a sensitive indicator of exocrine pancreatic dysfunction. *Pediat. Res.* **15**, 1351–5.

EDMONDS C. J. (1971) Absorption of sodium and water by human rectum measured by a dialysis method. *Gut* **12**, 356–62.

EDMONDS C. J. & GODFREY R. C. (1970) Measurement of electrical potentials of the human rectum and pelvic colon in normal and aldosterone-treated patients. *Gut* **11**, 330–7.

ERLANSON C. & BORGSTROM B. (1970) Tributyrine as a substrate for determination of lipase activity of pancreatic juice and small intestinal content. *Scand. J. Gastroenterol.* **5**, 293–5.

FRANK B. F. & KERN F. (1967) Menetrier's disease. *Gastroenterology* **53**, 953–60.

FROMM H. & HOFMANN A. F. (1971) Breath test for altered bile acid metabolism. *Lancet* **ii**, 621–5.

GREENBERGER N. J., SAEGH S. & RUPPERT R. D. (1968) Urine indican excretion in malabsorption disorders. *Gastroenterology* **55**, 204–11.

HADORN B. (1975a) The exocrine pancreas. In Anderson C. M. & Burke V. (eds.) *Paediatric Gastroenterology*, pp. 289–327. Blackwell Scientific Publications, Oxford.

HADORN B. (1975b) Assessment of pancreatic function. In Anderson C. M. & Burke V. (eds.) *Paediatric Gastroenterology*, pp. 661–6. Blackwell Scientific Publications, Oxford.

HAMBIDGE K. M. & WALRAVERS P. A. (1982) Disorders of mineral metabolism. *Clin. Gastroenterol.* **11**, No. 1, 87–117.

HARRIES J. T. & MULLER D. P. R. (1980) Basic causes of carbohydrate malabsorption. In Burman D. & Holton J. B. (eds.) *Inherited Disorders of Carbohydrate Metabolism*, pp. 21–37, MTP Press, Lancaster.

HEELEY A. F., HEELEY M. E., KING D. N., KUZEMKO J. A. & WALSH M. P. (1982) Screening for cystic fibrosis by dried blood spot trypsin assay. *Arch. Dis. Child.* **57**, 18–21.

HERBERT P. N., GOTTO A. M. & FREDRICKSON D. S. (1978) Familial lipoprotein deficiency. In Stanbury J. B., Wyngaarden J. B. & Fredrickson D. S. *The Metabolic Basis of Inherited Disease*, 4th Edition, pp. 544–88. McGraw-Hill, New York.

HILDEBRAND H., BORGSTROM B., BEKASSY A., ERLANSON-ALBERTSSON C. & HELIN I. (1982) Isolated co-lipase deficiency in two brothers. *Gut* **23**, 243–6.

HOFMANN A. F. (1977) The enterohepatic circulation of bile acids in man. *Clin. Gastroentero.* **6**, No. 1, 3–24.

IWATA T. & YAMASAKI K. (1964) Enzymatic determination and thin layer chromatography of bile acids in blood. *J. Biochem. (Tokyo)* **56**, 429–31.

KATZ M., LEE S. K. & COOPER B. A. (1972) Vitamin B12 malabsorption due to a biologically inert intrinsic factor. *New Engl. J. Med.* **287**, 425–9.

KERRY K. R. & ANDERSON C. M. (1964) A ward test for sugar in faeces. *Lancet* **i**, 981–2.

KREJS G. J., HENDLER R. S. & FORDTRAN J. S. (1980) Diagnostic and pathophysiologic studies in patients with chronic diarrhoea. In Field M., Fordtran J. S., Schultz S. G. *Secretory Diarrhoea*, pp. 67–83. Clinical Physiology Series; American Physiological Society, Bethesda, Maryland.

LAKE-BAKAAR G., MCKAVANAGH S., REDSHAW M., WOOD T., SUMMERFIELD J. A. & ELIAS E. (1979) Serum immunoreactive trypsin concentration after a Lundh meal. *J. Clin. Path.* **32**, 1003–8.

LAMABADUSURIYA S. P., PACKER S. & HARRIES J. T. (1975) Limitations of xylose tolerance test as a screening procedure in childhood coeliac disease. *Arch. Dis. Child.* **50**, 34–9.

LANZKOWSKY P. (1970) Congenital malabsorption of folate. *Am. J. Med.* **48**, 580–3.

LAUNIALA K., PERHEENTUPA J., PASTERNACK A. & HALLMAN N. (1968) Familial chloride diarrhoea–chloride malabsorption. *Modern Problems in Pediatrics* **11**, 137–49.

LEYLAND F. C., FOSBROOKE A. S., LLOYD J. K., SEGALL M. M., TAMIR I., TOMKINS R. & WOLFF O. H. (1969) Use of medium-chain tryglyceride diets in children with malabsorption. *Arch. Dis. Child.* **44**, 170–86.

MAKINO I. & SJOVALL J. (1972) A versatile method for analysis of bile acids in plasma. *Analyt. Lett.* **5**, 341–9.

MCCOLLUM J. P. K., MULLER D. P. R. & HARRIES J. T. (1977) Test meal for assessing intraluminal phase of absorption in childhood. *Arch. Dis. Child.* **52**, 887–9.

MCNEISH A. S. & HARRAN M. J. (1980) Clinical aspects of disordered carbohydrate metabolism. In Burman D., Holton J. B. & Pennock C. A. (eds.) *Inherited Disorders of Carbohydrate Metabolism*, pp. 39–57. MTP Press Ltd, Lancaster.

MILLA P. J. (1982) Disorders of electrolyte absorption. *Clin. Gastroenterol.* **11**, No. 1, 31–46.

MILLA P. J., AGGETT P. J., WOLFF O. H. & HARRIES J. T. (1979) Studies in primary hypomagnesaemia: evidence for defective carrier-mediated small intestinal transport of magnesium. *Gut* **20**, 1029–33.

MILLA P. J., ATHERTON D. A., LEONARD J. V., WOLFF O. H. & LAKE B. D. (1978) Disordered intestinal function in glycogen storage disease. *J. Inherit. Metab. Dis.* **1**, 155–7.

MITCHELL C. J., FIELD H. P., SIMPSON F. G., PARKIN A., KELLEHER J. & LOSOWSKY M. S. (1981) Preliminary evaluation of a single-day tubeless test of pancreatic function. *Br. Med. J.* **282**, 1751–3.

MITCHELL C. J., HUMPHREY C. S., BULLEN A. W., KELLEHER J. & LOSOWSKY M. S. (1979) Improved diagnostic accuracy of a modified oral pancreatic function test. *Scand. J. Gastroenterol.* **14**, 737–41.

MOYNAHAN E. J. & BARNES P. M. (1973) Zinc deficiency and a synthetic diet for lactose intolerance. *Lancet* **i**, 676–7.

MULLER D. P. R. (1982) Disorders of lipid absorption. *Clin. Gastroenterol.* **11**, No. 1, 119–40.

MULLER D. P. R. & GHALE G. K. (1982) The stability of pancreatic enzyme activities in duodenal juice following pancreatic stimulation by a test meal or exogenous hormones. *Ann. Clin. Biochem.* **19**, 89–93.

OGILVIE D., MCCOLLUM J. P. K., PACKER S., MANNING J., OYESIKU J., MULLER D. P. R. & HARRIES J. T. (1976) Urinary outputs of oxalate, calcium and magnesium in children with intestinal disorders. *Arch. Dis. Child.* **51**, 790–5.

RASMUSSEN H. & ANAST C. (1978) Familial hypophosphatemic (vitamin D-resistant) rickets and vitamin D-dependent rickets. In Stanbury J. B., Wyngaarden J. B. & Fredrickson D. S. *The Metabolic Basis of Inherited Disease*, pp. 1537–55. McGraw-Hill, New York.

ROCE G. A. (1964) Experience with the use of interrupted carmine red and continuous chromium sesquioxide marking with reference to human calcium, phosphorus and magnesium. *Gut* **5**, 274–9.

ROLLES C. J., KENDALL M. J., NUTTER S. & ANDERSON C. M. (1973) One-hour blood xylose screening test for coeliac disease in infants and children. *Lancet* **ii**, 1043–5.

RUDDELL W. S. J., MITCHELL C. J., HAMILTON I., LEEK J. P. & KELLEHER J. (1981) Clinical value of serum immunoreactive trypsin concentration. *Br. Med. J.* **283**, 1429–32.

SASAKI Y., IIO M., KAMEDA H., UEDA H., AOYAGI T., CHRISTOPHER N. L., BAYLESS T. M. & WAGNER H. N. (1970) Measurement of ^{14}C-lactose absorption in the diagnosis of lactase deficiency. *J. Lab. Clin. Med.* **76**, 824–34.

SCHWARTZ D. L., WHITE J. J., SAULSBURY F. & HALLER J. R. (1974) Gastrin response to calcium infusion: an aid to improved diagnosis of Zollinger–Ellison syndrome in children. *Pediatrics* **54**, 599–602.

SHELDON W. (1964) Congenital pancreatic lipase deficiency. *Arch. Dis. Child.* **39**, 268–71.

SILK D. B. A. (1972) Disorders of nitrogen absorption. *Clin. Gastroenterol.* **11**, No. 1, 47–72.

SIMMONDS W. J., KORMAN M. G., GO V. L. W. & HOFMANN A. F. (1973) Radioimmunoassay of conjugated cholyl bile acids in serum. *Gastroenterology* **65**, 705–11.

SKEGGS L. T. & HOCHSTRASSER H. (1964) Multiple automatic sequential analysis. *Clin. Chem.* **10**, 918–36.

SOEPARTO P., STOBO E. A. & WALKER-SMITH J. A. (1972) Role of chemical examination of the stool in the diagnosis of sugar malabsorption in childhood. *Arch. Dis. Child.* **47**, 56–61.

STERCHI E. E., GREEN J. R. & LENTZ E. (1981) Hydrolysis of N-benzoyl-L-tyrosyl-p-aminobenzoic acid by human small intestine. *Pediatric Res.* **15**, 1203.

Strauss E., Gerson C. D. & Yalow R. S. (1974) Hypersecretion of gastrin associated with short bowel syndrome. *Gastroenterology* **66**, 175–80.

Tarlow M. J., Hadorn B., Arthurton M. W. & Lloyd J. K. (1970) Intestinal enterokinase deficiency. *Arch. Dis. Child.* **45**, 651–5.

Teo N. H., Scott J. M., Neale G. & Weir D. G. (1980) Effect of bile on vitamin B12 absorption. *Br. Med. J.* **281**, 831–3.

Townes P. L., Bryson M. F. & Miller G. (1967) Further observations on trypsinogen deficiency disease: Report of a second case. *J. Pediatr.* **71**, 220–4.

Van de Kamer J. H., Huinink H. T. B. & Weyers H. A. (1949) Rapid method for the determination of fat in feces. *J. Biol. Chem.* **177**, 347–55.

Verner J. V. & Morrison A. B. (1958) Islet cell tumour and a syndrome of refractory water diarrhoea and hypokalaemia. *Am. J. Med.* **25**, 374–80.

Waldmann T. A. (1972) Protein-losing enteropathy and kinetic studies of plasma protein metabolism. *Semin. Nucl. Med.* **2**, 251–63.

Walker W. A., Lowman J. T. & Hong R. A. (1973) Measuring albumin turnover rates in patients with hypoproteinaemia. *Am. J. Dis. Child.* **125**, 51–4.

Wiggins H. S. (1967) Simple method for estimating trypsin. *Gut* **8**, 415–16.

Chapter 7

Liver Disorders

MICHAEL J. TARLOW

The pattern of liver disease in infants and children is very different from that in adults. Malignant disease, damage by alcohol or drugs, and gallstone related disease are all rare. Jaundice is a common phenomenon in the first few days of life. The newborn has a metabolically immature liver on which is imposed the excretion of a massive load of bilirubin. Any extra stress overwhelms the already taxed excretory capacity of the liver and leads to an unconjugated hyperbilirubinaemia. In the newborn this is particularly worrying since unconjugated bilirubin in high concentration has been linked with the particular pattern of brain damage termed kernicterus (possibly better named bilirubin encephalopathy).

Liver damage in the newborn (the neonatal hepatitis syndrome) is related to a wide range of different insults; few of these have yet been clearly defined. Of those that have, some are infective, but a different array of organisms to those causing liver damage in older patients, and some are metabolic. From this group must be separated those infants in whom the bile ducts have become totally obstructed and fibrosed by extrahepatic biliary atresia. This is possibly the result of viral damage to the ducts in the perinatal period. If these infants are to have any reasonable chance of survival, remedial surgery needs to be performed early, probably within the first eight weeks of life. This places pressure on the clinician and the pathologist to reach a clear diagnosis as soon as possible in patients with neonatal liver damage.

Some of the metabolic disorders causing severe liver damage in infancy may be treated by diet. These include galactosaemia and hereditary fructose intolerance. Despite their rarity, these and other treatable metabolic disorders assume a disproportionate place in the investigation and diagnosis of liver disease in infancy.

This chapter reviews the commoner or more important aspects of paediatric liver disease as they may involve the chemical pathologist. For more detailed surveys of these subjects, the reader is referred to one of the more specialised textbooks on the subject (Mowat 1979; Chandra 1979).

LIVER FUNCTION TESTS

Serum bilirubin, aspartate aminotransferase (AST), alanine aminotransferase (ALT) and alkaline phosphatase, together with a variety of other enzyme determinations, are routinely measured in most laboratories when 'liver function tests' are requested. Their importance usually lies in the detection of an abnormality of liver function and in monitoring its progress; they are seldom valuable in making a diagnosis of a specific type of liver disease. Other tests such as serum caeruloplasmin and alpha-1-antitrypsin help in making particular diagnoses. Ranges of normal values sometimes differ in infants and small children compared with older patients (Clayton *et al.* 1980).

Bilirubin

A physiological hyperbilirubinaemia in the neonate is expected and is discussed in more detail in succeeding paragraphs. After the first two weeks of life, serum bilirubin levels in infants and children should fall within the normal adult range.

Aspartate aminotransferase (AST) and alanine aminotransferase (ALT)

Serum levels may be raised in the early months of life and it is usual to accept as normal levels up to twice the upper limit of the adult range. They fall to within the adult range by the second year of life. Detailed studies have not been performed to define this elevation more precisely, or to account for it.

Alkaline phosphatase

Alkaline phosphatase concentrations are appreciably higher in normal infants and children than in adults. Since levels reported by any particular laboratory are critically dependent on the exact assay technique used, reference ranges tend to vary between different laboratories and each one needs to define its own age-related range.

Alkaline phosphatase isoenzymes

Alkaline phosphatase as measured, reflects the total activity of a variety of enzymes from different sources which include liver, bone, intestine and kidneys. Total alkaline phosphatase is therefore of poor specificity in defining liver disease in children. The elevated concentration found in liver disease in children is frequently associated with an elevation of the bone isoenzyme to a greater extent than the liver isoenzyme. This effect is thought to be due to impaired absorption and metabolism of vitamin D, which leads to the biochemical changes of rickets and the associated elevation of bone alkaline phosphatase.

Definition of the isoenzyme pattern of a raised serum alkaline phosphatase may therefore direct attention to the diseased organ. It is unlikely to be of appreciable diagnostic value in specifying the cause.

JAUNDICE IN THE NEWBORN

Physiological jaundice

Jaundice is a common feature of normal newborns. 10–15% of them have clinically detectable jaundice, and it is probable that all newborns develop chemical hyperbilirubinaemia at some stage during their first few days of life. This so called 'physiological' jaundice of the newborn is associated with the accumulation of unconjugated bilirubin (the indirect reacting bilirubin of the chemist's Van den Bergh test). Urine and faeces remain a normal colour, and the skin is tinged an orange-yellow, in distinction to the greeny-yellow of conjugated hyperbilirubinaemia. The causes of this physiological jaundice are unclear, but the following factors are considered to be relevant (Odell 1980).

1 Bilirubin is more rapidly produced from red cell breakdown in the newborn than in the older child. The rate of bilirubin production has been assessed at 6–8 mg/kg body weight per 24 hours in the normal neonate, which is at least twice the rate occurring later in childhood.

2 There is a probable deficiency of ligandin, the intracellular bilirubin binding protein that prevents bilirubin diffusing back into the circulation after being taken up by the liver.

3 Disturbance of perfusion of the neonatal liver may exist because of the prominence of haemopoietic tissue which distorts the normal liver architecture.

4 The hepatic conjugation of bilirubin to form its water soluble glucuronide may be impaired in the newborn. This is one of the early explanations for physiological jaundice in the newborn. Some evidence exists to support it as important in other primates, but there is little to make one think it is a problem in human infants.

5 An enterohepatic circulation of bilirubin exists. After excretion into the gut, conjugated bilirubin can be deconjugated either spontaneously in the alkaline medium, or by gut bacteria, and free bilirubin released. This is then reabsorbed across the mucosa and returned to the liver for further conjugation and excretion. It has been calculated that at birth the gut of the normal newborn contains between five and ten times the daily production of bilirubin from red cell breakdown. In older patients, any unconjugated bilirubin released in the gut is likely to be further metabolised to urobilinogen by bacteria. The neonate cannot do this until the initially sterile gut is colonised, and thefore reabsorption of bilirubin from the gut is quantitatively important in the newborn, especially in situations where gut transit is prolonged (e.g. in starvation or pyloric stenosis).

6 The newborn is subjected to a very high oestrogen concentration in the last trimester of pregnancy. This may suppress conjugation of bilirubin in the hepatocyte although direct evidence for this postulated role of oestrogens in man is lacking at present.

BILIRUBIN ASSAY

Bilirubin can be measured directly in the serum by virtue of its yellow colour, or it can be estimated after chemical transformation—usually by coupling to an azo dye. In addition, measurement of conjugated bilirubin as a proportion of the total is of value clinically, since its presence in excess suggests an obstruction or impairment of bile flow and may indicate hepatic damage or post-hepatic obstruction. Two groups of techniques are in common use.

Van den Bergh analysis

Bilirubin is directly coupled with a diazo reagent (diazotized sulphanilic acid) and the resulting colour read spectrophotometrically. Since conjugated (direct reacting) bilirubin couples rapidly with the azo dye while unconjugated (indirect reacting) bilirubin needs an accelerator (methanol or diphylline) to speed the reaction, an assessment can be obtained of both conjugated and unconjugated bilirubin by this method. Many modifications of this basic technique have been described. Measurement of direct reacting bilirubin by this technique is critically dependent on the exact assay conditions. These vary between laboratories, and since this is the only practical method of measuring direct bilirubin, comparisons between different laboratories may not always be valid.

Direct spectrometric analyses

These techniques directly measure the absorption of bilirubin containing serum, often at more than one wavelength and they correct for turbidity and other light-absorbing components such as haemoglobin and carotenoids. They are simple, rapid and use only small quantities of serum. However, they do not correct effectively for the presence of significant amounts of conjugated bilirubin. In most neonates with jaundice, this is not a problem, as only insignificant amounts of conjugated bilirubin are present. Only if this is known can one of these methods be used with confidence.

Transcutaneous bilirubinometry

Recently a number of instruments have been introduced for the measurement of spectral reflectance of human skin. They assess the yellow colour of a baby's forehead and give values which correlate well with serum concentrations of bilirubin using a diazo technique. The full scope of this technique remains to be assessed. Problems arise with infants of different racial backgrounds and in situations where phototherapy is used. They are more accurate than visual assessment of jaundice and seem likely to gain a place in the neonatal nursery as a screening tool to reduce the total number of serum assays required (Engel 1982; Hegyi 1982).

Accuracy of routine bilirubin measurement

Bilirubin measurement by any of the above techniques is inaccurate and provides only a rough guide to the true bilirubin level, especially when results between different laboratories are compared with one another. In a recent study, in which over 100 laboratories in Australia were compared, the 95% confidence interval for a standard sample ranged from 262 to 386 µmol/l. Similar results have been obtained in studies in other countries. Although the use of a common standard improved these results, they were still outside the limits desired by the clinician for informed decision making. No one technique for bilirubin estimation was consistently more accurate than any other. One reason for this problem is the difficulty in guaranteeing pure bilirubin standards for assay purposes. Another is the photolability of bilirubin. This physicochemical phenomenon, which helps the clinician (see below 'Phototherapy') can be the source of appreciable laboratory error unless adequate precautions are taken.

It is important for clinicians to appreciate that the apparent accuracy of bilirubin estimations may not be real. Important decisions should not be made on the basis of minor variations in bilirubin concentration which may be artefactual (Schreiner & Glick 1982; Watkinson *et al.* 1982).

Free and bound bilirubin

Unconjugated bilirubin in the circulation is almost entirely bound to serum albumin. Albumin possesses primary binding sites at which bilirubin is tightly bound and can only be released with difficulty, and secondary binding sites where bilirubin is loosely bound. Bilirubin can easily be dissociated from secondary binding sites by competition from other anions, including organic acids (e.g. salicylate, lactate), and drugs such as sulphasoxisole and heparin. A small amount of the total bilirubin (about one ten thousandth of the whole) exists as free bilirubin in solution and in equilibrium with the albumin bound bilirubin.

The 'free bilirubin' hypothesis

In laboratory studies, unconjugated bilirubin disrupts lipid membranes within the cell, uncouples oxidative phosphorylation, blocks the cytochrome chain and inhibits protein synthesis. This *in vitro* toxicity can be prevented by an excess of albumin. The 'free bilirubin' hypothesis states that free bilirubin unbound to serum albumin is the primary cause of this toxicity, and that it also produces the neurological damage associated with kernicterus in jaundiced newborns. Conjugated bilirubin is non-toxic as are the bilirubin degradation products produced by the action of light on the molecule.

Drugs, anions and 'free bilirubin'

A study of prophylactic antibiotics performed on neonates 25 years ago (Silverman *et al.* 1956) found a much higher incidence of subsequent brain damage in the group given a sulphonamide (sulphasoxisole) than in those given penicillin. Subsequent *in vitro* studies showed that the drug released bilirubin from its secondary binding sites on albumin. These studies provide some of the best evidence for the relevance of albumin binding in the prevention of brain damage due to bilirubin. Other substances competing with bilirubin for binding sites on albumin include salicylates and many other organic anions. Because of this phenomenon, many drugs are contraindicated in neonates with indirect hyperbilirubinaemia, and active treatment for hyperbilirubinaemia is started earlier if lactate, salicylate or other organic anions are present in abnormally high concentrations.

Kernicterus and the 'free bilirubin' hypothesis

Kernicterus is the yellow staining of the basal ganglia, cerebellum and brain nuclei with bilirubin. It is diagnosed pathologically at post-mortem and is traditionally associated with a particular pattern of clinically detectable neurological damage. For many years physicians have assumed that kernicterus occurred primarily as a result of raised levels of unconjugated bilirubin. It is certainly true that brain damage occurs in situations where the elevation of unconjugated bilirubin is the only appreciable biochemical or clinical abnormality. Crigler–Najjar syndrome (a rare inherited disorder of bilirubin metabolism) and experimental jaundice in the Gunn rat (which has an inborn defect of bilirubin conjugation) are examples of this. This pattern is possibly simulated clinically in patients with haemolysis due to blood group incompatibility, but without appreciable anaemia. In these situations it is likely that brain damage is the direct result of bilirubin encephalopathy. However, in smaller preterm infants who develop brain damage following jaundice, the pattern is not so clear cut. In this group, and in infants who are sick for other reasons, the connections between bilirubin levels and brain damage are not obvious. It has recently been shown that in experimental animals, the yellow staining of the brain structures may be dependent more on the loss of integrity of the blood–brain barrier than on the serum concentration of free bilirubin. Albumin-bound rather than free bilirubin produces this staining (Levine *et al.* 1982).

Loss of integrity of the blood–brain barrier occurs with cerebral anoxia or ischaemia. Many small neonates are known to suffer intracranial haemorrhages. These can now be readily detected by ultrasound studies and are associated both with brain damage and with hyperbilirubinaemia (the latter due to reabsorption of the blood clot). It may be that damage in these small infants attributed to kernicterus in the past is, at least in part, the result of intracranial bleeding. It that is true, the yellow staining of the brain is an index of the loss of integrity of the blood–brain barrier due to anoxia, and is a secondary rather than a primary phenomenon (Lucey 1982; Turkel *et al.* 1982; Levine *et al.* 1982). Nevertheless, it still seems sensible to recommend exchange transfusion or phototherapy for infants with high bilirubin levels, even if the direct relationship between these and brain damage is in question in some instances. Some brain damage previously attributed to kernicterus in small infants may therefore be primarily anoxic or multifactorial. This means that attempts to define levels of bilirubin that are toxic under all circumstances must be to some extent suspect (Cashore & Stern 1982).

Measurement of free bilirubin and albumin saturation

Many different techniques have been used to assess free bilirubin unbound to albumin in serum from jaundiced neonates. This has been pursued on the assumption that free bilirubin was always the primary factor in the development of kernicterus. Since modern theory suggests this may not invariably be the case, the clinical importance of developing an assay might now be less urgent than has been thought. Clinical chemists have described over 30 different assays for 'free bilirubin' or albumin-binding capacity. Many of them are tedious and time consuming. None is wholly satisfactory and few are in routine use in clinical neonatal units (Gitzelmann-Cumarasamy & Kuenzle 1979).

Among the more widely recognised of these tests are included: Sephadex gel filtration, horseradish peroxidase and the salicylate saturation test. As well as measuring free bilirubin, the reserve albumin binding capacity can be assessed by these tests. This is a theoretical concept in which the potential binding sites for bilirubin on albumin are measured. If the 'free bilirubin' hypothesis is true, these should correlate with the degree of protection against kernicterus.

Sephadex G-25 filtration (Kapitulnik *et al.* 1974)

In this test a small quantity (100 μmol) of test serum is passed down a Sephadex G-25 column. The process of gel filtration binds the free bilirubin, and a proportion of the bilirubin which is bound to plasma albumin. This bilirubin is separated from albumin and other serum components on the column, is eluted and its concentration assessed photometrically. The reserve binding capacity of plasma can be estimated by adding aliquots of bilirubin to the plasma and repeating the process of gel filtration and elution.

Levels of 'free bilirubin' found using this test are many times greater than the true 'free bilirubin', probably because the gel not only binds the 'free bilirubin' but also separates some of the loosely bound bilirubin from albumin.

No correlation has been shown between 'free bilirubin' levels assessed by this test and clinical evidence of bilirubin encephalopathy on follow-up.

Peroxidase method for 'free bilirubin' and for reserve albumin binding capacity (Jacobsen & Wennberg 1974)

This assay depends on the susceptibility of 'free bilirubin' to oxidation by ethyl hydroperoxide in the presence of horseradish peroxidase, while albumin-bound bilirubin is protected from oxidation. The initial oxidation velocity of bilirubin is proportional to the 'free bilirubin' concentration. This can be measured by the decrease in bilirubin concentration after a timed exposure to the oxidising agent. Reserve albumin binding capacity can be measured by *in vitro* titration of bilirubin, adding increasing amounts of bilirubin to serum samples.

There are however, many other substances in serum which can stimulate or inhibit horseradish peroxidase; these include vitamin C and many compounds with aromatic rings. Since the concentrations of these so-called 'effectors' will vary from person to person, and in the same individual on different occasions, levels of 'free bilirubin' obtained by this assay may be difficult to interpret or to reproduce.

Salicylate saturation test (Odell 1980)

This test measures the degree of saturation of serum albumin with bilirubin. It depends on the observation that albumin possesses both high-affinity and low-affinity binding sites for bilirubin. Bilirubin bound to a high-affinity site is not displaced *in vitro* addition of salicylate to jaundiced serum, but that bound to low-affinity sites is displaced and can be measured by a shift in spectral absorbance from 460 nm to a lower wavelength (440 nm). Reliable results in this test demand meticulous attention to detail, and even with extreme care, errors of up to 25% occur (Lee *et al.* 1978).

The salicylate saturation index may be a better predictor of long-term neurological damage in jaundiced infants than the total serum concentration of unconjugated bilirubin. It is probably the most widely recognised of the currently available tests for 'free bilirubin' and albumin binding.

Prevention of kernicterus

For many years techniques to prevent brain damage have concentrated on lowering the levels of unconjugated bilirubin in the blood, initially be exchange transfusion, and more recently by phototherapy and in some

Table 7.1 Concentrations of unconjugated bilirubin in serum (μmol/l) at which specific therapeutic measures might be considered.

	Phototherapy	Exchange transfusion
Term newborn, otherwise healthy	200	350
Pre-term newborn, otherwise healthy	170	250
Sick, pre-term infant	150	200

situations by the use of phenobarbitone (which increases hepatic conjugation and excretion of bilirubin).

Since the relationship of serum unconjugated bilirubin levels to kernicterus is uncertain, no clear indications for exchange transfusion can be given. The guidelines in Table 7.1 might be helpful in clinical practice. The reader is referred to a recent textbook of neonatal paediatrics for a more detailed discussion of the clinical factors involved in decision making (e.g. Avery 1981).

OTHER UNCONJUGATED HYPERBILIRUBINAEMIAS

The vast majority of jaundiced newborns have an unconjugated hyperbilirubinaemia and this seldom requires further investigation. However, in some instances there is an underlying pathological reason for the jaundice which may be especially prolonged or severe. The following should be considered as important factors suggesting jaundice is pathological rather than 'physiological' in origin, and that further investigations are warranted.

1 Appearance during the first 24 hours of life.
2 Persistence of jaundice for more than 14 days.
3 Elevation of total bilirubin levels above 250 μmol/l.
4 Conjugated hyperbilirubinaemia.
5 The presence of jaundice in association with other signs or symptoms of ill health.

Haemolytic states predominate amongst the pathological causes of unconjugated hyperbilirubinaemia. In Britain, blood group incompatibility is the commonest of these, especially Rh and ABO incompatibility. In other parts of the world, haemoglobinopathies such as thalassaemia and inherited disorders of red cell metabolism, such as glucose-6-phosphate dehydrogenase deficiency, become important. The investigation of these is within the realm of the haematologist rather than the clinical chemist and will not be considered further here.

There are however, non-haematological causes which need to be recognised if they are to be diagnosed and appropriate treatment given. These include hypothyroidism, breast milk jaundice and hereditary hyperbilirubinaemias.

Hypothyroidism

Prolonged neonatal jaundice is a well recognised feature of neonatal hypothyroidism (Aakerrén 1955), a condition which is almost always associated with an absent or grossly underdeveloped thyroid gland. The jaundice is probably associated with impaired synthesis of the enzymes involved in conjugation and excretion of bilirubin; this may be part of a more general effect of the thyroid hormone on protein synthesis (Gartner 1967). Since long-term brain development is dependent on thyroid hormone, recognition of the condition can mean the difference between a normal and a mentally retarded child. It is therefore mandatory to check thyroid function in all infants with prolonged unexplained unconjugated hyperbilirubinaemia (i.e. those jaundiced for more than 14 days). Jaundice may be the only clinical manifestation of congenital hypothyroidism.

Breast milk jaundice

Infants of some mothers develop prolonged unconjugated hyperbilirubinaemia despite remaining otherwise healthy. The bilirubin levels seldom rise above 350 μmol/l, and often reach their peak in the second or third week of life, which is well after the peak of physiological jaundice on the fourth or fifth day of life. The syndrome was originally attributed to the presence of a postulated sterol, 3-α, 20-β-pregnanediol, in the breast milk. This was thought to cause inhibition of bilirubin conjugation, but it has not been possible to verify either the persistent presence of the hormone in breast milk from affected mothers, or an effect of this hormone on the conjugation of bilirubin. A more recent suggestion, with better evidence, is the presence of high concentrations of non-esterified fatty acids (free fatty acids) in affected milk. These inhibit bilirubin glucuronyl transferase *in vitro*, and their high concentration in some milks may be associated with high activity of a specific breast milk lipase in these milks. Breast milk lipase hydrolyses triglycerides releasing free fatty acids. These fatty acids are either absorbed to reach the liver cell directly, or act in some way indirectly to impair bilirubin excretion (Poland 1981).

Whatever the cause of breast milk jaundice, it is harmless and need not lead to breast feeding being abandoned. If breast feeding is continued the jaundice resolves in time, and no long-term harmful effects have been recorded.

Phototherapy and bilirubin

It has been known for over 20 years that light reduces the serum bilirubin concentration of jaundiced infants. The mechanism is unclear, but it is probable that light induces an isomeric change in the ring structure of bilirubin so that its non-polar groups are re-arranged towards the centre of the molecule and polar groups towards the periphery. This renders the molecule more water soluble; excretion into the bile can then take place, although it is still unconjugated. It is known that in both the experimental animal and the human infant treated by phototherapy, the major excretory product in the bile is unconjugated bilirubin. Non-toxic breakdown products of bilirubin are also produced (McDonagh 1981).

The most effective wavelength to produce this effect is at the shorter end of the visible spectrum—'blue' light in the 400–500 nm waveband—and is produced by so-called 'daylight' fluorescent lighting. (Phototherapy has no effect on the plasma levels of conjugated bilirubin, which is anyway harmless in itself.)

Phototherapy itself presents certain potential hazards; these include skin burns, dehydration and eye damage associated with excessive or inappropriate exposure to the fluorescent lamps. In addition, diarrhoea is a common though poorly explained side effect and a generalised bronzing of the skin (the bronze baby syndrome), thought to be associated with polymerisation of circulating porphyrins, has been described.

There is no practical technique for measuring bilirubin degradation products in conventional assays. Some may react as does bilirubin itself in the Van den Bergh reaction. This may cause some difficulty in the interpretation of bilirubin concentrations in infants undergoing phototherapy. Failure to observe a fall in bilirubin concentrations during phototherapy does not necessarily mean the treatment has failed.

Indications for phototherapy are not yet universally agreed, and vary from unit to unit (see Table 7.1 for guidelines).

INBORN ERRORS OF BILIRUBIN METABOLISM

These are rare causes of jaundice, but it is important to recognise them. The Crigler–Najjar syndrome is treatable in some instances. The other causes of jaundice, if not correctly diagnosed, may lead to detailed and unnecessary investigations which may be painful, invasive or dangerous, or may appreciably curtail an individual's freedom to enjoy life.

Unconjugated hyperbilirubinaemias

Crigler–Najjar syndrome

This rare syndrome is associated with the partial or complete absence of the conjugating enzyme bilirubin UDP glucuronyl transferase in the liver. The disorder has been subdivided into two types:

Type I. Total absence of conjugating enzyme. The serum bilirubin rises inexorably in the first few days and weeks of life. Phototherapy is most effective in lowering the concentration of unconjugated bilirubin. Without this (and cholestyramine, an ion exchange resin which binds bile salts within the gut and stimulates bile flow) brain damage is likely to develop early in life. Even with treatment the prognosis is poor.

Type II. The conjugating enzyme is only partially absent. Lower levels of bilirubin (usually < 350 μmol/l) are found in the blood and kernicterus is rare. This anomaly can often be helped by treatment with a drug such as phenobarbitone, which stimulates the microsomal enzyme systems within the liver cell. This stimulation increases the total amount of conjugating enzyme present and lowers the serum concentration of bilirubin in some, but not all, patients.

In both Type I and II Crigler–Najjar syndromes, liver function tests and the liver histology are otherwise entirely normal. The diagnosis is made by assaying bilirubin UDP glucuronyl transferase activity. This can be performed by a specialised laboratory and a sample obtained by needle biopsy is usually sufficient.

Type I disease is inherited as an autosomal recessive, but Type II, as might be expected, is inherited as a dominant (Odell & Childs 1980).

Gilbert's disease (Gilbert's syndrome)

Gilbert's disease is a mild unconjugated hyperbilirubinaemia occurring without any other evidence of disturbed liver function. It is extremely common and it has been estimated that 1–6% of the population of the United Kingdom have this disorder. However it is almost certain that the elevation of serum bilirubin is due to a variety of different causes. In some individuals there is evidence of increased red cell haemolysis with a shortened red cell life span, and this is the cause of the mild jaundice. In others, no abnormality of red cell metabolism can be found. The elevation of serum bilirubin is mild (seldom greater than 100 μmol/l) and is unassociated with other biochemical, clinical or histological evidence of liver abnormality. A number of vague, non-specific symptoms such as lethargy and malaise have been attributed to this syndrome, but it is unlikely that it really produces any organic illness. In some individuals the jaundice may be precipitated by fasting or by the administration of intravenous nicotinic acid. Since fasting may be the unwelcome concomitant of many infections or other illnesses, it is not surprising that many of these patients present with jaundice in association with an acute infection. It is therefore important to demonstrate the absence of hepatic damage or of conjugated hyperbilirubinaemia, since otherwise the mistaken diagnosis of viral hepatitis or chronic liver disease may be suggested, and the patient treated by an unnecessarily restrictive regime.

Gilbert's disease has not been recognised as a problem in the neonate. It is probable that it does exist but is submerged within the many babies with physiological jaundice. It becomes recognisable in the older child and adolescent of 10 years and upward. Inheritance is said to be autosomal dominant but since several forms of the disorder almost certainly exist, and these have not yet been clearly delineated clinically, it is too early to make a definitive statement.

Classification into subgroups on the basis of plasma clearance of bromsulphthalein (BSP) and indocyanine (ICG) has been attempted (Berk *et al.* 1975), and three variants of the disorder defined on this basis. Electron microscopy of the liver cell has also been used to define subgroups (McGee *et al.* 1975).

Although the order should be recognised, the only treatment necessary is reassurance and the avoidance of unnecessary tests and restrictions. If the jaundice is sufficient to cause social embarrassment, a small dose of phenobarbitone which stimulates microsomal enzyme activity may be sufficient to clear it.

Diagnosis is made by demonstrating an unconjugated hyperbilirubinaemia unassociated with overt haemolysis or evidence of abnormal liver function or histology. Elevation of the serum bilirubin by fasting may be a useful clinical test (Owens & Sherlock 1973). Reduction in food intake to 400 calories/day for 72 hours results in a rise in serum bilirubin in both normal subjects and patients with Gilbert's syndrome. However, an increase of 100% or more in unconjugated bilirubin strongly suggests the patient has Gilbert's syndrome.

Conjugated hyperbilirubinaemia

Two inherited syndromes of conjugated hyperbilirubinaemia are recognised: the Dubin–Johnson sydrome and the Rotor syndrome (Odell & Childs 1980).

Dubin–Johnson syndrome

This is a rare disorder in which bilirubin is satisfactorily conjugated, but excretion into the bile is impaired. Serum bilirubin levels range up to about 100 μmol/l under normal circumstances, but may be much higher during illness, reaching 400 μmol/l or more. Bromsulphthalein (BSP) is also cleared inadequately by the liver. Following an intravenous dose, the initial clearance of BSP from the plasma (reflecting hepatic uptake of BSP) is normal. However, there is a secondary rise of plasma BSP between 60 and 120 minutes after administration, suggesting inadequate hepatic excretion of conjugated BSP and its secondary reflux into the blood. This peculiar pattern of BSP excretion is considered pathognomonic of the Dubin–Johnson syndrome. It does not occur with other dye tests in which the dye is excreted unchanged (e.g. indocyanine green, rose bengal or dibromosulphthalein).

Liver biopsy demonstrates the accumulation of a brown–black pigment in the liver; this is similar to melanin and lipofuscin, but its origin and exact chemical nature are unknown. Histology of the liver is otherwise normal.

The Dubin–Johnson syndrome is probably an asymptomatic and harmless disorder. Apart from the jaundice, excretion of some other substances sharing the route of excretion of bilirubin may be affected. These include the contrast material used by radiologists for cholecystography. If the biliary tree and gall bladder are not visualised on cholecystography, an abnormality may be suspected, and in the presence of conjugated hyperbilirubinaemia, surgical exploration undertaken to exclude biliary obstruction. The physician and pathologist should be aware of this possible pitfall in diagnosis.

In the neonate, the Dubin–Johnson syndrome may present as prolonged neonatal jaundice. If conventional liver function tests are normal, a BSP excretion test demonstrating the secondary rise in plasma BSP should enable the diagnosis to be made.

Inheritance is probably autosomal recessive. Heterozygotes can be detected because of an abnormal coproporphyrin excretion pattern in the urine. In normal individuals there is an excess of coproporphyrin III over coproporphyrin I in the urine but in patients and carriers of the Dubin–Johnson syndrome this pattern is reversed (Kondo *et al.* 1976), although the reason for this is not understood. The syndrome is particularly common among Jews of Iranian origin.

Rotor's syndrome

Rotor's syndrome is another rare syndrome in which concentrations of both conjugated and unconjugated bilirubin are raised in the serum. It differs from the Dubin–Johnson syndrome in that a BSP clearance test does not show a secondary rise, excretion of radiographic contrast material by the liver is normal, and there is no accumulation of dark pigment in the liver. It is likely that the defect lies in a reduced storage capacity of the liver cell for bilirubin rather than an excretory defect. Inheritance is apparently autosomal recessive.

NEONATAL HEPATITIS AND BILIARY ATRESIA

Neonatal hepatitis

The term 'neonatal hepatitis' denotes nothing more than inflammation of the liver in the newborn period in association with patent bile ducts. It does not presuppose an infective aetiology, and indeed it is rather unusual (in the United Kingdom) to find an infant with neonatal hepatitis associated with the recognised hepatitis viruses A or B. It is a relatively rare disorder (the incidence has been assessed at approximately 1:4000 live births). The aetiology may be any one of a wide range of different disorders (see Table 7.2) as similar clinical and histological damage may be caused by infective or by metabolic insults to the neonatal liver. While most causes are still 'idiopathic', the commonest which can be recognised is alpha-1-antitrypsin deficiency. This accounts for 20–25% of cases (Mowat *et al.* 1976).

Clinical picture

The liver damage presents clinically with hyperbilirubinaemia, usually detectable within the first week or two of life. Infants with neonatal hepatitis have a greeny-yellow jaundice in distinction to the orange-yellow

Table 7.2 Recognised causes of the 'neonatal hepatitis' syndrome.

Metabolic
Alpha-1-antitrypsin deficiency
Galactosaemia
Hereditary fructose intolerance
Hereditary tyrosinaemia
Niemann–Pick disease
Wolman's disease
Cystic fibrosis
Gaucher's disease
Infective
Antenatal infections
Rubella
Toxoplasma gondii
Cytomegalovirus
Herpes simplex
Hepatitis B
(? Non-A non-B hepatitis)
Syphilis
Perinatal and postnatal infections
Other bacterial infections especially those associated with bacteraemia
Others
Prolonged intravenous feeding
Chromosomal disorders (esp. trisomy 13 or 18)

colour of the more common unconjugated hyperbilirubinaemia. Even marginal elevations of conjugated bilirubin levels in the blood (above 35 μmol/l) should be taken as evidence of liver damage. The liver is usually enlarged, and splenomegaly may also be present. Aminotransferases are raised, usually only to a moderate degree. The stools may be of normal colour but are usually pale or even acholic. It is only with acholic stools that biliary atresia need be seriously considered in the differential diagnosis. A common pitfall is the failure to realise that so-called 'acholic' stools occurring in total biliary obstruction still have a tinge of green, associated with the diffusion of bile into the gut across the intestinal wall.

Growth may be inhibited in these babies, and their appetite is often poor. The clinical course is often unpredictable. When associated with a short-lived insult to the liver (such as a systemic bacterial infection) recovery is the rule; in other situations the clinician usually has to guess. In general, the more severe the initial liver disease, the worse the prognosis.

It is important to diagnose the minority of children with neonatal hepatitis who have a treatable problem. Occasional children with cystic fibrosis will present with this clinical pattern, as will children with galactosaemia and hereditary fructose intolerance. The specific diagnosis of these disorders is discussed elsewhere. The small group of babies with severe neonatal hepatitis, total cholestasis, and acholic stools, need to be differentiated from those with biliary atresia (Mowat *et al.* 1976).

Biliary atresia

Biliary atresia is a distressing illness in infants. The extrahepatic biliary tree is replaced by fibrous cords with no lumen. Untreated, it inevitably progresses to biliary cirrhosis and liver failure, and the child dies within a few years of birth. Biliary atresia was originally thought to be a developmental disorder (the term 'atresia' means failure to canalise and produce a lumen), but over the last 10–15 years, good evidence has been obtained that the disease is the result of damage to the infant's bile duct in the perinatal period. In the past, no form of surgery was effective except in those very few patients in whom only part of the extrahepatic biliary tree was occluded and when it was technically possible for the surgeon to bypass this. Over the last 20 years the Kasai operation (named after the Japanese surgeon who introduced it) has become accepted as a potentially life-saving procedure. The cord-like bile ducts are excised and an artificial conduit of jejunum is stitched to the liver. This links with the upper gut as a 'Roux-en-Y' anastomosis (see Fig. 7.1). This operation, however unphysiological it may appear, allows bile to flow and relieves the obstructed liver. This prevents the secondary toxic effects of damage by bile salt accumulation in the tissues. In the best hands, bile flow is obtained in over 70% of patients (Kasai *et al.* 1975). In many children, although the operation is an immediate success, the liver parenchyma is damaged and the child later develops cirrhosis, often with portal hypertension (Mustard *et al.* 1979).

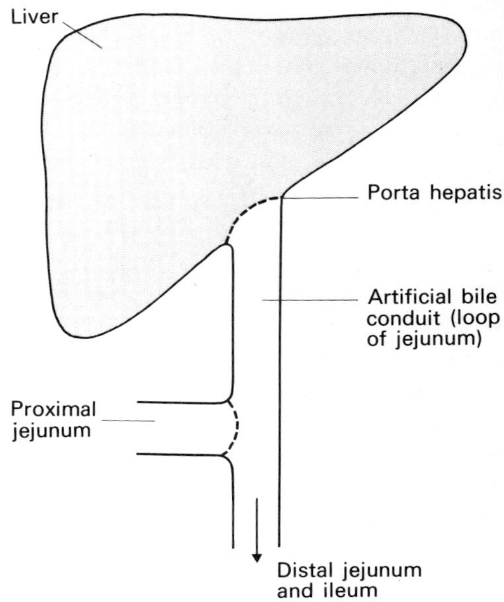

Fig. 7.1 The Kasai operation of porto-enterostomy for biliary atresia: the fibrosed bile ducts and gall bladder have been removed and replaced by a loop of jejunum.

For the operation to be successful it is now widely accepted that it should be performed before the infant has reached 60 days of age (Howard & Mowat 1977), which places considerable pressure on the clinician. He must clearly diagnose biliary atresia, and differentiate it from other causes of conjugated hyperbilirubinaemia, particularly severe neonatal hepatitis. The simplest and most useful differentiating feature is the stool colour.

The following are some of the other tests that have been used in an attempt to help in this difficult diagnostic procedure.

5′-nucleotidase

5′-nucleotidase is a phosphatase which hydrolyses nucleotides in which phosphate is attached to the pentose ring in the 5-position. Its concentration in the serum rises in obstructive rather than hepatocellular jaundice in adults (Goldberg 1973), and therefore it might be considered a particularly likely enzyme to differentiate biliary atresia from neonatal hepatitis.

In a group of 36 infants with liver disease Yeung (1972) found that the 9 children with biliary atresia had a mean 5′-nucleotidase level of 58 U/l, while the 27 with neonatal hepatitis had a mean level of 14 U/l (control mean 4.0). In his group of patients there was no overlap between the enzyme levels in the two pathological diagnoses. Conventional liver function tests (alkaline phosphatase, ALT and AST) did not differentiate between the two groups. Yeung has suggested that the high levels he found in biliary atresia were associated with bile duct proliferation, which is common in atresia but not in neonatal hepatitis, and is known from studies in adult liver disease to produce a high level of 5′-nucleotidase (Hobbs *et al.* 1968). However, appreciable overlap exists between the two groups, and the test, although helpful, should not be considered infallible in differentiating one from the other.

Gamma-glutamyl transpeptidase

Gamma-glutamyl transpeptidase is a microsomal enzyme in the liver cell, as well as occurring in association with other epithelial surfaces. It catalyses the transfer of gamma-glutamyl groups from donor peptides to other peptides and L-amino acids, and may play a role in amino acid and peptide transfer across cell membranes. Wright and Christie (1981) found gamma-glutamyl transpeptidase levels tended to be higher in their patients with biliary atresia than in those with neonatal hepatitis. Like 5′-nucleotidase measurement, gamma-glutamyl transpeptidase may therefore be helpful in suggesting one diagnosis rather than another. It should not be considered a reliable diagnostic tool.

ALPHA-1-ANTITRYPSIN DEFICIENCY

Alpha-1-antitrypsin is a serum glycoprotein. It travels with the alpha-1 fraction of serum globulins on electrophoresis, and it inhibits the activity of trypsin as well as a wide variety of other proteases. Its physiological role is probably as a regulator of proteolytic enzymes produced within the body so that it prevents the destruction of tissue by uncontrolled self digestion.

The first clinical correlation to be discovered in association with a deficiency of alpha-1-antitrypsin was pulmonary emphysema, and several years later a link with childhood liver disease was recognised by Sharp (1976).

Alpha-1-antitrypsin exists in at least 24 different allelic forms which are inherited co-dominantly. It is synthesised in the liver and the wide variety of molecules that can be produced are together known as the Pi (protease inhibitor) system. The individual variants of alpha-1-antitrypsin are characterised by their electrophoretic mobility, the fastest being known as PiA and the slowest PiZ. The commonest variant is PiM, which has a frequency between 0.86 and 0.99 depending on the particular population. The term alpha-1-antitrypsin deficiency is slightly misleading in that even the PiZZ phenotype does show some antitrypsin activity (10–20% of normal levels), although it has both a lowered concentration and a lowered enzymatic effect compared with normal.

Liver disease is associated particularly with the PiZZ phenotype, i.e. the Z gene has been inherited from both parents and the patient is therefore homozygous for the PiZ gene. Occasional patients with liver disease have been described who are heterozygous for PiZ (usually PiMZ, or PiSZ) and have liver disease, and a recent paper has suggested that the PiMZ patient may be more susceptible to liver damage than the general population.

The incidence of alpha-1-antitrypsin deficiency in western populations appears to be about 1:2000 live births.

Liver disease

Most children (80–90%) with PiZZ alpha-1-antitrypsin deficiency are asymptomatic. The small proportion who are symptomatic, present with prolonged neonatal jaundice, apparently indistinguishable clinically from other causes of neonatal hepatitis, which can be associated with a variable degree of liver damage. Mildly affected patients may recover with no evidence of permanent damage. Some patients at the other end of the scale progress rapidly to liver failure and soon die. Early reports suggested that when alpha-1-antitrypsin presented in the neonatal period, it relentlessly progressed to liver failure and early death. This is now known to be far too gloomy a prognosis.

Liver histology

All patients with the PiZZ phenotype, whether or not they have clinical evidence of liver damage, show accumulation of globules of alpha-1-antitrypsin within the cytoplasm of the cell. This has a characteristic red staining pattern with periodic acid–Schiff (PAS) stain. This PAS-positive material can be differentiated from glycogen within the cell since glycogen is digested out of the section with diastase, whereas alpha-1-antitrypsin globules are diastase-resistant. It is unusual to find alpha-1-antitrypsin globules in the liver in infants under three months of age, although they can be detected on electron microscopy. This means that in very young infants (the group in which the diagnosis is most likely to be considered) biochemical, rather than histological, diagnosis is usually necessary. It has been suggested that the histology of the liver during the early cholestatic phase of the illness may be a useful way of determining prognosis (Odievre *et al.* 1976).

It seems that the susceptibility of a small group of children to alpha-1-antitrypsin associated liver disease is linked with the presence of another trigger factor. The familial emphysema which occurs in some alpha-1-antitrypsin-deficient individuals is associated with cigarette smoking or with occupational exposure to lung irritants. It is probable that similar triggers precipitate liver disease in susceptible infants, but at the present time it is not known what these are.

Measurement of alpha-1-antitrypsin

For routine screening purposes, alpha-1-antitrypsin concentration is measured by single radial immunodiffusion (Mancini *et al.* 1965). This assay merely gives information about the total serum concentration of alpha-1-antitrypsin. To make a diagnosis of liver disease associated with alpha-1-antitrypsin deficiency, it is necessary to determine the type of alpha-1-antitrypsin. Only if the PiZ phenotype is present can alpha-1-antitrypsin deficiency be diagnosed confidently. Very occasionally, other variants have been associated with liver disease. If these are present in a patient with neonatal hepatitis, the relationship should be discussed with a member of a specialist reference laboratory before a definite diagnosis is made.

Phenotyping in alpha-1-antitrypsin deficiency

This is a specialised test best performed in reference laboratories or others with particular expertise. Specific techniques to determine the molecular type of protease inhibitor (Pi) present in the serum vary, but include acid starch gel electrophoresis and isoelectric focussing (Ward *et al.* 1977).

Other causes of abnormal alpha-1-antitrypsin concentrations

Alpha-1-antitrypsin is an acute phase reactant and therefore varies in serum concentration in association with a wide variety of disorders; it rises in infection and in association with chronic liver damage, and may be lower than expected in the neonatal period, and in babies with the respiratory distress syndrome.

In these patients the low serum levels of alpha-1-antitrypsin are usually associated with the common PiM variant of the protein, not linked with liver disease, and return to normal with physiological maturation or with recovery.

All infants and children with unexplained liver damage—even if it is only an elevation of serum aminotransferase levels—should be screened for alpha-1-antitrypsin deficiency.

Genetics

Alpha-1-antitrypsin deficiency is inherited as an autosomal recessive character. Siblings of affected patients should be screened and the parents should be checked for heterozygosity.

The disorder does not necessarily present in the same way in different members of the same family; one sibling may die of the disease, while another with an identical phenotype may be completely asymptomatic. However, there is an overall tendency for families to be either mildly or severely affected with the disease.

Other associations of alpha-1-antitrypsin deficiency

Apart from the well recognised association with emphysema, which may occasionally present in childhood, alpha-1-antitrypsin deficiency has been described in association with a fibrosing alveolitis and with a membrano-proliferative glomerulonephritis. Measurement of faecal alpha-1-antitrypsin has been used as a test for protein-losing enteropathy, since it is extremely resistant to proteolytic digestion and, following loss into the gut, could appear in the stools unchanged (Crossley & Elliott 1977). The reliability of this test is as yet unproven.

Lipoprotein X

Lipoprotein X is a low density lipoprotein which appears in the serum in cholestasis, but is not found in healthy individuals. It is almost certainly formed from a precursor lipoprotein which is normally secreted into the bile. In obstructive and cholestatic states this refluxes into the blood and binds to plasma albumin to form lipoprotein X (Manzato *et al.* 1976). It appears in the blood in both neonatal hepatitis and biliary atresia, but in the latter levels are rather higher. Its absence from the blood makes the diagnosis of biliary atresia untenable.

Poley *et al.* (1978) have modified this test by measuring serum lipoprotein X before and after a course of cholestyramine, given for 14–18 days. Cholestyramine is an ion-exchange resin which is given orally and is not absorbed. It binds bile salts reaching the gut and prevents their reabsorption. This leads to increased bile acid synthesis in the liver, and increased bile flow. In patients with interruption of the enterohepatic circulation, as in biliary atresia, this effect does not occur and cholestyramine does not influence bile flow. Serum levels of lipoprotein X, being derived from a biliary precursor lipoprotein, will diminish as bile flow increases. In biliary atresia, the lipoprotein X usually rises steadily with time, despite the administration of cholestyramine. Poley and his colleagues (1978) studied over 50 infants with liver disease. Twenty-seven of them had biliary atresia and all had lipoprotein X in the blood in concentrations between 0.87 and 11.4 g/l before cholestyramine. After cholestyramine, the concentration had risen in 23, remained the same in 2 and decreased slightly in 2. Of 28 infants with other liver disorders, 8 had no lipoprotein X in the serum; cholestyramine reduced the serum level of lipoprotein X in the remaining 20 by an average of 78%. The investigators concluded that their modified measurement of lipoprotein X was of value in the differentiation of neonatal hepatitis from biliary atresia.

The two week waiting period between measurements is nowadays considered a major drawback to the routine use of this test in clinical practice. Early surgery is mandatory for good clinical results in the management of biliary atresia and the liver may well deteriorate during the interval between admission and surgery, while this test is being carried out.

Alpha-fetoprotein (AFP)

AFP is the dominant serum protein of the early embryo, before adequate amounts of serum albumin develop. It is synthesised in the liver and released into the blood. It reaches a peak at about 3 months of gestation and then its concentration gradually falls away until it is only just detectable at birth in normal infants. It is thought to have an immunosuppressive role in the fetus and certainly possesses immunosuppressive properties, but details remain obscure (Keller & Tomasi 1981).

Its diagnostic value depends on its correlation with liver cell regeneration. Patients with neonatal hepatitis often have appreciable hepatocellular destruction and consequent regeneration with high serum levels of AFP; those with biliary atresia have less hepatocellular damage, and consequently a lower alpha-fetoprotein concentration.

The most detailed study of the diagnostic value of AFP in neonatal liver disease is that of Johnston *et al.* (1976). He and his colleagues concluded that high AFP concentrations (above 10 000 μg/l) were more likely to be associated with hepatitis than atresia, but found a high degree of overlap at lower levels, when the values were of little diagnostic significance. AFP levels did not correlate with standard liver function tests and gave no guide to prognosis of the disease, although high levels were associated with histological evidence of giant cell transformation in the liver. This latter is a change of unexplained significance in which large multinucleate parenchymal cells are formed in the liver. It is a particular feature of neonatal hepatitis, but is not specific or diagnostic, and may also occur in biliary atresia.

In alpha-1-antitrypsin deficiency with associated liver disease, Johnston *et al.* (1976) found rather low levels of AFP. This has not been adequately explained, but may be directly related to the underlying enzymatic abnormality in alpha-1-antitrypsin deficiency.

AFP in benign liver disease

A number of workers have found AFP to be raised in acute viral hepatitis during the recovery phase and have speculated that it is released by regenerating cells. Alpert and Feller (1978) have shown that the rapid regeneration of liver cells occurring with recovery after partial hepatectomy is not associated with any release of AFP. However, young partially hepatectomised rats (under 5 weeks of age) release AFP during the recovery phase, whilst adult rats subjected to the same procedure do not. It may therefore be that AFP is released from the neonatal liver more readily and in response to a wider variety of insults, than from the adult liver. It is secreted in high concentrations by malignant liver tumours, such as hepatoblastoma, and may be of great value in detecting their presence or their recurrence following surgery.

In older children AFP is released in liver injury associated with hereditary tyrosinaemia and viral hepatitis. The appearance of AFP in the serum has been used as a prognostic indicator in acute hepatic necrosis, suggesting that liver regeneration is beginning to take place. Its value in this remains speculative.

Radioactive rose bengal excretion

Rose bengal is a fluorescein derivative which can be labelled with ^{131}I. When injected intravenously, it is excreted by the liver into the bile. In total biliary obstruction less than 10% of an intravenously administered dose is excreted in the stool over a 72-hour period (usually less than 5%). This excretion test has been used to separate total biliary obstruction (biliary atresia) from other forms of liver disease, in which a higher proportion of rose bengal is excreted in the stool. In biliary obstruction degradation of the dye within the liver releases the radiolabel to be excreted in the urine. Contamination of the faecal specimens with urine therefore negates the value of the test. In routine practice the role of the rose bengal excretion test is limited by the time it takes, and the difficulty of obtaining uncontaminated stool specimens over a 3-day period in most neonatal and infant units. The test has proved to be of value in certain specialised units where enough highly trained staff are available to enable good stool collections to be made.

Oral cholestyramine has been used as an adjunct of this test in order to increase bile flow and increase the excretion of rose bengal in patients with any degree of bile duct patency. However, this demands a repeat rose bengal excretion test after 2 weeks of cholestyramine treatment. Modern concepts stress the necessity for early diagnosis of biliary atresia, and most paediatricians feel that this test takes too long to be valuable.

Liver biopsy

Despite the help that is provided by biochemical studies in the differential diagnosis of cholestatic jaundice in the neonate, liver biopsy is probably the most helpful test. This should be performed as early as possible in the course of investigation, as soon as a blood clotting screen has shown that a needle biopsy is safe. In the hands of a good histologist, this biopsy will be 95% accurate in separating biliary atresia from neonatal hepatitis. Any doubt left at this stage is best resolved by a laparotomy, since the success of surgical repair of biliary atresia is entirely dependent on early diagnosis, and this cannot be achieved with certainty by biochemical means.

Bile salts in paediatric liver disease

Serum levels of bile salts are elevated in patients with hepatocellular damage, and it has been suggested that these may be an even more sensitive indicator of liver disease than raised levels of ALT or AST. So far, measurement of total serum bile acid levels have not contributed significantly to the diagnosis of paediatric liver disease, although they may do so in the future.

Javitt (1974) suggested that there might be a consistently different pattern of serum bile acids in children with biliary hepatitis and those with intrahepatic cholestatic syndromes. He found predominantly chenodeoxycholic acid in the serum of patients with biliary atresia, and predominantly cholic acid in that of patients with other diseases. Cholestyramine administration had no effect on the serum bile acid concentration in the former group, but caused it to fall in the latter. Other workers were not able to confirm these findings and serum bile acid measurements, either quantitative or qualitative, have no part to play in the differential diagnosis of biliary atresia at present. Zellweger's syndrome, trihydroxycoprostanic acidaemia and cerebrotendinous xanthomatosis, are rare disorders associated with disturbed bile acid metabolism and are described later in this chapter.

HEPATITIS VIRUSES

In this section I will limit myself to a discussion of specifically paediatric aspects of viral liver disease of importance to the clinical chemist.

Two types of hepatitis viruses have been clearly defined and recognised antigenically and by electron microscopy: hepatitis A and hepatitis B viruses. In addition, other hepatitis viruses have been defined epidemiologically and have been named 'non-A non-B' hepatitis. This last group probably includes several distinct viruses.

Hepatitis A

This is the common form of infectious hepatitis in children. It is often subclinical and anicteric, but when clinical symptoms arise, it is associated with an acute illness with anorexia, jaundice and hepatomegaly. Type A hepatitis is not associated with the presence of viral particles in the blood and therefore laboratory personnel are not at special risk from handling these specimens.

Type A hepatitis resolves completely after the acute illness and is not associated with the development of chronic liver disease.

Hepatitis B

In the UK, hepatitis B is an unusual cause of infective hepatitis in infants and children. It is difficult to differentiate clinically from hepatitis A, although the incubation period is longer. Clinical features of the two types of liver disease may be very similar. In distinction to hepatitis A, both the non-infectious surface coat of the virus (Australia antigen, HBsAg) and the whole infective virus particle may be present in the blood of patients with hepatitis B, which should therefore be handled with careful precautions.

Hepatitis B is also associated with the development of chronic liver disease, especially chronic active hepatitis, in a small proportion of patients. In some of these the blood may contain hepatitis B surface antigen.

In distinction to hepatitis A, a carrier state exists for the hepatitis B virus. This can be identified in otherwise asymptomatic individuals by checking the blood for the presence of HBsAg. The incidence of the carrier state varies in different ethnic groups, being very high in South East Asians and in parts of Africa. In some areas HBsAg can be found in 70% or more of the population studied. In Britain the incidence is about 1:800 individuals.

Infants of carrier mothers

Infants of mothers who carry the HBsAg may acquire the antigen themselves, especially around the time of birth. In the vast majority of cases, the babies remain asymptomatic. Many of them clear the antigen from their blood over the first few months of life, but others go on to become chronic carriers themselves; this is especially likely to happen if the infants carry the 'e-antigen', an antigen particularly associated with the hepatitis B virus, but otherwise, as yet, poorly defined. Long-term carriers of hepatitis B antigen living in high risk areas have a much greater chance of developing hepatocellular carcinoma, the geographical incidence of which mirrors that of the hepatitis B surface antigen in carriers. Blood from infants of carrier mothers should be regarded as infective until this possibility has been excluded by virological examination.

Non-A non-B hepatitis

Epidemiological evidence suggests that most cases are caused by a single virus. Like hepatitis B, non-A non-B hepatitis is associated with a viraemia and blood from patients with an unidentified, but presumed viral, hepatitis should be regarded as infected. It is likely that a carrier state exists for non-A non-B hepatitis and that this can be transmitted vertically from mother to offspring. The whole picture will probably be clarified in the near future, with the development of a specific test for non-A non-B hepatitis. Until then, care should be taken with potentially infected samples.

CHRONIC HEPATITIS

Chronic hepatitis is usually defined by the presence of continuing liver damage (defined biochemically by abnormal liver function tests and histologically by the presence of inflammatory changes in the liver), for at least 6 months. This definition was derived from our knowledge of adult liver disease. We now know that in 6 months irreparable changes may have already occurred in the liver of affected children, and many paediatricians would investigate and make a presumptive diagnosis of chronic hepatitis in a child whose liver disease had continued for 2 months without clinical evidence of resolution.

Biochemical features suggesting the development of chronic liver disease following acute hepatitis, include the presence of a raised serum immunoglobulin concentration or the presence of autoantibodies, particularly antinuclear factor.

Causes of chronic liver disease in childhood

1 Chronic active hepatitis (may or may not follow viral hepatitis B infection).
2 Following neonatal hepatitis (either idiopathic or associated with alpha-1-antitrypsin deficiency).
3 Associated with chronic inflammatory bowel disease.
4 Associated with cystic fibrosis.
5 Associated with metabolic diseases, especially:
   (a) Wilson's disease,
   (b) galactosaemia,
   (c) hereditary fructose intolerance,
   (d) hereditary tyrosinaemia,
   (e) Niemann–Pick disease.

REYE'S SYNDROME

This is a syndrome of acute encephalopathy associated with fatty change in a variety of organs, particularly the liver. It occurs at any age between infancy and adult life, but especially in mid-childhood. The condition was first described 20 years ago, and since then well over a thousand cases have been recorded. Clinically, the patient, who was previously healthy, presents with symptoms of encephalopathy (impaired consciousness, clumsiness and neurological signs such as increased reflexes) progressing to coma. Often there has been a preceeding viral or bacterial infection such as chickenpox or gastroenteritis. The aetiology has been reviewed by DeLong and Glick (1982). The liver may be enlarged in some cases, but in many it will be of apparently normal size, and there will be nothing to differentiate patients with Reye's syndrome from other children with unidentified causes of impaired consciousness or coma. Jaundice is not present.

This condition is discussed further in Chapter 4.

INBORN ERRORS OF METABOLISM

Glycogen storage disorders

These are a group of disorders in which the breakdown of glycogen to glucose and its release from the liver is blocked at some point. There are only three common types of liver glycogen storage disease, type I (glucose-6-phosphatase deficiency), type III (debrancher enzyme deficiency) and type VI (deficiency of the phosphorylase system). The patients classified as having type VI disease may not be a homozygous group, since the activation of liver phosphorylase is a complex 'cascade' mechanism involving several enzymes, deficiency of any of which could produce clinical disease. It is difficult to separate these three common types of glycogen storage disease clinically; in all, the liver, but not the spleen, is enlarged due to accumulation of glycogen and of fat. In addition, hypoglycaemia may be a prominent clinical feature: lactic acidosis, hyperlipidaemia, ketosis and hyperuricaemia often coexist. However, there may be no clinical or biochemical features detectable apart from the enlarged liver. These disorders are discussed further in Chapters 3 and 17.

Diagnosis depends on the presence of suggestive clinical features, and needs to be confirmed by direct assay of the enzyme abnormality. In many cases this can be performed on white blood cells, but in the case of glucose-6-phosphatase deficiency, and of certain rarer types, it is necessary to assay enzyme activity directly in liver tissue. This can be obtained by needle biopsy of the liver. In experienced hands this is a very safe procedure, and the few milligrams of tissue obtained are usually perfectly adequate for a reference laboratory to measure the relevant enzyme. A wide range of physiological tests have been described to try and define the type of glycogen storage disease without biopsy (Fernandes *et al.* 1969). These include lactate measurements after glucose loading, galactose tolerance tests and the response of the blood sugar to glycogen. While these give valuable information they may be misleading, and should not be used as a substitute for liver biopsy, even though they are less invasive.

Type IV glycogen storage disease (deficiency of brancher enzyme) is rare and is different from the other types of glycogen storage disorder. It presents a separate clinical and biochemical picture, and is due to deficiency of brancher enzyme, which normally forms the branching 1,6-glucosyl limbs of glycogen. In its absence an abnormal glycogen consisting essentially of straight unbranched chains of 1,4-glucosyl units is formed. This is similar to the chemical structure of the plant polysaccharide amylopectin, and this disorder is alternatively known as amylopectinosis. Hypoglycaemia and acidosis are not particular features of the disease. Progressive liver damage however does occur, culminating in cirrhosis and portal hypertension. The few patients described have died early in life, primarily of liver disease. Abnormal polysaccharide accumulation has also been found within the brain.

Enzyme diagnosis can be made by direct assay of white cells for brancher enzyme; since some cases have been reported with normal concentrations of brancher enzyme in the white cells, but absent enzyme in the liver, it may be necessary to assay liver tissue if the white cells do not provide an answer.

Histologically type IV glycogen storage disease can be suspected since the abnormal glycogen accumulating in the liver stains with PAS (periodic acid–Schiff—a carbohydrate stain) but in distinction to normal glycogen is diastase resistant.

Galactosaemia (see also Chapter 3)

Classical galactosaemia is a rare disease with an incidence in our population of about 1:70000 live births (about 10 new cases in the UK each year). It is due to an enzymatic block in the conversion of galactose in the liver. This leads to the accumulation of the intermediate galactose-1-phosphate, which is thought to act as a tissue toxin causing multisystem damage (Donnell and Berggren 1975).

The clinical picture is one of neonatal hepatitis, the liver and spleen are enlarged, and the infant is jaundiced with a conjugated hyperbilirubinaemia and raised aminotransferases (ALT, AST). The gut cells are damaged, and vomiting, diarrhoea and failure-to-thrive are common. Damage to the renal tubular cells leads to a renal tubular syndrome with glycosuria, aminoaciduria, phosphaturia, acidosis and loss of renal concentrating ability. In addition, excess unmetabolised galactose is also lost in the urine. In the infant, very large quantities of galactose may be taken in daily, since all the baby's carbohydrate requirements are met by lactose in milk and half of this is galactose.

Brain damage and associated developmental delay is a common feature, but the development of cataracts is characteristic, probably associated with the accumulation of the sugar alcohol galactitol in the lens. These cataracts may develop very soon after birth and often need specialised ophthalmological examination for detection.

The diagnosis may be suspected by the combination of clinical signs, but needs confirmation by direct enzyme assay of galactose-1-phosphate uridyl transferase activity in red cells. While galactosuria may be expected in most cases, it is obviously necessary for galactose (as lactose) to be presented in the diet if it is to be lost in the urine, as there is very little endogenous synthesis of galactose in the body. Children admitted with galactosuria are often very sick and may not be taking milk orally. Testing the urine for reducing substances is not a reliable screening test (Mowat 1979).

DISORDERS OF FRUCTOSE METABOLISM

These include hereditary fructose intolerance (also known as essential fructosaemia, or fructose-1-phosphate aldolase deficiency) and fructose-1,6-diphosphatase deficiency (Fig. 7.2).

Hereditary fructose intolerance

Babies with hereditary fructose intolerance are well until they are given fructose or sucrose. They then develop the clinical syndrome of 'neonatal hepatitis' with progressive liver damage, conjugated hyperbilirubinaemia and raised aminotransferases (ALT, AST). In addition, acute loads of fructose may precipitate hypoglycaemia (Baker and Winegrad 1970); this latter effect is thought to be mediated by the sequestration of intracellular inorganic phosphate as fructose phosphates, making it unavailable for the synthesis of ATP. This limits energy within the cell for gluconeogenesis, and also directly inhibits the phosphorylase system of glycogen breakdown in the liver.

Because of this hypoglycaemia, fructose loading is a risky test to confirm a clinical suspicion of hereditary fructose intolerance. If it is performed at all, the intravenous route of administration should be used and only a small quantity (0.25 g/kg) of fructose given. In affected patients this produces marked hypoglycaemia and hypophosphataemia. Serum potassium may also fall. If liver disease is marked, several times this recommended dose may be needed to produce a clinical effect and this increases the risk.

Because of the dangers involved it is preferable to bypass this test and

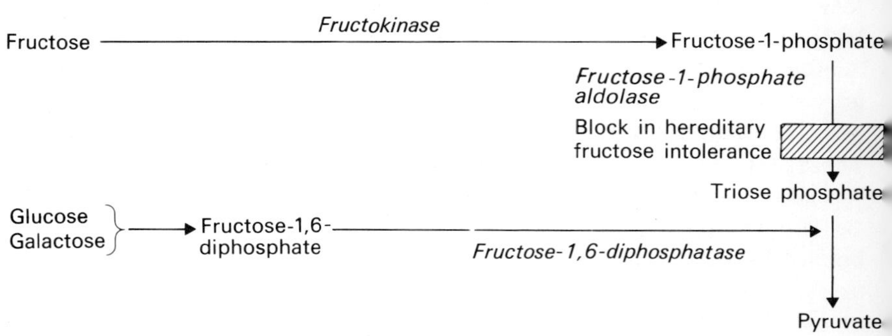

Fig. 7.2 Fructose metabolism in the liver.

proceed directly to liver biopsy, with direct assay of liver fructose-1-phosphate aldolase (Froesch 1975). This must be done in collaboration with a specialised laboratory where the test can be performed on a needle biopsy specimen containing only a few milligrams of liver tissue.

In addition to the liver damage, hereditary fructose intolerance is also associated with damage to the renal tubules and gut, with the accompanying clinical and biochemical features. Treatment with a diet free of fructose-containing foods usually produces a pleasing clinical response.

Hereditary fructose intolerance is inherited as an autosomal recessive.

Fructose-1,6-diphosphatase deficiency

Fructose-1,6-diphosphatase deficiency was first recognised in 1970 by Baker and Winegrad. It usually presents with the clinical picture of hypoglycaemia and acidosis on fasting, fructose ingestion, or infection. The liver is usually enlarged, but serum aminotransferase levels are only transiently elevated in the acute stage of the illness. Jaundice and severe derangement of liver function are not usually features. Administration of fructose or glycerol may provoke hypoglycaemia, but in general, these patients do not show the same clinical aversion to fructose as children with hereditary fructose intolerance. Diagnosis is made by direct assay of fructose-1,6-diphosphatase in a liver biopsy specimen, and the disorder can in this way be separated from glycogen storage diseases, which it otherwise resembles.

Hereditary tyrosinosis (hepato-renal tyrosinaemia, tyrosinosis type I)

This is a poorly defined clinical syndrome in which the underlying defect is not known. It affects many organs, but the primary effects are on the liver and the kidney. The liver is enlarged, with a clinical picture of neonatal hepatitis. Jaundice is often present and the transaminases are raised. Serum alpha-fetoprotein levels may be very high and progression to liver failure may occur. The kidneys show a renal tubular defect with a leak of glucose, phosphate, low molecular weight protein, aminoaciduria and a renal tubular acidosis. Rickets is frequently associated. The amino-aciduria, however, differs from that found in many other renal tubular disorders. As well as a generalised aminoaciduria, there is a specific renal tubular loss of tyrosine and methionine with high concentrations of the tyrosine metabolites p-hydroxyphenylpyruvic and p-hydroxyphenyllactic acids. The children characteristically have a 'cabbagey' smell, probably due to abnormal excretion of the methionine metabolite aphaketo-gammamethiobutyrate.

There is a wide clinical spectrum, from the acute and rapidly progressive to the chronic and slowly developing disease. Since both types may appear in the same family, it is probable that they are not genetically distinct. The condition is inherited as an autosomal recessive, and is presumably associated with a single enzyme defect. As yet, however, this has not been defined.

Treatment is symptomatic. Attempts are often made to lower the serum concentrations of tyrosine and/or methionine with diet, but as yet there is no convincing evidence that this alters the course of the disease.

Other storage disorders

These are a group of disorders in which the absence of an enzyme leads to the accumulation of a metabolite, usually in many other tissues as well as the liver. This is associated with progressive enlargement of the liver, impairment of function and eventual death. There is no specific treatment for this group of disorders, although active research at present holds out hope for enzyme replacement or a variety of transplant procedures in the future. Diagnosis is important for genetic counselling, and prenatal diagnosis may be possible.

Only those storage disorders which particularly effect the liver will be considered in this chapter.

Niemann–Pick disease

This is a lipid storage disorder in which sphingomyelin accumulates in the tissues. It may present in the neonatal period with a clinical picture very similar to other forms of neonatal hepatitis, or may present later in childhood. Biochemically the serum acid phosphatase is often markedly raised (this is also a feature of Gaucher's disease), probably as a reflection of the lysosomal damage occurring in association with sphingomyelin accumulation within the cell. The diagnosis can be made histologically, since characteristic 'foamy' macrophages are found in the tissues. Bone marrow or rectal biopsy are often more helpful than liver biopsy. The specific enzyme diagnosis may be made on fibroblasts or leukocytes.

Gaucher's disease

Gaucher's disease is probably the most common of the lipid storage disorders affecting the liver. The biochemical disorder is associated with the accumulation of glucocerebrosides in the tissues, and is due to the deficiency of a specific beta-glucosidase. Characteristic storages cells are present in many tissues. These consist of enlarged macrophages, containing 'wrinkled' cytoplasm. They can easily be distinguished from the 'foam cells' of Niemann–Pick disease and are of diagnostic importance. Biopsy of the bone marrow is one of the most useful diagnostic tests, although the cells can often be seen in other sites such as the liver or gut mucosa (rectal biopsy). Serum acid phosphatase may be greatly raised in this, as in other lysosomal storage disorders, and this may often be of help diagnostically.

The clinical picture is one of hepatosplenomegaly, with variable involvement of the central nervous system and viscera in different types of the disease. The patients are not jaundiced, although the serum aminotransferases (ALT, AST) may be raised.

The diagnosis is made on the clinical pattern of disease and presence of characteristic 'Gaucher' cells on tissue biopsy and can be confirmed by direct assay of beta-glucosidase in white blood cells or cultured fibroblasts.

Wolman's disease

This extremely rare disorder is associated with the storage of triglyceride and cholesterol esters in many organs, including the liver. The clinical pattern is one of organomegaly with failure to thrive. Liver function tests are abnormal, and some infants may be jaundiced. Characteristically, but for unknown reasons, adrenal calcification is found on abdominal X-ray. The underlying biochemical disorder is a lysosomal acid lipase deficiency. This can be detected in white blood cells and in cultured fibroblasts. The condition is said to be rapidly fatal in the first year of life, but it is probable that milder variants exist, associated with a better prognosis.

Zellweger's syndrome (cerebrohepatorenal syndrome)

This rare syndrome presents in infancy with hypotonia, a characteristic facial appearance, developmental problems, liver damage progressing to cirrhosis and the histological accumulation of fat, particularly cholesterol esters, in the white matter of the brain. An abnormal bile acid with a long side-chain is present. This was originally thought to be the underlying defect, but recently it has been demonstrated that these children have no detectable peroxisomes—subcellular organelles within the cell. This is the only disorder so far recognised associated with the absence of an organelle. It is genetically transmitted as an autosomal recessive (Goldfischer 1982).

RARE ERRORS OF BILE ACID METABOLISM

Trihydroxycoprostanic acidaemia

This rare disorder has only been described in two or three families. Children affected have had persistent liver disease with cholestasis from early infancy, and have usually died within the first few years of life. Very large quantities of trihydroxycoprostanic acid are present in their bile, serum, urine and faeces, and only small amounts of cholic acid.

Trihydroxycoprostanic acid is an intermediate in the pathway of synthesis of cholic acid from cholesterol. It is probable that the disorder is associated with an absent enzyme on this pathway (Hansen *et al.* 1975). Trihydroxycoprostanic acid is itself hepatotoxic, disrupting the rough endoplasmic reticulum of the liver in experimental animals.

Cerebrotendinous xanthomatosis

Patients with this rare disorder develop widespread xanthomas over the body, in the brain, lungs and in the tendons. Serum cholesterol levels are low. The bile contains a variety of abnormal C_{27} alcohols associated with the probable absence of mitochondrial C_{27}steroid,26-hydroxylase, a liver enzyme involved in the biosynthesis of bile acids from cholesterol (Oftebro *et al.* 1980). Liver function does not seem to be obviously affected.

Wilson's disease

Wilson's disease is a rare inborn error of metabolism in which excess copper is laid down in a number of tissues, but causes particular damage to the liver and brain. Since it takes several years for the copper deposition to produce appreciable damage, Wilson's disease is unrecognised in infancy or early childhood, and usually presents in late childhood or early adult life. Since the changes of Wilson's disease are treatable by using a chelating agent to reduce the body copper content of affected patients, the diagnosis assumes an importance quite out of proportion to the incidence of the disease. All children over 5 years of age with unexplained chronic or subacute liver damage, should have Wilson's disease excluded (Walshe 1975).

Clinical features

Wilson's disease classically presents with hepatic features. These may vary from asymptomatic liver enlargement to chronic or even fulminant hepatitis with jaundice and derangement of liver function tests. Neurological symptoms when present, include clumsiness, slurring of speech and poor school performance, and these often appear several years later than the liver features. Many patients also have a haemolytic anaemia, and renal tubular damage may lead to glycosuria, aminoaciduria and phosphaturia. Ophthalmological examination may reveal a typical brown ring around the edge of the cornea, the Kayser–Fleischer ring. Its presence is diagnostic, but its absence does not exclude the disease.

Biochemical tests

Characteristically, patients with Wilson's disease have a low serum caeruloplasmin with a raised urinary excretion of copper. Normal caeruloplasmin levels in children, as in adults, range down to 1.3 μmol/l. At least

80% of patients with Wilson's disease have levels below this range. Children with many other types of chronic liver disease may have high rather than low caeruloplasmin levels.

Urinary copper excretion, on a normal diet, does not exceed 0.47–0.63 μmol (30–40 μg) daily in normal children. In those with Wilson's disease, levels above this range are found, and this increases dramatically if the child is given penicillamine (0.5 g b.d.) as a chelating agent. Levels over 15.7 μmol (1000 μg) daily are usual, while normal patients excrete very much less.

Plasma copper is usually less than 15.7 μmol/l (100 μg/100 ml) in patients with untreated disease, while healthy individuals have levels of 15.1–22 μmol/l (90–140 μg/100 ml) and those with other forms of liver disease frequently have high levels.

While any one of these tests individually may occasionally be misleading, the serum caeruloplasmin is the most useful single test, and all three should ensure that the rare children with this treatable form of liver disease do not get overlooked.

The diagnosis can be confirmed by measurement of tissue copper levels in a liver biopsy which can also, less reliably, be stained for copper. Normal liver tissue contains up to 0.94 μmol/g dry weight (60 μg copper/g dry weight). Liver from patients with Wilson's disease almost always contains at least 3.9 μmol copper/g dry weight (250 μg copper/g dry weight). (High liver copper however, may also be found in other paediatric liver disorders.)

Treatment is with the toxic chelating agent penicillamine, and needs to be continued indefinitely. Since Wilson's disease is inherited as an autosomal recessive disorder, siblings of an affected child should be screened even if asymptomatic. Appropriate genetic advice must be given to the parents.

Indian childhood cirrhosis

This is a unique disorder of unclear aetiology which affects children from the Indian subcontinent (Tanner and Portmann 1981). It is rarely seen in Indians living outside that area. The clinical picture is one of progressive liver damage, jaundice, cirrhosis and liver failure occurring over a period of months rather than years. Children under 3 years are primarily affected. The diagnostic hallmarks of the disease are histological. Hyaline material (so called 'Mallory's hyaline') is prominent in the cytoplasm of the liver cell, extensive interstitial fibrosis occurs, cutting off individual cells or groups of cells, and regeneration is very limited. High concentrations of copper are found in the liver cell, and this may be important in its aetiology. Biochemical changes of hepatocellular damage are found. In addition, serum alpha-fetoprotein is usually markedly raised, a generalised hypergammaglobulinaemia is present, levels of haemolytic and C3 complement are low, and it is likely that circulating immune complexes are present. In addition, renal tubular damage may be detected, with an aminoaciduria and the presence of reducing sugars in the urine.

At present there is no specific treatment for Indian childhood cirrhosis.

REFERENCES

AAKERRÉN Y. (1955) Early diagnosis and early therapy in congenital cretinism. *Arch. Dis. Child.* **30**, 254–6.

ALPERT E. & FELLER E. R. (1978) Alpha fetoprotein (AFP) in benign liver disease. Evidence that normal liver regeneration does not induce AFP synthesis. *Gastroenterology* **74**, 856–8.

AVERY G. B. (1981) *Neonatology*, 2nd Edn. Lippincott, Philadelphia.

BAKER L. & WINEGRAD A. I. (1970) Fasting hypoglycaemia and metabolic acidosis associated with deficiency of hepatic fructose-1,6-diphosphatase activity. *Lancet* **ii**, 13–16.

BERK P. D., WOLKOFF A. W. & BERLIN N. I. (1975) Inborn errors of bilirubin metabolism. *Med. Clin. N. Am.* **59**, 803–16.

CASHORE, W. J. & STERN L. (1982) Neonatal Hyperbilirubinaemia. *Paed. Clin. N. Am.* **29**, 1191–203.

CHANDRA R. K. (1979) *The liver and biliary system in infants and children.* Churchill Livingstone, Edinburgh.

CLAYTON B. E., JENKINS P. & ROUND J. M. (1980) *Paediatric chemical pathology.* Blackwell Scientific Publications, Oxford.

CROSSLEY J. R. & ELLIOTT R. B. (1977) A simple method for diagnosing protein-losing enteropathies. *Br. Med. J.* **1**, 428–9.

DELONG R. & GLICK T. H. (1982) Encephalopathy of Reye's Syndrome. A review of pathogenetic hypotheses. *Pediatrics* **69**, 53–63.

DONNELL G. N. & BERGGREN W. R. (1975) The galactosaemias. In Raine D. N. (ed.) *The treatment of inherited metabolic disease*, pp. 91–114. MTP, Lancaster.

ENGEL R. R. (1982) Nuances or Nuisances for cutaneous bilirubinometry. *Pediatrics* **69**, 126–7.

FERNANDES J., HUIJING F. & VAN DER KAMER J. H. (1969) A screening method for liver glycogen diseases. *Arch. Dis. Child.* **44**, 311–17.

FROESCH E. R. (1975) Hereditary fructose intolerance and fructose-1,6-diphosphatase deficiency. In Raine D. N. (ed.) *The treatment of inherited metabolic disease*, pp. 151–69. MTP, Lancaster.

GARTNER L. (1967) The hormonal regulation of hepatic bilirubin excretion. In Bouchier I. and Billing B. (eds.) *Bilirubin metabolism*, p. 175. Blackwell Scientific Publications, Oxford.

GITZELMANN-CUMARASAMY N. & KUENZLE C. C. (1979) Bilirubin binding tests: living up to expectations? *Pediatrics* **64**, 375–8.

GOLDBERG D. M. (1973) 5'Nucleotidase: recent advances in cell biology, methodology and clinical significance. *Digestion* **8**, 87–99.

GOLDFISCHER S. (1982) The liver in Zellweger's Syndrome. VI International Congress of Liver Diseases. *Liver in Metaboic Diseases* (Congress report no. 35). Falk, Basel.

HANSEN R. F., ISENBERG J. N., WILLIAMS G. C., HACKEY D., SZCZEPANIK P., KLEIN P. D. & SHARP H. L. (1975) The metabolism of 3-alpha, 7-alpha, 12-alpha trihydroxy 5-beta cholestan-26-oic acid in two siblings with cholestasis due to intrahepatic bile duct anomalies. An apparent inborn error of cholic acid synthesis. *J. Clin. Invest.* **56**, 577–87.

HEGYI T. (1982) Transcutaneous bilirubinometry. *Pediatrics* **69**, 124–5.

HOBBS J. R., CAMPBELL D. M. & SCHEUER P. J. (1968) The clinical value of 5'nucleotidase assay in 6th international congress of clinical chemistry, Munich 1966. *Clinical Enzymology*, Vol. 2, p. 106. S. Karger, Basle.

HOWARD E. R. & MOWAT A. P. (1977) Extrahepatic biliary atresia. Recent developments in management. *Arch. Dis. Child.* **52**, 825–7.

JACOBSEN J. & WENNBERG R. P. (1974) Determination of unbound bilirubin in the serum of newborns. *Clin. Chem.* **20**, 783–9.

JAVITT N. B. (1974) Bile salts and liver disease in childhood. *Postgrad. Med. J.* **50**, 354–61.

JOHNSTON D. I., MOWAT A. P., ORR H. & KOHN J. (1976) Serum alpha fetoprotein levels in extrahepatic biliary atresia, idiopathic neonatal hepatitis and alpha 1-antitrypsin deficiency (PiZ). *Acta Pediat. Scand.* **65**, 623–9.

KAPITULNIK J., VALAES T., KAUFMANN N. A. & BLONDHEIM S. H. (1974) Clinical evaluation of sephadex gel filtration in estimation of bilirubin binding in serum in neonatal jandice. *Arch. Dis. Child.* **49**, 886–94.

KASAI M., WATANABE J. & OHI R. (1975) Follow up studies of long term survivors after hepatic portoenterostomy for 'non-correctable' biliary atresia. *J. Ped. Surg.* **10**, 173–82.

KELLER R. H. & TOMASI T. B. JR (1981) Alpha fetoprotein: a potential regulator of immune function. In Javitt N. B. (ed.) *Neonatal Hepatitis and Biliary Atresia*, pp. 119–29. Castle House Publications, Tunbridge Wells.

KONDO T., KUCHIBA K. & SHIMIZU Y. (1976) Coproporphyrin isomers in Dubin–Johnson syndrome. *Gastroenterology* **70**, 1117–20.

LEE K. S., GARTNER L. M. & VAISMAN S. L. (1978) Measurement of bilirubin-albumin binding. 1. Comparative analysis of four methods and four human serum albumin preparations. *Paed. Res.* **12**, 301–7.

LEVINE R. L., FREDERICKS W. R. & RAPAPORT S. I., (1982) Entry of bilirubin into the brain due to opening of the blood–brain barrier. *Pediatrics* **69**, 255–9.

LUCEY J. R. (1982) Bilirubin and brain damage—a real mess. *Pediatrics* **69**, 381–2.

MCDONAGH A. F. (1981) Phototherapy: A new twist to bilirubin. *J. Pediatrics* **99**, 909–11.

MCGEE J. O'D., ALLAN J. G., RUSSELL R. I. & PATRICK R. S. (1975) Liver ultrastructure in Gilbert's syndrome. *Gut* **16**, 220–4.

MANCINI G., CARBONARA A. O. & HEREMANS J. F. (1965) Immunochemical quantitation of antigens by single radial immunodiffusion. *Int. J. Immunochem.* **2**, 235–54.

MANZATO E., FELLIN R., BAGGIO G., WALCH S., NEUBECK W. & SEIDEL D. (1976) Formation of Lipoprotein X. *J. Clin. Invest.* **57**, 1248–60.

MOWAT A. P. (1979) *Liver disorders in childhood*. Butterworths, London.

MOWAT A. P., PSCHACHAROPOULOS H. T. & WILLIAMS R. (1976) Extrahepatic biliary atresia versus neonatal hepatitis. A review of 137 prospectively investigated infants. *Arch. Dis. Child.* **51**, 763–70.

MUSTARD R., SHANDLING B. & GILLAM J. (1979) The Kasai Operation (hepatic portoenterostomy) for biliary atresia—experience with 20 cases. *J. Ped. Surg.* **14**, 511–14.

ODELL G. B. (1980) *Neonatal hyperbilirubinemia*, pp. 35–51. Grune and Stratton, New York.

ODELL G. B. & CHILDS B. (1980) Hereditary hyperbilirubinemias. In Steinberg A. G. *et al.* (eds.) *Progress in Medical Genetics*, Vol. IV. W. B. Saunders, Philadelphia.

ODIEVRE M., MARTIN J. P., HADCHOUEL M. & ALAGILLE D. (1976) Alpha-1-antitrypsin deficiency and liver disease in children: phenotypes, manifestations and prognosis. *Pediatrics* **57**, 226–31.

OFTEBRO H., BJÖRKHEM I., SKREDE S., SCHREINER A. & PEDERSEN J. I. (1980) Cerebrotendinous xanthomatosis. A defect in mitochondrial 26-hydroxylation required for normal biosynthesis of cholic acid. *J. Clin. Invest.* **65**, 1418–30.

OWENS D. & SHERLOCK S. (1973) The diagnosis of Gilbert's syndrome: role of the reduced calorie intake test. *Br. Med. J.* **3**, 559–63.

POLAND R. L. (1981) Breast milk jaundice. *J. Pediatrics* **99**, 86–7.

POLEY J. R., CAPLAN D. B., MAGNANI H. N., ALAUPOVIC P., SMITH E. I., CAMPBELL D. P., BHATIA M., BURDELSKI M. & BOJANOVSKI D. (1978) Quantitative changes of serum lipoprotein X after cholestyramine administration in infants with cholestatic biliary tract and liver disease. *Eur. J. Clin. Invest.* **8**, 397–404.

SCHREINER R. L. & GLICK M. R. (1982) Interlaboratory bilirubin variability. *Pediatrics* **69**, 277–81.

SHARP H. L. (1976) The current status of alpha-1-antitrypsin, a protease inhibitor, in gastrointestinal disease. *Gastroenterology* **70**, 611–21.

SILVERMAN W. A., ANDERSON D. H., BLANC W. A. & CROZIER D. N. (1956) A difference in mortality rate and incidence of kernicterus among premature infants allotted to two prophylactic antibacterial regimens. *Pediatrics* **18**, 614–24.

TANNER M. S. & PORTMANN B. (1981) Indian Childhood Cirrhosis. *Arch. Dis. Child.* **56**, 4–6.

TURKEL S. B., MILLER C. A., GUTTENBERG M. E., MOYNES D. R. & HODGMAN J. E. (1982) A clinical pathologic reappraisal of kernicterus. *Pediatrics* **69**, 267–72.

WALSHE J. M. (1975) Wilson's disease (hepatolenticular degeneration). In Raine D. N. (ed.) *The treatment of inherited metabolic disease*, pp. 171–90. MTP, Lancaster.

WARD A. M., PICKERING J. D., FAGERHOL M. K. & MARTIN J. P. (1977) Pi (antitrypsin) typing methods: a comparison of acid starch gel electrophoresis and isoelectric focusing. *Hum. Hered.* **27**, 292–7.

WATKINSON L. R., ST JOHN A. & PENBERTHY L. A. (1982) Investigation into paediatric bilirubin analyses in Australia and New Zealand. *J. Clin. Path.* **35**, 52–8.

WRIGHT K. & CHRISTIE D. L. (1981) Use of Gamma-Glutamyl Transpeptidase in the diagnosis of biliary atresia. *Am. J. Dis. Child.* **135**, 134–6.

YEUNG C. Y. (1972) Serum 5'Nucleotidase in neonatal hepatitis and biliary atresia: preliminary observations. *Pediatrics* **50**, 812–14.

Chapter 8
The Use of Protein Assays

PAMELA G. RICHES & JOHN R. HOBBS

Biological fluids contain a variety of proteins each with a specific function or combination of functions. Proteins function as enzymes, antibodies, complement components, proteinase inhibitors, clotting factors, kinin precursors and transport substances for hormones, vitamins, metals, lipids and many other substances. Variations in protein patterns or amounts of specific proteins compared with an accepted reference range or reference pool can be attributed to genetic polymorphism, deficiency states or to disease activity.

During the last 30 years, the application of techniques of protein analysis to the investigation of biological fluids has provided information of value in the diagnosis and management of disease. Protein assays are routinely used in the examination of serum, urine, CSF, saliva and faeces.

Proteins are routinely analysed by a combination of qualitative and quantitative techniques. The use of specific protein profiling has been suggested but has not gained wide clinical use; simple listing of numerical results is of little value. The laboratory should provide an interpretation of the electrophoretic pattern and specific protein measurement with regard to relevant clinical information.

The most widely used technique for qualitative examination of serum proteins is electrophoresis on cellulose acetate membrane or agarose gel. Normal human serum is separated by electrophoresis at pH 8.6 into five principal fractions: albumin, alpha-1-globulin, alpha-2-globulin, beta-globulin and gammaglobulin (Fig. 8.1), most zones being composed of more than one protein.

The stained completed electrophoresis may be visually examined in comparison with a reference serum. Routine densitometric scanning of the stained strip has little application as a normal value for one fraction does not exclude the possibility of abnormal composition. Densitometry does, however, have a unique application in the estimation of paraprotein bands (although these are rare in children). Specific immunochemical estimations of paraproteins are unreliable.

Other techniques, such as polyacrylamide gel electrophoresis and isoelectric focussing, are finding increasing applications in specialised areas of protein analysis. Isoelectric focussing is routinely used for the identification of alpha-1-antitrypsin variants. Polyacrylamide gel electrophoresis is used for the analysis, with high resolution, of CSF, urine and other body fluids. The system has the added advantage that concentration

Table 8.1 Quantitative methods for specific protein determinations.

Method	Proteins commonly assayed	Approximate time for assay
Gel precipitation	Albumin, alpha-, beta- and gammaglobulins	24–72 hours
Electroimmunodiffusion (Rocket immuno-electrophoresis)	All proteins with fast anodic mobility*	3–16 hours
Nephelometry/turbidimetry	Albumin, alpha-, beta- and gammaglobulins	End point: 15 mins–3 hours Rate reaction: 30 secs–5 mins
Radioimmunoassay	Total IgE Specific IgE Alpha fetoprotein Serum ferritin Beta-2-microglobulin	24–48 hours

* Not recommended for C3, C4 and alpha-2-macroglobulin and immunoglobulins

of the fluid prior to electrophoresis is not always necessary. The higher resolution shown by systems such as polyacrylamide gel electrophoresis has less application in the routine analysis of serum as the resulting pattern is more difficult to interpret than the relatively simple pattern obtained with cellulose acetate or agarose.

Total protein and albumin are most frequently estimated by colorimetric assay. Albumin can be measured by specific immunochemical methods as is usual for the other serum proteins. Immunochemical assays in use are shown in Table 8.1 together with those proteins most commonly assayed by each technique and the approximate time for assay.

All immunochemical assays require very small sample volumes and are ideally suited, without modifications, to paediatric samples. Some of the more recent technological advances in immunochemical techniques, particularly in nephelometry and turbidimetry permit very rapid estimation of proteins, so that these can now help in emergency diagnostic problems.

Standardisation of immunochemical assays is not yet complete. Variations will occur with standards, reagents and methods. New methodology should not be introduced into a laboratory without consideration of these variables. Reference ranges should be established by a recommended method which will be in routine use. If commercial kits are used, then the manufacturers' normal range should be examined for acceptability.

The normal population ranges of many proteins show variation with ethnic groups and environmental influences. The normal ranges quoted throughout this chapter are for Caucasians domiciled in the UK. Similar values would be expected in all caucasians, except possibly those domiciled for long periods in third world countries.

SERUM PROTEINS

The concentration of total serum protein in the fetus steadily rises with progressing maturation, with levels of approximately 15.5 g/l in the second gestational month and approximately 55 g/l at birth.

There is an overall resemblance between the fetal and maternal electrophoretic pattern except for an additional band between the albumin and alpha-1 zone, which represents alpha-fetoprotein. The relative fractions show age-dependent quantitative differences from those of the adult. The establishment of reference ranges which may vary with age is obviously a pre-requisite for the interpretation of protein values in paediatric samples.

The comparative protein values of the neonate will differ when assessed against maternal or normal adult levels as pregnancy itself significantly changes levels of many serum protein fractions (Gitlin & Gitlin 1975). As reference ranges in most laboratories relate to a normal adult pool, the details of variations for specific proteins with age, phenotypic expression and pathology will be discussed with respect to normal adult ranges.

Electrophoretic patterns

The serum protein electrophoretic pattern in infancy is readily distinguishable from that of the adult (Fig. 8.1). The low total protein of the neonate is reflected in a low content of alpha-1- and alpha-2-globulins. The alpha-2-macroglobulin which is present in greater amounts than in the adult gives a sharp anodal front to the alpha-2 zone. Protein staining in the inter beta-gamma zone is reduced but the albumin and gamma zone is comparable to the adult. A faint band of alpha-fetoprotein between albumin and the alpha-1-globulin can be seen in some newborn premature infants. From birth to 2–6 months, there is an increase in the alpha and beta zone staining (the alpha-2 zone maintaning the sharp anodal appearance) with a reduction in the gammaglobulin zone. Thereafter there is a progressive increase in staining of the gammaglobulin zone and in the older child the pattern approaches that of the adult.

Significant changes in pattern are associated with nephrosis, liver disease, hypogammaglobulinaemia, abnormal production of immunoglobulin, alloalbuminaemia (bisalbuminaemia) and the acute phase reaction.

In nephrosis the serum electrophoresis pattern shows low albumin, alpha-1-globulin and transferrin bands (beta-globulin). On cellulose acetate a markedly elevated alpha-2 band is mainly due to increased lipoprotein with some contribution of alpha-2-macroglobulin. Depending on the severity of the renal lesion the gamma zone will either be normal or low,

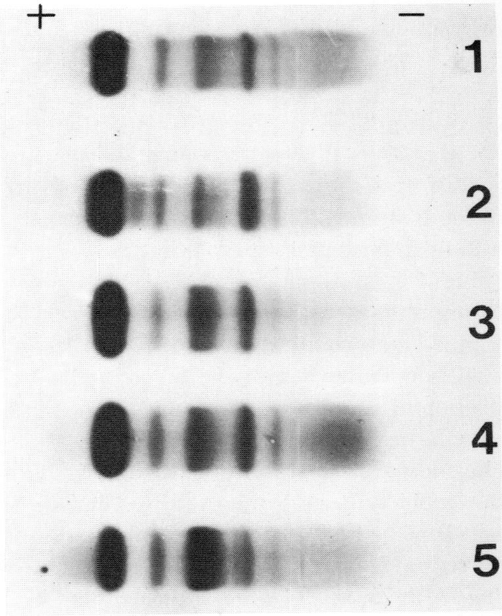

Fig. 8.1 Serum electrophoresis patterns. 1 Normal adult serum; 2 Serum of premature infant (note band of alpha-fetoprotein between albumin and alpha-1-globulin); 3 6-week-old infant (note the low gammaglobulin); 4 4-year-old child; 5 Typical pattern of serum from 3-year-old patient with nephrotic syndrome (note low albumin and transferrin bands and markedly raised alpha-2-globulin).

predominantly due to changes in IgG. A similar pattern is seen in protein-losing enteropathy.

In chronic liver disease there is a low albumin, raised alpha-1 and gamma zone with a normal alpha-2 zone which, however, fails to show reactive increases. The gammaglobulin increase which is attributable to increased synthesis is most marked in hepatic cirrhosis. The increase of IgA seen in many liver diseases, gives increased staining in the beta-gamma region due to migration of IgA to this region. In infective hepatitis the gammaglobulin tends to rise with increasing hepatocellular damage and is an unfavourable finding.

The albumin may be relatively unaffected in chronic aggressive hepatitis, the alpha-2 zone is raised and the gamma zone shows a marked increase from the onset of disease. Beta-gamma fusion is usually absent, the predominant increase in gamma being attributable to IgG.

In haemolytic jaundice there is marked reduction in the alpha-2 zone. Binding of haemoglobin to haptoglobin results in rapid metabolism of the complex and low circulating levels of haptoglobin, an alpha-2-globulin. Haemolysis *in vitro* also results in formation of a haemoglobin–haptoglobin complex which gives rise to a split alpha-2 zone on electrophoresis.

Hypogammaglobulinaemia with low values of all immunoglobulin classes or IgG results in a reduced gammaglobulin zone on electrophoresis. Hypogammaglobulinaemia involving only IgA or IgM, however, will result in an essentially normal electrophoretic pattern.

Paraproteins in childhood are rare and when found are usually transient. They may occur during infection, in lymphoid neoplasia (particularly after chemotherapy) and are frequently seen in children with bone marrow transplants. In these patients, reconstitution usually results in progress to the more usual polyclonal pattern.

Alloalbuminaemia results from inherited variants of albumin. Both fast and slow variants are reported. Alloalbuminaemia is of little clinical significance except where large changes of binding result. The total albumin is usually within normal ranges.

There is an increase in the hepatic synthesis of a number of plasma proteins as a result of trauma. Many of these proteins are alpha-1- and alpha-2-globulins and their increases led to the classic picture of raised alpha-1 and alpha-2 zones on electrophoresis.

Total protein

In the neonate variations in total protein are usually reflected in low levels of alpha- and beta-globulins. In the healthy newborn, the major fractions, except the immunoglobulins, reach steady-state levels by 6 months of age. Premature infants often show hypoproteinaemia involving all fractions and show a slow rate of increase in the first year of life.

Assay of total protein

The biuret reaction used overall for the total protein assay is an approximation in that one pure reference standard (bovine albumin) is compared with serum, which contains a mixture of proteins, so it is important that the method used conforms to that most widely accepted (Doumas *et al.* 1981).

Reference ranges

Newborn	45–65 g/l
Child	55–80 g/l

Alteration in disease

The isolated measurement of total protein in paediatric samples is of little value. Specific protein determinations are more helpful. Estimation as an indicator of hydration status has little application in the laboratory, where other biochemical measurements are available. Total protein is a necessary adjunct to electrophoresis and specific protein studies.

Albumin

Albumin is the most important protein in the control of water balance between the intra- and extra-vascular compartments, accounting for some

80% of the colloid osmotic pressure. It has a major role in the transport of fatty acids, tryptophan, haem, some hormones and cations including calcium. Since albumin binds calcium and bilirubin the level needs to be known to interpret total calcium results.

A recent monograph (Peters 1980) considers albumin in detail and provides an extensive bibliography.

Assay of albumin

In the UK bromocresol green binding is the most frequently-used method for assay of albumin. Variation in the time of absorbance reading can cause significant over-estimation of albumin. Specificity can be improved by estimating only the early reaction (less than 20 s). Bromocresol purple has been reported to be a more specific dye for the estimation of albumin.

Albumin may also be measured by specific immunochemical methods as is usual for other serum proteins. The human serum standard IFCC 74/1 has an assigned value of 40.3 g/l and is suitable as a primary standard (Hobbs *et al.* 1979).

Reference ranges

Albumin is synthesised from an early stage by the embryo and reaches adult levels by 20–26 weeks of gestation. The level may fall in the immediate neonatal period but thereafter should remain relatively stable, 30–45 g/l, throughout life.

Alterations in disease

Albumin values above normal are seen only with dehydration, excessive albumin therapy or artefactually with prolonged venous stasis. It is the low albumin values that are of most clinical significance. The fall in oncotic pressure as a consequence of hypoalbuminaemia results in oedema and values below 20 g/l almost always are associated with oedema.

The important transport functions of albumin have already been stated. Since the physiologically active calcium is the free fraction, as some 50% of the total calcium is bound to albumin, a low total calcium is seen in hypoalbuminaemia. The loss of albumin-binding capacity for many substances may increase their free levels, for example, reduced binding of drugs may result in toxic levels if these are administered in their normal doses.

Posture influences albumin levels, which may be 5–10 g/l lower in the recumbent than the upright position. Conditions associated with more significant changes in albumin are shown in Table 8.2.

Analbuminaemia results from a rare inherited deficiency in albumin synthesis. The condition is usually asymptomatic except for transient oedema in early life and persistent hyperlipidaemia.

Table 8.2 Conditions resulting in hypoalbuminaemia.

Mechanism	Disease
Genetic deficiency	Analbuminaemia
Severe changes in hydration	Dilutional hypoalbuminaemia
Fall in synthetic rate	Malabsorption Malnutrition Liver disease Acute phase reaction
Excessive protein loss	Nephrosis Protein-losing enteropathy Extensive burns Exudative skin diseases
Increased catabolism	Fever Malignancy Other rare conditions (familial hypercatabolic hypoproteinaemia)

Malabsorption may cause hypoalbuminaemia due to poor supply of amino acids and subsequent failure of synthesis. Prolonged disease results in generalised muscle and tissue wasting, osteoporosis and lowering of all protein values. The low albumin can result in oedema and low levels of circulating immunoglobulins predispose to infection.

Those situations which may give rise to malabsorption are summarised in Table 8.3.

In coeliac disease, low albumin values can precede the gut mucosal changes on re-introduction of gluten, so that precise monitoring of serum albumin is the most sensitive indicator for relapse or poor dietary control.

Table 8.3 Conditions leading to protein malabsorption.

Disease	Mechanism
Coeliac disease (gluten-sensitive enteropathy)	Reduction of absorptive surface
Milk and food allergy Crohn's disease	Inflammation
Fibrocystic disease of the pancreas Chronic pancreatitis Oesophageal atresia Pancreatic atresia Common bile duct atresia	Failure of digestion

Inflammatory conditions of the gastrointestinal tract result from a number of complex pathologies. Among these immunological hypersensitivity seems able to develop more readily in the immature and can play an important role. IgE antibodies to ingested food can be involved (see immunoglobulin section). The pathophysiological mechanisms can be similar to those of the immediate (type I) allergic reaction in other organs and consist of mucosal oedema, spasm of smooth muscle and increased secretion of mucus. It is also possible that type III (immune complex) and type IV (T-cell) reactions contribute in some patients.

Failure of digestion occurs in a number of conditions and predominantly affects large molecules. Fibrocystic disease of the pancreas (cystic fibrosis) is an inherited disorder presenting in early childhood. In addition to serum estimations, albumin in meconium up to three days after birth can indicate pancreatic insufficiency in affected infants (see faeces section).

Malnutrition is a general term referring to deficient dietary intake. Kwashiorkor which results primarily from a diet deficient in protein is characterised by the presence of oedema and a very low albumin value. In marasmus which results from calorie deficiency, there is no oedema and the serum albumin is less reduced or even normal.

It has been suggested that a fall in albumin is a relatively late event in protein depletion and other proteins have been investigated. Prealbumin has been found to be a very sensitive indicator, possibly related to its short half-life (2 days as opposed to 19 days for albumin). A fall in transferrin level, although a relatively sensitive marker, is of less value because the levels tend to rise in the iron-deficiency anaemia which may accompany malnutrition. Albumin remains the most regularly used serum protein in assessment of malnutrition. Proteins in malnutrition have recently been reviewed (Keyser 1979).

Protein-losing-enteropathy, although a rare syndrome, may be difficult to diagnose. Hypoalbuminaemia is frequently associated with low alpha-, beta- and gammaglobulin. The extent of protein loss is difficult to assess. Diagnosis may be confirmed by measuring the loss in the faeces of an intravenously-injected radioactive marker (dextran, albumin or polyvinylpyrrolidine) but since chromium is not reabsorbed (if released in the gut) intravenous chromate, which binds to albumin, offers a reliable test. Alpha-1-antitrypsin offers a non-invasive faecal screening test.

Protein losses of 5 g per day or greater, due to increased glomerular permeability, give rise to the classic features of the nephrotic syndrome: hypoalbuminaemia, oedema and hyperlipidaemia. Some idea of the sieving properties of the damaged glomerulus can be estimated by means of protein clearance ratio (see urine section). In children the most common cause of the nephrotic syndrome is 'minimal change' glomerulonephritis.

Albumin turnover in the nephrotic syndrome is greater than can be explained by measured urinary loss and results from increased protein catabolism induced by the fall of albumin levels; while some of this occurs in the renal cortex, it also occurs in the liver, gut and throughout the body.

Hypoalbuminaemia may also result from short survival without excessive loss due to accelerated endogenous catabolism. Such hypermetabolic states may affect albumin alone or many proteins due to defects in catabolic pathways. The defect may be a primary defect, as is seen in familial hypercatabolic hypoproteinaemia, or may occur with more complex diseases such as the Wiskott–Aldrich syndrome.

Hypoalbuminaemia is seen in liver disease and the acute phase response, the low albumin resulting from a significantly lowered synthetic capacity. In malignant disease acute phase response depresses synthesis, which is more especially seen in the prealbumin level, but there is also increased catabolism, particularly with fast-growing tumours which can be indicated by an unexplained low albumin, below 25 g/l.

Alpha-fetoprotein

Alpha-fetoprotein (AFP) is a glycoprotein synthesised by the fetal liver. Maximum circulating levels occur at about 20 weeks of gestation, levels then decrease and the newborn levels fall into the normal adult range by 8 months of age (Wu *et al.* 1981). The precise function of AFP is not known but it may be involved in the immune protection of the fetus from maternal rejection.

Assay of AFP

AFP is most usually assayed by radioimmunoassay. There is both an International Reference Preparation, IARC 72/225, and a British standard AFP preparation 72/227 (obtainable from the National Institute for Biological Standards and Control, Hampstead, London). The International Unit (IU) is similar in magnitude to 1 ng of current reference preparations.

Reference ranges of AFP (μg/l)

Neonate	50000–150000
1 month	100–10000
3 months	10–120
8 months–adult	below 10

Alterations in disease

After birth, raised AFP levels are found in association with:

1 Regenerating liver (a good prognostic feature following hepatitis).
2 Embryonal tumours.
3 Neonatal hepatitis.
4 Ataxia telangiectasia.
5 Tyrosinosis.
6 Congenital biliary atresia.

Serum levels in ataxia telangiectasia are rarely above 800 μg/l. Very high levels are seen in tyrosinosis (5000–5 × 10^6 μg/l). Elevated serum levels are seen both in congenital biliary atresia and neonatal hepatitis. Levels are usually higher, although not invariably, in the latter than in the former disorder.

Serum AFP levels and their use in childhood oncology are discussed elsewhere (Chapter 19).

Alpha-1-antitrypsin

Alpha-1-antitrypsin functions as a general proteinase inhibitor and shows a high degree of genetic polymorphism. The inheritance is autosomal co-dominant. The variants are classified by the Pi (proteinase inhibitor) system. Some 30 or more variants are recognised and allocated an alphabetical connotation formerly depending on electrophoretic mobility in acid conditions, and more recently characterised by isoelectric focussing. PiM is the most frequent allele. Letters earlier in the alphabet (e.g. PiF) refer to variants with more anodic mobility and those with later letters of the alphabet (e.g. PiZ) to variants with more cathodic mobility.

A few of the variants of alpha-1-antitrypsin (including the sub-types of PiM) are associated with deficiency. The circulating levels are shown in Table 8.4 together with the relative incidence within the UK population. The S and Z genes result in very low levels of serum alpha-1-antitrypsin. Evidence suggests that the protein is synthesised within the hepatocytes but the final stages of maturation involving glycosylation steps are defective leading to failure of secretion. A single amino acid substitution probably accounts for the defect. The gene associated with zero synthesis

Table 8.4 Some of the more common variants of alpha-1-antitrypsin.

Phenotype	% of MM value	Association % of population*
MM	100	88.5
MS	80	8.0
MZ	60	1.5
SS	60	0.35
SZ	35	0.22
ZZ	15	0.04
FM	100	0.66
FS	80	0.02
Z-	8	—

* Data from survey of 4565 people (Cook 1975). The Genetics of alpha-1-antitrypsin: A family study in England and Scotland. *Ann. Hum. Genet.* (*London*) **38**, 275–287.

is the null gene (Pi-). The genotypes PiZZ, PiZ-, PiSZ and Pi- are associated with the clinically important deficiency states.

Assay of alpha-1-antitrypsin

Alpha-1-antitrypsin may be assayed by gel immunoprecipitation or nephelometric techniques.

Reference ranges

In the first month of life the value falls from that seen in cord serum and then progressively increases to reach the adult level by 6 months of age (Table 8.5).

Alterations in disease

Low levels of alpha-1-antitrypsin are observed in the neonatal respiratory distress syndrome, which may result from capillary leakage into the interstitital compartment, and in severe protein losing conditions. Increased levels are more frequently seen as alpha-1-antitrypsin is one of the acute phase reactants.

Deficiency states are associated with lower lobe emphysema of early onset (20–30 years) and more importantly in the present context with transient cholestatic jaundice in the newborn proceeding in some cases to progressive juvenile cirrhosis. The neonatal hepatitis occurs in about 20% of PiZZ infants and presents within the first four weeks of life. The transaminases and alkaline phosphatase are raised and the histology is that of giant cell hepatitis. About one third of the patients progress to fatal juvenile cirrhosis.

There are two possible explanations for the liver disease associated with alpha-1-antitrypsin deficiency. The accumulation of alpha-1-antitrypsin may itself be a destructive process. Alternatively, a more probable explanation is that irreversible damage occurs as the result of an insult such as viral hepatitis. The situation in the liver is, therefore, analogous to that

Table 8.5 Reference range for alpha-1-antitrypsin (Ward & Allen 1983).

Age	No. tested	Mean g/l	Range g/l (5th to 95th percentile)
Cord serum	116	1.63	1.1–2.4
0–1 month	102	1.29	0.9–1.7
1–3 months	116	1.36	0.9–2.1
3–6 months	97	1.48	1.0–2.2
6 months to adult	701	1.50	1.1–2.1

of the lung, in that there is increased susceptibility to damage rather than inherently progressive changes. Other diseases have been associated with alpha-1-antitrypsin deficiency, including a number of connective tissue diseases. A convincing association with membranoproliferative glomerulonephritis has been shown in childhood cirrhosis.

Alpha-1-antitrypsin has an important role in counteracting the proteases spilt from activated leucocytes; not only its absence but also the allotypic variants result in less function, thus predisposing to more damage when there is excessive activation of leucocytes by immune complexes. This best explains the more severe effects of hepatitis, nephritis and lung infections. Where the immune complexes become excessive, the heterozygous MZ and the homozygous SS variations can predispose to such damage and this should be remembered in Still's disease where lung fibrosis is more likely to occur; indeed with such allotypes, prophylactic high dose corticosteroids should be used while immune complex levels are high.

Several reviews provide more extensive accounts of alpha-1-antitrypsin (Carrell & Owen 1979; Fagerhol & Cox 1981).

Caeruloplasmin

Caeruloplasmin is unlikely to function as a transport protein for copper since the binding of the copper is so tight. Alternatively it may function as an oxido-reductase and seems to play an important role in the binding of iron to apotransferrin. In addition there is evidence that caeruloplasmin may aid in maintenance of the homoeostasis of hepatic copper. Liver transplantation has shown that provision of a site of caeruloplasmin synthesis corrects both the serum level and the manifestations of Wilson's disease, including the clearing of cerebellar dysfunction. It is, of course, much easier to treat Wilson's disease by existing methods.

Assay of caeruloplasmin

Caeruloplasmin may be assayed by immunoprecipitation or nephelometric techniques.

Reference range

Caeruloplasmin is present in the neonate in lower levels than in the normal adult. Levels relatively similar to those of the adult are reached by 1 year of age. (Serum concentrations are given below.)

Age	g/l
0–4 months	0.08–0.23
4 months–1 year	0.12–0.35
1–10 years	0.20–0.40
10–13 years	0.15–0.23

Alterations in disease

The serum level may be raised in the acute phase. Low values are found in Wilson's disease, malnutrition, malabsorption and nephrosis.

The principal use of the measurement of caeruloplasmin in serum, together with serum copper and urine copper is in the diagnosis of Wilson's disease. Significant findings are:

1 Low levels of serum caeruloplasmin.
2 Low levels of total serum copper (95% of serum copper is associated with caeruloplasmin).
3 Raised levels of urinary copper.

The metabolic defect in Wilson's disease appears to be primarily deranged copper metabolism. Abnormal deposits of copper are found in tissues, particularly in brain, liver, renal tubules and the eye. Whereas neurological symptoms are more usual in the adult, children usually present with hepatic disturbances. Deposition of copper at the junction of cornea and sclera gives rise to the classic Kayser–Fleischer rings.

Transferrin

Transferrin is the specific iron-binding protein of plasma. The iron is carried in the ferric form to stores and to bone marrow. In stores iron is laid down as ferritin or haemosiderin. At least 20 variants of transferrin are documented, but so far only one variant is associated with a quantitative deficiency.

Assay of transferrin

Transferrin may be assayed directly by immunoprecipitation or nephelometric techniques. An indirect measurement of transferrin may be obtained by estimation of total iron binding capacity.

Reference range (immunochemical method)

The mean neonatal level of transferrin is in the range of 1.9 g/l. Postnatally levels are maintained in the healthy baby and reach the stable adult range of 2.0–4.0 g/l by 9 months of age.

Alterations in disease

A low transferrin level resulting from inherited deficiency (atransferrinaemia) is extremely rare and results in marked hypochromic anaemia during early childhood.

Low serum transferrin levels are also seen in malnutrition, malabsorption and protein-losing states. Raised values are associated with iron deficiency anaemia.

Ferritin

Transferrin levels do not always correlate well with total body iron stores, due to influences other than iron on transferrin levels. There is a direct relationship between serum ferritin levels and body iron stores. Iron deficiency is invariably associated with low serum ferritin levels. Ferritin is, therefore, the most reliable protein that can be measured for the assessment of iron status.

While the various tissue isoferritins and the embryonic variations have been of interest, the very wide range of reference ranges and the vast number of diseases that provoke elevations neutralise the value of assays for monitoring cancers, etc.

Acute phase proteins

In several recent studies the acute phase response has been evaluated as an aid to diagnosis of infection in the newborn. Septicaemia in the newborn may be life-threatening and early diagnosis and treatment are essential. Among the acute phase reactants, C-reactive protein (CRP) and alpha-1-acid glycoprotein appear to be the most sensitive indicators.

CRP is present normally at less than 10 mg/l and dramatic rises can be seen in infection. In a recent study (Sabel & Wadsworth 1979) 16 of 18 newborns with bacterial septicaemia showed raised levels. Raised CRP levels were also reported to be of definite value in the early diagnosis of bacterial meningitis but not in neonatal urinary tract infections or viral diseases.

Alpha-1-acid glycoprotein has also been reported to be raised in 85% of infants with severe bacterial infections (Sann *et al.* 1981).

Recently, there has been considerable interest in the serum amyloid A protein, which would appear to be the most sensitive of all acute phase reactants, including CRP. The clinical applications of the measurement of this protein in disease remain to be assessed.

Complement

Complement is a multi-component system of proteins in the blood and tissue fluids. Sequential activation leads to three main effects:

1 Release of peptides active in inflammation.
2 Attachment and digestion promotor for phagocytosis.
3 Membrane damage resulting in lysis.

Complement may be activated by either of two pathways, the classical and alternative. The complement components, in sequence of activation, together with their inhibitors and activators are shown in Table 8.6.

Assay of complement

Complement components may be investigated by total activity, quantitation of individual proteins by immunochemical methods, detection of products of activation (conversion products) or functional assays for individual components.

Total complement in serum may be assessed by haemolytic activity using a test system of sheep erythrocytes and rabbit antibody to sheep erythrocytes. Results are expressed as complement concentration giving 50% lysis (CH50). The test has a wide normal range and is insensitive, however it may be modified for the assay of individual components. Such assays are time consuming, expensive and are best performed in specialised laboratories.

Measurements of C3, C4, factor B and the C1 esterase inhibitor have the widest routine clinical applications. Techniques such as radial immunodiffusion, automated immunoprecipitation and laser nephelometry may be used. Conversion products may cause erroneous results. Assays should be done using fresh serum or EDTA plasma to minimise activation. Gels used in complement assays should contain EDTA to avoid activation by the gel.

Conversion products of C3 and factor B may be detected by crossed

Table 8.6 Complement components, control proteins and activators.

	Complement components	Control proteins (inhibitors)	Activators
Classical pathway	C1	C1 esterase inhibitor	Antigen–antibody complexes
	C4	C4 binding protein (C4 BP) + H	Aggregated Ig
	C2, C3, C5–9		Proteolytic enzymes
Alternative pathway	C3	I (previously C3b inactivator)	Bacterial lipopolysaccharide
	C5–C9	H	
	B		Endotoxin
	D		Aggregated IgG & IgA
	Properdin		

immunoelectrophoresis or more simply by immunofixation (Whicher *et al.* 1980).

Reference ranges

There is as yet no internationally accepted standard for complement components. Any laboratory performing such assays should establish normal ranges. The table below gives some indication of relative ranges in young children, together with the adult reference ranges of this laboratory.

Age	C3 (g/l)	C4 (g/l)
Cord blood	0.5–1.2	0.06–0.2
1–5 months	0.6–1.7	0.07–0.4
6–12 months	0.7–1.7	0.09–0.4
1–10 years	0.8–1.7	0.1–0.6
Adult	0.7–1.7	0.1–0.7

Complement and disease

Complement is involved in disease in a number of ways:
1 As a result of genetic deficiencies.
2 As a result of acquired deficiencies.
3 Secondary to activation by antigen–antibody complexes in the classical pathway or direct activation by endotoxin or other substances in the alternative pathway.

Tables 8.7 and 8.8 summarise the complement profile in those clinical situations where measurement is of value.

Deficiencies of complement components or activation results in subnormal values of various components. However, complement components are sensitive acute phase reactants and rising levels result in values within the normal range, even though activation is present. In this situation detection of conversion products is a more sensitive indicator of activation.

Immunoglobulins

The term immunoglobulin refers to that kind of protein in which specific antibody activity can be found. The immunoglobulin molecule consists of a basic subunit of two heavy chains joined to two light chains. In a given immunoglobulin, only one class of heavy chain (G, A, M, D, E) and one type of light chain (K and L) is found.

IgG is the major immunoglobulin and in normal serum accounts for 73% of the total immunoglobulin. Some 65% of IgG is in the extravascular,

Table 8.7 Genetic errors of complement.

Disease	Significant complement profile
Hereditary angio-oedema	Decreased C1 esterase inhibitor* Low C4 Normal C3 Low CH50
Increased incidence of infections with Gram-positive organisms	Inherited deficiency of C3 Inherited or acquired deficiency of C3b inactivator Inherited deficiency of C1q
Increased incidence of immune complex disease, e.g. SLE	Inherited deficiency of C1, C2 or C4
Recurrent or persistent infections with neisserian organisms	Inherited deficiency of C5, C6, C7 or C8
Partial lypodystrophy	Persistently low C3 C4 often normal Nephritic factors often present

* A non-functional C1 esterase inhibitor has been reported. Normal values are found by immunochemical assay but low values in a functional assay. However, in this laboratory we have yet to see such a case. The incidence in the UK is probably less than 10% of HAE patients. Non-functional C1 esterase inhibitor is falsely diagnosed on old samples and must be proved with absolutely fresh serum.

Table 8.8 Complement and acquired diseases.

Disease	Significant complement profile
Disseminated cytomegalovirus infection	Very low C4* Normal or raised C3
Gram-negative bacteraemia	Low C3 and Factor B with conversion products Normal C4
Gram-positive bacteraemia	Low C3 and C4 with conversion products
Sub-acute bacterial endocarditis	Low C3 and C4 with conversion products
SLE	Low C3 (or normal C3 with conversion products) Low C4
Post-streptococcal glomerulonephritis	Low C3 returning to normal after several months
Mesangiocapillary glomerulonephritis	Persistently low C3 C4 often normal Nephritic factors often present

* Probably due to IgM antibodies working directly through the classical pathway. C4 consumption out-stripping its replacement, whereas C3 acute phase response maintains the C3 level.

extracellular compartment and a major function is protection of the tissue spaces by the neutralisation of diffusible toxins.

About 20% of the circulating immunoglobulin pool consists of IgA and in the serum normally consists almost entirely of the basic 7S immunoglobulin. IgA is the major immunoglobulin in saliva and some 60% of all IgA is synthesised at the secretory surface of the gastrointestinal, respiratory and urinary tract. Secretory IgA is a dimer, whose formation is encouraged by the presence of a J-chain, also synthesised within the plasma cells. A secretory piece is added by the epithelial cells as the dimer is exported into the mucus, where it plays its major protective role. Secretory IgA reaches adult levels in the saliva by the age of six weeks. Serum 7S IgA is mainly derived from the bone marrow, slowly populated until adolescence by memory cells generated in the mucosa, and seems to play an important role in 'damping' and regulation of immune responses.

IgM is a pentamer and due to its high molecular weight (980 000) is 85% extravascular. It constitutes some 7% of the total circulating immunoglobulin pool and plays a major role in protection of the bloodstream by its great efficiency in directly fixing complement to rapidly inactivate organisms.

Reaginic antibodies in man are predominantly of the IgE class. These antibodies bind to cell membranes of mast cells and basophils. Reaction of IgE with specific antigen triggers the release of a number of amines (bradykinin, histamine, serotonin, etc.) which mediate the local anaphylactic reaction. Immediate hypersensitivity reactions of this type play an important role in defence against helminth infection. In regions where parasitic infestations are less common the adverse effects of IgE-mediated responses, i.e. atopy, are clinically of greater importance.

IgD is predominantly a membrane-bound immunoglobulin and its measurement in paediatric samples has as yet had no clinical application, although it seems an interesting marker of tonsillar and adenoidal B-cells.

Assay of immunoglobulins

IgG, IgA and IgM may be assayed by radial immunodiffusion, nephelometry or turbidimetry.

Total IgE is most often assayed by radioimmunoassay. This may be double antibody or solid-phase assay which is recommended. Enzyme-linked immunoassay may also be used. Specific IgE is similarly detected using radioimmunoassay or enzyme-linked immunoassay with specific antigens. The technique most widely used is the radioallergosorbent test (RAST).

The British Working Standard 67/99 for IgG, IgA and IgM and the British Standard for IgE 75/502 are available as primary standards from the National Institute for Biological Standards and Control, Hampstead, London NW3. The freeze-dried standard 67/99 is unsuitable as a standard in nephelometric techniques. The IFCC 74/1, a liquid preparation, suitable as a primary standard in nephelometric assays is available from the Protein Reference Unit, Westminster Hospital, London SW1P 2AR.

Reference ranges (IgG, IgA, IgM)

Of all the serum proteins the immunoglobulins show the widest variations with age. The full-term neonate has a plasma gammaglobulin concentration of about 9–12 g/l, which consists almost entirely of IgG derived from the maternal circulation by placental transfer. The fall of IgG in the first months of life can be attributed to the decay of this maternal IgG prior to the slow rise corresponding to production of gammaglobulin by the infant.

Age-related reference ranges for IgG, IgA and IgM are shown in Table 8.9. The mass concentration used to express immunoglobulin values was calculated from the assigned International Units of the standard 67/99 by standardised conversion factors. These are 86.6 μg/IU for IgG, 16.4 μg/IU for IgA and 8.67 μg/IU for IgM.

Alterations in disease. The measurement of serum immunoglobulins in clinical medicine is only of value when the observed pattern is related to

Table 8.9 Reference ranges for IgG, IgA and IgM.

Age	IgG (g/l)	IgA (g/l)	IgM (g/l)
Cord serum	5.2–18.0	below 0.02	0.02–0.2
0–2 weeks	5.0–17.0	0.01–0.08	0.05–0.08
2–6 weeks	3.9–13.0	0.02–0.15	0.08–0.4
6–12 weeks	2.1– 7.7	0.05–0.4	0.15–0.7
3–6 months	2.4– 8.8	0.1 –0.6	0.2 –1.0
6–9 months	3.0– 9.0	0.15–0.7	0.4 –1.6
9–12 months	3.0–10.9	0.2 –0.7	0.6 –2.1
1–2 years	3.1–13.8	0.3 –1.2	0.5 –2.2
2–3 years	3.7–15.8	0.3 –1.3	0.5 –2.2
3–6 years	4.9–16.1	0.4 –2.0	0.5 –2.0
6–9 years	5.4–16.1	0.5 –2.4	0.5 –1.8
9–12 years	5.4–16.1	0.7 –2.5	0.5 –1.8
12–15 years	5.4–16.1	0.8 –2.8	0.5 –1.9
15–45 years	5.4–16.1	0.9 –3.4	0.5 –2.0

Ranges are for levels in Caucasians (± 2 SD median by Log–Gaussian) determined in groups of at least 30 subjects. These are adjusted to the new g/l equivalents for International Standards (Reimer C. B., Smith S. J., Wells T. W., Nakamura R. M., Keitges P. W., Ritchie R. F., Williams G. W., Hanson D. J. & Dorsey D. B. (1982) *Am. J. Clin. Path.* 77, 12–19).

the clinical context and matched controls. The clinical situations in which serum immunoglobulins are of value are shown in Table 8.10.

In general infection provides a multiple antigenic challenge and typically all immunoglobulins show similar elevations. When the challenge is mainly directed at mucosal surfaces an IgA response may predominate, or where the bloodstream bears the main insult, IgM may predominate.

In paediatrics quantitation of immunoglobulins is requested most often

Table 8.10 Indications for determination of serum IgG, IgA and IgM.

Suspected intrauterine infection
Neonatal infection
Primary immune deficiency
Secondary immune deficiency
Transient immune deficiency of infancy
Monitoring of severe hypogammaglobulinaemia receiving gammaglobulin replacement therapy

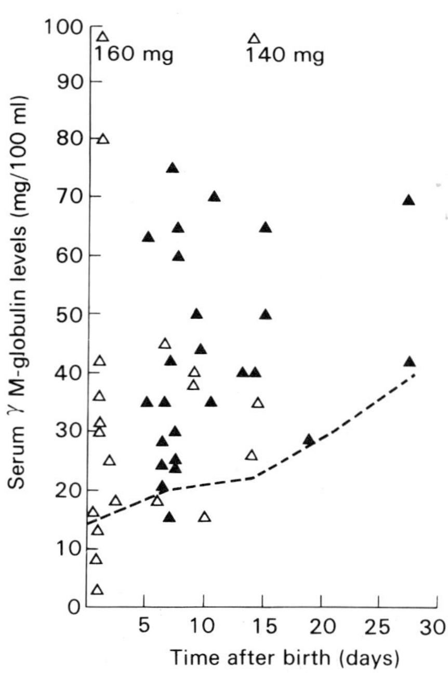

Fig. 8.2 Serum IgM levels in 45 neonates with confirmed infection. The broken line indicates the upper limit of normal for uninfected controls.

in the diagnosis of immunodeficiency states and diagnosis of intrauterine or neonatal infections. The neonate usually has very low levels of IgA and IgM. Raised IgM in cord serum may be associated with a number of intrauterine infections including congenital syphilis, toxoplasmosis, cytomegalovirus and rubella. Although during the first six weeks of life infection can be difficult to diagnose, elevations of IgM associated with infection are readily detected (Fig. 8.2). It is good practice to keep all cord sera for a limited time so that they may be screened for elevated immunoglobulin, which would indicate intrauterine stimulation or to provide a baseline for any subsequent elevation which might establish that postnatal infection has occurred.

After six weeks of age the normally rising IgM obscures the rises due to infection and IgM measurement in suspected infection is of less value.

Low immunoglobulin values may be associated with primary inherited deficiencies, or may occur transiently in infancy or secondarily to a number of other pathologies.

Primary immune deficiency

The primary immune deficiencies involving immunoglobulin are summarised in Table 8.11 in approximate order of frequency of occurrence.

Where another child is born into a known family, fetoscopy or cord samples have characteristic findings in Groups 3, 4, 6 and 12 in Table 8.10. Severe combined immune deficiency (SCID) broadly falls into two major categories: those with severe lymphopenia (less than 600×10^6/l) at birth who also have absent B-lymphocyte markers, e.g. Swiss-type; and those where B-cell markers are present and where occasionally a subset of helper T-cells (which may be identified by the monoclonal antibody OKT4) are not detectable but more often these may only be diagnosed after birth by functional assays. Deficiency of adenosine deaminase, of 5′-nucleoside phosphorylase and of carboxylase generally produce findings similar to the second group above, but of course can be directly diagnosed by enzyme assays. These assays are best done in a specialist centre.

Bruton's agammaglobulinaemia can be diagnosed by absence of membrane immunoglobulins on circulating lymphocytes. Ataxia telangiectasia may be identified by DNA challenge assay, where defective repair is seen.

Transient hypogammaglobulinaemia of infancy

This is usually classified as a primary deficiency but it is not sex-linked, occurs at a rate of about 4% of live births (1% suffer severe hypogammaglobulinaemia), and normal levels of circulating immunoglobulins are achieved, in most instances, after a slow start. The heterozygous state of genetic hypogammaglobulinaemia can account for only about 5% of the

Table 8.11 Primary immune deficiencies affecting immunoglobulin levels.

Disease		Immunoglobulin pattern	Clinical manifestations
1	Selective IgA deficiency	Both secretory and serum IgA absent[1]	Repeated infections at mucosal surfaces, 30% affected individuals are symptom-free
2	Selective IgM deficiency	IgM absent or low[2]	Septicaemia
3	Severe combined immune deficiency (SCID)	Maternal IgG, IgM variable, IgA absent	Recurrent infections without treatment do not usually survive childhood
4	X-linked agammaglobulinaemia (Bruton's)	Maternal IgG, IgM variable, IgA absent. Membrane immunoglobulin absent	Recurrent infections with extracellular pyogenic pathogens
5	Immune deficiency with thrombocytopenia and eczema (Wiskott–Aldrich)	Low IgM, frequently raised IgA. Deficient response to carbohydrate antigens. Absent isohaemagglutinins	ENT infections, thrombocytopenia and eczema
6	Ataxia telangiectasia	Variable but may be IgA deficient	Cerebellar ataxia appears around 3 years of age
7	IgG subclass deficiency	Borderline IgG levels. IgG_2 subnormal[3]	Pyogenic infections
8	IgA and IgM deficiency	Some IgG	Predominantly gut disease (lymphoid nodular hyperplasia)
9	IgA and IgG deficiency	Raised IgM	Recurrent infections. Tonsils and adenoids recur after removal. Hepatosplenamegaly
10	Other immunoglobulin deficiencies	Low or absent IgG or IgM. Functional deficiencies	Recurrent infections (staphylococcal, etc.)
11	Sporadic agammaglobulinaemia	All reduced. Membrane immunoglobulins present	Infections with varied organisms
12	Immunodeficiency and thymoma	All markedly reduced	Eosinophilia. Pure red cell aplasia in some cases
13	Transcobalamin II deficiency	All markedly reduced	Frequent bacterial infections

1 Exclude use of anti-epileptic drugs.
2 Exclude lymphoid neoplasia, cytotoxic drugs, etc.
3 Exclude protein-losing states, etc.

observed cases. Transient hypogammaglobulinaemia presents at 4–6 months of age when the infant's own IgG synthesis should normally be replacing the maternal IgG (Fig. 8.3). In most cases the mechanism of the 'delayed maturity' is largely unknown, however, a new common finding in these infants has been an excess of suppressor T-lymphocytes bearing the OKT8 marker. Contributing factors may also be transfer of alloantibodies or paraprotein from mother to fetus. Premature babies are also at risk due to incomplete transfer of maternal IgG (Fig. 8.3). The use of a single injection of prophylactic gammaglobulin for such infants does not delay subsequent maturation but does reduce both morbidity and mortality.

Secondary hypogammaglobulinaemia

This is 10–100 times more common than the primary forms. The clinical situations giving rise to secondary hypogammaglobulinaemia are shown in Table 8.12. In general, marrow disorders and hypercatabolism (including protein-losing states) result in predominantly low values of IgG and have less effect on IgA and IgM. In toxic suppressions and lymphoid neoplasia IgM is usually more affected than IgA, which in turn is affected more than IgG. The immune suppression associated with lymphoid neoplasia is not as marked as in adults, which may be due to these tumours being faster growing in children.

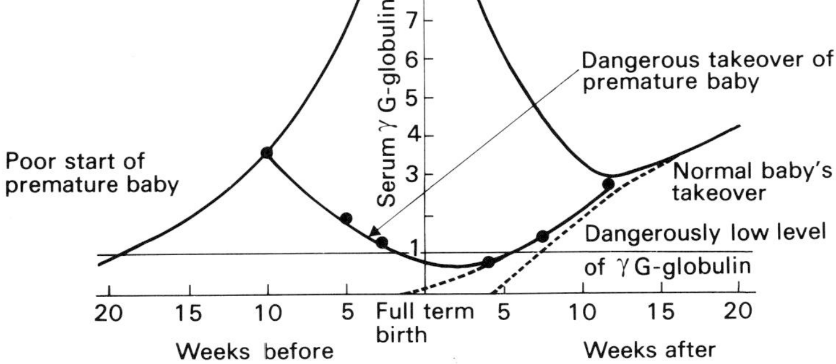

Fig. 8.3 Mechanism of hypogammaglobulinaemia in the premature infant. The maternal IgG may fall to low levels before the infant's own IgG takeover. Courtesy of *J. Clin. Path.* (Riches & Hobbs 1979).

Table 8.12 Clinical conditions giving rise to secondary hypogammaglobulinaemia.

Marrow disorders
Hypoplasia
Bony metastases
Myelosclerosis
Short survival of IgG
Nephrotic syndrome
Protein-losing enteropathy
Myotonic dystrophy
Toxic factors
Prolonged uraemia
Gluten-sensitive enteropathy
Cytotoxic and radiation therapy
Severe infection
Lymphoid neoplasia

Reference range for IgE

The normal values of serum IgE found in children are shown in Table 8.13. These values were originally derived in a Swedish study (Kjellman *et al.* 1976). Investigators in many countries have found similar ranges.

Alterations in disease. Primary deficiencies of IgE are recorded. Routine estimation of serum IgE is used for the detection of raised values. The conditions associated with raised values are shown in Table 8.14.

Most patients with allergic disease show raised values. An increased serum IgE level in childhood may predict future atopy and is a useful screening test, particularly for children of atopic parents. Where desired lymphocytes bearing IgE in cord blood samples can be counted and normally are less than 40×10^6/l, but for infants who become atopic they already exceed 80×10^6/l even though cord serum IgE is undetectable. This can identify those infants most worthy of preventative care.

The clinical situations in which it is useful to determine total or specific IgE are summarised in Table 8.15.

Specific IgE testing (RAST). The range of readily available antigens for detection of specific IgE by radioallergosorbent test (RAST) is steadily growing. The correlation with established procedures such as direct skin testing and bronchial or nasal provocation tests varies with the allergen being investigated. Correlation is high with some pollen allergens, 80% with skin prick and 90% with provocation testing. However, the house dust allergen correlation is much lower, at about 60%. This may reflect the incomplete extraction of antigens for the RAST.

Table 8.13 Total IgE reference values.

Age	Geometric mean (kU/l)	+1 SD (kU/l)
Below 3 months	3.2	11.0
4–12 months	3.8	11.0
1–2 years	8.3	29.2
2–3 years	14.4	42.5
4–5 years	15.3	52.2
6–7 years	17.8	56.2
8–10 years	18.0	63.4
11–12 years	13.0	44.8
13–14 years	20.9	70.3
Adult	25.6	81.0

Levels above mean +1 SD indicate a high probability of atopic disease (Fifield R., Protein Reference Unit, Cardiff Royal Infirmary, pers. comm.)

Table 8.14 Causes of increased serum IgE.

Allergic disease
Extrinsic asthma
Hay fever
Seasonal allergic rhinitis
Atopic eczema

Atopy-related disorders
Minimal change nephrotic syndrome
Interstitial nephritis due to drugs
Wiskott–Aldrich syndrome
Many T-lymphocyte deficiencies (primary or secondary)

Parasitic disorders
Tropical eosinophilia
Ascariasis
Echinococcosis
Hookworm
Intestinal capillariasis
Bilharziasis
Trichionosis
Schistosomiasis
Toxocara canis infestation

Table 8.15 Indications for determination of serum IgE.

Total IgE
Distinguishing IgE-mediated from non-IgE mediated disorders (rhinitis, bronchitis, asthma, dermatitis, chronic urticaria, food intolerance)
Indication of high risk for atopic diseases
Specific IgE
Severe dermatitis which excludes skin testing
Patients receiving symptomatic treatment (i.e. antihistamine)
Allergens that cannot be used for skin testing (toxic, insoluble, etc.)
High sensitivity where testing *in vivo* could be dangerous
Food allergies

A false-positive finding may indicate that low levels of specific IgE are not clinically important. Similarly RAST may be negative when provocation testing clearly indicates clinically significant hypersensitivity. In these cases it is possible that there is only regional synthesis, sufficient to sensitise mast cells but not to give detectable serum levels of specific IgE.

The value of RAST results with regard to food allergens is even more difficult to interpret. There is good correlation between testing *in vitro* and sensitivity to inhaled foods. However, it is rarely possible to implicate ingested foods unequivocally in the pathogenesis of chronic bronchial asthma. The problems are even more complex when dealing with suspected allergic gastroenteritis. Damaged mucosal surfaces may allow the passage of complex antigens, leading to detection of specific IgE to these antigens, which are unrelated to the antigen responsible for the allergic reaction. It must be noted that late asthmatic response may involve complex immune reactions, indicated by antibodies of IgG and IgM class which are not detected in the routine RAST. Undoubtedly RAST has a place in the battery of laboratory tests of allergy but as yet cannot fully replace skin and provocation testing. Measurement of specific IgE should not be requested either as a 'screen' for allergic disease or in conditions in which IgE-mediated mechanisms are not involved (e.g. contact dermatitis).

URINARY PROTEINS

With the exception of uromucoid, etc., originating in the renal tubular epithelium, and certain formed elements, e.g. casts, the urinary proteins are derived by a process of glomerular filtration from the plasma. Reabsorption of proteins by the renal tubules normally takes place.

Assay of urinary proteins

Assay of total protein may be made by dye-binding methods or turbidimetrically. Specific proteins can be estimated by any of the usual immunochemical methods although these often have to be used in a more sensitive range than for the corresponding serum proteins. The protein patterns are investigated by electrophoresis techniques.

Reference range

Normal subjects excrete up to 0.08 g of protein per day.

Alterations in disease

Glomerular proteinuria occurs with increased glomerular permeability. In the nephrotic syndrome the protein loss is in excess of 5 g/ day and may be massive. The protein loss results in hypoalbuminaemia, oedema and hyperlipoproteinaemia. The renal disease may be primary or secondary to other pathology. The serum electrophoretic pattern is characteristic (Fig. 8.1) and urine electrophoresis gives some indication of the severity of the lesion. A pattern predominantly of low molecular weight proteins (albumin, transferrin and alpha-1-antitrypsin) is associated with selective proteinuria. Increasing amounts of higher molecular weight proteins (gammaglobulin) are associated with less selective lesions. A more precise indication of selectivity may be obtained by immunochemical estimation of albumin and IgG in serum and urine. This differential protein clearance is expressed as a ratio, obviating the need for timed collection of urine. The clearance ratio is calculated as follows:

$$\frac{\text{Urine IgG}}{\text{serum IgG}} \times \frac{\text{serum-albumin}}{\text{urine albumin}}$$

A ratio of less than 0.16 is found with high selectivity and is associated with good response to steroid therapy and a favourable prognosis. The most common cause of proteinuria in children is minimal change glomerulonephritis, which shows high selectivity, i.e. a low clearance ratio.

Postural (orthostatic) proteinuria is more severe in the upright than the prone position. It appears to be glomerular in origin and is most common in adolescents and young adults. This type of proteinuria is usually harmless but renal diseases may occur after some years.

Tubular proteinuria is associated with low molecular weight proteinuria of less than 1 g/day. Concentration of urine followed by electrophoresis shows a characteristic pattern of low molecular weight alpha- and beta-globulins. The beta-2-microglobulin, assayed by radioimmunoassay of unconcentrated urine shows raised values. Renal tubular damage results from many causes but most frequently from pyelonephritis.

Severe haemolysis results in haemoglobinuria and severe muscle damage in myoglobinuria. In both these cases the urine will be red or brown in colour. The two may be distinguished by immunochemical methods.

CEREBROSPINAL FLUID (CSF)

The source of CSF proteins is primarily by filtration from the plasma across the blood–brain barrier. There is some selectivity on a molecular weight basis such that low molecular weight proteins pass more readily than high molecular weight proteins. During the circulation of the CSF there is re-absorption by venules and tissue cells so that CSF proteins tend to equilibrate increasingly with the plasma. The CSF protein concentration of the fluid from the lumbar sac may be three times more concentrated than that found within the ventricles. There is also a relative increase in concentration of higher molecular weight proteins in the lumbar fluid.

In neonates the blood–brain barrier is more permeable than in later life; the protein concentration of the CSF is greater, as is the proportion of higher molecular weight proteins.

There is some local production of immunoglobulin within the CSF, although normally this is a very small contribution to the total protein (about 1 mg per day).

Assay of CSF proteins

Assay of total protein may be made by dye-binding methods or turbidimetrically. Specific proteins of the CSF can be estimated by any of the usual immunochemical methods, although these often have to be used in a more sensitive range than for the corresponding serum proteins. The protein patterns are investigated by electrophoretic techniques, e.g. cellulose acetate membrane, agarose gel, starch gel or polyacrylamide gel.

Reference ranges (total protein, g/l)

Newborn	0.4–1.2
Up to 1 month	0.2–0.8
1 month–adult	0.15–0.45

Alterations in disease

In pathological conditions the factors contributing to increased protein levels are:

1 Transudation of serum which occurs due to local destruction of the barrier.

2 Local occurrence of plasma cells synthesising increased amounts of immunoglobulin.

3 Local destruction of immunoglobulins to produce fragments (including light chains).

The local production of IgG occurs in a number of neurological conditions, e.g. multiple sclerosis, sub-acute sclerosing panencephalitis, viral meningitis and viral encephalitis. The detection of this increased IgG synthesis can aid in the diagnosis of these conditions.

When the levels of IgG and albumin are estimated in relation to the serum levels in the same patient then normal CSF or CSF with raised total protein due to transudation fall into a narrow normal range. Intrathecal IgG synthesis causes ratios outside this normal range, as is illustrated in Fig. 8.4. The IgG synthesised in the conditions listed above also shows a characteristic pattern of cathodic IgG bands on electrophoresis (oligoclonal bands). Polyacrylamide gel electrophoresis, with its high resolving power is particularly useful for detecting these bands. Oligoclonal bands may

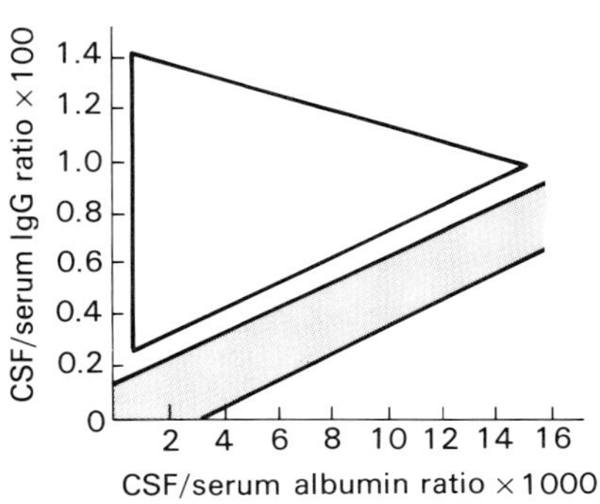

Fig. 8.4 The graph shows the relationship between serum and CSF IgG and albumin. Ratios falling within the stippled area are normal or result from a vascular leak. Ratios falling within the triangle suggest intrathecal IgG synthesis.

occur even when IgG levels are within the normal range. Serum should always be examined to eliminate this as the source of the oligoclonal bands.

SALIVA

The principal application of protein measurements in saliva is to establish the absence of IgA.

Normal range for salivary IgA

6 weeks–adult 0.03–0.2 g/l.

Alterations in disease

Neonates should reach adult salivary IgA levels by six weeks of age and the absence of salivary IgA at this age can give an early indication of immune deficiency. Mixed saliva shows a wide variation in volume and protein content and other protein measurements are of little value.

FAECES

The estimation of protein in faeces or meconium is complicated by protein digestion by both gut and bacterial proteases.

There are only two instances in which protein measurements may be of value. These are as an early screening test for cystic fibrosis and in the estimation of protein loss from the gut.

The protein in meconium probably originates from ingestion of amniotic fluid prior to birth. Most of the protein is degraded by the gut proteases and only two proteins, alpha-1-antitrypsin and alpha-1-antichymotrypsin, which are resistant to proteolysis, are found in the meconium of healthy infants. In pancreatic insufficiency most of the protein remains undegraded and a high level of other proteins, e.g. albumin, results. The ratio of albumin: alpha-1-antitrypsin is, therefore, much higher in cystic fibrosis than in normals. The stability of alpha-1-antitrypsin also makes this a useful, non-invasive test for protein-losing conditions.

Assay of proteins in faeces or meconium

Aqueous extracts of faeces or meconium (1% w/v) in 0.9% sodium chloride are assayed by the usual immunochemical techniques.

Reference ranges

An albumin: alpha-1-antitrypsin ratio of greater than 2.0 is suggestive of pancreatic insufficiency.

Forty normal faecal samples assayed in this laboratory gave alpha 1-antitrypsin values in the range 0.01–0.2 mg/g wet weight of faeces with a mean value of 0.07 mg/g.

REFERENCES

CARRELL R. W. & OWEN M. C. (1979) α_1-Antitrypsin: Structure, variation and disease. *Essays Med. Biochem.* **4**, 83–119.

COOK P. J. L. (1975) The genetics of alpha-1-antitrypsin: A family study in England and Scotland. *Ann. Hum. Genet.* (*London*) **38**, 275–87.

DOUMAS B. T., BAYSE D. D., CARTER R. J., PETERS T. & SCHAFFER R. (1981) A candidate reference method for determination of total protein in serum. 1. Development and validation. *Clin. Chem.* **27**, 1642–50.

FAGERHOL M. K. & COX D. W. (1981) The Pi polymorphism: Genetic, biochemical and clinical aspects of human α_1-antitrypsin. *Adv. Hum. Genet.* **11**, 1–62.

GITLIN D. & GITLIN J. D. (1975) Fetal and neonatal development of human plasma proteins. In Putnam F. W. (ed.) *The Plasma Proteins*, Vol. II, pp. 263–319. Academic Press, New York.

HOBBS J. R., HARBOE N., ALPER C. & JOHANSSON B. G. (1979) The human serum standard IFCC 74/1. *Clin. Chim. Acta* **98**, 179F–86F.

KJELLMAN N.-I. M. (1976) Predictive value of high IgE levels in children. *Acta Paediat. Scand.* **65**, 465–71.

KEYSER J. (1979) *Human Plasma Proteins*. John Wiley & Sons.

PETERS T. (1980) *Albumin. An overview and bibliography*, pp. 1–34. Miles Laboratories Inc.

RICHES P. G. & HOBBS J. R. (1979) Mechanisms in secondary hypogammaglobulinaemia. *J. Clin. Path.* **32**, suppl. (Roy-Coll. Path.) **13**, 15–22.

SABEL K. G. & WADSWORTH CH. (1979) C-reactive protein (CRP) in early diagnosis of neonatal septicaemia. *Acta Pediatr. Scand.* **68**, 825–31.

SANN L., BIENVENU J., LAHET C., DIVRY P., COTTE J. & BETHENOD M. (1981) Serum orosomucoid concentration in newborn infants. *Eur. J. Pediatr.* **136**, 181–5.

WARD A. M. & ALLEN R. (1983) in prep.

WHICHER J., HIGGINSON J., RICHES P. G. & RADFORD S. (1980) Clinical applications of immunofixation: detection and quantitation of complement activation. *J. Clin. Path.* **33**, 781–5.

WU J. T., BOOK L. & SUDAR K. (1981) Serum alpha fetoprotein (AFP) levels in normal infants. *Pediatr. Res.* **15**, 50–2.

BIBLIOGRAPHY

PUTNAM F. W. (ed.) *The Plasma Proteins*. Vol. 1, 2 (1975) Vol. 3 (1977). Academic Press, New York.

HOBBS J. R. (1971) Immunoglobulins in clinical chemistry. *Adv. Clin. Chem.* **14**, 219–313.

Use and abuse of laboratory tests in clinical immunology: Critical consideration of eight widely used diagnostic procedures. Report of an IUIS/WHO Working Group (1981) *Clin. Exp. Immunol.* **46**, 662–74.

SOOTHILL J. F. (1981) The role of immunology in paediatrics. *Rec. Adv. Paed.* **6**, 237–47.

Chapter 9
Plasma Lipid Disorders

JUNE K. LLOYD

Requests for estimations of plasma lipids and lipoproteins in children are relatively uncommon and such analyses form only a small part of the workload of a Department of Chemical Pathology. In General Hospitals with paediatric units figures for work done on children are not usually kept separately, but in the author's hospital, a Teaching Hospital with a childrens department serving a total population of about 250,000, it has been estimated that out of some 2700 estimations of plasma cholesterol and triglyceride annually, about 2% relate to children, and even in a large Childrens Hospital providing both a district and a regional service a relatively small number of estimations a year are performed as routine investigations*. In both hospitals, however, additional studies are undertaken as part of research programmes, but even these do not involve the main hospital laboratory in a large amount of extra work. Nevertheless increasing awareness of the importance of hyperlipidaemia in childhood as a risk factor for coronary heart disease in adult life may result in such laboratories receiving more requests in the future for studies on children, and biochemists being asked to advise on appropriate investigations.

Recent advances in both methodology and in understanding of metabolic pathways have been considerable and the data on which to base concepts of normality have become more secure. Knowledge of lipoprotein structure and metabolism is essential for the interpretation of analytical results and can contribute to understanding the clinical manifestations of disease states, and planning treatment. Conversely, detailed study of some of the disorders still has much to teach us about the functions of lipoproteins.

BIOCHEMISTRY AND METABOLISM

Although lipoproteins were originally classifed according to their mobility on electrophoretic separation it has become more customary to define them in relation to their density, and terminology based on this will in general be used. Table 9.1 gives the average composition of the major classes of lipoproteins. Within each class there is a spectrum of molecular size, and

---

* Data kindly provided by Professor J. A. Owen, St George's Hospital, London and Dr K. Griffiths, Birmingham Children's Hospital.

Table 9.1 Composition of lipoproteins.

	Density g/ml	Percentage composition (approx.)				Apoproteins[1]	
		Protein	Cholesterol[2]	Phospholipid	Triglyceride	Major	Minor
Chylo-microns	< 0.95	1–2	3–7	3–6	80–95	A B C	E
VLDL (pre-beta)	< 1.006	6–10	20–30	15–20	45–65	C B	E
LDL (beta)	1.019–1.063	20–25	50–60	18–24	4–8	B	
HDL (alpha)	1.063–1.21	40–50	18–25	26–32	2–7	A	C E

1 Apoproteins may have several alleles designated A–I, A–II etc.
2 Includes cholesterol esters.
VLDL = very-low-density lipoprotein; LDL = low-density lipoprotein; HDL = high-density lipoprotein.

the fatty acids esterified to cholesterol, triglyceride and phospholipid vary greatly both in respect of their chain length and degree of unsaturation.

Fig. 9.1 illustrates current concepts of lipoprotein metabolism in over-simplified form. Ingested triglyceride is transported out of the intestinal absorptive cell in the form of chylomicrons which, in addition to the triglyceride, contain small amounts of cholesterol (both unesterified on the surface, and esterified within the core), phospholipid, and a mixture of apoproteins. Within the circulation chylomicrons, together with very-low density lipoproteins (VLDL) carrying endogenous triglyceride from the liver, are exposed to extrahepatic lipoprotein lipases at the endothelial surfaces of adipose tissue and muscle (and probably other tissues) which hydrolyse the triglyceride prior to incorporation of fatty acids into the cell for energy or storage purposes. As a result the chylomicrons and VLDL particles shrink in size and increase in density. They also lose some of their surface apoproteins, cholesterol and phospholipids by transfer to high-density lipoproteins (HDL). The remnants of chylomicrons are taken up by specific receptors in the liver and degraded. The small amount of dietary cholesterol entering by the chylomicron route is rapidly cleared from the circulation by this process and thus meals do not for practical purposes affect plasma cholesterol concentration.

Hydrolysis of VLDL triglyceride results in a particle of intermediate density (IDL) whose triglyceride is further hydrolysed by a hepatic lipase. At the same time interchange occurs between the cholesterol of IDL and HDL; the unesterified cholesterol in HDL is esterified by the action of the enzyme lecithin–cholesterol acyl transferase (LCAT) and incorporated into IDL. Removal of all apoprotein apart from apoB, and the loss of most of

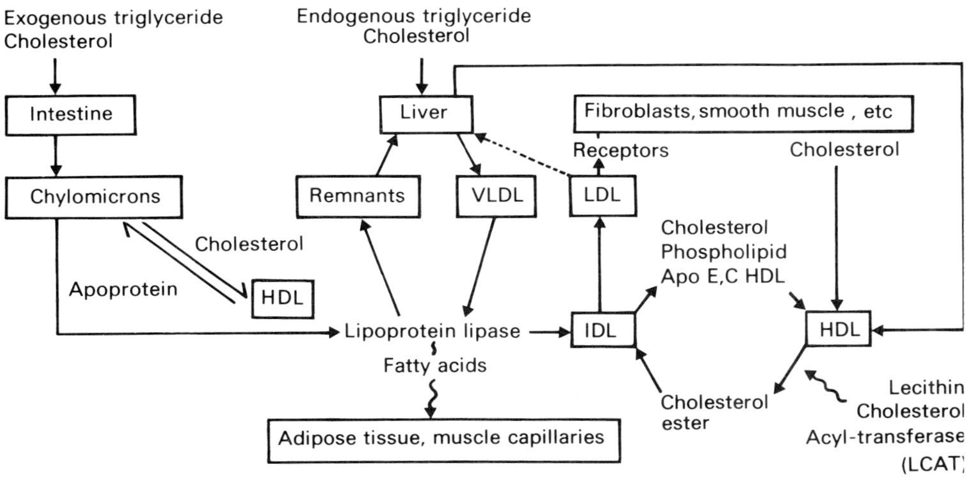

Fig. 9.1 Schematic representation of lipoprotein metabolism and interrelationships.

its triglyceride, results in the conversion of IDL into low-density lipoprotein (LDL).

LDL, carrying about two-thirds of its cholesterol in unesterified form, is bound by specific receptors present on the surface of fibroblasts and smooth muscle cells (probably also of many other cells in the body). Receptor-bound IDL is then taken into the cell where the cholesterol ester is hydrolysed by a lysosomal acid-lipase. Free cholesterol thus liberated is used for membrane synthesis and also regulates both the amount of endogenous synthesis of cholesterol by the cell, and the formation of further surface receptors. HDL is synthesised in the liver and intestine and probably takes up the surplus cholesterol resulting from cell death and membrane turnover. It appears to function primarily as a reservoir for the exchange of free cholesterol and phospholipid between lipoproteins, and for interchange between certain apoproteins. The close interrelationship that exists between the lipoprotein classes is such that in any disorder affecting lipoprotein metabolism some derangement can be expected in all lipoproteins. An excellent review of lipoprotein metabolism, and particularly of lipoprotein receptors and their role in the regulation of plasma cholesterol, has been written by Brown *et al.* (1981).

SAMPLING AND INTERPRETATION

Venous blood is usually taken for plasma lipid analyses although capillary blood is equally acceptable if the laboratory methods have been adapted for small samples. Because plasma triglyceride concentrations fluctuate during lipid absorption, and because plasma cholesterol concentrations are slowly influenced by the total quantity and nature of dietary fat, samples for lipid analysis should ideally be obtained in the fasting state and when the subject has been known to have been eating his usual diet over the preceding two weeks. In practice, however, a random sample is acceptable for plasma cholesterol estimation and, because the clearing of dietary fat in children is normally a very efficient process (the half-time for clearance of chylomicrons is only 4–5 minutes, but the absorptive process of course continues for much longer), it may also be possible to interpret the triglyceride value in such a sample. Nevertheless, if it is likely that the patient has a primary abnormality of lipid metabolism requiring precise diagnosis, blood should be obtained after a fast of not less than 8 hours. Specific instructions are advisable: 'nothing to eat or drink except water after your last meal on the night before the test'. For milk-fed babies the last milk feed should be at midnight, with blood being taken at 8.00 am. If this degree of fasting is unacceptable, a fruit drink or other small amount of carbohydrate in the form of bread and jam is unlikely to interfere with the interpretation of results.

Confirmation of abnormal results is a wise measure, especially if the values are in the 'borderline range'. In the primary lipoprotein disorders treatment is not likely to be required urgently (an exception is familial hyperchylomicronaemia with an acute abdominal crisis, see p. 253), follow-up will be lengthy, and two estimations before treatment is started will not only confirm the diagnosis but also provide a more secure baseline from which to judge the efficacy of therapy.

Lipoprotein studies are generally made either qualitatively by electrophoretic separation or quantitatively by estimating the amount of cholesterol present in HDL after precipitation of the low-density lipoproteins. Such analyses, coupled with inspection of the plasma (if necessary stored in the refrigerator at 4 °C overnight), knowledge of the timing of the sample in relation to the last meal, and an understanding of the composition of the lipoproteins (Table 9.1), should enable an estimation of the nature of any lipoprotein abnormality to be made. Table 9.2 indicates the deductions regarding the lipoprotein status that may be so derived. Only rarely will studies of fatty acid composition, apoprotein distribution, or associated enzyme concentrations be required for diagnosis and management; if necessary, these more specialised determinations are best carried out in centres with special interest and expertise in lipid metabolism. Identification of an abnormal lipoprotein pattern does not of course by itself provide a diagnosis; this can only be established after consideration of the clinical and family data. The typing system based on the electrophoretic appearance (Beaumont *et al.* 1970) should be used with caution when reporting results. The Type II pattern if reported as 'type II hyperlipidaemia' may be misinterpreted by the clinician as indicating that the patient has familial hypercholesterolaemia.

Table 9.2 Interpretation of some plasma lipid and lipoprotein analyses.

Appearance									
Turbidity at room temp.	Cream layer at 4 °C	Total lipids		Lipoprotein electrophoresis					
		C	T	Chylo	β	Pre-β	α	HDL–C	Interpretation
0	0	N	N	nil	N	±	N	N	Normal
±	±	N	±	±	N	±	N	N	Probably non-fasting sample
+	0	N	+	nil	N	+	N	N(−)	Slight increase VLDL
+ +	0	N(+)	+ +	nil	N	+ +	−	−	Increase VLDL
+ +	0	+ +	+ +	nil	broad-beta		−	−	Increase IDL
+ + +	+ + +	+ +	+ + +	+ + +	−	−	−	−	Pathological increase chylomicrons
0	0	+ + (+)	N(±)	nil	+ +	±	N(−)	N(−)	Increase LDL
0	0	−	N	nil	−	−	N(−)	N(−)	Decrease LDL
0	0	− −	− −	nil	absent		− −	− −	Absent LDL, VLDL
+ +	(+)	−	+	+	−	±	absent	− −	Absent HDL
0	0	+(+)	N	0	N	N	+(+)	+(+)	Increase HDL

N = normal range; + or − = degree of increase or decrease; C = cholesterol; T = triglyceride; LDL = low-density lipoprotein; VLDL = very-low-density lipoprotein; IDL = intermediate-density lipoprotein; HDL = high-density lipoprotein; HDL–C = high-density lipoprotein–cholesterol.

NORMAL VALUES

Knowledge of what should be regarded as 'normal' for plasma cholesterol and triglyceride concentrations in children is still imperfect. Values vary between communities and are influenced by a combination of genetic and environmental factors. The degree of change with age, and especially at puberty, has been inadequately studied and ethical considerations have tended to limit the number of investigations performed on children. Even more importantly the relationship between 'normal' values, defined statistically in relation to the values obtaining in a population, and the values desirable for health, or indicative of disease, has not been evaluated.

The largest epidemiological studies have been made in the USA, and these provide ranges for plasma cholesterol, triglycerides and lipoproteins from birth through childhood to adolescence, which are probably applicable to children in industrialised countries. Table 9.3 gives values derived from the Bogalusa heart study (Berenson 1980). Figures from other studies in the USA are comparable but levels of cholesterol in the UK and Australia are probably somewhat higher.

When considering the significance of values in an individual against the background of the population data the following points may be relevant:

1 At birth plasma cholesterol concentrations do not predict levels in later

Table 9.3 'Normal' values[1] for plasma lipids and lipoproteins (median, 5th and 95th centiles, mmol/l).

Age	Total cholesterol	Total triglyceride	HDL cholesterol
Newborn (cord blood)	1.8 (1.1–2.6)	0.4 (0.1–0.9)	0.9 (0.3–1.5)
6 months	3.4 (2.3–4.9)	1.0 (0.6–1.9)	1.3 (0.6–2.2)
1 year	3.9 (2.5–4.9)	0.9 (0.5–1.8)	1.3 (0.6–2.2)
2–14 years	4.1 (3.1–5.4)	0.7 (0.4–1.4)	1.7 (0.8–2.6)

HDL = High-density lipoprotein.
1 Values derived from Bogalusa heart study (USA) (Berenson 1980); values in UK may be somewhat higher (Clayton *et al.* 1980).

infancy or childhood; although the reason for this may be due in part to the fact that a smaller proportion of total cholesterol is in the LDL fraction at birth than at any other age, even LDL values at birth lack correlation with those at 6 months of age. The type of feeding and the changes that occur as the composition of the diet alters may account for some of the variability.

2 During the first year of life, and especially during the first 6 months, concentrations of plasma lipid and lipoprotein rise rapidly. Thereafter there is relatively little change with age up to puberty, such change being largely confined to a slight increase in total and HDL cholesterol.

3 After 1 year of age (and perhaps as early as 6 months) 'tracking' (staying within the same centile band) of plasma cholesterol can be demonstrated. In the Muscatine study (Clarke *et al.* 1978) the 6-year correlation coefficient for cholesterol was 0.61. Children with levels in the upper decile are most likely to remain in this position. Tracking of plasma triglyceride is much less marked, with a correlation coefficient of only about 0.3.

4 Up to 5 years of age lipid and lipoprotein levels do not differ significantly between the sexes. From 5 to 14 years girls tend to have higher triglyceride, LDL and VLDL levels, and lower HDL levels, than do boys.

5 During puberty variations occur which may make interpretation in individuals difficult. Total cholesterol and LDL levels tend to fall and triglyceride and VLDL to rise. Sometime during puberty and in later adolescence LDL and VLDL concentrations in boys become higher and HDL levels become lower than in girls, thus approaching the adult pattern.

6 At all ages diet and other environmental factors may need to be considered (see 'hyperlipoproteinaemia of affluence' p. 259).

HYPERLIPOPROTEINAEMIA

The most common abnormality found on analysis of plasma lipids is some form of hyperlipoproteinaemia. This may be a secondary manifestation in another disease state or may represent a primary genetically determined disturbance. Table 9.4 summarises the hyperlipoproteinaemias which may present in childhood according to the predominant lipoprotein affected.

Primary hyperlipoproteinaemias

The classification based on the lipoprotein pattern which was originally devised for the primary hyperlipoproteinaemias (Beaumont *et al.* 1970) has achieved considerable familiarity and is still useful in certain instances. As understanding of the basic defects improve, however, terminology based on this classification should be replaced by more precise terms.

All the primary disorders can occur in children but most will be uncommon findings in chemical pathology laboratories: some because the disease is itself rare (lipoprotein lipase deficiency, hyperchylomicronaemia type I), some because although fairly common in adults the disorder is infrequently expressed during childhood (hypertriglycerideaemia type IV; combined hyperlipidaemia type IIb; type III; type V), and some because children rarely have clinical abnormalities and therefore plasma lipids are not estimated (hypercholesterolaemia type IIa). Detailed accounts of the disorders may be seen in Stanbury *et al.* (1983).

Familial lipoprotein lipase deficiency (hyperchylomicronaemia type I)

This disorder is rare, is inherited as an autosomal recessive, the basic defect being deficiency of lipoprotein lipase activity, and may present clinically at any time during childhood with eruptive xanthomata, unexplained hepatosplenomegaly or acute attacks of abdominal pain. It may also first be recognised in the laboratory when plasma is received for investigation of some other disorder. Impaired hydrolysis of chylomicrons and VLDL due to lack of lipoprotein lipase activity results in accumulation of these lipoproteins with gross turbidity of the plasma and an obvious cream layer on standing. The biochemical defect is fully expressed at birth and the lipoprotein abnormalities are evident as soon as the infant starts to ingest fat. It is not known why the clinical manifestations are so variable in expression between children, and even in the same child at different ages.

Analysis of the lipids shows greatly increased triglycerides, usually in the range of 40–100 mmol/l. Cholesterol concentrations are also high because the degree of hyperchylomicronaemia is sufficient to make the small proportion of cholesterol in the molecule significant in terms of total plasma values. If the concentrations of triglyceride and cholesterol are expressed as mg/dl the triglyceride : cholesterol ratio is of the order of 10 : 1;

Table 9.4 Hyperlipoproteinaemias.

Main lipoprotein abnormality	Lipoprotein type[1]	Plasma lipids Cholesterol	Triglyceride	Primary disorders	Secondary disorders
Increased chylomicrons	I	++	++++	Familial lipoprotein lipase deficiency	Diabetes mellitus (rare)
Increased LDL	IIa	++(+)	N	Familial hypercholesterolaemia	Hypothyroidism, nephrotic syndrome, obstructive jaundice[2], hepatic glycogenoses (rare), acute porphyria, cholesterol ester storage disease
Increased LDL	IIb	++(+)	+	Familial combined hyperlipidaemia	
Increased IDL	III	++	++	Familial type III hyperlipidaemia	Hypothyroidism (rare)
Increased VLDL	IV	N or +	++	Familial hyper-triglyceridaemia	Diabetes mellitus, nephrotic syndrome, renal dialysis, hepatic glycogenoses, obstructive jaundice, lipodystrophy, idiopathic hypercalcaemia
Increased VLDL and chylomicrons	V	++(+)	++(+)	Familial type V hyperlipidaemia	Diabetes mellitus, obstructive jaundice
Increased HDL	–	+	N	Familial hyperalphalipo-proteinaemia	

1 WHO classification (Beaumont *et al.* 1970).
2 In obstructive jaundice there may be an accumulation of lipoprotein X.
LDL = low-density lipoprotein; IDL = Intermediate density lipoprotein, abnormal in composition and electrophoretic and ultracentrifugal characteristics; VLDL = very-low-density lipoprotein.

about 50% of the cholesterol is unesterified, reflecting the composition of chylomicrons. Lipoprotein electrophoresis shows considerable accumulation at the origin often tracking forwards to obscure the beta and pre-beta areas. Estimations of LDL and HDL show these lipoproteins to be reduced in concentration and this feature becomes even more noticeable after treatment.

The diagnosis can usually be made on the plasma findings together with rapid clearing of the hyperchylomicronaemia (usually within 4–5 days) on withdrawal of dietary fat, and the exclusion of any other disorder known to cause secondary hyperchylomicronaemia. Lipoprotein lipase activity can be estimated after heparin stimulation but if this is done the test must be carried out before (or within a few days of) starting a low-fat diet; such investigations are best undertaken by a specialised laboratory.

Investigation of the parents (heterozygotes) may show mild lipoprotein abnormalities but is not often rewarding. Siblings should be studied, as symptomless individuals with the disorder may thus be detected.

Treatment is by reduction of dietary fat of all types. In children with abdominal crises (acute pain, often with vomiting, which is thought to be due to pancreatitis and in which plasma amylase levels may be raised) *all* fat should be withdrawn. In other cases a virtually fat-free diet is usually given until the plasma is optically clear, and initial treatment can be easily monitored by observing the decline in plasma turbidity. For subsequent management the degree of reduction in dietary fat required in the long-term depends on the individual child and should be low enough to prevent symptoms. Medium-chain triglycerides may prove a helpful adjunct to the diet as they do not utilise the chylomicron route of absorption. Laboratory monitoring of plasma cholesterol and triglyceride usually shows that triglyceride levels remain moderately elevated but cholesterol is within, or even below, the normal range. The disorder is not thought to be associated with premature atherosclerosis and attempts to maintain plasma triglyceride within the normal range are probably not justified.

Familial hypercholesterolaemia (FH, type IIa)

This is the most common of the familial hyperlipoproteinaemias to be expressed during childhood and is inherited as an autosomal dominant with a gene frequency perhaps as great as 1 in 500. The basic defect is thought to be a deficiency of LDL receptors at cell surfaces (Brown *et al.* 1981).

The homozygous form is very rare, presenting during early childhood with tendon and skin xanthomata, and causing premature atherosclerosis with clinical evidence of cardiac dysfunction towards the end of the first decade and death from myocardial infarction either in the second or third decades. Plasma cholesterol levels are extremely high (of the order of 17–26 mmol/l), triglycerides are normal and HDL is often low. Electrophoresis shows a dense beta band. The diagnosis must be confirmed by demonstrating the heterozygous state in both parents or, if a parent has already died, the disorder should be shown to be present in both parental families.

A child who has tendon or skin xanthomata, plasma lipid and lipoprotein levels compatible with the homozygous state, no evidence of any other disease that might cause hyperlipidaemia, and whose parents have entirely normal plasma lipids may have the even rarer condition of *pseudohomozygous hypercholesterolaemia* (Mishkel 1976). The basic defect in this disorder has not been established but LDL receptor function appears to be normal.

In both the homozygous and pseudohomozygous disorders, intermittent arthralgia may occur and the erythrocyte sedimentation rate may be raised. These findings, coupled with cardiac murmurs, have sometimes resulted in an erroneous diagnosis of rheumatic carditis.

The heterozygous form of FH is usually suspected in a child because a parent is known to have the disorder, or premature ischaemic heart disease has occurred in a first degree relative. Clinical manifestations in childhood are rare but skin xanthomata and corneal arcus may occur. The lipoprotein abnormality is fully expressed in childhood and the diagnosis usually presents no problem, with plasma cholesterol being unequivocally high at around 8–12 mmol/l, triglycerides normal, HDL normal or low, and lipoprotein electrophoresis showing a prominent beta band. Difficulties may arise, however, in interpreting cholesterol concentrations at or only just above the normal range. There is no absolute cut-off point for either plasma cholesterol or LDL cholesterol above which the diagnosis is certain, and it is essential that family studies are undertaken and the disease demonstrated in a first degree relative. Under these circumstances the studies of Leonard *et al.* (1977) showed that, at least in the UK, the plasma cholesterol concentration at which minimal misclassification occurred was 6.77 mmol/l. Even with values at this level uncertainties may still exist especially if the child has already been receiving a modified fat diet. In all cases in which there is doubt, and even in those in which the diagnosis appears clear, repeat estimations are advisable before the definitive diagnosis is made. Children examined as part of a family study who are found to have unequivocally normal values can be reassured and do not require further follow-up. If one parent is known to be heterozygous the diagnosis of FH can probably be made at birth by examination of cord blood plasma using LDL cholesterol values rather than total cholesterol (Kwiterovich *et al.* 1973). There appears, however, to be no major therapeutic advantage in establishing the diagnosis at birth and it can be delayed to the end of the first year, by which time the child is usually taking a mixed diet. Neither screening the general population for hypercholesterolaemia at birth nor in later childhood is a satisfactory method for detecting FH.

Treatment of FH in children is undertaken in the hope that by lowering plasma cholesterol and LDL concentrations the premature development of atherosclerosis may be prevented or at least delayed. There is as yet no evidence that treatment is effective in these respects. Plasma cholesterol may be lowered by means of diet or drugs or a combination of both. Once treatment has been started regular estimations of plasma lipids should be made both to ensure the efficacy of the treatment procedure in lowering plasma cholesterol, and to check compliance. Initially monthly measurements are desirable but after treatment has been well established 3–4 monthly checks may suffice. In general compliance with treatment in the long term is not good (West *et al.* 1980) and it is unwise to relax on follow-up estimations.

The drug therapy most likely to be used for children is an ion exchange resin such as cholestyramine. Patients on long term treatment with this drug should be receiving additional folic acid and have their folate, vitamin D status, and clotting function regularly checked. If nicotinic acid is used, periodic tests of liver function should be carried out.

Homozygous FH generally responds poorly to diet and drug therapy and more radical measures such as regular plasmapheresis, ileal by-pass operation, and portocaval shunting are under investigation.

Some patients with pseudohomozygous hypercholesterolaemia appear to be very sensitive to treatment with a rapid fall in plasma cholesterol and LDL to normal on diet combined with small doses of cholestyramine.

Familial combined hyperlipidaemia (type IIb)

This disorder is now considered to be distinct from FH; the basic defect remains unknown. Although probably inherited in an autosomal dominant manner it is seldom fully expressed during childhood. Investigation of an affected adult shows moderate hypercholesterolaemia with a proportionately smaller rise in triglyceride, and lipoprotein electrophoresis shows a prominent beta band with some increase in pre-beta and a diminished alpha band. Thus both LDL and VLDL are raised and HDL is somewhat depleted. Studies of the offspring of 33 affected parents showed that of the 35 children under the age of 13 years, 16 (45%) had some abnormality of plasma lipoproteins but in only 2 children was it similar to the parental pattern; in the remainder there was either a small rise in LDL only (type IIa) or VLDL only (type IV) (Glueck *et al.* 1973a). Children in such families probably need regular follow-up into adult life.

Familial type III hyperlipoproteinaemia

In this condition, which is inherited as an autosomal dominant, IDL accumulates in the plasma and results in elevation of both cholesterol and triglyceride with a 'broad beta' band on electrophoresis. The latter finding is not, however, always present and if the disorder is suspected ultracentrifugal studies are indicated. Neither clinical nor biochemical expression during childhood is common; clinical suspicion may be alerted by the presence of xanthomata in the palmar creases. The disorder is associated with premature atherosclerosis causing both ischaemic heart disease and peripheral vascular disease. Obesity is often present and the lipoprotein abnormality responds to both weight-reduction and to clofibrate.

Familial hypertriglyceridaemia (type IV)

This disorder, characterised by an increase in VLDL with variable but often quite marked hypertriglyceridaemia and normal or only slightly elevated plasma cholesterol, has a lipoprotein pattern (increased pre-beta lipoprotein) which is relatively common in adults and often associated with carbohydrate-rich diets, the taking of contraceptive pills, or alcohol ingestion. Inheritance is probably as an autosomal dominant but the lipoprotein abnormality is rarely found in children before puberty, and becomes more common during adolescence (Glueck *et al.* 1973b). Children of affected parents should therefore be examined both at the time of parental diagnosis or concern and, if the lipid pattern is normal, again during adolescence and early adult life.

Familial type V hyperlipoproteinaemia

In this condition (which may represent more than one basic disorder) the lipoprotein pattern is complex with an increase of chylomicrons and VLDL. There is thus marked hypertriglyceridaemia and hypercholesterolaemia with chylomicrons and pre-beta lipoprotein on electrophoresis. The clinical features resemble those of primary lipoprotein lipase deficiency (type I), and attacks of acute abdominal pain due to pancreatitis are a particular feature. The primary disorder appears to be very rare during childhood and has to be distinguished from lipoprotein lipase deficiency. Post heparin lipolytic activity (adipose tissue lipoprotein lipase) is, however, normal and, as the condition is inherited as a dominant, family studies should help to clarify the diagnosis. Treatment is essentially similar to that for lipoprotein lipase deficiency.

Familial hyperalphalipoproteinaemia

With the more widespread measurement of HDL cholesterol concentrations it has become clear that hypercholesterolaemia in some children is due to an increase in HDL rather than in LDL. The strong inverse correlation of HDL with coronary heart disease in adults highlights the importance of determining whether raised total plasma cholesterol concentrations are due to increased HDL, so that inappropriate treatment and unnecessary anxiety are avoided.

Criteria for the diagnosis of familial hyperalphalipoproteinaemia (Glueck *et al.* 1978) include plasma HDL cholesterol levels greater than 1.8 mmol/l, absence of diseases to which hyperalphalipoproteinaemia might be secondary, and detection of a similar pattern in a first degree relative. Neither the mode of inheritance nor the influence of shared dietary or other environmental factors is yet clearly established. The condition appears to be associated with reduced coronary heart disease and longevity.

Secondary hyperlipoproteinaemias

Raised concentrations of plasma cholesterol and/or triglyceride associated with a variety of different lipoprotein patterns may occur as a secondary manifestation in a number of diseases (Table 9.4), and in any child with hyperlipoproteinaemia this possibility must be excluded, although in most instances the underlying condition will have already been diagnosed. Occasionally however, hyperlipidaemia may be the initial feature, and its presence may influence further investigations and management.

The disorders most commonly associated with secondary hyperlipoproteinaemia in childhood are poorly-controlled diabetes mellitus, hypothyroidism, the nephrotic syndrome, the hepatic glycogenoses, and obstructive liver diseases.

Diabetes mellitus

The poorly-controlled diabetic child may have hyperlipidaemia with almost any type of lipoprotein pattern, although types I, IV and V are most usual. The degree and nature of the hyperlipoproteinaemia does not necessarily correlate with the degree of glycaemic control though the most gross forms of hyperlipidaemia are generally associated with ketoacidotic states. In the presence of gross accumulations of chylomicrons or VLDL plasma electrolyte results may be fallacious, with apparent very low levels (Frier *et al.* 1980) and if treatment is based on these misleading figures serious consequences can result. In any markedly lipaemic plasma the electrolytes should only be measured in the aqueous phase, the lipids having been previously extracted with a solvent such as carbon tetrachloride.

Plasma lipid estimations are in general not helpful in the assessment of overall diabetic control and there would appear to be no indication for performing either cholesterol or triglyceride measurements routinely in the management of childhood diabetes. Nevertheless in some individuals the finding of abnormally turbid plasma or hypercholesterolaemia will prompt reappraisal of insulin control. Occasionally diabetes mellitus and a primary hyperlipoproteinaemic state will co-exist and family studies should be done if hyperlipidaemia persists in spite of improved control. The role of hyperlipoproteinaemia in the pathogenesis of diabetic vascular complications remains controversial.

Hypothyroidism

Hypercholesterolaemia due to raised LDL levels is common, though not invariable, in untreated hypothyroidism in children, but rarely found in infancy. Occasionally levels of IDL are raised resulting in hypertriglyceridaemia in addition to hypercholesterolaemia, and the plasma is then turbid. The lipoprotein pattern reverts to normal with treatment; unusually low plasma cholesterol can be an indication of excess thyroxine administration.

Renal disease

Hypercholesterolaemia is a classic feature of the nephrotic syndrome. Plasma LDL is increased and VLDL levels may also be raised, with associated hypertriglyceridaemia. There is a strong negative correlation between plasma cholesterol and albumin concentrations, but no correlation between plasma triglyceride and albumin. Plasma lipoproteins return to normal with control of proteinuria and restoration of plasma albumin levels. Measurement of plasma lipids does not contribute to diagnosis or management and need not be performed routinely.

Hyperlipoproteinaemia may occur in other forms of renal disease but is less well documented. Increase in VLDL with moderate hypertriglyceridaemia and some hypercholesterolaemia has been reported in children undergoing regular haemodialysis (El-Bishti *et al.* 1977). It is suggested that plasma cholesterol and triglyceride concentrations should be measured regularly in such children, and in those with chronic renal insufficiency likely to require dialysis, in order that dietary manipulation aimed at reducing the hyperlipidaemia may be attempted. Premature atherosclerosis is known to be associated with chronic renal disease in adults and occurs in adults on long-term haemodialysis.

Hepatic glycogenoses

Raised levels of VLDL with turbidity of the plasma and hypertriglyceridaemia are commonly found in hepatic glycogen storage disease due to glucose-6-phosphatase, debrancher, or phosphorylase deficiencies. Hypercholesterolaemia and some increase in LDL may also occur. The finding of turbid plasma and hyperlipidaemia in a child with hepatomegaly but no evidence of obstructive liver disease should prompt investigations for glycogen storage disease. Hyperlipidaemia is not a feature of the lipid storage disorders, with the exception of cholesterol ester storage disease in which hypercholesterolaemia is common (Assman & Fredrickson 1983).

Obstructive liver disease

Hypercholesterolaemia and hypertriglyceridaemia associated with raised levels of LDL and/or VLDL and reduced concentrations of HDL are common in children with obstructive liver disease. When liver failure supervenes, however, all lipoprotein levels decline.

In intrahepatic biliary atresia an unusual type of hyperlipoproteinaemia may be found due to accumulation of so-called lipoprotein X. This abnormal lipoprotein contains high proportions of cholesterol and phospholipid and most of its cholesterol is in the unesterified form. Investigations show gross hypercholesterolaemia and hyperphospholipidaemia, and after electrophoresis the lipoprotein bands are found to stain very poorly, because both unesterified cholesterol and phospholipid have little affinity for lipid dyes. LCAT activity is reduced, as are levels of HDL. Xanthomata associated with this type of hyperlipoproteinaemia have a rather characteristic flat appearance and occur on the palms and soles and the mucous membranes, as well as other skin areas.

'Hyperlipoproteinaemia of affluence'

In most industrialised countries plasma cholesterol and LDL levels in children are higher than in poorer societies and this is probably due, at least in part, to a greater consumption of saturated fat in the richer nations. Between countries values vary and the somewhat higher mean levels of cholesterol in children in the UK and Australia compared with the USA may reflect differing dietary patterns. Even in apparently homogenous populations the proportion of children with cholesterol concentrations above the arbitrary upper limit of normal is different, and in some studies as many as 20% of children have been designated as 'hypercholesterolaemic'. The chance finding of a cholesterol level somewhat above the normal range in a child need not imply an underlying disease process nor provoke extensive investigations. Although the 95th centile for plasma cholesterol in children, given in Table 9.3, is 5.4 mmol/l, many apparently healthy children in the UK will be found to have values above this, and levels up to about 6.5 mmol/l should not necessarily cause concern. Measurement of HDL cholesterol should probably be done in these children and some are likely to have high values (see hyperalphalipoproteinaemia, p. 257). For those with higher LDL concentrations some reduction in plasma cholesterol by dietary change may be beneficial, but there is at present no justification for advocating this on a widespread basis, nor for population screening of serum cholesterol in childhood. An excellent review of the problems involved in the prevention of coronary heart disease through an attack on hypercholesterolaemia in childhood has recently been written by Berwick *et al.* (1980).

HYPOLIPOPROTEINAEMIA

As with hyperlipoproteinaemia, hypolipoproteinaemia may be due to a primary genetic disorder or be a secondary manifestation of another disease (Table 9.5). Hypolipoproteinaemia affecting the low density lipoprotein species and therefore presenting as hypocholesterolaemia on routine analysis is common in most disorders of fat malabsorption in children; cholesterol concentrations may be reduced to the order of around 2.0–2.5 mmol/l and levels rise as the underlying abnormality is treated. Hypocholesterolaemia may also occur in hyperthyroidism, in hepatic failure, and in association with certain chronic haemolytic disorders (e.g. thalassaemia); the mechanism in the latter conditions is not understood. Reduction in HDL is a fairly common non-specific finding in many disease states and very low levels may be found in liver disease.

Table 9.5 Hypolipoproteinaemias.

Main lipoprotein abnormality	Plasma lipids Cholesterol	Triglyceride	Primary disorders	Secondary disorders
Absent LDL, VLDL, Chylos	– –	– –	Abetalipoproteinaemia; Homozygous hypobetalipoproteinaemia	
Absent LDL, VLDL	– –	N	Normotriglyceridaemic abetalipoproteinaemia	
Low LDL	–	N	Familial hypobetalipoproteinaemia	Fat malabsorption, thyrotoxicosis, chronic anaemia
Absent HDL	–	+(+)	Familial high-density lipoprotein-deficiency (Tangier disease)	
Low HDL	N or + +	N or + +	Familial LCAT deficiency	Liver disease, non-specific

LDL = low-density lipoprotein; VLDL = very-low-density lipoprotein; HDL = high-density lipoprotein; LCAT = lecithin – cholesterol acyltransferase.

Primary hypolipoproteinaemias

Primary disorders causing hypolipoproteinaemia, although rare, are of great scientific interest and the investigation of patients with such disorders has been responsible for many advances in knowledge of lipoprotein metabolism and function. In most of the conditions the original diagnosis was made by the assiduous follow-up of an unusually low plasma cholesterol level or other lipid abnormality found on 'routine' investigation or requested with a quite different disease in mind!

Deficiency of betalipoprotein

Since absence of betalipoprotein (abetalipoproteinaemia) was first reported in 1960 it has become clear that there are at least two, and probably three, distinct genetic disorders in which synthesis of Apo B and hence LDL is impaired.

Abetalipoproteinaemia is a recessively inherited disorder characterised clinically by fat malabsorption and failure to thrive which is present from birth, and a pigmentary retinopathy and ataxic neuropathy which develop in later childhood and adolescence. Investigation reveals total absence of LDL, VLDL and chylomicrons from the plasma, acanthocytosis of the red cells, and vacuolation of the intestinal epithelial cells on jejunal biopsy. Presentation to the clinical chemistry laboratory is likely to be either because no other cause has been detected for fat malabsorption, or acanthocytosis has been reported, or because of unusual ophthalmological or neurological features. Acanthocytosis, although always arousing the suspicion of abetalipoproteinaemia, is not pathognomonic for this disorder (Herbert *et al.* 1983).

Examination of the plasma shows completely clear serum, even in the post-absorptive state, very low total cholesterol (usually below 1.3 mmol/l), virtually undetectable triglycerides, and absence of chylomicron, pre-beta and beta bands on electrophoresis, with a reduced alpha band. Confirmation of the diagnosis should be obtained by immunochemical estimation of LDL, and family studies (parental lipoproteins) are necessary to exclude homozygous hypobetalipoproteinaemia (*vide infra*). Acanthocytes, if not already visualised, should be looked for in a fresh wet preparation of blood and intestinal biopsy will show the typical appearances. Because of the fat malabsorption and lack of LDL, fat-soluble vitamins are depleted and estimations of plasma vitamin A, D, E and K (by prothrombin time) are required. Levels of vitamins A and E are likely to be very low and prothrombin time prolonged. Increased peroxide haemolysis of the red cells is a sensitive test of vitamin E function at cellular level.

Treatment comprises reduction of fat intake and administration of large doses of vitamins A and E, with the probable need for additional vitamin

K. As there is now increasing evidence that the neurological sequelae may be alleviated by prolonged therapy with vitamins A and E (Illingworth *et al.* 1980; Muller *et al.* 1983), plasma vitamin levels should be monitored regularly, although levels of vitamin E will always be low because of lack of the carrier protein (LDL).

Familial hypobetalipoproteinaemia is a dominantly inherited disorder which in its homozygous form appears clinically and biochemically indistinguishable from abetalipoproteinaemia, although there is a suggestion that the long-term neurological sequalae may be less marked (Herbert *et al.* 1983).

In the heterozygous state LDL is reduced to about half its normal concentration; plasma cholesterol levels are of the order of 1.7–2.2 mmol/l. Many individuals may be asymptomatic with normal fat absorption, normal red cell morphology and no neuro-ophthalmological features. The diagnosis is then established from the plasma lipoprotein pattern, the finding of a similar abnormality in at least one first degree relative, and the absence of other disease to explain the hypolipoproteinaemia. In some patients, however, failure to synthesise chylomicrons results in malabsorption; in others acanthocytosis, or neurological abnormalities may be found. Possibly in the latter cases one or other of the alleles of Apo B may be selectively affected (*vide infra*).

Normotriglyceridaemic abetalipoproteinaemia is a newly described disorder (Malloy *et al.* 1981) in which plasma LDL and VLDL are absent but chylomicrons can be formed and fat absorption is normal. Acanthocytosis is not a feature. The finding of severe hypocholesterolaemia and absent LDL in the presence of turbid plasma after a meal containing fat, with chylomicrons visible as a cream layer on standing, is therefore an indication for more detailed studies of lipoprotein metabolism. It is postulated that the molecular defect in this condition is selective deletion of a hepatic form of Apo B (B-100) with retention of the intestinal form (B-48). No evidence is yet available regarding the mode of inheritance of this disorder.

Deficiency of alphalipoprotein

Familial high-density lipoprotein deficiency (Tangier disease) is inherited as an autosomal recessive and characterised clinically by accumulation of cholesterol (mainly in esterified form) in the reticuloendothelial system (especially the tonsils), by corneal opacities, and by an unusual peripheral neuropathy. Any of these clinical features may prompt examination of plasma lipids which will show low levels of total cholesterol (around 2.0–2.5 mmol/l), modest hypertriglyceridaemia, and virtual absence of HDL. Ultracentrifugal and immunochemical studies should be undertaken to confirm the HDL deficiency, and family studies may reveal reduced HDL levels in the heterozygotes. Other causes of LDL deficiency must be excluded, especially LCAT deficiency (*vide infra*) and liver diseases.

Familial deficiency of lecithin-cholesterol acyl transferase (LCAT)

This rare, recessively inherited disorder, presents clinically with proteinuria, a mild normochromic anaemia, and corneal opacities (Glomset *et al.* 1983). The renal lesion progresses to renal failure and most patients have presented in adult life, though presumably the defect is detectable in childhood. Plasma levels of total cholesterol and triglyceride are variable, HDL is very low, and electrophoresis shows a very faint alpha band and a poorly staining beta band. The latter is explained by the fact that most of the cholesterol is in the unesterified form. If LCAT deficiency is suspected determination of unesterified cholesterol is an important initial investigation. Further studies include determination of the enzyme activity and characterisation of the HDL abnormality.

REFERENCES

Assman G. & Fredrickson D. S. (1983) Acid lipase deficiency Wolman's disease and cholesterol ester storage disease. In Stanbury J. S. *et al.* (eds.) *The Metabolic Basis of Inherited Disease*, 5th edition, pp. 803–20. McGraw-Hill, New York.

Beaumont J. L., Carlson L. A., Cooper G. R., Fejfar Z., Fredrickson D. S. & Strasser T. (1970) Classification of hyperlipidaemias and hyperlipoproteinaemias. *Bull. WHO* **43**, 891–915.

Berenson G. S. (1980) *Cardiovascular Risk Factors in Children*. Oxford University Press, Oxford.

Berwick D. M., Cretin S. & Keeler E. (1980) *Cholesterol, Children and Heart Disease*. Oxford University Press, Oxford.

Brown M. S., Kovanen P. T. & Goldstein J. L. (1981) Regulation of plasma cholesterol by lipoprotein receptors. *Science* **212**, 628–35.

Clarke W. R., Schrott H. G., Leaverton P. E., Connor W. E. & Lauer R. M. (1978) Tracking of blood lipids and blood pressures in school-age children: The Muscatine Study. *Circulation* **58**, 626–34.

Clayton B. E., Jenkins P. & Round J. M. (1980) *Paediatric Chemical Pathology, Chemical tests and reference ranges*. Blackwell Scientific Publications, Oxford.

El-Bishti M., Counahan R., Jarrett R. J., Stimmler L., Wass U. & Chantler C. (1977) Hyperlipidaemia in children on regular haemodialysis. *Arch. Dis. Child.* **52**, 932–6.

Frier B. M., Steer C. R., Baird J. D. & Bloomfield S. (1980) Misleading plasma electrolytes in diabetic children with severe hyperlipidaemia. *Arch. Dis. Child.* **55**, 771–5.

Glomset J. A., Norum K. R. & Gjone E. (1983) Familial lecithin-cholesterol acyl transferase deficiency. In Stanbury J. S. *et al.* (eds.) *The Metabolic Basis of Inherited Disease*, 5th edn., pp. 643–55. McGraw-Hill, New York.

Glueck C. J., Fallat R., Buncher C. R., Tsang R. & Steiner P. (1973a) Familial combined hyperlipoproteinaemia: Studies in 91 adults and 95 children from 33 Kindreds. *Metabolism* **22**, 1403–28.

GLUECK C. J., KELLY K., MELLIES M. J., GARTSIDE P. S. & STEINER P. M. (1978) Hypercholesterolaemia and hyperalphalipoproteinaemia in school children. *Pediatrics* **62**, 478–87.

GLUECK C. J., TSANG R., FALLAT R., BUNCHER C. R., EVANS G. & STEINER P. (1973b) Familial hypertriglyceridemia: Studies in 130 children and 45 siblings of 36 index cases. *Metabolism* **22**, 1287–309.

HERBERT P. N., ASSMAN G., GOTTO A. M. & FREDRICKSON D. S. (1983) Familial lipoprotein deficiency. In Stanbury J. S. *et al.* (eds.) *The Metabolic Basis of Inherited Disease*, 5th edn., pp. 589–622. McGraw-Hill, New York.

ILLINGWORTH D. R., CONNOR W. E. & MILLER R. G. (1980) Abetalipoproteinaemia: Report of two cases and review of therapy. *Arch. Neurol.* **37**, 659–62.

KWITEROVICH P. D., LEVY R. I. & FREDRICKSON D. S. (1973) Neonatal diagnosis of familial type II hyperlipoproteinaemia. *Lancet* **i**, 118–22.

LEONARD J. V., WHITELAW A. G. L., WOLFF O. H., LLOYD J. K. & SLACK J. (1977) diagnosing familial hypercholesterolaemia in childhood by measuring serum cholesterol. *Br. Med. J.* **1**, 1566–8.

MALLOY M. J., KANE J. P., HARDMAN D. A., HAMILTON R. L. & DALAL K. B. (1981) Normotriglyceridemic abetalipoproteinemia: absence of the B-100 apolipoprotein. *J. Clin. Invest.* **67**, 1441–50.

MISHKEL A. (1976) Pseudohomozygous and pseudoheterozygous type II hyperlipoproteinaemia. *Am. J. Dis. Child.* **130**, 991–3.

MULLER D. P. R., LLOYD J. K. & WOLFF D. H. (1983) Vitamin E and Neurological function. *Lancet* **i**, 225–8.

STANBURY J. S., WYNGAARDEN J. B., FREDRICKSON D. S., GOLDSTEIN J. L. & BROWN M. S. (Eds.) (1983) *The Metabolic Basis of Inherited Disease*, 5th edition, part 4, Disorders of Lipoprotein and Lipid Metabolism, pp. 589–731. McGraw-Hill, New York.

WEST R. J., LLOYD J. K. & LEONARD J. V. (1980) Long-term follow up of children with familial hypercholesterolaemia treated with Cholestyramine. *Lancet* **ii**, 873–5.

Chapter 10
Diabetes Mellitus

NEVILLE R. BELTON & JAMES W. FARQUHAR

Diabetes mellitus is defined by the World Health Organisation Expert Committee (1980) as a state of chronic hyperglycaemia which may result from many environmental and genetic factors, often acting jointly. It involves an imbalance between insulin and those factors which oppose it and leads to abnormalities of carbohydrate, protein and lipid metabolism. No single causal factor has been identified and it may be that diabetes mellitus is a heterogeneous group of disorders. At the present time causative factors identified include viral infection, autoimmunity, heredity and nutrition. The insulin secreting β cells of the pancreas may be attacked directly by infection, but it is probable that in genetically susceptible children (Sacks *et al.* 1980) characterized by particular HLA genotypes (e.g. DRW4) an abnormal immune process, which may be triggered by an intercurrent viral infection, results in the destruction of the pancreatic β cells. The likelihood of an immune aetiology gains support from the histological features seen in the limited number of post-mortem pancreatic specimens that have been available, and from the presence in patients' blood of pancreatic islet cell antibodies (ICA) either before, or more commonly at the time of, and for a year or so after diagnosis (Gorsuch *et al.* 1981). The aetiology of diabetes has been reviewed by Farquhar (1979), Gamble (1980) and Cahill and McDevitt (1981).

The diagnosis of insulin-dependent diabetes mellitus (IDDM) in childhood is usually straightforward because of the acute onset of hypoinsulinaemia leading to the rapid occurrence of the classic symptoms of thirst, polyuria and weight loss. There are many disorders which feature a gradual loss of carbohydrate tolerance in children (Farquhar 1983), the most common of which is probably cystic fibrosis; these disorders are however easily distinguishable from IDDM.

Insulin

The β cells of the islets of Langerhans in the pancreas normally secrete preproinsulin which is a single chain insulin precursor. From this chain a polypeptide segment is cleaved to produce proinsulin. Proinsulin is a complex spiral molecule consisting of an A and a B chain connected by two disulphide bridges and a C-peptide chain. Prior to insulin release the C-peptide chain is cleaved and both C-peptide and active insulin are then released in equimolar amounts.

Both insulin and C-peptide can be measured by specific radioimmunoassays. The concentration of C-peptide in the blood is a useful measure of endogenous insulin production, as levels are not influenced by injected insulin or by circulating insulin antibodies. Using the C-peptide assay it has been possible to show that newly diagnosed young patients may improve with treatment, increasing their endogenous insulin production, a phenomenon known as 'the honeymoon period' (Menchini *et al.* 1980).

The action of insulin in a patient may be impaired by the development of insulin antibodies and in this respect porcine insulin has proved less antigenic than the bovine-derived product (Bloom *et al.* 1979), as it is closer to human insulin in structure. It is hoped that the use of human insulin produced by recombinant DNA techniques will from 1982 eliminate antibody formation.

Most body cells carry insulin receptor sites on their membranes. At these sites insulin forms a complex which gives rise to complex intracellular events which are outside the scope of this chapter. The function of insulin may be interrupted either by poor receptor affinity or by a reduction in the number of receptor sites; this is particularly common in obese patients. Antibodies to receptors have also been described (Flier *et al.* 1976). Receptor number and function can be improved by dietary restriction and weight loss; this subject has been extensively reviewed by Olefsky (1981).

BIOCHEMICAL ASSESSMENT

On presentation, assessment of a ketotic child should be made as soon as possible. Blood should be drawn for glucose, potassium, sodium, urea, and osmolality testing. The osmolality and/or haematocrit will provide a baseline for rehydration. While in theory the acid–base state of a shocked ketotic child should be assessed with an arterial blood sample, this may carry an increased risk of thrombosis in diabetics and has never been our (JWF) practice (in a large paediatric series extending over 30 years with only one death, which was associated with an inexplicable hyperlipidaemia). Initial emergency establishment of a near normal blood volume with saline and, rarely, plasma, makes possible adequate assessment of acid–base and potassium status. Urine should be collected for glucose and ketones and both blood and urine should be cultured for possible infection.

Heavy glycosuria (2% glucose or more) with or without ketones, and a blood glucose exceeding 10 mmol/l are usually diagnostic. In this situation a standard glucose tolerance test is very rarely necessary and may in fact be dangerous if the blood glucose exceeds 20 mmol/l. However, glucose tolerance tests are useful in the rare maturity onset diabetes in the young and possibly when screening affected families or in the prospective assessment of the HLA identical sibling of a child with insulin-dependent diabetes. It may also be helpful to assay insulin in difficult diagnostic problems (Grant 1967; Belton 1978) and in the assessment of pancreatic destruction in cystic fibrosis even before diabetic symptoms become clear.

During initial therapy blood glucose and electrolytes should usually be measured at frequent intervals. For blood glucose, indicator sticks, with or without reflectance meters, are useful, but when blood is collected for electrolyte measurement, glucose should always be measured on the same specimen. Hyperlipidaemia can lead to spuriously low plasma glucose results (Frier *et al.* 1980) and an obviously lipaemic specimen should be commented on by the staff of the chemical pathology department. It has been reported that paracetamol (acetaminophen) can interfere with glucose estimations using the YS1 23AM glucose analyser (Fleetwood & Robinson 1981), but this interference was not observed when a standard glucose method was used (Trinder 1969).

Glycosylated haemoglobin

In the late 1950s Schroeder and his co-workers chromatographically separated several fast-moving haemoglobin components from red cell haemolysates (Schnek & Schroeder 1961). These were designated HbA_1a, HbA_1b, HbA_1c, HbA_1d and HbA_1e, according to the order in which they were eluted. It was subsequently found that these fast components, in particular HbA_1c, were increased in diabetics (Huisman & Dozy 1962; Rahbar 1968; Rahbar *et al.* 1969).

It is now known that HbA_1c is structurally identical to haemoglobin A, having the same amino acid composition, except that a glucose molecule is added to the N(amino) terminus of the beta chains of normal haemoglobin A. It is attached by a non-enzymatic reaction which involves the formation of a Schiff base (aldimine) between the aldehyde of glucose and the terminal valine of the β chain. While the red cell is circulating the relatively unstable aldimine undergoes an Amadori rearrangement to the stable and relatively irreversible ketoamine (Bunn *et al.* 1975). Evidence suggests that there is a continuous formation and accumulation of glycosylated haemoglobin during the 120 day life span of the red cells and that the rate of synthesis of HbA_1c is related to blood glucose concentration. This supports the idea that a single HbA_1c determination reflects the average glucose concentration during the preceding few weeks. HbA_1c is used as an index of diabetic control. Although it is slower to reflect changes (Koenig *et al.* 1976), it is almost certainly better and more reliable than a blood glucose value obtained at the clinic and much better than daily urine test records by the patient, since it eliminates the common experimental error known to exist among diabetic patients, as well as their non-compliance. The times of urine and blood tests are arbitrary and provide only very small windows on the day's glycaemic control. The remaining merit in traditional urine and blood specimens lies in their indication of acute changes and the need for daily adjustments to the treatment programme. The main problems involved in the use of HbA_1c as a clinical index lie in the complexity of the many different methods available for its measurement and the lack of suitable reference material. Some methods measure HbA_1 while others measure HbA_1c alone. This has led to uncertainty in the interpretation of results and problems in the comparison of data from different laboratories and clinic populations. Coupled with this there is a lack of clear information as to what factors *in vivo* affect haemoglobin glycosylation.

The methodology most widely used at the present time is cation exchange chromatography. Originally relatively long (30 cm) columns were used, but shortening of the columns to 8–10 cm has both reduced the size of the sample required and the time needed in which to complete the assay. Methods like those of Trivelli *et al.* (1971) and Gabbay *et al.* (1977) have now been incorporated into commercially-available micro-column kits. Although these are relatively simple to use they have major drawbacks, as reproducibility is strongly temperature-dependent and the columns may vary from batch to batch (Johnson *et al.* 1980).

Speed and simplicity have been the criteria for the development of more recent methods. Kynoch and Lehmann (1977), Davis and Nicol (1978), Welch and Boucher (1978), Baron *et al.* (1980) and Baumgart *et al.* (1980) have all published rapid microcolumn methods for the measurement of HbA_1. In all of these methods the red cell haemolysate is prepared using buffer, potassium cyanide and in some cases sonication, or freezing and thawing. Elution of fast haemoglobins (HbA_1) followed by the major haemoglobin components, using different buffers, enables the percentage of HbA_1 to be measured. Changes in pH, ionic strength of the buffers, temperature and types of resin used, can all affect the results. In the absence of reliable and stable glycosylated haemoglobin standards many laboratories use blood samples from laboratory staff or pooled blood samples for quality control.

Electrophoresis on commercially-produced agar gel films followed by scanning densitometry is now widely accepted as a reliable alternative to ion exchange (Read *et al.* 1980). This is now the method used routinely in the Child and Adolescent Diabetic Unit of the Royal Hospital for Sick Children. Isoelectric focussing (IEF) with quantitation by high resolution microdensitometry has also been used to measure HbA_1c (Spicer *et al.* 1978; Stickland *et al.* 1982).

A colorimetric method for glycosylated haemoglobin has been described by Fluckiger and Winterhalter (1976) and Davis *et al.* (1978) have published a method using high-performance liquid chromatography (HPLC), but at the present time IEF is probably the most accurate and it is specific for HbA_1c. Electrophoresis is however still widely used for routine measurements, although it measures total HbA_1. It is probably more reliable than microcolumn chromatography where such careful control of pH, temperature, etc. is required (Welch 1981).

Attempts have been made to minimise variation and improve specificity in the measurement of glycosylated haemoglobin by pretreatment of blood samples. These efforts are particularly aimed at removing the unstable intermediate known to be formed *in vivo* and *in vitro* in response to short-term fluctuations in blood glucose (Svendsen *et al.* 1980). Methods used include incubation of the red cells in saline (Welinder & Svendsen 1980), washing and dialysis of the red cells (Goldstein *et al.* 1980) or the addition of glycerol to stabilise the samples before analysis (Spicer *et al.* 1978). Mortensen showed that whole blood samples are stable for 5 days in the absence of stabilising agent, and haemolysates for 9 days when stabilised with potassium cyanide (Mortensen 1980). Baumgart *et al.* (1980) found that blood stored in heparin tubes containing acid-citrate-dextrose (ACD) solution maintained their estimated HbA_1c values for 31 days. Svendsen *et al.* have reviewed this subject and recommend incubation of the red cells in a low glucose medium at 4 °C for 7 days to eliminate the unstable fraction. The role of red cell turnover rate in influencing levels of glycosylated haemoglobin with particular reference to thalassaemia and sickle cell anaemia was discussed in a BMJ editorial (1980b).

Reference ranges for glycosylated haemoglobin will vary between laboratories but are generally considered to be 4–8% of the total HbA. Quoted reference ranges from large series are 5.98–8.45% (Baron *et al.* 1980) using a column method and 5.4–7.6% (Mortensen *et al.* 1982) using IEF.

The use of glycosylated haemoglobin as a time-averaged index of blood glucose levels is still slightly controversial, much of this controversy having arisen because of the earlier problems of methodology discussed above. Only some of the more recent literature regarding its clinical use is reviewed here.

Studies by Fraser *et al.* (1979) showed the importance of duration of hyperglycaemia in newly diagnosed diabetics. The usefulness of glycosylated haemoglobin investigations in children as a measure of the duration of hyperglycaemia was confirmed by Williams and Savage (1979), but many doubts have been expressed regarding the interpretation of results because of the presence of unstable components and the lack of a satisfactory reference method. In the RHSC clinic, progressive improvements in methodology have yielded a monitoring system which now seems to be very reliable. A significant correlation between HbA_1c and 24-hour urinary glucose excretion has been reported (Heinze *et al.* 1979) but these workers found no correlation between HbA_1c levels and symptoms, sporadic blood and urine determinations, or ketonuria. Using a microcolumn technique, Daneman *et al.* (1980) showed that in a group of insulin-dependent diabetic children glycosylated haemoglobin levels fell as their condition was brought under control with insulin. A mean nadir of 8.2% from an original pretreatment mean of 13.6% was achieved after periods on treatment ranging from 3 weeks to 6 months, these changes being consistent with clinical evidence of control. Mortensen used an IEF method to demonstrate a seasonal variation in HbA_1c, lowest values being found in June and July. This was consistent with improved metabolic control during the summer months (Mortensen 1980; Mortensen *et al.* 1982).

Most clinicians treating diabetic children, particularly those dealing with adolescents, would probably be happy if the glycosylated haemoglobin results for their patients remained less than 10%; unfortunately many do not and results may increase to 20% or more when compliance and therefore control are poor. Realisation that such alarming figures were possible led to new approaches to the problem of management and a satisfactory response in HbA_1c values has followed.

Despite problems of methodology glycosylated haemoglobin does provide a satisfactory index of blood glucose over a 4–6 week period and if it can be accurately measured it will give a better indicator of overall plasma glucose control than do tests on isolated blood and urine samples.

CLINICAL CARE

Coma in children with diabetic ketoacidosis should be a rare event in well-educated communities served by knowledgeable physicians, since symptoms will be recognised early and metabolism corrected by simple means. The following description of immediate management should therefore be needed less and less.

Immediate management

Details of therapy vary among paediatricians; in this chapter we have emphasised our own practice which has been established over many years (Farquhar 1983).

The immediate aim is to provide available insulin and at the same time to replace fluid and electrolytes. In the past high doses of intravenous or intramuscular insulin were given; it is now practice to achieve a more physiological insulin level by continuous intravenous infusion containing relatively low concentrations of insulin (Kaufman *et al.* 1975). Infused in this way it produces a smooth predictable fall in blood glucose and is less likely to cause dramatic swings in potassium concentrations. We give a solution containing 1 unit of soluble insulin per ml of normal saline together with a starting intravenous bolus of 0.5 units of soluble insulin per year of age. The infusion is continued at 2.5 units hourly for children under 8 years of age and 5.0 units hourly for those of 8 years or over. This broad division of therapeutic dose is modified in the context of the blood glucose concentration, the flow rate being increased if the fall is too slow and vice versa. The same effect may be achieved by using the same solution to deliver 0.1 unit of insulin/kg per hour (Tamborlane & Genel 1978) modifying the flow-rate according to the response.

The hyperglycaemia should not be corrected too quickly in severely acidotic children. A blood glucose concentration of 12 mmol/l may be achieved smoothly over 4–6 hours. The flow rate of insulin is then adjusted to maintain normoglycaemia and should not be discontinued until fluids and food are being taken securely by mouth. A fresh insulin solution should be substituted every 6 hours to ensure continued potency, but no other precaution is necessary.

When constant rate infusion pumps are not available, insulin may be given intramuscularly; the initial dose is then 0.25 units/kg body weight followed by 0.1 units/kg each hour until the blood glucose concentration falls to 12 mmol/l. The dose may then be reduced and the time interval increased (Onur *et al.* 1979).

The subcutaneous route is never used now in moderate or severe diabetic ketoacidosis but intraperitoneal administration through a dialysis cannula has been used experimentally (Greenwood *et al.* 1979). In mild, early diagnosed cases however subcutaneous insulin is adequate.

Fluid and electrolytes

In addition to the administration of insulin to reduce hyperglycaemia therapy must aim to:

1 Replace fluid.
2 Provide potassium.
3 Eliminate hydrogen ions.

Fluid replacement. We usually commence intravenous infusions with 0.154 M (0.9%) saline, 20 ml/kg body weight being given in the first hour. A shocked, poorly perfused child may well be hyperosmolar and uraemic; here, initial infusion should be 10–15 ml/kg body weight of low molecular weight dextran or plasma. Calculation of the subsequent fluid requirements is then made using the child's weight and a clinical assessment of the degree of dehydration (see Table 10.1; Baum 1981).

The practice of giving proportionally more fluid in the first 6 hours should be avoided, especially in very young children because of the risk of cerebral oedema, which can be fatal (Duck *et al.* 1976). A change to 'half strength' (0.077 M) saline containing 5% dextrose is made when the blood glucose concentration falls below 11 mmol/l.

In the second 24 hours fluid replacement may be continued or the intravenous line may be discontinued if vomiting has ceased, nausea has subsided and the child is eating normally.

Potassium. Total body potassium is always reduced in diabetic ketoacidosis (Soler *et al.* 1972), although the plasma potassium may not be very low. Potassium, as potassium chloride, is usually given after the first hour, and supplementation should be continued in parallel with fluid therapy during the second 24 hours of treatment. Daily oral potassium should be given

Table 10.1 Guide to fluid intake in first 24 hours of treatment of severe diabetic ketoacidosis.

Age in years	Average weight for age of healthy children (kg)	Deficit due to 10% dehydration (ml)	Fluid per kg needed for maintenance (ml)	Total volume of fluid for maintenance (ml)	Volume needed for deficit and maintenance (ml)	Volume to be given in first hour (ml)	Volume per hour from 1–24 hours (ml)
1	10	1000	120	1200	2200	200	85
5	18	1800	100	1800	3600	360	140
10	30	3000	75	2250	5250	600	200
15	50	5000	50	2500	7500	1000	280

for 7 days after the intravenous therapy has been discontinued, the exact dose being varied according to the child's consumption of good sources of potassium, such as fresh fruit juice, and as indicated by the plasma potassium concentrations. Because major losses of phosphate in the urine occur in diabetic ketoacidosis, the use of potassium phosphate rather than potassium chloride has been advocated, particularly during the first 6–8 hours of therapy. Use of the phosphate should be discontinued after 8 hours to avoid precipitating hypocalcaemia (Drash 1977). Polyionic solutions containing phosphate and other ions such as calcium and magnesium have long been used in initial therapy for ketosis and may still have a place in treatment (Bruck 1979).

Bicarbonate is not always given routinely. Following the administration of insulin, ketone bodies are oxidised via the citric acid cycle to carbon dioxide and water, thus generating an endogenous source of bicarbonate (Kaye 1975; Lightner *et al.* 1977). Insulin will also inhibit lipolysis and organic acid production (Zimmet *et al.* 1970). Thus less bicarbonate is needed in diabetic ketoacidosis than in comparable metabolic acidoses of other origins. When used only one-third of the dose normally given to correct acidosis is administered because of the endogenous bicarbonate production already mentioned.

The progress of therapy should be followed by frequent estimations of relevant blood biochemistry.

Personal practice

Having described the range of desirable treatment it is fair to confess that a flawless record has been achieved in some hundreds of cases at the RHSC over many years using rapid restoration of normovolaemia with saline, the *partial* correction of metabolic acidosis with bicarbonate, the addition of potassium chloride and then of glucose and the progressive correction of dehydration over 24 hours. Indeed the record was free from mortality when correction of dehydration and acidosis was monitored by clinical observation alone, although the caring team felt less secure than now. For them, the provision of indicator sticks with which to monitor blood glucose movement was the greatest single advance. These points are made for the comfort of many who do not have an efficient 24-hour laboratory service but who have the clinical medical and nursing skills.

Intermediate management

Insulin

The intermediate 24–48 hours between acute intravenous therapy and maintenance treatment require only the simplest manipulation of insulin, using as many subcutaneous doses as may be needed, guided by blood glucose values using paper strip tests. Thus, for example, the previous 24 hour total of intravenous soluble insulin may be reduced by one third (60 units $\times 2/3 + 40$ units). Half of this dose is then given in the morning and what is necessary of the balance to maintain normoglycaemia is given at lunch and in the evening. During the next 24 hours a proportion of intermediate insulin is added (say 1/3) and a simple blood profile before and between meals makes subsequent adjustment simple. Recovering β cell secretion may reduce the subsequent insulin needs and daily readjustment may be necessary in this 'honeymoon period'.

Fluid and food

The vomiting and indeed the abdominal pain of ketoacidosis is quickly stopped by metabolic correction and oral fluids should then be introduced. Appetite however is poor for some days and light diet (with little attention to carbohydrate control) should be made as attractive as possible. To a body which has been seriously depleted of calories and carbohydrates for some weeks, further restriction is undesirable and it is wise to provide what is necessary to repair and replenish (see below). Potassium may be best given at this stage in such natural sources as fresh orange juice, since pharmaceutical preparations may irritate the recovering bowel.

Maintenance

Treatment should aim at restoring normality and that means restoring insulin as and when necessary. Soluble insulin should be used and by providing it to cover the balance of energy intake and output, the diabetic child's metabolism is corrected. Proof of this has been provided by use of the closed-loop artificial pancreas. Multiple injections of soluble insulin, even if given into the systemic and not the portal circulation, can approach this standard of control when an emotionally stable, intelligent, educated patient calculates soluble insulin needs on the basis of food, intake, energy-output and blood glucose concentrations. Unfortunately from the diabetic point of view only, diabetic children have unpredictable appetite and activity, show varied response to daily injections and the simple rules of eating and depend for years on even those parents who are fortunate enough to be stable and to understand.

Assured normoglycaemia day and night, year in and year out, is unlikely to be achieved by the majority until a process as free from conscious co-operation as are the β-cells themselves is available to them. In the meantime the best control will be achieved by imposing as few controls as possible, by families sharing together in a healthy diet and by approaching as closely as possible the achievement of normal insulin concentrations in response to events throughout each 24-hour period.

Insulin

A plethora of insulin preparations now exists but children should have soluble insulin as the most effective, and a preparation of intermediate duration of action as back-up. All should be highly purified and as 'human' as possible, but it is true that the more expensive so-called human insulin preparations have not yet shown measurable advantages over *less* expensive highly purified porcine insulins (Parr *et al.* 1982; Clark *et al.* 1982). Their further development and lower cost however may be expected. The modification of insulin to delay its absorption from the injection site and so to prolong its action made possible the survival, with well-being, of diabetics and a reduction in the daily number of injections. Only in the past decade has the wisdom of this pharmaceutical effect been questioned. Such products have made the childhood years of young diabetics more tolerable but there is a suspicion that for many this advantage may have been gained at the expense of optimal glycaemic control and at increased risk of ultimate handicap.

Use. The practice of giving insulin subcutaneously once or twice a day at fixed times and then eating to prevent hypoglycaemia is false and unnatural. It is therefore likely to be disobeyed openly or surreptitiously. A natural day, punctuated at appropriate times by healthy meals, is made possible by selecting the best insulin or mixture of insulins to be given by injection at times most likely to achieve normoglycaemia for the greatest proportion of 24 hours. Ideally this is best done by using multiple subcutaneous injections of soluble insulin. An intermediate acting insulin is used to meet the needs between soluble insulin peaks and the results are monitored to achieve the best 'basic daily pattern'. This is then adjusted on unusual days (food, exercise, infection, etc.). Physicians have appreciated for a long time that the times quoted by insulin manufacturers for peak insulin activity and for total duration of activity provide only a rough guide to expected performance. Sometimes an injection has little early effect and an unexpectedly great late one. Alternatively there is a dramatic early effect followed by escape from control much too soon. Recent work on absorption kinetics using marked insulin has shown convincingly the wide variation in absorption rate. This relates to blood flow and muscular activity among other factors and it fits the blood glucose profile (Lauritzen *et al.* 1979, 1980, 1981b; Williams *et al.* 1981). Absorption rates may also vary from patient to patient at the same site and between sites in the same patient (Berger *et al.* 1979). Accelerated absorption is also claimed to result from the hyperaemia of sauna bathing (Cüppers *et al.* 1980) and although this is not a children's pastime, the principle may apply when sun-bathing or being very hot in bed or in a prolonged hot bath.

Inevitably at the time of writing this chapter the type of insulin used, the number of injections given daily and the times selected in relation to the day's programme represent a compromise between the desire to achieve optimal normoglycaemia and the need to avoid confusing the inexperienced or disturbing the emotionally unstable. Soluble insulin exerts an early effect which passes in a few hours. The effect of intermediate insulins is delayed for an hour or two, then achieves its maximum effect and fades in the course of 12–14 hours.

Morning and evening injections of insulin are now the general rule for diabetic children in many clinics. It is often an unnecessary routine or counter-productive before the age of 10 years but thereafter it is desirable in the context of social life in North West Europe. Other cultures with other patterns of daily life need to develop their own (Farquhar 1981a).

Newer means of giving insulin. New delivery systems have been tested in adults and children in recent years but none is yet applicable to all children at home and at school in the community. They are either 'closed' or 'open' systems. The 'closed loop' system is an artificial pancreas which samples the blood glucose continuously from one blood vessel, and which is then programmed to deliver the correct amount of insulin into another so that the blood glucose is held in the normal range. Such machines are very expensive, large and of value only under selected hospital conditions. The 'open loop' system is essentially a small electric pump which delivers insulin subcutaneously and continuously and which has a variable flow rate permitting more to be delivered when food is taken and less between meals. Most are worn on belts or in holsters but smaller models which could be worn like a wrist watch or implanted are being developed. The blood glucose is then intermittently sampled by the patient and the flow-rate adjusted to maintain normoglycaemia. The following references represent the present development of the open loop system: Arky (1979); Pickup *et al.* (1979); Tamborlane *et al.* (1979); Lancet editorial (1979); Rizza *et al.* (1980); Pickup and Keen (1980); Barnett *et al.* (1980); Barbosa *et al.* (1981); Pickup *et al.* (1981); Schade *et al.* (1981); Hanuschak (1981).

While there is little doubt that in the miniaturisation, simplification and low-cost production of pumps capable of maintaining normoglycaemia lies the main hope for young diabetics (other than perhaps the remoter possibility of β cell transplants), the immediate future for all diabetic children involves the intermittent subcutaneous injection of insulin. Some still believe that delayed-action insulins can achieve as satisfactory results (Holman & Turner 1981), while others simply exclude the pump and give natural (soluble; regular) insulin intermittently through a subcutaneous butterfly needle or cannula (Slama *et al.* 1980; Nelson *et al.* 1980; Cohen *et al.* 1981). Such a use of soluble insulin would certainly overcome the finding that it is the more variable absorption-rate of delayed-action insulins which is largely responsible for erratic control. The use of a jet (without needle) injector for children probably offers no advantage (Worth *et al.* 1980).

Food

Family eating habits, personal likes and dislikes and economic factors make a nonsense of attempting to impose on diabetic children inflexible diets with fixed proportions of protein, fat and carbohydrate. Where such are prescribed and enforced under hospital experimental conditions, the period is for most but an irrelevant fantasy in the hard reality of home, school and community life. Protein for the family is limited not only by taste but by expense and the diabetic child has an equal share. The merits of polyunsaturated fats compared with the saturated are still debated and at least it does no harm to use corn oil for frying, soft vegetable fat for cooking and soft margarine for spreading. More elaborate attempts to manipulate fat intake in the community are unlikely to be sustained.

The tradition of carbohydrate restriction lingers on from pre-insulin times when severe limitation prolonged life a little. There is no evidence that a low total daily carbohydrate intake is beneficial and quite a bit to suggest that it is harmful. The need is for healthy carbohydrate and fibre intake so that, for so long as intermittent insulin injections are needed, the post-prandial glucose curves are lowered and lengthened, so avoiding explosive hyperglycaemic peaks and urinary loss. A rough guide to carbohydrate needs at different ages is given in Table 10.2. The total and distribution are then adjusted according to needs, weight-for-height, cultural pattern of meals, etc.

The present western trend, away from low-residue convenience foods and toward wholefoods with a high fibre (and therefore residue) content has much to commend it as food for the family including the diabetic. This helps reduce the 'them and me' persecution complex of young diabetics. In furtherance of this policy we avoid the use of the term 'diabetic diet' and speak of a 'whole or good food system' (Farquhar 1981b; Farquhar 1982).

The magnitude of the glycaemic curve decreases as the nature of the carbohydrate moves from the immediately assimilable (glucose) toward the crude fibrous fibre of uncooked grain and the viscous fibre of beans. There is an obvious difference between the glucose curve and insulin response following the ingestion of cooked and raw starch (Collings *et al.* 1981).

Table 10.2 Principles of deciding the *initial* carbohydrate intake of the diabetic child (100 grams + (age in years) × 10).

		Allocation of grams by meals					
Age in years	Grams of CHO	B'fast	Snack	Midday meal	Snack	Evening meal	Snack
3	130	20	10	40	10	40	10
4	140	30	10	40	10	40	10
5	150	30	15	40	15	40	10
6	160	30	20	40	20	40	10
7	170	30	20	50	20	40	10
8–9	180–190	40	20	50	20	40	10
10	200	50	20	50	20	50	10
12	220	50	20	60	20	50	20

If 'still hungry' (and not fat) increase the total and adjust individual meals.
If 'too much' decrease the total and adjust individual meals.
Evening intake may be increased by adjusting *evening* insulin—especially for pubertal children and adolescents. Time and blend of evening insulin to be fitted to social programme.

Guar (a gel-forming carbohydrate) has been incorporated as a powder or granules or already baked into a crisp-bread in an attempt to flatten the curve, improve control and limit sugar loss (Cohen & Martin 1979; Goulder 1979; Jenkins *et al.* 1979b, 1980a–d; Heaton *et al.* 1981). Leguminous seeds such as soya beans and lentils also lower the glycaemic curve and the effects of guar and of legumes is additive (Jenkins *et al.* 1980c, 1980d). The effects of fibre (or guar) extend beyond the control of blood glucose. Thus Jenkins *et al.* (1979a) claimed a lowering effect on 3-hydroxybutyric acid; Rivellese *et al.* (1980) showed a reduction of atherogenic lipoproteins and Aro *et al.* (1980) found that guar significantly reduced the urinary calcium loss.

Leguminous foods may be acceptable in a family's daily diet but guar certainly is not, especially in the quantities required to exert a significant effect. Canivet *et al.* (1980) have raised the theoretical possibility that a fibre diet in older diabetics with visceral neuropathy may cause bezoar formation. Further developments seem likely in the search for natural foods which have a rate-limiting effect on carbohydrate absorption. Dietary advice already available to diabetics shows carbohydrate values of foods and an index of whether the foods are 'fast' or 'slow' in changing blood glucose concentrations. The British Diabetic Association provides valuable printed tables of carbohydrate portions, or exchanges each of which is equivalent to 10 grams of glucose. Its new (1982) edition of *Countdown* lists the glucose equivalent of many familiar foodstuffs and uses the familiar traffic-light colours on pages to indicate wise and unwise choices. Efforts are being made to indicate the 'fast' and 'slow' carbohydrates so that diabetic (and indeed normal) cooking may favour the latter. In short, whole grain produce (e.g. rough wheaten bread, brown rice, Scottish rough oat-cake and oatmeal porridge) and legumes are desirable and are likely also to be more sustaining than fast-refined carbohydrate, e.g. in preparation for sport. Highly-refined carbohydrate is valuable in treating (rather than preventing) hypoglycaemia. The role of fibre has also been discussed by Baum (1981).

Sweeteners

The use of substitute sweeteners may be necessary initially or for some time for diabetic children. Saccharin, although banned in the United States remains freely available in Britain. Cyclamates were formerly used also until being prohibited and now aspartamate, a dipeptide of aspartic acid and phenylalanine is being launched as a further alternative sweetener; however some caution is advised in its use with patients who have liver disease (Uribe 1982). For diabetics it seems likely that the health risks, if any, of using such sweeteners are less than the alternatives of doing without them.

Sorbitol and fructose have also been proposed as sweetening agents in the diet of patients with diabetes mellitus but may cause diarrhoea when given orally. The use of intravenous fructose has been advocated in diabetic ketoacidosis because of four theoretical advantages over glucose. It is metabolised more rapidly, its utilisation is not dependent on insulin, it is antiketogenic and it is less damaging to veins. However it causes lactic acidosis, exacerbates the hypophosphataemia associated with insulin therapy and enhances water and electrolyte losses. Its use is therefore not recommended (Woods & Alberti 1972; Peaston 1973; Sestoft 1979).

Exercise

Careful considerations are given to insulin and energy intake but energy output, particularly exercise, is less easily prescribed or measured. Exercise nevertheless is as desirable for diabetics as it is for non-diabetics. Many aspects of exercise and its metabolic effects are considered in a supplement of *Acta Paediatrica Scandinavica* (Ludvigsson 1980a; Dahl-Jorgensen *et al.* 1980). Wahren and Hagenfeldt (1980) demonstrated that in well-controlled insulin-treated patients with mild hyperglycaemia and no or minimal ketonaemia, the utilization of free fatty acids (FFA), blood glucose and glycogen by working muscles is similar to that of healthy subjects, and exercise is accompanied by a fall in blood glucose concentrations. In contrast, patients in poor metabolic control tend to respond to exercise with further rises in FFA, ketone and glucose concentrations in the blood, reflecting augmented rates of tissue lipolysis and hepatic ketogenesis. Nevertheless regular physical exercise does appear to improve and correlate with the degree of metabolic control (Ludvigsson 1980b) and will reduce insulin requirements. Pedersen *et al.* (1980) suggest that this improved glucose tolerance and diminished insulin requirement due to physical exercise may be due to increased insulin binding. Holm and Björntorp (1980) agree that physical training increases peripheral insulin sensitivity and reduces insulin requirements. Dahl-Jorgensen *et al.* (1980) showed that HbA_1 concentrations were significantly reduced from 15.1% to 13.8% following a 5 month one-hour-twice-weekly supervised exercise programme in a group of juvenile diabetics.

In summary one must agree with Larsson (1980) that physical exercise does improve metabolic control and it may postpone the appearance of microangiopathy, but ketosis must be removed and optimal insulin status achieved before any exercise and particularly acute exercise is encouraged and taken (Sherwin & Koivisto 1981).

MONITORING CONTROL

The belief is growing that normoglycaemia protects the diabetic from those complications which commonly convert, over the course of 20 or more years, an inconvenient disorder into a personal and perhaps family tragedy. It is also obvious that urine testing is not only messy and awkward but quite incapable of telling the diabetic and the doctor to what extent a steady state of normoglycaemia has been achieved. The *traditional urine tests* are usually conducted with Clinitest* (5 drops of urine and 10 drops of water to which is added a Clinitest tablet) or a paper strip such as Ketodiastix*. The latter has the advantage of convenience but may be less accurate. The Clinitest may also yield a false underestimate where excessive glucose is present and is then best repeated using 2 drops of urine, for which modification the manufacturer provides an alternative colour chart. They should be conducted on second voidings before breakfast, lunch, evening meal and bedtime, but very rarely are, and were designed to measure quite different parameters. Even a trace of glycosuria ($\frac{1}{4}$%) means that the renal threshold of 10 mmol/l has been exceeded and that the upper limit of normal has been surpassed to an extent exceeding 2 mmol/l or 36 mg per cent.

Glycosuria of $\frac{3}{4}$–1% is always bad news since it implies that even higher blood glucose concentrations (12–14 mmol/l) have occurred. Even now we employ outdated criteria of success, e.g. in any one month in which 4 tests have been done daily at traditional times, at least half of all tests should be negative, two thirds should contain $\frac{1}{2}$% glucose or less and no more than one fifth should contain 2% glucose or more. Many physicians/paediatricians are even less demanding. Mostly these traditional tests were designed to show whether the patient was approaching hypoglycaemia immediately before meals or at bedtime. Parents were often counselled to let the child

---

* Ames Division, Miles Laboratories, England.

'show a trace of glucose' or to give a child, negative at bedtime, an extra snack to prevent hypoglycaemia. The traditional objectives therefore to prevent diabetic ketoacidosis and hypoglycaemia were achieved by a programme which is the antithesis of ensuring normoglycaemia. There would be merit in testing second voidings $1\frac{1}{2}$ hours after starting each meal to assess peak glycaemia. However, like all self-monitoring systems the patient can, and commonly does, provide a false record and second voidings by small children or teenagers in a busy household are commonly a figment of the physician's imagination.

Were estimations of 24-hour urinary glucose loss likelier to be more honest, to be compatible with a normal day, to be practicable over many years and to be successful in the community after hospital discharge, then they would have merit. The day's loss could be fractionated, urine being collected from say 0800–1400 hours, 1400–2000 hours and 2000–0800 hours. Such results could guide manipulations of insulin dosage. There may rarely be little Lord Fauntelroys who can cooperate immaculately with such control, but none live in the Edinburgh area!

The possibility that urine test results may be seriously misinterpreted by diabetics with defective colour vision (Thompson *et al.* 1979) has proved to be untrue in our clinic and in the experience of Graham *et al.* (1980). Perhaps as Graham *et al.* (1980) suggest the problem results from advancing age or retinopathy rather than from defective colour vision and this may explain its significance in a children's clinic where patients and parents are young. Where such a problem does exist the Hypoguard* firm can supply an electric meter, made for the blind diabetic, which will read the colour of the urine tests and provide the answer as an audible signal.

Such systems may of course be accurately conducted in hospital and are used to achieve initial control as a more acceptable routine than daily *blood glucose profiles*. We have used the latter for many years, blood being drawn hourly or at less frequent intervals from an indwelling venous cannula. This technique enables the physician to spot the waves and troughs on the 24 hour clock and to manipulate the insulin dose and blend appropriately (Werther *et al.* 1980; Griffin *et al.* 1980).

Because of the vagaries of human child behaviour the hospital profile does not necessarily reflect blood glucose control at home. The application of paper strip tests to home use provides a method of substituting a quick clean convenient blood test for urine tests. Young children may be upset by frequent finger pricking even with the help of a device such as an Autolet† but from the age of 10 years many would prefer to be regarded by their friends as public heroes rather than as urine-testers. Stick tests may be read by eye and many physicians teach their patients and parents to do so since *reflectance meters* must be bought privately or lent for limited

---

* Hypoguard Ltd., Dock Lane, Melton, Woodbridge, Suffolk 1P12 1PE.

† Autolet, Owen Mumford Ltd., Medical Division, Brook Hill, Woodstock, Oxon. OX7 1TU.

periods by hospital clinics. Many people are also inept in their use. The new models are fairly robust and have the advantage of constancy of reading irrespective of whether it is day or night. A variety of stick tests is available for blood glucose estimation. All are expensive although a price-war among manufacturers is anticipated (Diabetes Bulletin 1982). They work just as well if halved longitudinally (Jones *et al.* 1980) and clever people cut each into three: so economising in cost and blood. We much prefer the BM test‡ which needs no washing and can be carried out anywhere.

Some of the artefacts found in paper-strip tests may be due to glucose-contamination of the fingers from glucose-loaded urine or from glucose tablets (Kinmonth 1981), an observation made by ourselves 20 years ago with laboratory samples during experimental work on glucose–insulin relationships in the newborn. The patient and the parent/nurse/doctor should have freshly washed hands when taking finger-prick blood samples.

Children capable of such testing and trained to do it well may keep a daily blood profile. One of us (JWF) encourages patients to test on 2–3 days per week; one of which should be at the weekend. On one day the tests should be before breakfast, before lunch, and before the evening meal. On another day tests should be $1\frac{1}{2}$ hours after breakfast, lunch and evening meal with a late test at about 2200 hours. Occasional tests should be done at 0300 hours. This system may be backed up with urine tests on other days. Until patients are willing to test throughout each day such a system permits 'glycaemic biopsies' at the peak of the post-prandial curves and at the troughs midway between meals. Blood testing of this kind is also useful to parents or others (authorised by parents) to check blood glucose concentrations when hypoglycaemic symptoms are suspected to be simulated or late at night when, following a negative late urine test, parents are anxious about possible impending hypoglycaemia.

Whether the glucose concentration is being estimated in urine or blood, *the addition of a test for ketones* is of fundamental importance since it may be life-saving especially in otherwise unreliable patients. Its use should be restricted to urine samples containing 2% glucose or to urine obtained from patients whose blood glucose exceeds 14 mmol/l. Where neither blood nor urine are being tested or where the recorded results conflict with the clinical features of thirst, polyuria, weight loss, dehydration, etc., a ketone test may confirm the patient's non-compliance and imminent danger.

‡ BM-test Glycemie, The Boehringer Corporation (London) Ltd., Bell Lane, Lewes, East Sussex BN7 1LG.

COMPLICATIONS

Short-term

Ketoacidosis

Diabetic ketoacidosis is, of course, by far the most important short-term complication of diabetes and represents a return of hypoinsulinaemia relative to the current demand as a result of which the clearance of glucose and ketones is impaired and there are increases of glucose and ketones in blood as well as increased glucagon (Miles *et al.* 1980).

Whatever the mechanism may be, the most common and most easily recognised cause in an established insulin-dependent diabetic is infection. Heavier glycosuria, greater hyperglycaemia, ketonaemia and ketonuria may be recognised before the infection itself is recognised, e.g. in the late incubation period of mumps or chickenpox. Any infection, from florid measles to a dental root abscess, may unstabilise the diabetic child.

The possibilities of impotent insulin or of withheld insulin injection must be considered and so must the development of allergy or of destruction by a local insulinase. The latter may be treated with a proteinase inhibitor, aprotinin (Pickup *et al.* 1980). Poor control over a period of time may deplete the number of insulin receptors available (Eaton *et al.* 1981), but they can be made available again by effecting good control and by exercise. This may account in part for the remarkable results experienced among young diabetics admitted to residential care at 'Cruachan', Barnardo's hostel in Edinburgh. They have often been unable to attend school for many months and yet, within a week, can become regular in school attendance and much more sensitive to insulin. Psychosocial factors play a very big part in causing recurrent ketoacidosis in some children, mostly young adolescents.

Lastly, of course, there are side-effects of other medication—for example the use of salbutamol by asthmatic diabetics (Gündŏgdu *et al.* 1979).

The temporary changes in osmotic pressure within the vitreous humour of the eye as a result of dehydration explains the short-term loss of visual acuity in ketoacidosis. It may take some weeks to recover, so that the diabetic child's return to school may be associated with some difficulty in reading. Acuity should not be re-tested and spectacles certainly should not be prescribed until this phase has passed.

Re-admission of a diabetic child in ketoacidosis is usually confined to those with a primary vomiting disease or severe infection. Most others can be dealt with in the community where appropriate services exist (Farquhar & Campbell 1980).

Hypoglycaemia

Hypoglycaemia causes more anxiety to parents than does ketoacidosis, and more than it deserves. It usually represents an involuntary and temporary excess of insulin due either to the dose given or to delayed food intake, increased activity or unusual exposure to cold (wet, wind, inadequate clothing). In the early days after diagnosis it may represent β cell recovery, endogenous insulin secretion and the combined effect of that and the daily exogenous dose. Rarely it may represent a failure of counter-regulatory hormones, e.g. the so-called Houssay phenomenon following infarction of the anterior pituitary gland or the destruction of the adrenal cortex by an autoimmune process.

The clinical result of interplay of insulin dose and of counter-regulatory hormones has been known as the *Somogyi effect*. The accepted theory has been that an excessive insulin effect provokes hypoglycaemic symptoms and that with or without therapeutic intervention (glucose, glucagon) the patient responds with increased secretion of adrenaline, cortisol, glucagon and growth hormone. The resulting over-reaction causes hyperglycaemia and heavy glycosuria which may be misinterpreted as insulin deficiency. In consequence the physician increases the insulin dose or alters the blend of short and intermediate action insulins. Hypoglycaemia recurs and an upward spiral of insulin dose results. There is no doubt about the counter-regulatory reaction. It may be observed whenever an insulin hypoglycaemia test is conducted in the investigation of very short stature, but the ability of diabetics to respond normally has been questioned (*Lancet* Editorial 1980a). So-called 'brittle' diabetics may be displaying a similar phenomenon.

More recently however another cause of such glycaemic swings has been proposed by Gale *et al.* (1980). Hypoglycaemia of less than 2 mmol/l was found between 2300 and 0300 hours in a group of patients who did not respond with raised counter-regulatory hormones. Surprisingly the catecholamine response was not measured, because they assumed that the blood glucose concentration depends on the amount of free insulin available and that an insulin concentration in excess of 20 mU/ml causes hypoglycaemia in most people. Yet we know that a beta adrenoceptor stimulant like salbutamol can generate hyperglycaemia in asthmatic diabetics (Gündŏgdu *et al.* 1979).

In short the authors suggest that too rapid absorption of injected insulin may cause hypoglycaemia and that the subsequent hyperglycaemia represents hypoinsulinaemia consequent upon discharge from the site. This concept is supported by Faber *et al.* (1980) who correlated the peripheral blood glucose concentrations with the rate of disappearance of insulin from injection sites. They too suggest that the Somogyi effect represents too rapid absorption followed by decreased absorption or total discharge from the site. The concept has been challenged by Lyen *et al.* (1980) who are impressed by the effect of catecholamines and of beta adrenergic blockade in hypoglycaemia. They also ask the pertinent question about why a reduction in insulin dose should *help* brittle Somogyi effect diabetics if the hyperglycaemic counter swing results from too rapid exhaustion of the dose given.

In general the hypoglycaemia of diabetic children causes little anxiety by day and is easily treated, but the children should wear identity discs. Delayed recognition of nocturnal hypoglycaemia however does cause concern although measurable lasting damage is rare. There are very exceptional non-ketoacidotic deaths in diabetic children. One of us (JWF) has experienced two such; both were found unconscious in the morning and died quickly. No cause could be proved for either and neither was hypoglycaemic on admission. Severe prolonged hypoglycaemia with convulsions and anoxia during the night could not be excluded in either and gross brain destruction at autopsy did not assist diagnosis. A *Lancet* editorial (1980c) pointing out that a fasting blood glucose concentration does not represent overnight control, claimed that 25% of diabetic children are hypoglycaemic at night and that most are asymptomatic. If this is so then it is a remarkably benign process apart from an unfortunate few whose social background may account for the non-intervention of parents.

A simple laboratory method of seeking retrospective proof of hypoglycaemia was proposed by Moore *et al.* (1979). They measured the urinary cortisol and creatinine in the first morning urine sample and proposed a ratio

$$\frac{\text{cortisol (nmol/l)}}{\text{creatinine (mmol/l)}} \times 10^{-6}$$

and concluded that values exceeding 55×10^{-6} suggested hypoglycaemia. The work was criticised by Darlow *et al.* (1980) who claimed that the ratio was too easily disturbed because cortisol concentrations are more labile than in adults. There are workers, most of whom have already been quoted above, who believe that continuous overnight subcutaneous infusion is better or worse than the use of a late evening dose of intermediate insulin. Certainly at present it would be a brave or foolish man who entrusted all diabetic children to the unsupervised mercy of subcutaneous infusion.

Lastly, there is the increasing or recently recognised phenomenon of the wilful injection of insulin as a means of manipulating the adult environment. It is relatively common, but may rarely cause death. Insulin is readily available to diabetics and even under hospital conditions one of us (JWF) spent much time excluding problems ranging from the Houssay phenomenon to APUD cell carcinoma before an adolescent girl's persistent secret overdosage was proved. Her serum C-peptide concentration was extremely low although she had hyperinsulinaemia. The insulin source therefore was exogenous and a padlock on the insulin refrigerator door ended a long and dramatic illness.

The association of anorexia nervosa and diabetes in adolescents can also cause hypoglycaemia and dwindling insulin needs, but the clinical picture is obvious and is in striking contrast to the self-overdosing with insulin and the consequent correction of hypoglycaemia. The latter produces obesity and a typical puffy appearance.

Long-term

The joy of finding a cure for young diabetics 60 years ago was sadly tempered 20 years later with the discovery that Fate had suddenly restored life with one hand and then slowly withdrawn it with the other. The aetiology of retinopathy, nephropathy, peripheral and autonomic neuropathy and atherosclerosis generates apparently endless debate and research. The grim fact that diabetes mellitus from childhood or adolescence is the most common cause in Britain of being registered as blind between the most responsible ages of 30 and 45 years spurs appeals for research funding.

The early belief that 'good control' could prevent complications disintegrated by the 1950s. It has been restored by an appreciation of the fact that neither physicians nor their patients have achieved or even defined good control until the past few years, during which time methods of achieving it have been found.

Pathological thickening in the capillary basement membrane is characteristic of diabetic microangiopathy. The changes particularly demonstrated in the eye, kidney and muscle underlie the clinical retinopathy, nephropathy and neuropathy respectively which are commonly associated with diabetes. The kidneys from diabetics show basement membrane thickening with arteriosclerosis and nodular necrosis characteristic of Kimmelstiel–Wilson nodules (De Bats *et al.* 1982).

The glomerular glycoprotein contains increased total hexoses, mannose and galactose and decreased sialic acid and glucose were found; in the component amino acids there are increases in hydroxylysine, proline and glycine and decreases in lysine, histidine and leucine (De Bats *et al.* 1982). These authors observed that the changes were similar to, but greater than, the chemical changes which occur with aging in man. Similar biochemical data in diabetics has been previously described by Beisswenger and Spiro 1970; Kefalides 1974; Sato *et al.* 1975; Westberg 1976 and Canivet *et al.* 1979, although Beisswenger and Spiro found increased rather than decreased glucose concentration in the glomerular basement membrane. Capillary basement thickening has also been found in myocardium (Fischer *et al.* 1979) and accumulation of glycoprotein and lipids in small vessels of the heart (Regan *et al.* 1977). Studies ranging from that of civil servants in Whitehall (Sayegh & Jarrett 1979) to Pima Indians (Pettit *et al.* 1980) strongly suggest that retinopathy is much more likely to occur in those whose fasting blood glucose exceeds about 7.5 mmol/l and whose 2 hour blood glucose concentration during an oral test exceeds 11.1 mmol/l. Using oscillatory potential, macula recovery time, posterior vitreous fluorophotometry, glomerular filtration rate and urinary albumin excretion, Lauritzen *et al.* (1981a) claim to have provided the first controlled evidence that strict metabolic control improves or stabilises pathophysiological changes in the eyes and kidneys of insulin-dependent diabetics.

At a much simpler level and indirectly, one of us (JWF) has consistently argued over at least 20 years that the most rewarding, quickly disposable and ultrasensitive human animal is the placenta. Steadily improving control over the past 30 years has enabled obstetricians to bring pregnant women closer to term and normal delivery and away from placental failure, fetal distress and perinatal mortality. While many factors have contributed to the outcome of diabetic pregnancy, none has been as significant as good control.

Naturally this knowledge has led to increasing pressure on children and parents to achieve perfect control as represented by consistent normoglycaemia. This is largely responsible for adverse emotional reaction in diabetic children and the physician's art is to find the best way of exerting this long-term pressure with the minimum of disturbance.

Other factors such as hypertension (Knowler *et al.* 1980) may coexist and predispose young diabetics to complications, just as they seem to play some part also in the outcome of diabetic pregnancy. It is surprising that retinopathy and nephropathy do not develop at the same pace since both are due to changes in the basement membranes of their capillaries (Bodansky *et al.* 1981).

There is a suggestion that insulin-dependent diabetics developing retinopathy have increased red cell concentrations of glyceraldehyde reductase, which correlates neither with age nor HbA_1c (Crabbe *et al.* 1980).

Early warning of impending microvascular disease is desirable and apart from bad prognostic signs like frequent episodes of ketoacidosis, poor growth and weight, heavy glycosuria and a consistently high HbA_1c, a number of tests have been devised. Thus a transcutaneous Pco_2 electrode (Ewald *et al.* 1981) may show a poorer transcutaneous Pco_2 change in response to heat-induced reactive hyperaemia of skin in diabetic children than in matched controls. Renal complications may be most sensitively measured by elevation of beta-2-microglobulinaemia and indeed it is suggested that when this is detectable the progress to nephropathy is inexorable (Jones 1981).

It is suggested however that microalbuminuria may be reversed by high quality control (Viberti *et al.* 1979) even if the ultimate prognosis for nephropathy is unknown.

Recently attention has been drawn to the increased occurrence of limited joint mobility and multiple joint contractures, primarily of the fingers (Beacon *et al.* 1981; Rosenbloom *et al.* 1981; Knowles 1981) in young diabetics. It has been suggested that these increase in severity with the diabetes and may even be a herald of other connective tissue complications and an early indication of microvascular changes. Dupuytren's disease is said to affect about 42% of diabetics and only 13% of controls (Heathcote *et al.* 1981).

REFERENCES

ARKY R. A. (1979) The engineering of blood sugar. *New Engl. J. Med.* **300**, 618–19.

ARO A., UUSITUPA M., KORHONEN T. & SIITONEN O. (1980) Dietary fibre and calcium excretion in diabetics. *Br. Med. J.* **281**, 871.

BARBOSA J., MENTH L., EATON J., SUTHERLAND D., FREIER E. F. & NAJARIAN J. (1981) Long-term ambulatory, subcutaneous insulin infusion versus multiple daily injections in brittle diabetic patients. *Diabetes Care*, **4**, 269–74.

BARNETT A. H., ROBINSON M. H., HARRISON J. H. & WATKINS P. J. (1980) Minipump: method of diabetic control during minor surgery under general anaesthesia. *Br. Med. J.* **1**, 78–9.

BARON M. D., SHENOUDA F. S. & SÖNKSEN P. H. (1980) Micro-column method for HbA_1 determination. *Lancet* **1**, 1114–16.

BAUM J. D. (1981) Clinical management of diabetes mellitus. In Brook C. G. D. (ed.) *Clinical Paediatric Endocrinology*, pp. 616–36. Blackwell Scientific Publications, Oxford.

BAUMGART R., FRANCE M. W., UACONAILL E. D., MARNAGHAN S., PEREIRA M. & DRURY M. I. (1980) An assessment of a microcolumn chromatographic method for the measurement of haemoglobin A_1. *Ann. Clin. Biochem.* **17**, 101–4.

BEACON R., KENNEDY L., CARSON D. & CAMPBELL S. L. (1981) Joint contractures in Type I (Insulin-Dependent) Diabetes. *Diabetologia* **21**, 503.

BEISSWENGER P. J. & SPIRO R. G. (1970) Human glomerular basement membrane: chemical alteration in diabetes mellitus. *Science* **168**, 596–8.

BELTON N. R. (1978) Biochemical and Physiological tables and function tests. In Forfar J. O. & Arneil G. C. (eds.) *A Textbook of Paediatrics*, 2nd ed., pp. 1708–37. Churchill Livingstone, Edinburgh.

BERGER M., HALBAN P. A., GIRARDIER L., SEYDOUX J., OFFORD R. E. & RENOLD A. E. (1979) Absorption kinetics of subcutaneously injected insulin. Evidence for degradation at the injection site. *Diabetologia* **17**, 97–9.

BLOOM S. R., ADRIAN T. E., BARNES A. J. & POLAK J. M. (1979) Autoimmunity in diabetics induced by hormonal contaminants of serum. *Lancet* **1**, 14–17.

BODANSKY H. J., VIBERTI G. C., DRURY P. L., MACKINTOSH D., KEEN H. & CUDWORTH A. G. (1981) Diabetic Microvascular Disease: A heterogenous problem. *Diabetologica* **21**, 504.

British Diabetic Association (1982) *Countdown*. London.

BRUCK E. (1979) The New Dogma—Therapy for diabetic acidosis. *Am. J. Dis. Child.* **133**, 24.

BUNN H. F., HANEY D. N., GABBAY K. H. & GALLOP P. M. (1975) Further identification of the nature and linkage of the carbohydrate in hemoglobin A_1c. *Biochem. Biophys. Res. Commun.* **67**, 103–9.

CAHILL G. F. & MCDEVITT H. O. (1981) Insulin-dependent diabetes mellitus: the initial lesion. *New Engl. J. Med.* **304**, 1454–65.

CANIVET B., CREISSON G., FREYCHET P. & DAGEVILLE X. (1980) Fibre, diabetes and the risk of bezoar. *Lancet* **2**, 862.

CANIVET J., CRUZ A. & MOREAU-LALANDE H. (1979) Biochemical abnormalities of the human diabetic glomerular basement membrane. *Metabolism* **28**, 1206–10.

CLARK A. J. L., ADENIYI-JONES R. O., KNIGHT G., LEIPER J. M., WILES P. G., JONES R. H., KEEN H., MACCUISH A. C., WARD J. D., WATKINS P. J., CAULDWELL J. M., GLYNNE A. & SCOTTON J. (1982) Biosynthetic human insulin in the treatment of diabetes. A double-blind crossover trial in established diabetic patients. *Lancet* **2**, 354–7.

COHEN M., BERMAN G. M. & ZIMMET P. (1981) Are reflectance meters necessary for home blood glucose monitoring? *Br. Med. J.* **282**, 742.

COHEN M. & MARTIN F. I. R. (1979) Guar crispbread in the diabetic diet. *Br. Med. J.* **1**, 616–17.

COLLINGS P., WILLIAMS C. & MACDONALD I. (1981) Effects of cooking on serum glucose and insulin responses to starch. *Br. Med. J.* **282**, 1032.

CRABBE M. J. C., PECKAR C. O., HALDER A. B. & CHENG H. (1980) Erythrocyte glyceraldehyde—reductase levels in diabetics with retinopathy and cataract. *Lancet* **2**, 1268–70.

CÜPPERS H. J., BERCHTOLD P. & BERGER M. (1980) Sauna-induced acceleration in insulin absorption. *Br. Med. J.* **2**, 307.

DAHL-JORGENSEN K., MEEN H. D., HANSSEN K. F. & AAGENAES Ø. (1980) The effect of exercise on diabetic control and haemoglobin A_1 (HbA_1) in children. *Acta Paed. Scand. Suppl.* **283**, 53–6.

DANEMAN D., TSALIKIAN E., HENGSTENBERG F., BECKER J. & DRASH A. L. (1980) Glycosylated haemoglobin in children with insulin-dependent diabetes mellitus. *Diabetologia* **19**, 423–6.

DARLOW B. A., ABBOTT G. D., SCANDRETT M. S. & STEER B. (1980) Early morning urinary cortisol/creatinine ratios, an unreliable test for nocturnal hypoglycaemia in juvenile diabetics. *Lancet* **2**, 266–7.

DAVIS J. E., MCDONALD J. M. & JARETT L. (1978) A high-performance liquid chromatography method for haemoglobin A_1c. *Diabetes* **27**, 102–7.

DAVIS R. E. & NICOL D. J. (1978) A rapid simplified method for routine measurement of glycosylated haemoglobin. *Lancet* **2**, 350–1.

DE BATS A., RHODES E. L., GORDON A. H. & PARKE D. V. (1982) Biochemical differences in human glomerular basement membrane related to diabetes and age. *Ann. Clin. Biochem.* **19**, 17–21.

Diabetes Bulletin (1982) **18**, 1. "Strips price-war for Christmas?"

DRASH A. L. (1977) The treatment of diabetic ketoacidosis. *J. Ped.* **91**, 858–60.

DUCK S. C., WELDON V. V., PAGLIARA A. S. & HAYMOND M. W. (1976) Cerebral edema complicating therapy for diabetic ketoacidosis. *Diabetes* **25**, 111–15.

EATON R. P., GALAGAN R., KAUFMAN E., ALLEN R. C., RUSSELL L. & MILLER F. (1981) Receptor depletion in diabetes mellitus: correction with therapy. *Diabetes Care* **4**, 299–304.

EDITORIAL (1978) Glycosylated haemoglobin and diabetic control. *Br. Med. J.* **1**, 1373–4.

EDITORIAL (1979) New insulin-delivery systems for diabetics. *Lancet* **1**, 1275–7.

EDITORIAL (1980a) Diabetic control at night. *Lancet* **2**, 297–8.

EDITORIAL (1980b) Haemoglobin A_1 and diabetes: a reappraisal. *Br. Med. J.* **281**, 1304–5.

EDITORIAL (1980c) Home blood-glucose monitoring: revolution, revelation or rip-off? *Lancet* **2**, 187–8.

EWALD U., TUVEMO T. & ROOTH G. (1981) Early reduction of vascular reactivity in diabetic children detected by transcutaneous oxygen electrode. *Lancet* **1**, 1287–8.

FABER O. K., BINDER C. & LAURITZEN T. (1980) In search of the Somogyi effect. *Lancet* **2**, 701.

FARQUHAR J. W. (1979) Juvenile Diabetes Mellitus: possibility of prevention. *Arch. Dis. Child.* **54**, 569–80.

FARQUHAR J. W. (1981a) Childhood diabetes mellitus in the tropics. In Hendrickse R. G. (ed.) *Paediatrics in the Tropics: Current Review*, pp. 200–19. Oxford University Press, Oxford.

FARQUHAR J. W. (1981b) *The Diabetic Child.* Churchill Livingstone, Edinburgh.

FARQUHAR J. W. (1982) *Diabetes in Your Teens.* Churchill Livingstone, Edinburgh.

FARQUHAR J. W. (1983) Diabetes mellitus. In Forfar J. O. & Arneil G. C. (eds.) *A Textbook of Paediatrics*, 3rd ed. Churchill Livingstone, Edinburgh.

FARQUHAR J. W. & CAMPBELL M. L. (1980) Care of the diabetic child in the community. *Br. Med. J.* **281**, 1534–7.

FISCHER V. W., BARNER H. B. & LESKIW M. L. (1979) Capillary basement laminar thickness in diabetic hyman myocardium. *Diabetes* **28**, 713–19.

FLEETWOOD J. A. & ROBINSON S. M. A. (1981) Paracetamol interference in glucose measurement by YSI analyser. *Br. Med. J.* **283**, 438–9.

FLIER J. S., KAHN R., JARRETT D. B. & ROTH J. (1976) The immunology of the insulin receptor. *Immunol. Commun.* **5**, 361–73.

FLUCKIGER R. R. & WINTERHALTER K. H. (1976) In vitro synthesis of haemoglobin A_1c. *FEBS Letters* **71**, 356–60.

FRASER D. M., SMITH A. F., GRAY R. S., BORSEY D. Q., SINCLAIR M. E., CLARKE B. F. & DUNCAN L. J. P. (1979) Glycosylated haemoglobin concentrations in newly diagnosed diabetics before and during treatment. *Br. Med. J.* **1**, 979–81.

FRIER B. M., STEER C. R., BAIRD J. D. & BLOOMFIELD S. (1980) Misleading plasma electrolytes in diabetic children with severe hyperlipidaemia. *Arch. Dis. Child.* **55**, 771–5.

GABBAY K. H., HASTY K., BRESLOW J. L., ELLISON R. C., BUNN H. F. & GALLOP P. M. (1977) Glycosylated haemoglobins and long-term blood glucose control in diabetes mellitus. *J. Clin. Endocrin. Metabol.* **44**, 859–64.

GALE E. A. M., KURTZ A. B. & TATTERSALL R. B. (1980) In search of the Somogyi effect. *Lancet* **2**, 279–82.

GAMBLE D. R. (1980) The Epidemiology of insulin-dependent diabetes with particular reference to the relationship of virus infection to its etiology. *Epidemiol. Rev.* **2**, 49–70.

GOLDSTEIN D. E., PETH S. B., ENGLAND J. D., HESS R. L. & DA COSTA J. (1980) Effects of acute changes in blood glucose on HbA_1c. *Diabetes* **29**, 623–8.

GORSUCH A. N., SPENCER K. M., LISTER J., MCNALLY J. M., DEAN B. M., BOTTAZZO G. F. & CUDWORTH A. G. (1981) Evidence for a long prediabetic period in Type I (insulin-dependent) diabetes mellitus. *Lancet* **2**, 1363–5.

GOULDER T. J. (1979) Guar and diabetes. *Lancet* **1**, 612.

GRAHAM K., KESSON C. M., KENNEDY H. B. & IRELAND J. T. (1980) Relevance of colour vision and diabetic retinopathy to self-monitoring of blood glucose. *Br. Med. J.* **281**, 971–3.

GRANT D. B. (1967) Fasting serum insulin levels in childhood. *Arch. Dis. Child.* **42**, 375–8.

GREENWOOD R. H., DAVIES C. J., SENATOR G. B., O'BRIEN I. & HAYES T. M. (1979) Intraperitoneal insulin in diabetes. *Lancet* **2**, 312.

GRIFFIN N. K., SPANOS A., JENKINS P. A., TURNER R. C., WERTHER G. & BAUM J. D. (1980) 24-hour metabolic profiles in diabetic children. *Arch. Dis. Child.* **55**, 112–17.

GÜNDOĞDU A. S., BROWN P. M., JUUL S., SACHS L. & SÖNKSEN P. H. (1979) Comparison of hormonal and metabolic effects of salbutamol infusion in nomal subjects and insulin requiring diabetics. *Lancet* **2**, 1317–21.

HANUSCHAK L. N. (1981) Control of blood sugar in insulin-requiring diabetes. *New Engl. J. Med.* **304**, 908.

HEATHCOTE J. G., COHEN H. & NOBEL J. (1981) Dupuytren's Disease and diabetes mellitus. *Lancet* **1**, 1420.

HEATON K. W., HARTOG M., MANHIRE A. & HENRY C. L. (1981) Dietary fibre and diabetes. *Lancet* **1**, 1157.

HEINZE E., KOHNE E., MEISSNER C., BEISCHER W., TELLER W. M. & KLEIHAUER E. (1979) Haemoglobin A_1c (HbA_1c) in children with long standing and newly diagnosed diabetes mellitus. *Acta Paed. Scand.* **68**, 609–12.

HOLM G. & BJÖRNTORP P. (1980) Metabolic effects of physical training. *Acta Paed. Scand. Suppl.* **283**, 9–14.

HOLMAN R. R. & TURNER R. C. (1981) Insulin infusion pump. *Lancet* **2**, 41.

HUISMAN T. H. J. & DOZY A. M. (1962) Studies on the heterogeneity of hemoglobin. V. Binding of hemoglobin with oxidized glutathione. *J. Lab. Clin. Med.* **60**, 302–19.

JENKINS D. J. A., HOCKADAY T. D. R., WOLEVER T. M. S., NINEHAM R., GOFF D. V., HAISMAN P., CHARNOCK R., TAYLOR R. H. & BACON S. (1979a) Dietary fibre and ketone bodies: reduced urinary 3-hydroxybutyrate excretion in diabetics on guar. *Br. Med. J.* **2**, 1555.

JENKINS D. J. A., TAYLOR R. H., NINEHAM R., GOFF D. V., BLOOM S. R., SARSON D. & ALBERTI K. G. M. M. (1979b) Combined use of guar and acarbose in reduction of postprandial glycaemia. *Lancet* **2**, 924–7.

JENKINS D. J. A., TAYLOR R. H., WOLEVER T. M. S., BACON S., HOCKADAY T. D. R. & REYNOLDS D. J. (1980a) Guar crispbread in the diabetic diet. *Br. Med. J.* **281**, 62.

JENKINS D. J. A., WOLEVER T. M. S., NINEHAM R., SARSON D. L., BLOOM S. R., AHERN J., ALBERTI K. G. M. M. & HOCKADAY T. D. R. (1980b) Improved glucose tolerance four hours after taking guar with glucose. *Diabetologia* **19**, 21–4.

JENKINS D. J. A., WOLEVER T. M. S., TAYLOR R. H., BARKER H. M., FIELDEN H. & JENKINS A. L. (1980c) Effect of guar crispbread with cereal products and leguminous seeds on blood glucose concentrations of diabetics. *Br. Med. J.* **281**, 1248–50.

JENKINS D. J. A., WOLEVER T. M. S., TAYLOR R. H., GHAFARI H., JENKINS A. L., BARKER H. & JENKINS M. J. A. (1980d) Rate of digestion of foods and postprandial glycaemia in normal and diabetic subjects. *Br. Med. J.* **281**, 14–7.

JOHNSON M. W., DOBREA G. M., BENDEZU R. & WIELAND R. G. (1980) Temperature dependence of the chromatographic assay of haemoglobin A_1 and application of a temperature controlled assay to clinical evaluation of diabetic control. *Clin. Chim. Acta* **104**, 319–28.

JONES G. R., BIRTWELL J., JONES M. K. & HAYES T. M. (1980) Home blood glucose monitoring. *Br. Med. J.* **281**, 1358.

JONES R. (1981) $Beta_2$-microglobulinaemia as index of diminishing renal function in diabetics. *Br. Med. J.* **282**, 2060.

KAUFMAN I. A., KELLER M. A. & NYHAN W. L. (1975) Diabetic ketosis and acidosis: the continuous infusion of low doses of insulin. *J. Ped.* **87**, 846–8.

KAYE R. (1975) Diabetic ketoacidosis—the bicarbonate controversy. *J. Ped.* **87**, 156–9.

KEFALIDES N. A. (1974) Biochemical properties of human glomerular basement membrane in normal and diabetic kidneys. *J. Clin. Invest.* **53**, 403–7.

KINMONTH A.-L. (1981) Home blood glucose monitoring: a sticky artefact. *Br. Med. J.* **282**, 272–3.

KNOWLER W. C., BENNETT P. H. & BALLINTINE E. J. (1980) Increased incidence of retinopathy in diabetics with elevated blood pressure. A six-year follow-up study in Pima Indians. *New Engl. J. Med.* **302**, 645–50.

KNOWLES H. B., JR. (1981) Joint contractures, waxy skin, and control of diabetes. *New Engl. J. Med.* **305**, 217–19.

KOENIG R. J., PETERSON C. M., JONES R. L., SUADE K. C., LEHRMAN M. & CERAMI A. (1976) Correlation of glucose regulation and haemoglobin A_1c in diabetes mellitus. *New Engl. J. Med.* **295**, 417–20.

KYNOCH P. A. M. & LEHMANN H. (1977) Rapid estimation ($2\frac{1}{2}$ hours) of glycosylated haemoglobin for routine purposes. *Lancet* **2**, 16.

LARSSON Y. (1980) Physical exercise and juvenile diabetes—summary and conclusions. *Acta Paed. Scand. Suppl.* **283**, 120–2.

LAURITZEN T., BINDER C. & FABER O. K. (1980) Importance of insulin absorption, subcutaneous blood flow and residual beta-cell function in insulin therapy. *Acta Paed. Scand. Suppl.* **283**, 81–5.

LAURITZEN T., FABER O. K. & BINDER C. (1979) Variation in ^{125}I-insulin absorption and blood glucose concentration. *Diabetologia* **17**, 291–5.

LAURITZEN T., FROST-LARSEN K., CHRISTIANSEN J. S. & PARVING H. H. (1981a) Pathophysiological changes in eyes and kidneys during strict metabolic control in Type I (insulin-dependent) diabetes. *Diabetologia* **21**, 510.

LAURITZEN T., PRAMMING S., GALE E., BINDER C. & DECKET T. (1981b) ^{125}I-NPH Insulin absorption and its clinical implications. *Diabetologia* **21**, 8.

LIGHTNER E. S., KAPPY M. S. & REVSIN B. (1977) Low-dose intravenous insulin infusion in patients with diabetic ketoacidosis: biochemical effects in children. *Pediatrics* **60**, 681–8.

LUDVIGSSON J., LARSSON V. & SVENSSON P. G. (1980a) Attitudes towards physical exercise in juvenile diabetics. *Acta Paed. Scand. Suppl.* **283**, 106–11.

LUDVIGSSON J. (1980b) Physical exercise in relation to degree of metabolic control in juvenile diabetics. *Acta Paed. Scand. Suppl.* **283**, 45–9.

LYEN K. R., FINEGOLD D. & BAKER L. (1980) In search of the Somogyi effect. *Lancet* **2**, 700–1.

MENCHINI M., MESCHI F., LAMBIASE R., PUZZOVIO M., DEL GUERCIO M. J. & CHIUMELLO G. (1980) C-peptide response to arginine stimulation in diabetic children. *J. Ped.* **96**, 362–6.

MILES J. M., RIZZA R. A., HAYMOND W. & GERICH J. E. (1980) Effects of acute insulin deficiency on glucose and ketone body turnover in man. *Diabetes* **29**, 926–30.

MOORE R. A., SMITH R. F. & ASPLIN C. M. (1979) Simple test for nocturnal hypoglycaemia in diabetic patients. *Lancet* **1**, 409–10.

MORTENSEN H. B. (1980) Quantitative determination of hemoglobin A_1c by thin-layer isoelectric focusing. *J. Chromatogr.* **182**, 325–33.

MORTENSEN H. B., VESTERMARK S. & KASTRUP K. W. (1982) Metabolic control in children with insulin dependent diabetes mellitus assessed by hemoglobin A_1c. *Acta Paed. Scand.* **71**, 217–22.

NELSON J. D., MARLISS E. B., ALBISSER A. M. & ZINMAN B. (1980) Role of continuous component in subcutaneous 'open loop' insulin delivery. *Lancet* **1**, 1383–6.

OLEFSKY J. M. (1981) Insulin resistance and insulin action. An *in vitro* and *in vivo* perspective. *Diabetes* **30**, 148–62.

ONUR K., LALA V. R., JUAN C. S. & AVRUSKIN T. W. (1979) Glucagon suppression with low-dose intramuscular insulin therapy in diabetic ketoacidosis. *J. Ped.* **94**, 307–11.

PARR J. H., ABRAHAM R. R., DAVIE M. W. J., DORNHORST A. & WYNN V. (1982) Whither the latest insulins. *Lancet* **2**, 831.

PEASTON M. J. T. (1973) Dangers of intravenous fructose. *Lancet* **1**, 266.

PEDERSEN O., BECK-NEILSON H. & HEDING L. (1980) Increased insulin receptors after exercise in patients with insulin-dependent diabetes mellitus. *New Engl. J. Med.* **302**, 886–92.

PETTIT D. J., KNOWLER W. C., LISSE J. R. & BENNETT P. H. (1980) Development of retinopathy and proteinuria in relation to plasma glucose concentrations in Pima Indians. *Lancet* **2**, 1050–2.

PICKUP J. C., BILOUS R. W. & KEEN H. (1980) Aprotinin and insulin resistance. *Lancet* **2**, 93–4.

PICKUP J. C. & KEEN H. (1980) Continuous subcutaneous insulin infusion: a developing tool in diabetes research. *Diabetologia* **18**, 1–4.

PICKUP J. C., KEEN H., PARSONS J. A., ALBERTI K. G. M. M. & ROWE A. S. (1979) Continuous subcutaneous insulin infusion: improved blood-glucose and intermediary-metabolite control in diabetics. *Lancet* **1**, 1255–8.

PICKUP J. C., KEEN H., VIBERTI G. C. & BILOUS R. W. (1981) Patient reactions to long-term out patient treatment with continuous subcutaneous insulin infusion. *Br. Med. J.* **282**, 766–8.

RAHBAR S. (1968) An abnormal hemoglobin in red cells of diabetics. *Clin. Chim. Acta* **22**, 296–8.

RAHBAR S., BLUMENFELD O. & RANNEY H. M. (1969) Studies of an unusual hemoglobin in patients with diabetes mellitus. *Biochem. Biophys. Res. Commun.* **36**, 838–43.

READ A., TIBI L. & SMITH A. F. (1980) Assessment of a simple electrophoretic method for measuring HbA_1. *Clin. Chim. Acta* **108**, 487–91.

REGAN T. J., LYONS M. M., AHMED S. S., LEVINSON G. E., OLDEWURTEL H. A., AHMAD M. R. & HAIDER B. (1977) Evidence for cardiomyopathy in familial diabetes mellitus. *J. Clin. Invest.* **60**, 884–99.

RIVELLESE A., RICCARDI G., GIACCO A., PACIONI D., GENOVESE S., MATTIOLI P. L. & MANCINI M. (1980) Effect of dietary fibre on glucose control and serum lipoprotein in diabetic patients. *Lancet* **2**, 447–50.

RIZZA R. A., GERICH J. E., HAYMOND M. W., WESTLAND R. E., HALL L. D., CLEMENS A. H. & SERVICE F. J. (1980) Control of blood sugar in insulin-dependent diabetes: comparison of an artificial endocrine pancreas, continuous subcutaneous insulin infusion and intensified conventional insulin therapy. *New Eng. J. Med.* **303**, 1313–18.

ROSENBLOOM A. L., SILVERSTEIN J. H., LEZOTTE D. C., RICHARDSON K. & MCCALLUM M. (1981) Limited joint mobility in childhood diabetes mellitus indicates increased risk for microvascular disease. *New Engl. J. Med.* **305**, 191–4.

SACHS J. A., CUDWORTH A. C., JARAQUEMADA D., GORSUCH A. N. & FESTENSTEIN H. (1980) Type I Diabetes and the HLA-D Locus. *Diabetologia* **18**, 41–3.

SATO T., MUNAKATA H., YOSHINGA K. & YOSIZAWA Z. (1975) Comparison of the chemical composition of glomerular and tubular basement membranes obtained from human kidneys of diabetics and non-diabetics. *Clin. Chim. Acta* **61**, 145–50.

SAYEGH H. A. & JARRETT R. J. (1979) Oral glucose-tolerance tests and the diagnosis of diabetes: results of a prospective study based on the Whitehall survey. *Lancet* **2**, 431–3.

SCHADE D. S., EATON R. P., CARLSON G. A., BAIR R. E., GAONA J. I., LOVE J. T., URENDA R. S. & SPENCER W. J. (1981) Future therapy of the insulin diabetic patient—the implantable insulin delivery system. *Diabetes Care* **4**, 319.

SCHNEK A. G. & SCHROEDER W. A. (1961) The relation between the minor components of whole normal human adult hemoglobin as isolated by chromatography and starch block electrophoresis. *J. Am. Chem. Soc.* **83**, 1472–8.

SESTOFT L. (1979) Fructose and the dietary therapy of diabetes mellitus. *Diabetologia* **17**, 1–3.

SHERWIN R. S. & KOIVISTO V. (1981) Keeping in step: does exercise benefit the diabetic? *Diabetologia* **20**, 84–6.

SLAMA G., GARREL D. & TCHOBROUTSKY G. (1980) Multiple daily insulin injections through subcutaneously implanted needle. *Lancet* **1**, 1078.

SOLER N. G., BENNETT M. A. & DIXON K. (1972) Potassium balance during treatment of diabetic ketoacidosis. *Lancet* **ii**, 665–7.

SPICER K. M., ALLEN R. C. & BUSE M. G. (1978) A simplified assay of haemoglobin A_1c in diabetic patients by use of isoelectric focusing and quantitative microdensitometry. *Diabetes* **27**, 384–8.

STICKLAND M. H., PERKINS C. M. & WALES J. K. (1982) The measurement of haemoglobin A_1c by isoelectric focussing in diabetic patients. *Diabetologia* **22**, 315–17.

SVENDSEN P. A., CHRISTIANSEN J. S., SØEGAARD U., WELINDER B. S. & NERUP J. (1980) Rapid changes in chromatographically determined haemoglobin A_1c induced by short-term changes in glucose concentration. *Diabetologia* **19**, 130–6.

SVENDSEN P. A., LAURITZEN T. & NERUP J. (1981) The connection between HbA_1c and mean blood glucose in Type I (insulin-dependent) diabetics. *Diabetologia* **21**, 514.

TAMBORLANE W. V. & GENEL M. (1978) Discordant correction of hyperglycaemia and ketoacidosis with low-dose insulin infusion. *Pediatrics* **61**, 125–7.

TAMBORLANE W. V., SHERWIN R. S., GENEL M. & FELIG P. (1979) Reduction to normal of plasma glucose in juvenile diabetes by subcutaneous administration of insulin with a portable infusion pump. *New Engl. J. Med.* **300**, 573–8.

THOMPSON D. G., HOWARTH F., TAYLOR H., LEVY I. S. & BIRCH J. (1979) Defective colour vision in diabetes: a hazard to management. *Br. Med. J.* **1**, 859–60.

TRINDER P. (1969) Determination of glucose in blood using glucose oxidase with an alternative oxygen acceptor. *Ann. Clin. Biochem.* **6**, 24–7.

TREVELLI L. A., RANNEY H. M. & LAI H. T. (1971) Hemoglobin components in patients with diabetes mellitus. *New Engl. J. Med.* **284**, 353–7.

URIBE M. (1982) Potential toxicity of a new sugar substitute in patients with liver disease. *New Engl. J. Med.* **306**, 173–4.

VIBERTI G. C., PICKUP J. C., JARRETT R. J. & KEEN H. (1979) Effect of control of blood glucose on urinary excretion of albumin and $beta_2$ microglobulin in insulin-dependent diabetes. *New Engl. J. Med.* **300**, 638–41.

WAHREN J. & HAGENFELDT L. (1980) Free fatty acid and ketone body metabolism during exercise in diabetes. *Acta Paed. Scand. Suppl.* **283**, 39–44.

WELCH S. G. (1981) Changes in glycosylated haemoglobin after oral glucose load. *Br. Med. J.* **283**, 1403.

WELCH S. G. & BOUCHER B. J. (1978) A rapid micro-scale method for the measurement of haemoglobin $A_1(a+b+c)$. *Diabetologia* **14**, 209–11.

WELINDER B. S. & SVENDSEN P. A. (1980) Heterogeneity of the haemoglobin A_1c-band in isoelectric focusing. *Diabetologia* **19**, 465–7.

WERTHER G. A., JENKINS P. A., TURNER R. C. & BAUM J. D. (1980) Twenty-four hour metabolic profiles in diabetic children receiving insulin injections once or twice daily. *Br. Med. J.* **2**, 414–8.

WESTBERG N. G. (1976) Biochemical alterations of the human glomerular basement membrane in diabetes. *Diabetes Suppl.* **2**, 920–4.

WHO Expert Committee on Diabetes Mellitus (1980) 2nd Report. Technical Report Series 646, p. 286. World Health Organisation, Geneva.

WILLIAMS G., CLARKE A. J. L., COOKE E., BOWCOCK S., PICKUP J. C. & KEEN H. (1981) Local changes in subcutaneous blood flow around insulin injection sites measured by photoelectric plethysmography. *Diabetologia* **21**, 516.

WILLIAMS M. L. & SAVAGE D. C. L. (1979) Glycosylated haemoglobin levels in children with diabetes mellitus. *Arch. Dis. Child.* **54**, 295–8.

WOODS H. F. & ALBERTI K. G. M. M. (1972) Dangers of intravenous fructose. *Lancet* **2**, 1354–7.

Worth R., Anderson J., Taylor R. & Alberti K. G. M. M. (1980) Jet injection of insulin: comparison with conventional injection by syringe and needle. *Br. Med. J.* **281**, 713–14.

Zimmett P. Z., Taft P., Ennis G. C. & Sheath J. (1970) Acid production in diabetic acidosis: a more rational approach to alkali replacement. *Br. Med. J.* **3**, 610–12.

Chapter 11
Growth and its Problems

JAMES M. TANNER

Problems concerning growth make up a considerable part of the work-load of general paediatricians, and a part that is increasing now that the public has become generally aware that there are some forms of growth disorder which are amenable to treatment. Correct differential diagnosis is of the utmost importance; the parents have to be told as soon as possible whether their child has:

1 An untreatable disorder such as achondroplasia or Bloom's syndrome, in which case long-term medical and social support must be organised.

2 A treatable disorder such as growth hormone deficiency or coeliac disease, in which case long-term treatment must be explained and embarked upon, or

3 No disorder at all, but a family pattern of short stature, or delayed growth or both combined, in which case long-term and persistent reassurance may be needed, allied in some cases to short-term, palliative 'treatment'.

The chemical pathologist plays a part in this differential diagnosis, as does the cytogeneticist, but it is a part complementary to that of the auxologist who will normally be a general paediatrician. The auxologist (auxology is the study of growth, from the Greek *auxein*—to grow) must have a detailed understanding of how to measure infants and children, calculate velocity of growth, use and correctly interpret the plots on distance (e.g. the height attained at a given age) and velocity charts, how to assign a skeletal maturity score and a bone age to a hand–wrist radiograph and make a prediction of adult height from present height and bone age, and compare it with the range of heights within which that particular child would normally fall, given the heights of his parents.

The paediatrician who wishes to handle growth disorders with competence will require proper measuring instruments and the relevant growth charts, and will need to make himself familiar with the technique of assessing skeletal maturity. (It is the uniform and regrettable experience of paediatricians and auxologists all over the world that their radiologist colleagues, with only rare exceptions, have shown a surprising but persistent disinclination to involve themselves at more than an amateur level in this type of work.) Lastly, the general paediatrician, when he has done his best to sort out a patient's problem, may be impelled to seek more specialist advice or obtain more specialist treatment. Both are readily available in the UK, because the shortage of human growth hormone and the problems associated with obtaining it from cadavers, caused a national organization handling this treatment to be set up in 1959, and this has grown into a network of 20 Growth Disorder Clinics, covering the UK and Republic of Ireland on a more or less regional basis.

Amongst the textbooks concerned with growth, the simplest to begin with is *Foetus Into Man* (Tanner 1978) which is equally appropriate for the biochemist and the physician. There are excellent books by Marshall (1977) and Smith (1977), both rather more medically oriented, and a good manual of growth charts by Buckler (1979). More general discussions, covering the biology of growth as well as growth disorders, are those edited by Barltrop (1977) and Tanner (1981b) and at the encyclopaedic level there is the three-volume handbook *Human Growth* (Falkner & Tanner 1978–9). There is also a comprehensive chapter in Forfar and Arneil's (1983) *Textbook of Paediatrics*.

The auxology laboratory needs a stadiometer, a sitting height table, an infantometer for supine length and crown–rump measurement in the under-twos, a skinfold caliper and a tape measure (an anthropometer is optional). All can be obtained from Holtains (Crosswell, Nr. Crymmych, Pembrokeshire).

Charts of height, weight, sitting height and leg length, height velocity, weight velocity, head circumference, skinfolds and parent-allowed-for height standards (see below) are obtainable from Castlemead Publications (40 Greyfriars, Ware, Herts, SG14 1LH).

TECHNIQUES OF MEASUREMENT

Height is measured as shown in Fig. 11.1. The child stands on the footplate of the stadiometer, with heels together and head in the Frankfurt plane (i.e. with the lower border of the orbit in the same horizontal plane as the external auditory meatus). The top of the stadiometer, being counter-balanced, rests lightly on his head. The child then stretches upwards fully, aided by relaxing the shoulders and by the anthropometrist applying a gentle upward pressure on the mastoid processes and encouraging him verbally. The recorder sees that the heels do not come off the ground, holding them down if necessary. The height is read from a digital counter attached to the stadiometer. This technique minimizes variation in posture due to tiredness or boredom, which may otherwise amount to 1–2 cm. A good anthropometrist can repeat his reading to 0.3 cm or less on 95% of occasions, and two well-trained observers should seldom differ by more than 0.4 cm.

The accuracy of measurement is particularly important where calculation of increments or velocities of growth are concerned since there are two errors involved, and these may summate. Casual observers without

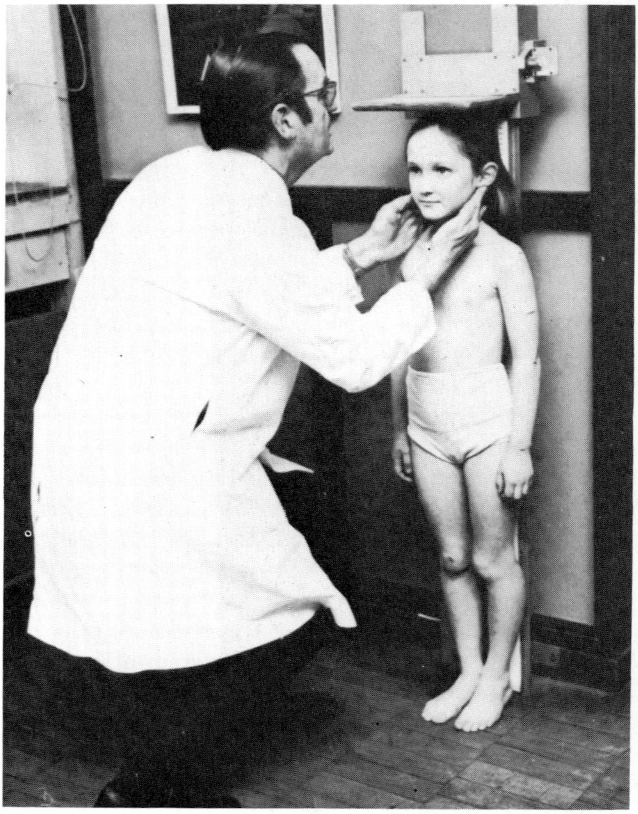

Fig. 11.1 Measurement of height (courtesy of R. H. Whitehouse).

special training may differ in their measurement by 1.5 cm, and two summating errors of this magnitude introduce large errors into a yearly velocity and make a three-monthly velocity meaningless.

Sitting height is taken as shown in Fig. 11.2. The subject is positioned on the sitting-height table so that the head is in the Frankfurt plane, the back is straight, the thighs horizontal and comfortably positioned, so that the tendons of the biceps femoris are approximately one inch clear of the table. The feet are supported on the footboard or metal bar incorporated in the apparatus, and the hands rest comfortably in the subject's lap. To ensure that the subject's back is fully extended the observer may run his index finger up the spine, applying pressure to the sacral and lumbar regions, thus causing the subject to sit up as a reflex action. Having lowered the headboard, which, as in the measurement of stature, may have a small weight placed on top of it, gentle upward pressure is applied to the mastoid processes and the subject is instructed to 'sit tall' or 'sit up, take a deep breath, and relax'.

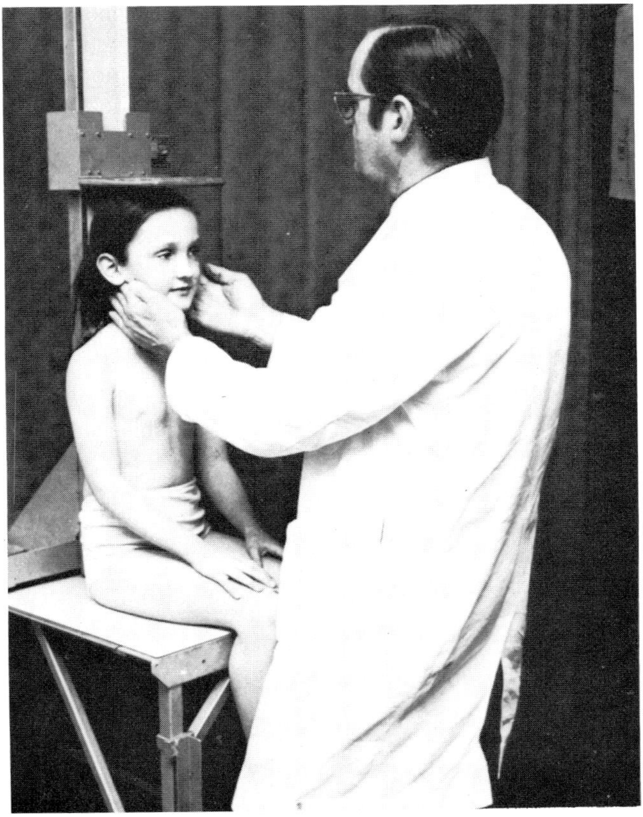

Fig. 11.2 The laboratory measurement of sitting height (courtesy of R. H. Whitehouse).

The measurement value is read by the observer who is standing in front of the table looking horizontally at the counter. As in the measurement of stature, the technique of stretching the subject reduces the effect of diurnal variation and aids reproducibility of measurement (Whitehouse *et al.* 1974).

Methods for measuring supine length and crown–rump length are illustrated and described on the back of the sitting height–subischial leg length charts obtainable from Castlemead.

The skinfold measurement is also described and illustrated on the cover of the skinfold standard charts. There are two sites for which standards are available; over the triceps on the back of the left arm, and on the left side of the back just below the angle of the scapula. Skinfold thickness is a better guide to the amount of fat that a child has than simple weight, or even weight for height, and the degree of fatness is important in the differential diagnosis of short stature, for example in distinguishing

Silver–Russell syndrome (low skinfolds) for growth hormone deficiency (high skinfolds).

Growth charts

There is no point in measuring a child unless the measurement is plotted on a chart and compared with the centiles for normal children. Growth charts may be presented either in centiles or standard deviations; those presented in centiles are much more useful in clinical work, as parents can be told the status of their child in terms they can understand, e.g. '5 per cent of all British children are less tall than yours'. A child's height should be reported to the nearest fifth centile, interpolations being made by eye between the given lines. If a child's height falls below the third centile, it is convenient to say by how many centimetres. (This is the situation in which a standard deviation scale is useful, but this requires tables of means and standard deviations for each age—they are available in Tanner *et al.* 1966.) For weight, centiles should always be used, since the non-Gaussian distribution of weight makes standard deviations meaningless.

Fig. 11.3 shows an example of the height for age ('height distance') chart, with successive plots of a growth hormone deficient boy treated with human growth hormone (hGH). The shaded areas represent the 97th and 3rd centiles for *population* charts, which are not the same as charts used for following individual subjects (see discussion in Tanner 1983) and should be used as the reference centiles only on the first occasion a child is seen (as in a survey). If a child's height falls within the shaded area on the first occasion then on the second, normally six months later, it should be closer, if not within, the clear lines of centiles. At the bottom of the chart are horizontal boxes for plotting the child's progress through the stages of puberty. The stages, and the way in which the plotting is done, will be found in any of the textbooks cited above.

At the top right hand margin of the chart, the vertical bar labelled F and M represents the 'parental height target'. The heights of the parents should always be *measured* (comparison with reported heights shows the latter to be often grossly inaccurate) and the *centiles* for father (F) and mother (M) are plotted. (To plot mother's centile on a boy's chart add 13 cm or 5 inches; to plot father's centiles on a girl's chart subtract the same.) Ninety-five per cent of the population of sons born to this couple will, if normal, reach the height range corresponding to the vertical bar (mid-parent centile ± 8.5 cm; see Tanner 1978, p. 168). Comparison of the height prediction (see Tanner *et al.* 1975) with the target height is a prime diagnostic point in dealing with children with short stature.

Fig. 11.4 shows a height velocity chart. Rate of growth is more important from a physiological point of view than height attained, since it indicates what is happening in the recent past; growth attained, or 'distance', integrates the whole course of growth, and considered alone

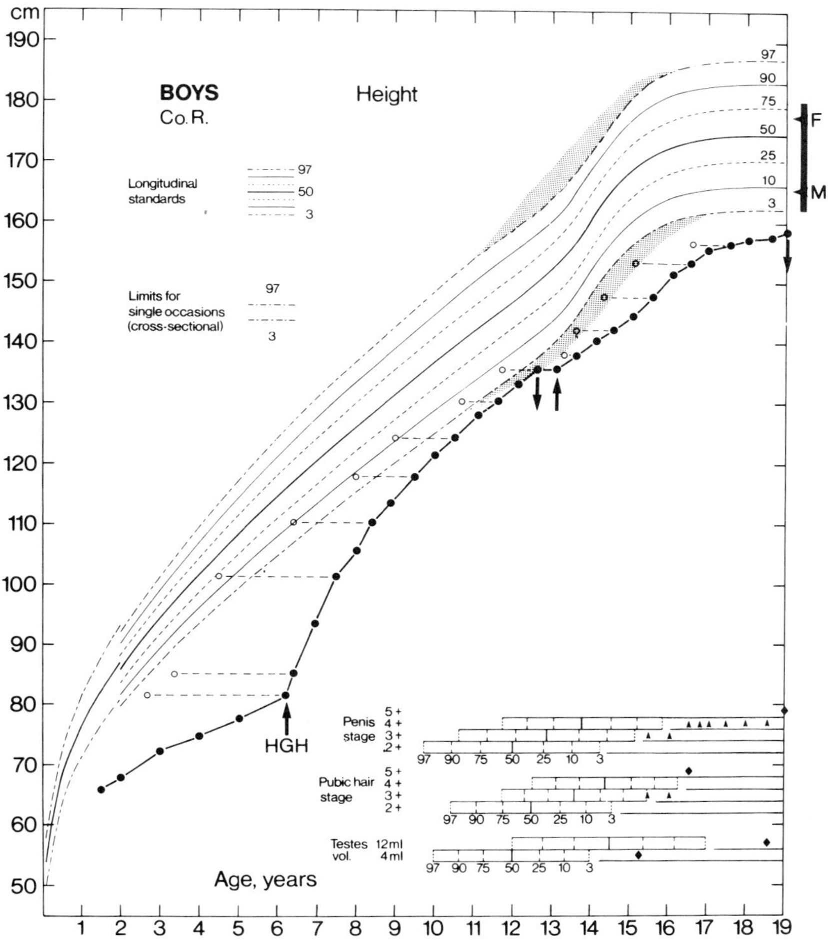

Fig. 11.3 Standards of height attained at given age, boys (from Tanner & Whitehouse 1976). Plot is of a growth hormone-deficient boy treated with hGH. Solid circles are height for chronological age, open circles are height for bone age.

may give little indication of recent events; growth velocity should be estimated and plotted. Note that the centiles in the chart are appropriate for whole-year periods only; children grow considerably faster, on average, in the spring than in the autumn, so multiplying a six-month increment by two leads to error in assigning a centile. Centiles for three- and six-monthly periods however, will be found in Marshall (1971). The clear centiles are those which characterise boys whose *tempo* of growth is average, that is, who have their adolescent growth spurt at the average time. The shaded areas to left and right characterise cohorts of boys with their spurts two standard deviations (of age at peak) earlier, and two

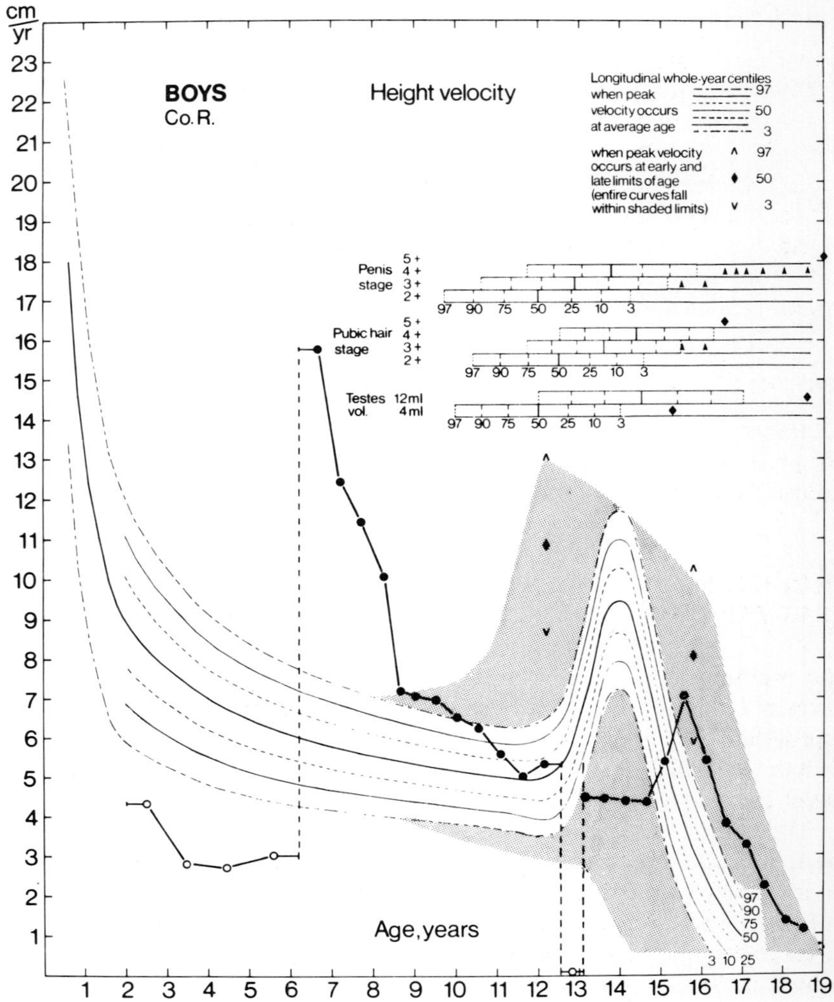

Fig. 11.4 Standards of whole-year height velocity, boys (from Tanner & Whitehouse 1976). Plot is of a growth hormone-deficient boy treated with hGH. Each successive whole-year value of height velocity is plotted.

standard deviations later. The plot is of the same boy as in Fig. 11.3. Each successive point represents a whole-year value of height velocity covering the six months before and the six months after the point (annual growth velocities are calculated successively, like the more familiar annual inflation rates). It is much easier to use age in years and decimals than in years and months when calculating velocity. Tables of decimal years are given in the textbooks.

Skeletal maturity

Standards for skeletal maturity and bone age, are given in Tanner *et al.* (1975). It should be remembered that the standardising group for these was a large sample of lowland Scottish children, representing all social classes, X-rayed in the 1950s. In some areas of England, especially the South, children are nowadays on average some three or four months advanced over these standards. Average British children however are still considerably less advanced than the very well-off American children who served as the standard for the older Greulich–Pyle atlas method of assigning bone age, which is therefore less appropriate, at least in Europe.

Prediction of adult height

Tables for the prediction of adult height are given in Tanner *et al.* (1975). The predicted height should be compared with the 'target height', obtained as described above.

PROCEDURE FOR INVESTIGATING A CHILD WITH SHORT STATURE

The procedure suggested below, based on the practice at the Growth Disorder Clinic of the Hospital for Sick Children, Great Ormond Street, is appropriate for children over the age of twelve to eighteen months; we exclude the first-year 'failure-to-thrive' patients, whose problems are somewhat different.

The child is referred because the parents, the general practitioner or the school doctor consider him or her to be suffering from short stature. He is usually the smallest in the class, and often has a younger brother (still worse, a younger sister) who surpasses him in height. If the child is in the older age group, the complaint of lack of pubertal development (which includes the expectation of a height spurt) is added to that of shortness; now the other boys or girls are 'shooting up while he/she isn't growing at all'. (It has to be remembered that the *minimum* velocity is indeed experienced just before the adolescent spurt begins.)

At the initial visit the child's height and weight are measured and sitting height and skinfolds too if possible. The bone age is assessed and the parents' heights are measured. There is, of course, the usual clinical examination and history-taking and the latter should include enquiry as to age of menarche in mother and whether father was an early, middling or late maturer. The diagnosis of Silver–Russell syndrome (see Tanner *et al.* 1975), or other recognisable pattern of malformation (Smith 1976) may be made at this visit. Few of these syndrome diagnoses can be verified by laboratory tests at present, but the possibility of Turner's syndrome (often without any stigmata whatever, especially in the mosaic forms) and Bloom's syndrome (rash on light exposure may be the only clue) should be remembered and karyotyping done. Any otherwise undiagnosed girl with short stature should be karyotyped, with at least 20 cells counted, but this should usually be done after the other diagnoses have been ruled out. If the leg length and sitting height are discordant a chondrodystrophy is likely and a skeletal survey should be done. But in hypochondroplasia both trunk and legs are often affected (the legs more, however) and the radiological diagnosis is notoriously uncertain. The same applies to the situation where the trunk is clearly abnormally short on measurement. Routine determinations of thyroxine and thyroid-stimulating hormone may be done at this visit to exclude hypothyroidism, which in the juvenile form may produce practically no symptoms other than reduced growth. A lateral X-ray of the skull is used to rule out the presence of a craniopharyngioma.

The second visit is after 3 months. At this time the bone age has been determined, and the prediction of adult height made and compared with the parental target, and these matters are discussed with the parents. The height velocity, though calculated, is not given much weight, the interval being too short. At this point a tentative differential diagnosis is made between constitutional growth delay (the most common single diagnosis in growth disorder clinics), growth hormone deficiency (total or partial and perhaps transient) and other situations causing short stature. The latter include psychosocial stress producing inhibition of growth hormone secretion, which in principle is reversible on hospitalisation, undernutrition which is often for psychosocial reasons too (Davis *et al.* 1978), iatrogenic short stature due to excessive or incorrect treatment of some other condition with corticosteroids (see Preece 1976) including those applied to inflamed and vasodilated skin after the evening bath, asymptomatic coeliac syndrome (see Groll *et al.* 1980), asymptomatic intestinal disease and asymptomatic renal disease. Enough patients are diagnosed by jejunal biopsy to make that procedure a routine part of the investigation where other tests fail to make the diagnosis, but it is usually left until the growth hormone tests have been performed. Plasma creatinine may be determined at this stage.

GROWTH HORMONE TESTS AND THE DIFFERENTIAL DIAGNOSIS

Auxology and GH tests play equal and complementary parts in the hardest differential diagnosis, that of growth delay from growth hormone deficiency (GHD). The problem arises because idiopathic isolated GHD can occur in a partial form, and even in a transient form, this latter usually in the years immediately preceding a (much delayed) puberty (see Tanner 1981a).

Though a GH test may be scheduled after, or even before, the 3 month visit where clinical and auxological data make a diagnosis of total isolated GHD or of panhypopituitarism likely, in cases where the differentiation from growth delay assumes importance, the GH test is best done after the 6-month velocity is known. Tests for the ability of the hypothalamic–pituitary axis to secrete GH are of two sorts; physiological and pharmacological.

Physiological tests

Under ordinary circumstances of daytime rest, the amount of GH in the blood, of an adult or a child, is negligible. Thus GH levels taken on so-called random blood specimens are useless. Growth hormone is secreted in a pulsatile manner; some bursts of secretion occur for no obvious reason, but exercise, emotion and sleep are rather regularly accompanied by GH peaks. The first and third can be used as physiological stimuli.

Exercise test

As a screening test, exercise can be used by any paediatrician or biochemist who can run up three flights of stairs with the child in tow, since the exercise, to be effective, has to be fairly strenuous! If this produces a level of GH over 15 mU/l then the child does not have GHD, but if the level fails to reach 15 mU/l, then the test has simply given no information, and another, more searching test must be used. The exercise test should really be done more accurately of course, using an exercise bicycle. The child exercises for 10 minutes, the friction wheel being tightened to increase the resistance to pedalling to as much as the child can reasonably tolerate. The exercise should induce breathlessness. Blood should be taken before exercise and at suitable intervals after it, usually 0, 10, 20 and 30 minutes; the peak value is the important one (see Lacey *et al.* 1973). There is some evidence that the test is improved (in the sense of giving a higher percentage of reactors amongst normal children) by fasting the children for 8 hours prior to exercise and then giving propranolol, 20 mg orally to children weighing less than 20 kg, 40 mg to others (Shanis & Moshang 1976).

Sleep

Bursts of GH secretion occur in most or all normal children during sleep, and following blood levels as closely as possible during an ordinary night's sleep is probably the best test available for a child's ability to secrete GH under the normal circumstances of life. It is seldom done, however, because of administrative and logistic problems and because it involves an over-night stay in hospital, in contrast to the other tests, all of which can be done on an out-patient basis except in the very young or very ill child. Blood is taken every half hour during the first $5\frac{1}{2}$ hours of sleep. Either the peak value, or the area under the curve of $\frac{1}{2}$ hour values may be used; the peak should be above 15 mU/l to rule out GHD (Bierich & Potthoff 1979; Mace *et al.* 1972). Howse *et al.* (1977) used the more sophisticated technique of connecting a Watson–Marlowe flow inducer to an indwelling arm vein catheter and an automatic sample collector. Blood was collected in 15 minute aliquots and the mean of the 5 hour values from 22.30–03.30 hours calculated. This mean value varies with age, decreasing from age 6 to 11 (in boys) and then rising fairly sharply to age 15. The children studied had growth delay and the rise began a year or so before the adolescent growth spurt. Howse *et al.* give a range for ages 6–15 based on these short stature-with-growth-delay patients.

These sleep tests need further validation, but both, and perhaps especially the second, deserve to be introduced much more widely than at present. Sleep tests have the great advantage over pharmacological tests that they measure what happens normally; in principle it is perfectly possible for a patient to secrete growth hormone in response to a pharmacological stimulus yet be unable to do so under the normal circumstances of life.

Pharmacological tests

There are numerous pharmacological tests for GH secretion. The most widely used, and probably the best, is the insulin tolerance test (ITT, also called insulin stimulation test, insulin hypoglycaemia test). Glucagon, L-dopa and arginine have also been used as stimulators, alone or combined with propranolol. Prior priming with sex steroids is also used.

The ITT (details in appendix below) is the test demanded by the Health Service Human Growth Hormone Committee in the UK. It has the advantage that it combines a test of growth hormone secretion with a test, equally necessary in all patients, of cortisol secretion. It is also compatible with the thyrotrophin releasing hormone (TRH) test, another that is mandatory, the solutions of TRH and insulin being injected together. A luteinising hormone releasing hormone (LHRH) test can be also added, and this is done in most of the UK growth disorder clinics. Its usefulness has never been validated and it remains at present a research tool. If it could distinguish between those patients who have true isolated GHD and will secrete gonadotrophins and enter puberty normally but late, and those who will turn out to lack gonadotrophins as well as GH and who need treatment to induce puberty, that would be very valuable: no statistics have yet been published showing it can do this. It is worth emphasising, too, that a single-bolus injection of LHRH is also a pharmacological test, for in normal life LHRH is secreted in pulses at rather set intervals and the intervals are of paramount physiological importance.

In North America, two tests are demanded by the National Pituitary Agency, and a frequent protocol is an ITT followed at the same session by a 30-minute intravenous infusion of arginine 0.5 g/kg (see Savage *et al.* 1978). There is little compelling evidence to suggest that the combined test has any advantage over a simple ITT if this is done using the strict protocol outlined below. The disadvantage of the ITT lies precisely in the need for this protocol to be followed and in particular for the blood sugar to be dropped sharply and to quite low levels. In inexperienced hands the test is said to be dangerous, but we have encountered no difficulty, nor any unwished-for or prolonged hypoglycaemias during over a thousand tests. However, these have all been done by three or four experienced paediatric registrars; the test is not one to be left to the first-week house officer. In children with a history of spontaneous hypoglycaemic episodes the ITT should be done, if at all, very circumspectly, initially using low doses of insulin. In very young children it is awkward, in that an intravenous line needs to be established and this may be difficult without a cut-down. It is particularly in young children that the alternative tests, using subcutaneous or intramuscular injection or oral administration, are useful.

In the *glucagon test* 0.5 mg or 0.1 mg/kg body weight is injected intramuscularly and blood GH estimated at intervals thereafter (Vanderschueren-Lodeweyckx *et al.* 1974). It may perhaps be improved by pretreatment of the patient with propranolol by mouth, 20 mg in a child weighing less than 20 kg, 40 mg in others, given 2 hours before the glucagon (Parks *et al.* 1973; Rochiccioli *et al.* 1976). In some patients glucagon produces the side-effects of nausea and vomiting.

More recently a combined *propranolol–laevodopa* test has been used. In this laevodopa is administered by mouth following overnight fasting, in a dose of 125 mg for subjects weighing less than 10 kg, 250 mg for subjects weighing 10–30 kg and 500 mg for subjects weighing over 30 kg; at the same time propranolol is administered as above. Blood samples are taken at intervals which should include 60 and 90 minutes (Fass *et al.* 1979). Laevodopa also causes nausea and vomiting in some patients.

Primed insulin tolerance test

There exists a group of children, mostly in the immediately pre-pubertal age range, who do not respond to the ITT, or who produce only 'partial' responses and who are growing slowly, even for their bone age. Despite this, they increase their growth rate to normal as soon as sex hormone secretion starts and they attain normal adult heights (Eastman *et al.* 1971). These children, when retested after puberty, have uniformly normal GH secretion; they do not require hGH treatment. In such cases the primed ITT is very useful (Martin *et al.* 1979). In boys 100 mg of testosterone esters is given intramuscularly 4 days before the ITT, which is then done in the usual way. In girls 200 μg of oral ethinyl oestradiol is given each day for

3 days before the ITT. We now use this test routinely as a second test when the first, the regular ITT, yields a 'partial' result, and we also use it as the first test in children with a bone age of 10 years or over.

Somatomedin test

Growth hormone does not act directly on the cartilage cells; it causes secretion of somatomedin C (SMC) primarily by the liver, and it is this which causes growth of cartilage. Accordingly, it would be very convenient to have an assay for SMC in blood, both for use in diagnosis and to monitor the effects of administration of hGH. At present, however, only cumbersome and highly specialised bioassays are available in the UK, and those only in 1 or 2 laboratories. A radioimmunoassay has long been sought and is now available from the Nicols Institute in Los Angeles and in one or two North American departments (see Copeland *et al.* 1980). Its development is especially important in that recently it has been claimed (Rudman *et al.* 1981) that there exist cases in which the radioimmunoassay of GH gives a false answer, assayable but biologically inactive GH being present. These cases, which, it is said, are more common than idiopathic GH deficiency and just as amenable to treatment, can only be discovered by SMC assay before and after a few days treatment with hGH. The subject is under discussion at the time of writing.

APPENDIX

Technique of ITT test

The patient should fast overnight and on the morning of the ITT have only water to drink. For the duration of the test he should rest on a bed and 50% dextrose, hydrocortisone, oxygen and aspiration equipment should be readily available (for although highly unlikely the patient may have a hypoglycaemic convulsion). A medical officer should remain with the patient for the duration of the test.

A 21 bufferfly cannula is then inserted into a vein and fixed into position for the duration of the test. A three-way stopcock is attached to the butterfly and the needle kept patent with normal saline. A ligature is applied proximal to the needle and fasting blood samples collected for growth hormone, glucose and cortisol determinations.

Soluble insulin is administered intravenously in a dose of 0.1 unit/kg body weight. As the amount of insulin to be used is generally rather small it is convenient, and safer, to prepare a solution of insulin in normal saline of 1 unit/ml. For example for a child of 16 kg the dose of insulin would be 1.6 units. If the patient is suspected of having panhypopituitarism the insulin dosage should be reduced to 0.05 units/kg body weight.

When the insulin has been injected the butterfly should be washed through with heparinised saline. Samples for GH and glucose are then collected at 20, 30, 45, 60 and 90 minutes and for cortisol at 45 and 90 minutes. If severe symptomatic hypoglycaemia develops (and if it is going to do so, it generally occurs between 20 and 45 minutes after the insulin is administered) the patient should be given 50% dextrose intravenously; a dose of approximately 1 ml/kg is usually adequate to correct the hypoglycaemia. Occasionally the response to this is poor and it is then advisable to give 100 mg of hydrocortisone intravenously (*not* further dextrose); sampling should continue if possible.

On completion of the test a small bolus of 50% dextrose may be given and following removal of the butterfly the child should receive a meal.

The ITT is only considered valid when the blood glucose drops to below 2.2 mmol/l (40 mg/100 ml) or when the sugar level falls to below 50% of the fasting value.

REFERENCES

BARLTROP D. (1977) *Paediatrics and Growth.* Unigate Paediatric Workshops: The Fellowship of Postgraduate Medicine, London.

BIERICH J. R. & POTTHOFF K. (1979) Spontaneous secretion of growth hormone in children with constitutional delay of growth and adolescence and with early normal puberty. *Monatsschr. Kinderheilkd.* **127**, 561–5.

BUCKLER J. M. H. (1979) *A Reference Manual of Growth and Development.* Blackwell Scientific Publications, Oxford.

COPELAND K. C., UNDERWOOD L. E. & VAN WYCK J. J. (1980) Induction of immunoreactive somatomedin-C in human serum by growth hormone–dose-response relationships and effect on chromatographic profiles. *J. Clin. Endocrinol. Metabolism* **50**, 690–7.

DAVIS D. R., APLEY J., FILL G. & GRIMALDI C. (1978) Diet and retarded growth. *Br. Med. J.* **1**, 539–42.

EASTMAN C. J., LAZARUS L., STUART M. C. & CASEY J. H. (1971) The effect of puberty on growth hormone secretion in boys with short stature and delayed adolescence. *Aust. NZ J. Med.* **2**, 154–9.

FALKNER F. & TANNER J. M. (1978–9) *Human Growth:* 3 vols. Baillière Tindall & Cox, London.

FASS B., LIPPE B. M. & KAPLAN S. A. (1979) Relative usefulness of 3 growth hormone stimulation screening tests. *Am. J. Dis. Child.* **133**, 931–3.

FORFAR J. O. & ARNEIL G. C. (1983) *Textbook of Paediatrics.* 3rd Edn. Churchill Livingstone, Edinburgh.

GROLL A., CANDY D. C. A., PREECE M. A., TANNER J. M. & HARRIES J. T. (1980) Short stature as the primary manifestation of coeliac disease. *Lancet* **ii**, 1097–9.

HOWSE P. M., RAYNER P. H. W., WILLIAMS J. W., RUDD B. T., BERTRANDE P. V., THOMPSON C. R. S. & JONES L. A. (1977) Nyctohemeral secretion of growth hormone in normal children of short stature and in children with hypopituitarism and intrauterine growth retardation. *Clin. Endocrinol.* **5**, 347–59.

LACEY K. A., HEWISON A. & PARKIN J. M. (1973) Exercise as a screening test for growth hormone deficiency in children. *Arch. Dis. Child.* **48**, 508–12.

MACE J. W., GOTLIN R. W. & BECK P. (1972) Sleep related human growth hormone (GH) release: A test of physiologic growth hormone secretion in children. *J. Clin. Endocrinol.* **34**, 339–41.

MARSHALL W. A. (1971) Evaluation of growth rate in height over periods of less than a year. *Arch. Dis. Child.* **46**, 414–20.

MARSHALL W. A. (1977) *Human Growth and its Disorders*. Academic Press, New York.

MARTIN L. G., GROSSMAN M. S., CONNOR T. B., LEVITSKY L. L., CLARK J. W. & CAMITTA F. D. (1979) Effect of androgen on growth hormone secretion and growth in boys with short stature. *Acta Endocrinol.* **91**, 201–11.

PARKS J. S., AMRHEIN J. A., VAIDYA V., MOSHANG T. & BONGIOVANNI A. M. (1973) Growth hormone responses to propranolol-glucagon stimulation: A comparison with other tests of growth hormone reserve. *J. Clin. Endocrinol.* **37**, 85–92.

PREECE M. A. (1976) The effect of administered corticosteroids on the growth of children. *Br. Postgrad. Med. J.* **52**, 625–30.

ROCHICCIOLI P., ENJEAUME P., DUTAU G., RIBOT G. & AUGIER D. (1976) Stimulation of growth hormone secretion using propranolol-glucagon. Results, statistical study and comparison in 71 children, with the test using insulin-arginine. *Arch. Fr. Pediatr.* **33**, 453–65.

RUDMAN D., KUTNER M. H., BLACKSTON R. D., CUSHMAN R. A., BAIN R. P. & PATTERSON J. H. (1981) Children with normal-variant short stature: Treatment with human growth hormone for six months. *New Engl. J. Med.* **305**, 123–31.

SAVAGE D. C. L., SWIFT P. G. F., JOHNSTON P. G. B., GOLDIE D. J. & MURPHY D. (1978) Combined test of anterior pituitary function in children. *Arch. Dis. Child.* **53**, 301–4.

SHANIS B. S. & MOSHANG T. (1976) Propranolol and exercise as a screening test for growth hormone deficiency. *Pediatrics* **57**, 712–14.

SMITH D. W. (1976) *Recognizable Patterns of Human Malformation*, 2nd edn. W. B. Saunders Company, Philadelphia.

SMITH D. W. (1977) *Growth and Its Disorders*. W. B. Saunders Company, Philadelphia.

TANNER J. M. (1978) *Foetus Into Man*. Harvard University Press, Cambridge, Mass.

TANNER J. M. (1981a) Short stature of pituitary origin: the clinical state of the art. In J. L. Gueriguian (ed.) *Insulins, Growth Hormone and Recombinant DNA Technology*, pp. 133–59. Raven Press, New York.

TANNER J. M. (1981b) Catch-up growth in man. *Br. Med. Bull.* **37**, 233–8.

TANNER J. M. (1983) Growth charts and pubertal staging. In R. A. Donald (ed.) *A Guide to the Diagnosis of Endocrine Disorders*. Dekker, New York.

TANNER J. M., LEJARRAGA H. & CAMERON N. (1975) The natural history of the Silver–Russell syndrome: A longitudinal study of 39 cases. *Pediatr. Res.* **9**, 611–23.

TANNER J. M. & WHITEHOUSE R. H. (1976) Clinical longitudinal standards for height, weight, height velocity and weight velocity and the stages of puberty. *Arch. Dis. Child.* **51**, 170–9.

TANNER J. M., WHITEHOUSE R. H., MARSHALL W. A., HEALY M. J. R. & GOLDSTEIN H. (1975) *Assessment of Skeletal Maturity and Prediction of Adult Height*. Academic Press, New York.

TANNER J. M., WHITEHOUSE R. H. & TAKAISHI M. (1966) Standards from birth to maturity for height, weight, height velocity and weight velocity: British children 1965. *Arch. Dis. Child.* **41**, 454–71; 613–35.

Vanderschueren-Lodeweyckx M., Wolter R., Malvaux P., Eggermont E. & Eeckels R. (1974) The glucagon stimulation test: Effect on plasma growth hormone and on immunoreactive insulin, cortisol and glucose in children. *J. Pediatrics* **85**, 182–7.

Whitehouse R. H., Tanner J. M. & Healy M. J. R. (1974) Diurnal variation in stature and sitting height in 12–14-year old boys. *Ann. Hum. Biol.* **1**, 103–6.

Chapter 12
Endocrine Disorders

DAVID B. GRANT

This chapter outlines the laboratory investigation of thyroid, adrenal, and 'intersex' disorders. More detailed discussions of these disorders can be found in several recently published textbooks on paediatric endocrinology (Hung *et al.* 1978; Frasier 1980; Brook 1981; Job & Pierson 1981).

THYROID DISORDERS

Embryology and physiology

The development of fetal thyroid function and the changes which take place after birth have been fully reviewed by Fisher & Klein (1981). In summary, the anlage of the thyroid gland develops as a bud of endodermal tissue between the first and second pharyngeal arches. By the seventh week of fetal life the developing gland has taken up its definitive position in front of the trachea and has been joined by components from the fourth branchial arches which contribute to the lateral lobes; by the 12th week its characteristic structure with formation of colloid vesicles is evident. There is good evidence that the maternal thyroid has no significant role in maintaining the concentration of thyroid hormones in the fetal circulation as the placenta is impermeable to thyroxine (T_4) and triiodothyronine (T_3). About 18–20 weeks after conception the concentrations of T_4 and thyrotrophin (TSH) in the fetal circulation begin to increase, suggesting that the hypothalamic-pituitary control of thyroid function has begun to develop, and a normal rise in plasma TSH after thyroid-releasing hormone stimulation has been described in infants born after 26–28 weeks gestation.

The biosynthesis of thyroid hormones has been fully reviewed by DeGroot & Niepomniszcze (1977) and can be summarised as follows: circulating iodine is 'trapped' by the thyroid cells and oxidised by a membrane-bound peroxidase system to allow iodination of the tyrosine residues in thyroglobulin, a high molecular weight glycoprotein. The biologically active thyroid hormones (L-thyroxine—T_4; triiodothyronine —T_3) are formed by coupling of iodotyrosines in the thyroglobulin molecule, which is then stored in the extracellular colloid. Both the trapping of iodine and the peroxidase activity are stimulated by TSH, which also controls the cellular uptake of thyroglobulin-carrying colloid in proteolytic vesicles and the subsequent release of thyroid hormones into the circulation. Under normal conditions, approximately 85% of the iodinated material which is released is in the form of T_4 and most of the remainder as T_3. Small amounts of other iodinated compounds (reverse T_3—rT_3; mono- and di-iodinated tyrosine—MIT and DIT; iodinated histones) are also released but most of this material is de-iodinated within the cells and the iodine recycled for further hormone synthesis. Inhibition of thyroid activity by iodine excess (Wolff–Chaikoff effect) is well-recognised in adults and it appears that the fetal or neonatal thyroid gland is particularly sensitive to iodine excess.

Most of the circulating T_4 and T_3 are bound to plasma proteins. Approximately 60% is carried by thyroxine-binding globulin (TBG), 30% by thyroxine-binding pre-albumin, and 10% by serum albumin; only very small proportions of the T_4 (0.03%) and T_3 (0.3%) are in the biologically available free forms. As discussed below, this binding is modified by a number of drugs, with the result that the total level of hormone may be reduced or increased with little or no change in the concentration of free hormone.

The mechanisms whereby thyroid hormones exert their metabolic effects have been reviewed by Sterling (1979). These include effects on nuclear transcription, mitochondrial activation, and the transport processes across cell membranes. It is generally agreed that T_4 is deiodinated to T_3 in the peripheral tissues, and that the latter is the active thyroid hormone. rT_3, a hormone with relatively little biological activity, seems to be the main product of T_4 deiodination in fetal life. The physiological significance of this difference between fetal and post-natal thyroxine metabolism is still not completely clear.

One of the most important functions of thyroid hormones in late fetal life and early infancy is related to brain development and the importance of untreated congenital hypothyroidism as a cause of mental retardation or neurological handicap is well known.

In later childhood the main clinical importance of thyroid hormones lies in their role as factors which are necessary for normal growth, and the importance of hypothyroidism as a cause of short stature has long been known.

Changes in serum thyroid hormone levels during childhood

After birth. Almost immediately after birth there is a marked rise in serum TSH which reaches a peak 1–2 hours after delivery and is usually complete after 12–24 hours. This rise in serum TSH secretion probably accounts for some of the striking changes in circulating thyroid hormone levels which take place during the first few days of life.

Total thyroxine levels in cord blood plasma are in the upper normal range for adults but within 12 hours of delivery they show an almost

Table 12.1 Plasma thyroxine levels (nmol/l) ±SD in full-term and premature infants during the first 3 months of life (after Cuestas 1978).

	Estimated gestational age (weeks)				
Age	30–31	32–33	34–35	36–37	37–40
Cord blood	84±13	97±27	86±15	97±36	106±23
12–72 hours	148±27	159±41	160±40	200±34	245±27
3–10 days	99±23	110±25	129±31	164±32	205±39
11–20 days	97±23	107±21	135±23	144±37	157±26
21–45 days	101±19	103±22	120±17	147±54	156±19
46–90 days	124±22 (30–37 weeks)				135±25

two-fold rise to values which are well above the normal adult range. These high levels are sustained for some time after delivery but by the age of 6 weeks the levels have returned to values similar to those in cord blood. A very similar pattern is shown by free-T_4 (Abuid *et al.* 1974). These changes in plasma T_4 are also found in premature infants (Cuestas 1978) (Table 12.1).

Even more striking changes are found for plasma T_3 which is very low in cord blood when compared with adult standards but which shows a three- or four-fold increase in the first 12 hours after delivery. Unlike plasma T_4, relatively high plasma T_3 values persist well into the first year of life. Free-T_3 values show an almost identical pattern to total T_3 levels (Abuid *et al.* 1974).

rT_3 levels in cord serum are much higher than those seen in adult life but they fall progressively after delivery and are within the adult range after about one week.

Thyroglobulin can be detected in cord plasma at concentrations which are greater than those in the maternal circulation. Like T_4 and T_3, thyroglobulin levels rise after delivery (Pezzino *et al.* 1981), probably as a result of increased thyroid hormone secretion.

Plasma thyroxine-binding globulin concentrations in cord blood are higher than those in adult life. However, they do not change during the early neonatal period and while increased hormone binding capacity probably accounts for the relatively high T_4 levels in cord blood it does not account for the striking changes in T_4 and T_3 which take place after delivery.

During later childhood. After 3–4 weeks of age until the onset of puberty the mean plasma T_4 level is higher than the mean adult level, but individual values usually fall within the normal adult range (Ryness 1972); there is very little information on free-T_4 values during this period.

Unlike T_4, T_3 levels during the first 6–12 months of life are often well above the normal adult range.

Laboratory tests of thyroid function

The different methods which are used to assess thyroid function are listed in Table 12.2.

Radioimmunoassay or competitive protein binding methods for measuring total *plasma* T_4 have been widely available for several years. Radioimmunoassay methods are also available for estimating plasma total T_3, rT_3, and thyroglobulin (Tg), as well as iodotyrosines (MIT and DIT), but with the exception of T_3 assay these latter methods are largely research tools. Immunoassay methods can also be used to estimate thyroxine-binding globulin (TBG) concentrations.

Plasma T_4 and T_3 values do not always reflect thyroid activity as they are also dependent on the thyroid-hormone binding capacity of the plasma. Examples of this are the low plasma T_4 values seen in familial TBG deficiency, the nephrotic syndrome, or during treatment with phenytoin, anabolic steroids, or high doses of glucocorticoids. Conversely, increased binding occurs as a result of pregnancy or oestrogen therapy leading to high plasma T_4 levels.

To offset this difficulty, several indirect tests have been devised to assess the concentration of 'free hormone'. Plasma samples are incubated with trace amounts of radioactive T_3 and resin, to estimate 'T_3 resin uptake'. T_3 is distributed between the resin and the free binding sites in the plasma. Thus, a low T_3-resin uptake is found when the T_4 levels are low or when binding capacity is increased, and vice versa, a high T_3-resin uptake indicates high T_4 levels or reduced binding capacity.

The product of $T_4 \times T_3$-resin uptake gives the free thyroxine index (FTI), thus in theory allowing for variations in hormone binding capacity. Similarly, the effective thyroxine ratio (ETR) is obtained by comparing the

Table 12.2 Direct and indirect tests of thyroid function.

Plasma
T_4, T_3, free T_4, free T_3 (reverse T_3, thyroglobulin, T_2, etc.)
Resin T_3 uptake (RT_3U), free thyroxine index (FTI), effective thyroxine ratio (ETR), thyroxine binding globulin (TBG)
TSH: basal or after TRH stimulation
Antithyroid antibodies
Long-acting thyroid stimulator (LATS) and human specific thyroid stimulating antibody (HTSI)
Thyroid scan
Localization of gland (^{99m}Tc or ^{123}I)
Investigation of biosynthetic defect

results obtained by adding the patient's serum or a control serum to the assay system. These indirect methods are now being replaced by radioimmunoassay methods to measure free-T_4 and free-T_3; these methods initially included a time-consuming dialysis stage to separate the free hormones, but immunoassay methods, using a tracer which is not bound by serum proteins, are likely to simplify estimation of free T_4 and T_3. T_4 (or T_3) can also be estimated in dried blood spots on filter paper for screening purposes.

TSH assay can also be used to assess thyroid function and may be particularly useful in patients with abnormal T_4-binding. Elevated basal levels of TSH or an exaggerated response to injection of TRH usually suggest primary hypothyroidism.

Other thyroid function tests

Tests for the presence of antibodies to thyroid components and tests to elucidate the different types of enzyme defects of thyroxine biosynthesis are considered below.

Clinical applications

Suspected hypothyroidism

The causes of hypothyroidism are listed in Table 12.3.

The most common cause of congenital hypothyroidism is hypoplasia or aplasia of the thyroid. Most cases are sporadic but families with two affected siblings have been described. The next most common cause is one of the various enzyme defects of thyroid hormone biosynthesis; these make up about 15% of the cases. Autoimmune thyroiditis is the third important cause of hypothyroidism, particularly in girls. Such cases usually present after the age of 5 and there is often a positive family history. Other rare causes of primary hypothyroidism include antithyroid medication during pregnancy, excessive iodine intake, or previous surgery for hyperthyroid-

Table 12.3 Causes of hypothyroidism.

Primary hypothyroidism
Thyroid dysgenesis (aplasia, hypoplasia, ectopic gland)
Defective thyroxine biosynthesis
Autoimmune thyroiditis (Hashimoto's thyroiditis)
Other (anti-thyroid drugs, iodine deficiency, excess iodine intake, previous surgery, etc.)
Secondary and tertiary hypothyroidism
TSH deficiency
TRH deficiency

ism. Iodine deficiency is now a very rare cause of congenital or acquired hypothyroidism in the Western world.

Secondary hypothyroidism (pituitary) or tertiary hypothyroidism (hypothalamic) are less common and may occur as an isolated defect of TSH secretion or, more commonly, as part of a more generalised anterior pituitary failure—for example, after breech delivery or surgery for craniopharyngioma.

Hypothyroidism due to insensitivity of the thyroid gland to TSH or due to resistance of the peripheral tissues to thyroid hormones are very rare and have only been described in a handful of cases.

Laboratory diagnosis of hypothyroidism. In most cases the laboratory diagnosis of primary hypothyroidism is straightforward and rests upon the demonstration of below normal T_4 values in plasma or serum. However, errors in the diagnosis can arise in the following situations:

1 When inappropriate reference ranges are used. This is mainly a problem in the newborn period when a plasma T_4 value which appears to be normal when judged by adult standards is below the lower limits of normal for the neonatal period and in fact indicates hypothyroidism.

2 When the thyroxine-binding capacity of the serum proteins is abnormal. The classic examples of this are the very low T_4 levels which are found in familial sex-linked TBG deficiency. Perhaps a more common example would be the child with suspected hypothyroidism who is on treatment with phenytoin or some other drug which reduces thyroxine binding. While the indirect tests of free thyroxine concentration (FTI, ETA) will usually indicate that abnormal hormone-binding is present, they are not always entirely satisfactory for determining whether or not the patient is euthyroid. It is anticipated that the more widespread introduction of assays for free-T_4 and T_3 will be particularly helpful in these patients.

3 When only very mild or 'compensated' hypothyroidism is present. In these cases, the plasma T_4 may lie just within the normal range and the plasma T_3 may be normal.

In all of these situations TSH assay can be particularly helpful in clarifying the true situation; a normal value essentially excludes primary hypothyroidism while a TSH level which is quite definitely elevated (say > 20 mU/l) makes the diagnosis almost certain. However, there are a few conditions in which TSH assay may be misleading. One is coeliac disease in which a low serum T4 may be associated with an exaggerated TSH response to TRH; another is pseudohypoparathyroidism which is often associated with slightly elevated basal TSH values in the presence of apparently normal thyroid function.

TSH assay is of rather limited value in the diagnosis of pituitary or hypothalamic hypothyroidism. In both conditions the basal levels are low and indistinguishable from normal. However, in the hypothalamic form there is a definite but delayed rise in TSH after TRH injection, while in the pituitary form there is no appreciable rise in TSH.

There is still controversy over the place of isotope scans of the thyroid gland in the diagnosis of hypothyroidism in early childhood, particularly as the result will have very little bearing on initial treatment and may be difficult to interpret if carried out in a department which has little experience of thyroid scanning in small infants. However, the scan may show an unsuspected ectopic gland or indicate that the patient has one of the inherited disorders of thyroxine biosynthesis (see below). This information allows the clinician to give a clearer explanation as to the cause of hypothyroidism and more accurate advice as to the risk of recurrence in future pregnancies. Either ^{123}I or ^{99m}Tc pertechnetate can be used but ^{123}I has the advantage of giving a smaller dose of irradiation. Unfortunately, this isotope is not always immediately available and in this situation the use of ^{99m}Tc is probably justified.

Screening tests for congenital hypothyroidism. It is now widely recognised that it is possible to detect congenital hypothyroidism soon after birth by routine screening of all newborn (Fisher *et al.* 1979; De Lange *et al.* 1981). Two different methods of screening are in general use; both use either blood samples collected between the 6th and 14th days of life (often as dried blood spots on filter paper) or samples of cord blood. In the first method, which is largely used in North America, T_4 concentration is estimated and all the samples which give values below the lowest 3–5% of the normal range are reassayed for TSH. Samples which give an elevated TSH value almost certainly indicate primary hypothyroidism while those with normal TSH values may indicate secondary hypothyroidism, or more likely, the clinically unimportant condition of TBG deficiency. The second method, which is largely used in European screening programmes, relies on the estimation of TSH alone. While this method will not detect a small proportion of cases with secondary hypothyroidism, it has the great advantage of giving fewer false-positive results. It also allows the identification of 'compensated cases' who usually have large ectopic glands or biosynthetic defects and in whom the plasma T_4 level may be within normal limits in early infancy. This ability to identify 'mild' cases probably explains why the incidence of congenital hypothyroidism detected by TSH screening has usually been higher than that with T_4/TSH screening. It should however be noted that spuriously elevated TSH screening values have been reported from France, probably as a result of previous immunisation of the mother with vaccine grown on fetal rabbit lung. This led to the production of maternal antibodies which could cross the placenta to the fetus and interfere with the TSH radioimmunoassay.

Most screening programmes rely on relatively imprecise methods and in suspected cases of hypothyroidism it is important to confirm the diagnosis by quantitative measurement of T_4 and TSH on plasma samples before treatment is begun. However, in cases with clinical evidence of hypothyroidism such as slow feeding, lethargy or delayed bone maturation, treatment should not be delayed and thyroxine can be given as soon as a blood sample has been obtained for definitive tests of thyroid function.

It is now clear that hypothyroidism may be transient in a small proportion of cases diagnosed by neonatal screening. For this reason, temporary withdrawal of treatment after the first year is probably warranted in cases who show no clinical features of hypothyroidism at the time of initial diagnosis. In order to keep the period without treatment to a minimum, T_3 (10–20 μg/day) can be substituted for the usual T_4 therapy for 2–3 weeks before withdrawing medication. As the T_3 is rapidly cleared from the circulation, estimation of T_4 and TSH can be carried out a week after stopping treatment.

Monitoring treatment in hypothyroidism. Plasma T_4 and TSH assays are now extensively used to monitor treatment in cases of hypothyroidism. In general, a dose of L-thyroxine is chosen which will keep the plasma T_4 in the normal range and at the same time suppress TSH to normal values. However, it should be noted that in some infants with congenital hypothyroidism full TSH suppression may take weeks or even months (Schultz *et al.* 1980). In such cases it is probably acceptable to have mildly elevated TSH values, providing plasma T_4 levels are in the upper part of the normal range.

Hyperthyroidism

Hyperthyroidism (Graves' disease) is relatively uncommon in early childhood but is not particularly rare in girls in their mid and late teens. It seldom presents diagnostic difficulties as the classic symptoms of tachycardia, exophthalmos and goitre are usually obvious. Exceptions are in mild or early cases where the symptoms may be those of erratic behaviour or poor school progress, and in very young children when poor weight gain and persistent diarrhoea suggest the presence of a gastrointestinal disorder.

Confirmation of the diagnosis is usually straightforward and depends on demonstration of elevated serum thyroid hormones. In many cases a single assay of plasma T_4 will suffice but when this gives borderline results estimation of plasma T_3 will usually clarify the diagnosis. In a small proportion of cases T_4 values are well within normal while T_3 is markedly elevated. This state of 'T_3-toxicosis' is usually found in adults, particularly those with functioning thyroid adenomas. Only rarely is it necessary to confirm the diagnosis of thyrotoxicosis by estimating plasma TSH after injection of TRH; release of TSH is absent in hyperthyroidism while a normal rise in TSH is considered to exclude the diagnosis. However, it should be noted that there is very little published data on the place of this test in very young children. These methods have largely replaced techniques such as estimating the uptake of radioiodine by the thyroid and determining whether or not it can be blocked by giving triiodothyronine (T_3 suppression test).

Neonatal hyperthyroidism. This is due to maternal Graves disease and is caused by the passage of thyroid stimulating immunoglobulins (TSI) from the mother to the fetus. It can occur even when the maternal Graves disease has been controlled by surgery or antithyroid drugs and to some extent can be predicted by estimating the TSI titres in the maternal circulation. Signs of hyperthyroidism do not usually develop for some days after delivery—this delay may be partly related to the protective effect of antithyroid drugs which have been given to the mother and partly to the preferential production of rT_3 which is a feature of fetal thyroid function and which changes to T_3 production soon after birth. The diagnosis can usually be confirmed quite easily by demonstrating plasma T_4 and T_3 levels which are even higher than those normally expected in the newborn period.

Goitre

The causes of generalized enlargement of the thyroid gland are given in Table 12.4. All are relatively uncommon during childhood (Reiter *et al.* 1981). The most important cause—Graves disease—has already been considered.

Chronic lymphocytic thyroiditis (Hashimoto's thyroiditis) is characterized by firm, and sometimes asymmetrical enlargement of the thyroid: a small lymph gland can often be felt just above the isthmus. The clinical diagnosis can be confirmed by estimation of serum antibody titres to thyroid microsomes. Antithyroglobulin antibody titres may not be elevated in a

Table 12.4 Causes of goitre.

Hyperthyroidism (Graves' disease)
Thyroiditis
Chronic lymphocytic thyroiditis (Hashimoto disease)
Acute and subacute thyroiditis
Disorders of thyroid biosynthesis
Iodide trapping defect
Peroxidase 'organification' defect
Iodotyrosine coupling defect
Iodotyrosine deiodinase defect
Defects in thyroglobulin synthesis
Simple or colloid goitre (puberty, iodine deficiency)

small but significant proportion of cases. Tests of thyroid function can range from those of severe hypothyroidism to those of thyrotoxicosis. Most often they are normal or merely show a low-normal T_4 and a slightly elevated TSH suggesting 'compensated' hypothyroidism. Radioiodine uptake may be enhanced during the early stages and a positive perchlorate discharge test is sometimes seen in these cases.

Acute and subacute thyroiditis are uncommon and are characterized by pain and tenderness over the gland. Thyroid antibodies are usually negative, and thyroid function tests normal. However, high T_4 values may occur at the onset as a result of hormone release from the damaged gland. During recovery T_4 may be low and TSH elevated but thyroid function eventually returns to normal.

Disorders of thyroid biosynthesis. A number of different types of metabolic defect in thyroid hormone biosynthesis have been described. They are listed in Table 12.4. They are all associated with some enlargement of the gland and cause clinical and biochemical features of hypothyroidism of varying degrees of severity.

Estimation of radioactive iodine uptake by the gland is probably the single most useful investigation. With the exception of the 'trapping' type of defect, in which there is no iodine uptake by the thyroid (or the salivary glands), all the biosynthetic defects cause increased radioiodine uptake. In the 'organification' type, which may be due to peroxidase deficiency, administration of perchlorate will result in a rapid discharge of unbound iodine from the gland. The dehalogenase type leads to rapid iodine turnover due to loss of iodotyrosines from the gland wheras iodine turnover is slow in the iodoprotein type of defect in which thyroglobulin is replaced by an abnormal iodoprotein. In these latter cases radioimmunoassay for thyroglobulin may be helpful by demonstrating undetectable levels in the plasma. Further investigation of these different metabolic defects calls for sophisticated techniques such as estimation of radioactive thyroid products in the circulation following administration of radioiodine, estimation of iodoproteins in the urine, and metabolic studies on thyroid tissue obtained by biopsy. Fortunately all these disorders respond to treatment with L-thyroxine and the above rather complicated investigations are rarely called for on clinical grounds.

DISORDERS OF THE ADRENAL CORTEX

Embryology and Physiology

The adrenal cortex is derived from mesoderm between the root of the mesentery and the germinal ridges, and can be recognized about 6 weeks after conception: the close relationship of the adrenal anlage to the germinal ridge probably accounts for the occurrence of 'adrenal rests' in the testes, and less often, the ovaries. An outer layer of small cells which will become the definitive cortex and an inner mass of large cells which will develop into the fetal cortex become evident, and the gland grows rapidly until, 4 months after conception, it is larger than the fetal kidney. This increase in size is largely due to growth of the fetal zone. In anencephaly, the fetal cortex is poorly developed, indicating that its growth is at least partly determined by the fetal pituitary gland.

Shortly after birth the fetal cortex undergoes involution and has all but disappeared by the age of 6 months. During this period the cells of the definitive cortex increase and by the age of 2 years the outer two layers of the adult cortex are evident (zona glomerulosa and zona fasciculata). The zona reticularis is incomplete but it increases in thickness during childhood to reach the adult pattern around the time of puberty.

It has been known for a number of years that the fetal adrenal cortex has a complicated relationship with placental steroid metabolism (Diczfalusy 1969). In brief, steroid sulphatase and 3β-hydroxysteroid dehydrogenase activities in the fetal cortex are low. In contrast, the placenta has 3β-hydroxysteroid dehydrogenase and steroid sulphatase activity but is lacking in sulphate transferase. As a result, the fetal cortex is unable to synthesise cortisol from dehydroepiandrosterone and relies on placental progesterone as a substrate for cortisol synthesis. Conversely, dehydroepiandrosterone sulphate and 16α-hydroxydehydroepiandrosterone sulphate are used by the placenta as substrates for oestrone and oestriol synthesis. The importance of these latter compounds is still not clear. However, hypoplasia of the fetal cortex—for example in anencephaly—and absence of placental sulphatase lead to very low maternal oestriol excretion, but have little if any effect on the course of the pregnancy.

The pattern of adrenal steroid synthesis changes with involution of the fetal cortex after birth. The definitive zona fasciculata produces cortisol and the zona glomerulosa synthesises aldosterone. C19 steroids such as dehydroepiandrosterone and androstenedione are mainly formed in the zona reticularis and the progressive rise of these steroids which takes place towards the end of the first decade seems to mirror growth of the zona reticularis.

Cortisol production is stimulated by pituitary corticotrophin (ACTH) while aldosterone is largely regulated by the renin–angiotensin system but is also stimulated by ACTH. There is still controversy as to whether androgen secretion is controlled by ACTH or an as yet unidentified pituitary 'androgen-stimulating hormone' (ASH).

The best-documented physiological roles of cortisol are in the maintenance of blood glucose by facilitating gluconeogenesis and in the maintenance of blood pressure. Less clearly documented actions are in combatting 'stress' and suppressing inflammatory responses. Aldosterone promotes reabsorption of sodium from the distal renal tubules in exchange for potassium, and thus regulates sodium excretion in the urine. The physiological role of adrenal androgens is still controversial.

ADRENAL FUNCTION TESTS

In the past urinary steroid analysis formed the basis of most tests of adrenal function, but with the development of radioimmunoassay methods estimation of plasma steroid levels has, to some extent, replaced urine steroid analysis in the investigation of adrenal disorders.

There are still relatively few control data which have been collected in normal children of different ages using these newer laboratory methods.

Analysis of basal steroid production or excretion

Urine. Until fairly recently urinary steroid analysis was carried out using chemical or fluorometric methods on aliquots from timed collections of urine. These methods are still used in many laboratories and normal values

Table 12.5 Urinary steroid excretion at different ages.

Age	Excretion rate (μmol/24 h)
17-hydroxycorticoid [1]	
5 months to 1 year	4–24
2–5 years	4–38
5–10 years	2–22
10–14 years	14–50
17-oxosteroids [1]	
Neonates	up to 7
4 weeks–1 year	up to 2
1–5 years	up to 6
6–10 years	up to 10
11–17 years	up to 32
Urinary free cortisol [2]	
4–12 years	107 ± 33 nmol/m² per day
Cortisol production rate [3]	
4 months–20 years	32.5 ± 6.3 μmol/m² per day

[1] Clayton *et al.* (1980)
[2] Franks (1973)
[3] Kenny *et al.* (1966)

obtained in children of different ages are given in Table 12.5. However, in many centres these methods have been replaced by gas–liquid chromatography, which has the advantages of specificity and allowing the quantitation of individual steroids or metabolites.

Estimation of steroid production rates are possible if an isotope-labelled steroid is administered and its recovery and specific activity are measured in the urine. Methods for estimating cortisol production rates have been available for a number of years (Table 12.5). However, it should be noted that measurement of steroid production rates is largely a research tool and is rarely needed for clinical diagnosis.

Plasma. It is now possible to assay plasma cortisol and most of its precursor steroids, aldosterone, androgens, and oestrogens by specific radioimmunoassay methods. However, care is still needed in interpreting the results obtained with these different assays. The conditions under which basal samples are collected must often be standardised. For example, there is marked diurnal variation in plasma cortisol concentration after the first 1–2 years of age and random values obtained during the evening may be misleadingly low.

In addition, age and puberty-related control values are required for the correct interpretation of results for many adrenal and gonadal steroids. Such control data has been published by a number of workers and some representative results are given in Tables 12.6 and 12.7.

Non-specific fluorometric methods for measuring plasma steroids are particularly unreliable in children and have now been almost entirely replaced by radioimmunoassays.

Table 12.6 Mean values (±1 SD) for plasma dehydroepiandrosterone sulphate (DHAS), dehydroepiandrosterone (DHA), androstenedione (A), testosterone (T), oestrone (E1) and oestradiol (E2) in males at different stages of pubertal development.

Puberty stage	DHAS[1] (μmol/l)	DHA[1] (nmol/l)	A[2] (nmol/l)	T[2] (nmol/l)	E1[3] (pmol/l)	E2[3] (pmol/l)
1	0.25±0.19[1]	3.00±1.07[1]	0.44±0.40[3]	0.18±0.16[3]	39±10	27±10
1	1.06±0.81[2]	7.13±1.73[2]	0.68±0.51[4]	0.37±0.48[4]		
2	1.60±0.91	10.63±3.18	1.49±0.79	1.85±1.84	57±11	38±11
3	1.88±1.25	13.97±3.46	1.76±0.70	8.32±3.36	79±7	58±19
4	2.64±1.42	13.01±3.43	3.59±2.02	18.9±9.64	123±23	80±29
5	3.16±2.09	18.83±3.91	—	—	116±26	76±19

[1] De Peretti and Forest (1978)
[2] Korth-Schutz *et al.* (1978)
[3] Sippell *et al.* (1980)

1 Aged 6–9 years
2 Aged 9–11 years
3 Bone age < 7 years
4 Bone age > 7 years

Table 12.7 Mean values (with SD or range) for plasma dehydroepiandrosterone sulphate (DHAS), dehydroepiandrosterone (DHA), androstenedione (A), testosterone (T), oestrone (E1), and oestradiol (E2) in females at different stages of pubertal development.

Puberty stage	DHAS[1] (μmol/l)	DHA[1] (nmol/l)	A[2] (nmol/l)	T[2] (nmol/l)	E1[3] (pmol/l)	E2[3] (pmol/l)
1	0.38 + 0.32[1]	3.60 + 1.32[1]	0.63 + 0.42[3]	0.20 + 0.08[3]	50 (10–107)	30 (4–73)
1	1.10 + 0.48[2]	6.00 + 2.36[2]	1.28 + 0.82[4]	0.25 + 0.06[4]		
2	1.25 + 0.56	11.40 + 2.08	2.07 + 1.46	0.45 + 0.24	79 (37–122)	61 (37–88)
3	1.82 + 1.37	14.82 + 5.58	2.39 + 1.45	0.61 + 0.37	112 (55–159)	93 (26–220)
4	2.21 + 1.24	17.04 + 4.16	4.89 + 1.95	1.36 + 0.71	133 (59–285)	172 (77–312)
5	3.08 + 1.08	22.00 + 9.37	—	—	227 (167–285)	407 (124–624)

[1] De Peretti and Forest (1978)
[2] Korth-Schutz *et al.* (1978)
[3] Sippell *et al.* (1980)

1 Aged 6–9 years
2 Aged 9–11 years
3 Bone age < 7 years
4 Bone age > 7 years

Stimulation tests

ACTH. In the short ACTH test up to 250 μg tetracosactryn is given by either intramuscular or intravenous injection. The plasma cortisol level 30 minutes later is generally above 700 nmol/l and the increment in plasma cortisol is > 280 nmol/l (Barnes *et al.* 1972). The long ACTH test has been largely superseded by the short test described above, but is still occasionally of value in patients with very mild enzyme defects of steroid biosynthesis, particularly if they have already been treated with steroids, as the prolonged ACTH stimulation may bring to light the abnormal pattern of steroid secretion. In this test urinary (and plasma) steroids are measured after twice-daily injections of 20 IU ACTH gel given for 3 days. Urine 17-hydroxysteroid excretion should normally be at least 70 μmol/24 hours on the third day of ACTH injections (Clayton *et al.* 1980).

Insulin. This test, like the glucagon and metyrapone tests described below, assesses the function of the hypothalamic–pituitary–adrenal axis and is useful in suspected secondary adrenal insufficiency. The test depends on the release of ACTH as a result of insulin-induced hypoglycaemia and the subsequent rise in plasma cortisol. Sixty minutes after injection of 0.1 U/kg insulin the plasma cortisol level is normally over 600 nmol/l (Barnes *et al.* 1972).

Growth hormone secretion can be investigated at the same time as cortisol secretion in the insulin test, and by adding TRH and luteinizing hormone-releasing hormone (LHRH) to the insulin injection it is possible to investigate secretion of growth hormone, ACTH, TSH and the gonadotrophins in a single test procedure.

Glucagon. This test avoids the hazard of serious hypoglycaemia and is often preferred to the insulin test in young children. Plasma cortisol is measured at intervals for $2\frac{1}{2}$ hours after intramuscular injection of glucagon (100 μg/kg): a value over 500 nmol/l is considered normal (Vanderscheuren-Lodeweyckx *et al.* 1974).

Metyrapone. This agent inhibits adrenal 11-hydroxylase and blocks cortisol synthesis. The rise in ACTH which follows this leads to an increase in plasma 11-deoxycortisol and urine 17-hydroxysteroids.

In the standard form of the test, 300 mg metyrapone/m^2 is given every 4 hours for a total of six doses and normally the urinary 17-hydroxysteroids show at least a three-fold increase above basal levels. This test is unpleasant and often unreliable in small children but the short metyrapone test described by Leisti (1977) appears to avoid these disadvantages. A single dose of metyrapone (1 g/m^2) is given between 0800 and 0900 hours and plasma 11-deoxycortisol is estimated 3 hours later. A value over 130 nmol/l would normally be expected.

Suppression tests

Dexamethasone. The low-dose dexamethasone suppression test was at one time used extensively to differentiate the high urine steroid excretion of obesity from that due to Cushing's syndrome. In adults, either 1.0 mg dexamethasone is given at 2400 hours and plasma cortisol estimated 8 hours later, or 0.5 mg dexamethasone is given 6-hourly for 3 days and urinary 17-hydroxysteroids estimated during the last 24 hours. Normally, plasma cortisol will be below 150 nmol/l in the first test and urinary 17-hydroxysteroids completely suppressed during the second test.

The high-dose dexamethasone test has been used to differentiate between pituitary-dependent Cushing's syndrome and Cushing's due to adrenal tumour or ectopic ACTH secretion. In adults, 2 mg dexamethasone are given 6-hourly for 3 days. In pituitary-dependent Cushing's, there is significant ($> 50\%$) suppression of urinary 17-hydroxysteroids whereas no suppression is expected in cases of adrenal tumour or ectopic ACTH secretion. However, these suppression tests have not been widely standardised in young children, particularly with regard to the dose of dexamethasone, and have at times given misleading results (Streeten *et al.* 1975).

Other tests of adrenal function

ACTH. Estimation of plasma ACTH by radioimmunoassay may be of value as an indirect guide to adrenal function; persistently elevated basal values suggest the presence of defective cortisol secretion.

Plasma renin activity (PRA). Demonstration of elevated PRA can provide a very sensitive index of impaired aldosterone secretion, providing that other causes of salt loss, such as vomiting and diarrhoea, can be excluded.

Electrolyte balance. Restriction of sodium intake may be very useful in bringing to light minor degrees of aldosterone deficiency. Normally, urinary sodium loss will mirror the reduced sodium intake and inappropriately high urinary sodium levels, particularly if associated with mild hyponatraemia, point to impaired sodium reabsorption by the kidneys.

Visualisation of the adrenal glands. The presence of an adrenal tumour can often be most easily demonstrated by ultrasound examination or a computerised tomographic (CT) scan. However, in difficult cases, aortography may still be necessary. Adrenal scan with selenocholesterol can also be helpful in determining the presence of a tumour, with suppression of uptake by the other gland. In general, these methods are unhelpful in the diagnosis of adrenal hypofunction.

HYPOADRENAL STATES

The causes of impaired adrenocortical function are listed in Table 12.8.

Adrenal hypoplasia or atrophy

There are a number of different causes of adrenal atrophy or hypoplasia; all are uncommon in childhood. The majority are associated with both cortisol deficiency and aldosterone deficiency but only one of these features

Table 12.8 Causes of adrenal hypofunction.

Inherited
Congenital adrenal hyperplasia (recessive) (see Table 12.9)
Congenital adrenal hypoplasia
  cytomegalic : (X-linked)
  atrophic : (recessive)
Familial glucocorticoid insufficiency (recessive)
Addison–Schilder's disease (X-linked)

Sporadic
Addison's disease
Autoimmune, 'idiopathic', tuberculosis

'Secondary'
Iatrogenic
ACTH deficiency: panhypopituitarism, pituitary–hypothalamic tumours, etc.

may be evident at the time of initial diagnosis. The presence of cortisol deficiency can often be easily confirmed by demonstrating a subnormal rise in plasma cortisol after tetracosactryn injection coupled with low urinary 17-hydroxycorticoid excretion. It is rarely necessary to carry out a prolonged ACTH stimulation test but in borderline cases demonstration of a markedly elevated plasma ACTH level may provide the most sensitive index of impaired cortisol secretion. However, when measuring plasma ACTH it is important to avoid stress as this can also cause elevation of plasma ACTH.

Hyponatraemia and hyperkalaemia, associated with inappropriate sodium loss in the urine, point to the diagnosis of aldosterone deficiency, although these changes are also found in some renal salt-wasting conditions. However, in mild aldosterone deficiency it may be necessary to monitor plasma and urinary sodium during a low-sodium diet to confirm the diagnosis. In these cases estimation of PRA can be very helpful; a raised value is probably the most sensitive index available of depleted body sodium. As with ACTH, the conditions under which samples for PRA assay are collected can be very important—in particular, whenever possible the subject should be kept supine for at least an hour before the sample is taken.

Laboratory investigation is usually of relatively little value in differentiating the different causes of adrenal hypofunction from one another. However, it may be possible to demonstrate positive anti-adrenal antibody titres in some patients, particularly those with other autoimmune disorders such as hypoparathyroidism or diabetes mellitus. There are no specific tests for the different types of familial adrenal insufficiency, i.e. X-linked cytomegalic adrenal hypoplasia, recessive atrophic adrenal hypoplasia, and recessive isolated cortisol deficiency. The true nature of these conditions may only be appreciated when a positive family history comes to light but the demonstration of persistently normal aldosterone secretion strongly suggests the diagnosis of familial cortisol deficiency, particularly if the patient has absent tear secretion or achalasia of the cardia, which are also present in some families. In general, the adrenal insufficiency of 'Addison–Schilder's disease' (familial adrenoleucodystrophy) is relatively mild and the illness is dominated by the severe and progressive neurological disturbance which characterises the disorder.

Enzyme defects of adrenal steroid biosynthesis

At least seven different inherited disorders of adrenal steroid biosynthesis are known (Fig. 12.1). While defective cortisol production is a feature of most of them, this is often largely offset by adrenal hyperplasia and symptoms of cortisol deficiency may be mild or only come to light during periods of infection or stress.

The main clinical features of these disorders are either abnormal

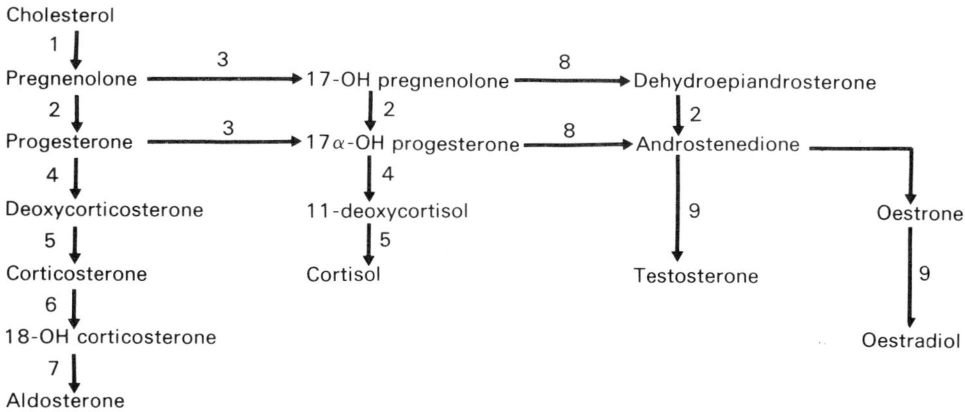

Enzyme deficits causing:

Defective cortisol biosynthesis	Defective aldosterone biosynthesis	Defective testosterone biosynthesis
1 20–22 desmolase		1
20-hydroxylase	2	2
22-hydroxylase	4	3
2 3β-hydroxysteroid dehydrogenase	5	8 17–20 desmolase
3 17α-hydroxylase	6 18-hydroxylase	9 17β-hydroxysteroid
4 21-hydroxylase	7 18-hydroxysteroid dehydrogenase	
5 11β-hydroxylase		

Fig. 12.1 Outline of cortisol, aldosterone and testosterone biosynthetic pathways and known enzyme deficits.

Table 12.9 Types of congenital adrenal hyperplasia.

Type	Virilisation in girls	Incomplete masculinisation in boys	Salt loss	Hypertension	Urinary oxosteroids
21-hydroxylase	Yes	No	Yes/No	No	Elevated
11β-hydroxylase	Yes	No	No	Yes	Elevated
3β-hydroxysteroid dehydrogenase	Yes	Yes	Yes	No	Elevated
17α-hydroxylase	No	Yes	No	Yes	Very low
20-22 desmolase	No	Yes	Yes	No	Very low

patterns of virilisation or feminisation, increased urinary salt loss, or hypertension (Table 12.9).

Congenital adrenal hyperplasia (CAH) due to 21-hydroxylase deficiency is the most common type and is characterised by virilisation which is usually evident at birth in girls but which may not become apparent in boys until later childhood; salt loss is a feature in approximately 50% of cases. Until the last decade, urinary steroid analysis was used to confirm the diagnosis —17-oxosteroids, pregnanetriol, and the 11-oxygenation index (Clayton *et al.* 1971) are all increased. However, these tests were inconvenient and sometimes unreliable during the first weeks of life, and they have been largely replaced by measurement of plasma 17α-hydroxyprogesterone (17-OHP), either by competitive protein binding (Barnes & Atherden 1972) or more often by radioimmunoassay (Hughes *et al.* 1979). A basal plasma 17-OHP concentration above 20 nmol/l virtually confirms the diagnosis. However, immediately after birth, 17-OHP is elevated in normal infants and as a result cord blood or samples taken within a few hours of birth should not be used for the diagnosis of 21-hydroxylase deficiency.

In cases where steroid therapy was begun before the diagnosis of 21-hydroxylase deficiency had been confirmed, estimation of plasma 17-OHP or urine steroids after ACTH stimulation can be used to establish or refute the diagnosis. During these tests dexamethasone should be given in place of the usual steroid therapy.

Plasma 17-OHP assay is also used to monitor treatment in a number of centres but care must be taken in the timing of samples for this purpose, as there is marked variation in plasma concentration both as a result of circadian changes in ACTH secretion and as a result of ACTH suppression immediately after a dose of glucocorticoid. Some authors have suggested using plasma testosterone, androstenedione or DHA assay to offset this problem, but there is still disagreement on the merits of these assays for monitoring treatment (Grant *et al.* 1977).

Plasma renin assay can be of considerable value in the management of the 21-hydroxylase type of CAH, particularly in assessing the dosage of mineralocorticoid therapy: a high value suggests that insufficient mineralocorticoid is being used.

Radioimmunoassay methods have been described for measuring 17-OHP in saliva (Hughes & Read 1982) and in dried blood spots (Sólyom 1981). These may be useful in obtaining multiple samples to assess the efficacy of treatment while the blood spot assay makes neonatal screening for CAH possible.

There have now been several reports indicating that the 17-OHP concentration in amniotic fluid is elevated when the fetus is a homozygote for 21-hydroxylase deficiency (Frasier *et al.* 1975) and it appears that this may allow prenatal diagnosis of the condition in 'at risk' pregnancies. Extensive control data for amniotic fluid steroid concentrations at different stages of gestation have been given by Sippell *et al.* (1981). It is also clear that investigation of the fetal and family HLA haplotypes can also be used in prenatal diagnosis, as the gene for regulating 21-hydroxylase is located close to the HLA loci on the short arm of chromosome number 6 (Pollack *et al.* 1979). In addition, estimation of plasma 17-OHP after infusion of tetracosactryn can be used to identify heterozygotes for 21-hydroxylase deficiency with a fair degree of accuracy (Lejeune-Lenain *et al.* 1980).

11-Hydroxylase deficiency causes virilisation without salt loss, and hypertension develops in later childhood because of overproduction of 11-deoxycorticosterone. Urinary 17-oxosteroids, pregnanetriol, tetrahydrodeoxycortisol, and the 11-oxygenation index are all raised but urinary studies are being replaced by plasma 11-deoxycortisol assay as a diagnostic test. As a result of improved laboratory methodology, it is becoming clear that 11-hydroxylase deficiency is not as uncommon as has been believed, and it appears that in the past this type of CAH was often misdiagnosed as non-salt-losing 21-hydroxylase deficiency.

3β-Hydroxysteroid dehydrogenase deficiency produces genital ambiguity in both boys and girls; most cases have severe salt loss but incomplete forms without salt wasting have been described. Urine pregnanetriol and pregnendiol are elevated, and high levels of plasma pregnenolone, 17-hydroxypregnenolone, and DHA are likely to be found.

20-22 Desmolase (or *22-hydroxylase*) *deficiency* result in lipoid adrenal hyperplasia. Adrenal and gonadal steroid production is blocked and as a result there is usually severe salt loss and complete absence of virilisation in males. The diagnosis is confirmed by demonstrating very low urinary and plasma steroid levels.

17-Hydroxylase deficiency leads to imperfect masculinisation in males and lack of pubertal development in females as a result of failure of androgen and oestrogen synthesis. Hypertension subsequently develops because of overproduction of corticosterone.

Salt loss as a result of selective *defects of aldosterone biosynthesis* have been well documented. These cases have no virilisation or impaired cortisol production and have been shown to be due to 18-hydroxylase or 18-dehydrogenase defects. Urine analysis by GLC has been required to pin-point the nature of these metabolic blocks. (Salt loss in the neonate is also discussed on p. 52.)

Secondary adrenal insufficiency

Previous steroid medication is the most common cause of secondary adrenal insufficiency during childhood. Less often it occurs as part of a more general hypothalamic–pituitary disturbance, for example, combined with growth hormone deficiency after breech delivery or after surgery for craniopharyngioma. The clinical features are those of cortisol deficiency—hypoglycaemia is probably the most important. Aldosterone secretion is normal and salt loss does not occur.

As in primary adrenal insufficiency, plasma cortisol and urinary 17-hydroxycorticoids are low. Plasma cortisol responses to insulin hypoglycaemia and glucagon are impaired but these tests do not clearly differentiate between primary and secondary adrenal underfunction. While there is rarely difficulty in distinguishing between the two on clinical grounds, in cases of doubt plasma ACTH assay can be very helpful. A high value indicates primary adrenal disease while low values which do not rise after insulin or metyrapone stimulation indicate a hypothalamic–pituitary deficit in ACTH secretion.

End-organ insensitivity to steroid hormones

The only clear-cut example of this is pseudohypoaldosteronism, a condition which, in at least some cases, is recessively inherited. It presents as a salt-losing state in early infancy and in some cases this is part of a more general abnormality of sodium transport in other tissues; for example, sweat glands and intestinal mucosa. PRA is raised, as may be plasma aldosterone. In addition, there may be marked resistance to synthetic mineralocorticoids such as 9α-fluorohydrocortisone.

HYPERADRENOCORTICAL STATES

Excess production of adrenal steroids are considered under four headings: Cushing's syndrome, virilising disorders, feminising disorders, and conditions associated with excessive aldosterone secretion.

Cushing's syndrome

Cushing's syndrome is very uncommon during childhood; the causes are listed in Table 12.10. In young children adrenal tumour is the most likely cause, while in adolescents pituitary-dependent adrenal hyperplasia (Cushing's disease) is more likely. Cushing's syndrome due to ectopic ACTH production or to steroid production by a non-adrenal tumour is very rare.

In cases with florid clinical features of Cushing's syndrome, investigation may be fairly straightforward and demonstration of persistently-elevated plasma cortisol levels, and increased excretion of 17-hydroxycorticoids and urinary free cortisol, readily confirms the diagnosis. However, in some cases it is more difficult to demonstrate that cortisol secretion is abnormal. In these cases loss of the normal diurnal pattern of cortisol secretion has been the most reliable test in a number of published series. As noted above, the overnight and 3-day low-dose dexamethasone tests have not been very

Table 12.10 Causes of Cushing's syndrome.

Adrenal tumour: adenoma or carcinoma
Bilateral adrenal hyperplasia
Pituitary-dependent
Ectopic ACTH secretion
Nodular hyperplasia of early infancy
Iatrogenic
Corticotrophin therapy
Corticosteroid therapy

well standardised in children and have sometimes given misleading results. In very difficult cases, formal measurement of cortisol production rate may be helpful.

Once the diagnosis of excessive steroid production has been established it is necessary to differentiate between adrenal tumour and bilateral adrenal hyperplasia. High dose dexamethasone usually results in at least a 50% reduction in urine steroids in Cushing's disease but normally has no effect in cases due to tumour or ectopic ACTH production. However, exceptions to this have been reported and the test is not completely reliable. Metyrapone or ACTH tests may be helpful in difficult cases; an exaggerated response suggests the presence of Cushing's disease. In adults, plasma ACTH determination is often recommended. Very high values have been reported in ectopic ACTH secretion, high-normal values in Cushing's disease, and low values in tumour cases. However, our own experience with ACTH assay in this situation has been disappointing and low levels have been found in pituitary-dependent cases.

Abnormal virilisation

The causes of abnormal virilisation are shown in Table 12.11. In girls, the differential diagnosis lies between adrenal tumour, CAH (in a minority of girls this is not evident at birth), and premature pubarche. The latter is due to normal but early secretion of adrenal androgens and does not cause enlargement of the clitoris. Very rarely virilisation may be due to an androgen-secreting ovarian tumour. In boys, testicular Leydig-cell tumours and ectopic secretion of human chorionic gonadotrophin (hCG), together with precocious puberty, must be added to the list.

Determining the cause of virilisation is usually fairly straightforward but in some cases, particularly in boys with small testicular tumours, diagnosis may be difficult. As a first step it is important to exclude 21-hydroxylase and 11β-hydroxylase deficiencies by estimation of plasma 17-OHP and compound-S, and by analysis of urinary steroid excretion. In

Table 12.11 Causes of virilisation during childhood and adolescence.

Both sexes
Congenital adrenal hyperplasia
Adrenal tumour: adenoma or carcinoma
Cushing's disease
Premature 'adrenarche'
Girls
Polycystic ovary disease (Stein–Leventhal)
Ovarian tumour
Boys
Ectopic hCG production: hepatoma, teratoma
Testicular tumour
Precocious puberty

very mild cases, estimation of plasma steroids after ACTH stimulation may clarify the diagnosis. If these tests are negative, and if plasma androgens and urine oxosteroids are elevated, a high-dose 3-day dexamethasone suppression test will usually help differentiate between adrenal tumour and premature pubarche, as steroid values are suppressed in the latter condition. A discrepancy between a high plasma testosterone and only slightly elevated urinary oxosteroid secretion suggests the presence of a testicular tumour.

Abnormal feminisation

Moderate breast enlargement in boys is common during early puberty and is associated with a high ratio of plasma oestradiol to plasma testosterone. Occasionally it is a manifestation of Klinefelter's syndrome, and very rarely it may bring to light undiagnosed male pseudohermaphroditism. Breast development after the neonatal period but before puberty should always be considered abnormal. It has been described in the 11-hydroxylase type of CAH but can also occur as a result of an oestrogen-secreting adrenal tumour or hCG production by a testicular or hepatic tumour. In all these disorders there is usually virilisation in addition to the breast enlargement.

In girls, premature breast enlargement is usually due to precocious puberty or an ovarian follicular cyst or tumour, usually a granulosa cell tumour. A small number of cases with feminisation due to an oestrogen-secreting adrenal tumour have also been described (Bhettay & Bonnici 1977).

Investigation of plasma oestrogens, urinary 17-oxosteroids, and plasma gonodotrophin levels after luteinizing hormone releasing hormone (LHRH) stimulation will help differentiate these different causes of feminisation from one another.

Excess aldosterone secretion

Aldosterone-secreting adrenal adenomas are very rare in childhood and usually present with hypertension or hyokalaemia (Kafrouni *et al.* 1975). PRA is very low and this serves to distinguish them from Bartter's syndrome which is due to excessive renin production with secondary hyperaldosteronism. Excess aldosterone secretion as a result of adrenal hyperplasia is very uncommon in children and the clinical and biochemical features are the same as those for an aldosterone-producing tumour. There is still relatively little information on plasma and urinary aldosterone values in these disorders.

INTERSEX CONDITIONS

Embryology and physiology

The embryological development of the gonads and accessory structures has been well reviewed by a number of authors (Imperato-McGinley & Peterson 1976; Wachtel 1979; Josso 1981).

Until about 6 weeks after conception the gonads in both sexes are represented by the primordial gonadal ridges which project into the coelomic cavity—these are made up from mesenchyme and germ cells which have migrated through the tissues from the yolk sac. In males, the first signs of testicular differentiation become apparent about 6 weeks after conception. This is now thought to be under the influence of HY-antigen which in turn is probably under the control of gene(s) on the short arm of the Y chromosome. In females, ovarian differentiation becomes apparent 11–12 weeks after conception. The presence of a 46XX karyotype is not essential for early development of the ovaries but is required for normal maturation before birth.

Differentiation of the male external genitalia is entirely dependent on the function of the fetal testes. About 10–12 weeks after conception there is a marked increase in the number of Leydig cells in the testes which is associated with an increase in testosterone secretion. Initially this is under the control of placental hCG but in the latter part of gestation fetal pituitary LH takes over this role. The increased concentrations of fetal plasma testosterone lead to increased levels of 5α-dihydrotestosterone in the peripheral tissues and in turn this causes masculinisation of the external genitalia, with closure of the urogenital sinus. Testosterone diffusion down the Wolffian ducts prevents their involution—these structures form the vas deferens and accessory structures. At the same time the Sertoli cells of the fetal testes produce 'anti-Müllerian hormone', a polypeptide which causes regression of the Müllerian structures.

In males, fetal plasma testosterone falls to low levels towards the end of gestation, but after delivery there is a further period of increased testosterone secretion, which is under the control of pituitary LH. This increased secretion lasts for 3–4 months and while its physiological significance is still uncertain it may play a role in ensuring testicular descent. After the first months of life there is little or no spontaneous testosterone secretion by the Leydig cells until the onset of puberty, when plasma levels rise progressively to reach adult values in the later stages of puberty.

In complete contrast to the pattern in males, female genital development proceeds normally in the absence of the ovaries, and differentiation of the Mullerian ducts to form the fallopian tubes, uterus, cervix and upper vagina is not dependent on active hormone secretion by the ovaries.

Secretion and metabolism of testosterone

The metabolic steps in the formation of testosterone from cholesterol are shown in Fig. 12.1. As described below, a number of different enzyme defects which interfere with testosterone production have been recognised. In addition, testosterone acts as a prohormone in many tissues and is reduced to 5α-dihydrotestosterone which is the active hormone. Some dihydrotestosterone is secreted by the testes but most is formed by the action of intracellular 5α-reductase. The dihydrotestosterone is bound by intracellular receptors and the receptor–steroid complexes are transported to the nuclei of the cells where they induce RNA synthesis, which in turn regulates protein synthesis. A variety of disorders due to inherited abnormalities of the receptor system have been described.

Laboratory investigation of testosterone metabolism

With the development of radioimmunoassay methods it is now possible to estimate the plasma concentrations of testosterone and dihydrotestosterone, as well as those of precursors such as DHA, 17-OHP and androstenedione. However, as there is little or no spontaneous testosterone secretion in boys before the onset of puberty, it is usually necessary to use hCG stimulation in order to assess Leydig cell potential.

The testosterone response to hCG stimulation is biphasic, and Forest *et al.* (1980) have recommended the use of 2–4 doses of hCG given at 4-day intervals. However, in most of the published studies hCG (1000–5000 IU) has been given for 2–5 consecutive days.

There has been considerable variation in the post-hCG testosterone concentrations reported by different authors, probably as a result of differences in the assays used to quantitate plasma testosterone, but it is clear that the results are also related to the pubertal status of the subjects: pre-pubertal boys show lower post-hCG values than those found in pubertal and adult males. Representative values published by Winter *et al.* (1972) are given in Table 12.12.

Table 12.12 Plasma testosterone levels (±1 SD) after one and three injections of 2000 IU hCG in males at different stages of pubertal development (after Winter *et al.* 1972).

Age range (years)	Puberty stage	Plasma testosterone (nmol/l)		
		Day 0	Day 1	Day 4
5.5–9.0	1	0.6±0.1	2.1±0.5	6.0±0.9
9.1–11.0	2	1.1±0.2	6.1±1.4	9.2±1.7
11.1–14.0	3	3.3±0.9	14.8±4.5	21.0±3.7
14.1–16.0	4	10.0±2.0	25.3±6.6	41.6±4.9
16.1–20.0	5	16.4±1.7	28.1±6.1	41.9±3.6

The hCG test can also be used to investigate the plasma concentrations of androgen precursors and 5α-dihydrotestosterone, and can thus be of value in demonstrating a defect in testosterone biosynthesis or impaired 5α-reductase activity (Pang *et al.* 1979). However it should be noted that the ratios between plasma testosterone and precursors may be related to the duration of hCG stimulation. For example, Walsh *et al.* (1976) found a negligible rise in plasma androstenedione after three injections of hCG whereas Forest (1979) obtained a two- to four-fold rise after 7 doses of hCG.

A number of laboratories have developed methods for studying tissue 5α-reductase activity and dihydrotestosterone binding to intracellular receptors, using either homogenates of skin or skin fibroblasts grown in tissue culture (Griffin & Wilson 1980). However, these methods are still largely research tools and are not generally available for routine investigation of patients.

Disorders of sexual differentiation and development

In males, abnormal fetal Leydig cell function or abnormal testosterone metabolism in the genital tissues will lead to imperfect masculinisation with either genital ambiguity (male pseudohermaphroditism) or genital underdevelopment (micropenis). The recognised causes of male pseudohermaphroditism are listed in Table 12.13.

Conversely, exposure of a female fetus to excess androgen will result in masculinisation of the external genitalia (female pseudohermaphroditism). The most important cause of female pseudohermaphroditism is CAH, but other causes are listed in Table 12.13.

Genital ambiguity can also arise as a result of abnormal gonadal differentiation—for example, hermaphroditism or mixed gonadal dysgenesis (Table 12.13).

Table 12.13 Causes of 'intersex' disorders.

Male pseudohermaphroditism
Impaired testosterone secretion
Inborn errors of biosynthesis
20-22 desmolase deficiency
3β-hydroxysteroid dehydrogenase deficiency
17α-hydroxylase deficiency
17-20 desmolase deficiency
17α-hydroxysteroid dehydrogenase deficiency
Leydig cell hypoplasia/receptor deficiency
Impaired testosterone metabolism
5α-reductase deficiency
'Complete' and 'incomplete' androgen insensitivity
Absent receptors for dihydrotestosterone (DHT)
Unstable receptors for DHT
Normal receptors for DHT
Other forms of male pseudohermaphroditism
Associated with other congenital anomalies
Iatrogenic
Persistent Müllerian structures
Female pseudohermaphroditism
Virilisation by fetal androgens
Congenital adrenal hyperplasia
21-hydroxylase type
11β-hydroxylase type
3β-hydroxysteroid dehydrogenase type
Other causes of fetal androgen overproduction
Fetal virilisation by maternal androgens
Iatrogenic fetal virilisation
Associated with other dysmorphic conditions
Abnormal gonadal differentiation
Turner's syndrome
XX and XY gonadal dysgenesis
Mixed gonadal dysgenesis (XO/XY)
'Agonadism' and micropenis with rudimentary testes
Hermaphroditism
XX males
Klinefelter's syndrome

Laboratory investigation of genital ambiguity

The presence or absence of palpable gonads in an infant with abnormal external genitalia will largely determine which are the most appropriate biochemical investigations.

If there are no palpable gonads the most likely diagnosis is that of female pseudohermaphroditism due to CAH and if there are symptoms of salt loss this diagnosis is virtually certain. Assay of plasma 17-OHP levels (or plasma 11-deoxycortisol), together with estimation of urine 17-oxosteroid excretion, or the urine 11-oxygenation index may be all that is needed to establish the diagnosis. If these tests are negative, full karyotype analysis is required to confirm the presence of female pseudohermaphroditism.

If gonads are palpable in both the labio-scrotal folds the most likely diagnosis is that of male pseudohermaphroditism and this is almost certain if the karyotype shows a 46XY pattern. Estimation of the plasma testosterone and dihydrotestosterone responses to hCG will usually be normal in such cases, indicating the presence of abnormal testosterone metabolism by the genital tissues (Savage *et al.* 1978). In a small proportion of cases, there may be a poor rise in 5α-dihydrotestosterone in the face of a normal testosterone response, suggesting a diagnosis of 5α-reductase deficiency. Alternatively, there may be an impaired or negligible rise in plasma testosterone, indicating the presence of impaired Leydig cell function or a specific defect in testosterone biosynthesis. Analysis of plasma androgen precursors may help to determine the particular enzyme which is deficient.

The presence of a single gonad raises the possibility of either hermaphroditism or mixed gonadal dysgenesis due to a 45X/46XY karyotype. Full chromosome analysis with laparotomy and gonadal biopsy are usually indicated in these cases, and biochemical investigation is of little help in establishing the diagnosis. However, measurement of plasma testosterone after hCG stimulation may be helpful in predicting whether spontaneous virilisation is likely to take place at the time of puberty.

Investigation of micropenis or anorchia

Gonadotrophin deficiency is probably the most common cause of marked underdevelopment of the penis and scrotum and estimation of plasma LH and follicle-stimulating hormone (FSH) after intravenous injection of LHRH (also called Gn–RH, see Chapter 13) is the most useful investigation in boys with underdeveloped external genitalia. However, after the neonatal period the presence of an impaired or absent response can be difficult to interpret in prepubertal boys, as gonadotrophin secretion is normally low during this period. Primary testicular dysplasia can also be associated with micropenis and in these cases plasma LH and FSH levels are likely to be increased. Estimation of plasma testosterone after hCG may be of little help in differentiating between these two conditions as a poor response is common in boys with gonadotrophin deficiency. However, the test can be very useful in deciding which surgical treatment is indicated in boys with bilateral cryptorchidism. An absent or severely impaired plasma testosterone response to hCG stimulation strongly suggests bilateral anorchia, while a significant rise in plasma testosterone indicates the presence of intra-abdominal testicular tissue and the need for surgical exploration.

REFERENCES

Abuid J., Klein A. H., Foley T. P. & Larsen P. R. (1974) Total and free triiodothyronine and thyroxine in early infancy. *J. Clin. Endocrinol.* **39**, 263–8.

Barnes N. D. & Atherden S. M. (1972) Diagnosis of congenital adrenal hyperplasia by measurement of plasma 17-hydroxyprogesterone. *Arch. Dis. Child.* **47**, 62–5.

Barnes N. D., Joseph J. M., Atherden S. M. & Clayton B. E. (1972) Functional tests of adrenal axis in children with measurement of plasma cortisol by competitive protein binding. *Arch. Dis. Child.* **47**, 66–73.

Bhettay E. & Bonnici F. (1977) Pure oestrogen secreting feminizing adrenocortical adenoma. *Arch. Dis. Child.* **52**, 241–3.

Bidlingmaier F., Wagner-Bernack M., Butenandt O. & Knorr D. (1973) Plasma estrogens in childhood and puberty under physiologic and pathologic conditions. *Pediat. Res.* **7**, 901–7.

Brook C. G. D. (1981) *Clinical Paediatric Endocrinology*. Blackwell Scientific Publications, London.

Clayton B. E., Edwards R. W. H. & Makin H. L. J. (1971) Congenital adrenal hyperplasia and other conditions associated with a raised urinary steroid 11-oxygenation index. *J. Endocrinol.* **50**, 251–65.

Clayton B. E., Jenkins P. & Round J. M. (1980) *Paediatric chemical pathology: clinical tests and reference ranges*. Blackwell Scientific Publications, London.

Cuestas R. A. (1978) Thyroid function in healthy premature infants. *J. Pediat.* **92**, 963–7.

Degroot L. J. & Niepomniszcze H. (1977) Biosynthesis of thyroid hormone: Basic and clinical aspects. *Metabolism* **26**, 665–718.

Delange F., Illig R., Rochiccioli P. & Brock-Jacobsen B. (1981) Progress report 1980 on neonatal thyroid screening in Europe. *Acta Paed. Scand.* **70**, 1–2.

De Peretti E. & Forest M. G. (1978) Pattern of plasma dehydroepiandrosterone sulfate levels from birth to adulthood: evidence for testicular production. *J. Clin. Endocrinol.* **47**, 572–7.

Diczfalusy E. (1969) Steroid metabolism in the human foeto-placental unit. *Acta Endocr. (Kbvh)* **161**, 649–64.

Fisher D. A., Dussault J. H., Foley T. P., Klein A. H., La Franchi S., Larsen P. R., Mitchell M. L., Murphey W. H. & Walfish P. G. (1979) Screening for congenital hypothyroidism: results of screening one million North American infants. *J. Pediatr.* **94**, 700–5.

Fisher D. A. & Klein A. H. (1981) Thyroid development and disorders of thyroid function in the newborn. *New Engl. J. Med.* **304**, 702–12.

Forest M. G. (1979) Pattern of the response of testosterone and its precursors to human chorionic gonadotrophin stimulation in relation to age in infants and children. *J. Clin. Endocrinol.* **49**, 132–7.

Forest M. G., David M., Lecoq A., Jeune M. & Bertrans J. (1980) Kinetics of the HCG-induced steroidogenic response of the human testis: III Studies in children of the plasma levels of testosterone and HCG: Rationale for testicular stimulation test. *Pediat. Res.* **14**, 819–24.

Franks R. C. (1973) Urinary 17-hydroxycorticosteroid and cortisol excretion in childhood. *J. Clin. Endocrinol.* **36**, 702–5.

Frasier S. D. (1980) *Pediatric Endocrinology*. Grune & Stratton, New York.

Frasier S. D., Thorneycroft I. H., Weiss B. A. & Horton R. (1975) Elevated amniotic fluid concentration of 17α-hydroxyprogesterone in congenital adrenal hyperplasia. *J. Pediat.* **86**, 310–12.

GRANT D. B., DILLON M. J., ATHERDEN S. M. & LEVINSKY R. J. (1977) Congenital adrenal hyperplasia: renin and steroid values during treatment. *Eur. J. Pediatrics*. **126**, 89–96.

GRIFFIN J. F. & WILSON J. D. (1980) The syndromes of androgen resistance. *New Engl. J. Med.* **302**, 198–209.

HUGHES I. A. & READ G. F. (1982) Simultaneous plasma and saliva steroid measurements as an index of control in congenital adrenal hyperplasia (CAH). *Hormone Res.* **16**, 142–50.

HUGHES I. A., RIAD-FAHMY D. & GRIFFITHS K. (1979) Plasma 17OH-progesterone concentrations in newborn infants. *Arch. Dis. Child.* **54**, 347–9.

HUNG W., AUGUST G. P. & GLASGOW A. M. (1978) *Pediatric Endocrinology*. Henry Kimpton, London.

IMPERATO-MCGINLEY J. & PETERSON R. E. (1976) Male pseudohermaphroditism: the complexities of male phenotypic development. *Am. J. Med.* **61**, 251–72.

JOB J.-C. & PIERSON M. (1981) *Pediatric Endocrinology*. John Wiley & Sons, New York.

JOSSO N. (1981) In: The Intersex Child. *Pediatric and Adolescent Endocrinology*, Vol. 8. Karger, London.

KAFROUNI G., OAKES M. D., LURVEY A. N. & DE QUATTRO V. (1975) Aldosteronoma in a child with localization by adrenal vein aldosterone: Collective review of the literature. *J. Pediat. Surg.* **10**, 917–24.

KENNY F. M., PREEYASOMBAT C. & MIGEON C. J. (1966) Cortisol production rate: II Normal infants, children and adults. *Pediatrics* **37**, 34–42.

KORTH-SCHUTZ S., VIRDIS R., SAENGER P., CHOW D. M., LEVINE L. S. & NEW M. I. (1978) Serum androgens as a continuing index of adequacy of treatment of congenital adrenal hyperplasia. *J. Clin. Endocrinol.* **46**, 452–8.

LEISTI S. (1977) Evaluation of 3-hour metyrapone test in children and adolescents. *Clin. Endocrinol.* **6**, 305–20.

LEJEUNE-LENAIN C., CANTRAINE F., DUFRASNES M., PREVOT F., WOLTER R. & FRANCKSON J. R. M. (1980) An improved method for the detection of heterozygosity of congenital virilizing adrenal hyperplasia. *Clin. Endocrinol.* **12**, 525–35.

PANG S., LEVINE L. S., CHOW D., SAGIANI F., SAENGER P. & NEW M. I. (1979) Dihydrotestosterone and its relationship to testosterone in infancy and childhood. *J. Clin. Endocrinol.* **48**, 821–6.

PEZZINO V., FILETTI S., BELFIORE A., PROTO S., DONZELLI G. & VIGNERI R. (1981) Serum thyroglobulin levels in the newborn. *J. Clin. Endocrinol. Metabolism* **52**, 364–6.

POLLACK M. S., LEVINE L. S., PANG S., OWENS R. P., NITOWSKY H. M., MAURER D., NEW M. I., DUCHON M., MERKATZ I. R., SACHS G. & DUPONT B. (1979) Prenatal diagnosis of congenital adrenal hyperplasia (21-hydroxylase deficiency) by HLA typing. *Lancet* **i**, 1107–8.

REITER E. O., ROOT A. W., RETTIG K. & VARGAS A. (1981) Childhood thyromegaly: recent developments. *J. Pediatrics* **99**, 507–18.

RYNESS J. (1972) The measurement of serum thyroxine in children. *J. Clin. Path.* **25**, 726–9.

SAVAGE M. O., CHAUSSAIN J. L., EVIAN D., ROGER M., CANLORBE P. & JOB J.-C. (1978) Endocrine studies in male pseudohermaphroditism in childhood and adolescence. *Clin. Endocr.* **8**, 219–31.

SCHULTZ R. M., GLASSMAN M. S. & MACGILLIVRAY M. H. (1980) Elevated threshold for thyrotrophin suppression in congenital hypothyroidism. *Am. J. Dis. Child.* **134**, 19–20.

SIPPELL W. G., DÖRR H. G., BIDLINGMAIER F. & KNORR D. (1980) Plasma levels of aldosterone, corticosterone, 11-deoxycorticosterone, progesterone, 17-hydroxyprogesterone, cortisol, and cortisone during infancy and childhood. *Pediat. Res.* **14**, 39–46.

SIPPELL W. G., MÜLLER-HOLVE W., DÖRR H. G., BIDLINGMAIER F. & KNORR D. (1981) Concentrations of aldosterone, corticosterone, 11-deoxycorticosterone, progesterone, 17-hydroxyprogesterone, 11-deoxycortisol, cortisol, and cortisone determined simultaneously on human amniotic fluid throughout gestation. *J. Clin. Endocrinol. Metabolism* **52**, 385–92.

SÓLYOM J. (1981) Blood spot 17alpha-hydroxyprogesterone radioimmunoassay in the follow-up of congenital adrenal hyperplasia. *Clin. Endocrinol.* **14**, 547–53.

STERLING K. (1979) Thyroid hormone action at the cell level. *New Engl. J. Med.* **300**, 117–23; 173–7.

STREETEN D. H. P., FAAS F. H., ELDERS M. J., DALOKOS T. G. & VOORHESS M. (1975) Hypercortisolism in childhood. Shortcomings of conventional diagnostic criteria. *Pediatrics* **56**, 797–803.

VANDERSCHEUREN-LODEWEYCKX M., WOLTER R., MALVAUX P., EGGERMONT E. & EECKELS R. (1974) The glucagon stimulation test: effect on plasma growth hormone and on immunoreactive insulin, cortisol, and glucose in children. *J. Pediatrics* **85**, 182–7.

WACHTEL S. S. (1979) The genetics of intersexuality: clinical and theoretic perspectives. *Obstet. Gynaecol.* **54**, 671–85.

WALSH P. C., CURRY N., MILLS R. C. & SIITERI P. K. (1976) Plasma androgen response to hCG stimulation in prepubertal boys with hypospadias and cryptorchidism. *J. Clin. Endocrinol.* **42**, 52–9.

WINTER J. S. D., TARASKA S. & FAIMAN C. (1972) Hormonal response to HCG stimulation in male children and adolescents. *J. Clin. Endocrinol.* **34**, 348–53.

Chapter 13
Precocious and Delayed Puberty

HANK K. A. VISSER

Puberty is usually taken to be the process of maturation of the sexually immature child into the sexually mature adolescent. Maturation of gonadal secretory function and the development of secondary sex characteristics during adolescence are the final phase of a continuous process that starts in early life. In this view sexual maturation begins with differentiation of the gonads during early embryonic life and continues until fertility is achieved.

Gonadal function is regulated by a neuro-endocrine system: the hypothalamic–pituitary–gonadal axis. The 'messenger' linkage between the central nervous system and the gonads are the pituitary gonadotrophins luteinizing hormone (LH) and follicle stimulating hormone (FSH). Release of both gonodotrophins is mediated through the gonadotrophins-releasing hormone (Gn-RH)* a decapeptide synthesised in the medial basal part of the hypothalamus and transmitted to the anterior pituitary gland by the hypothalamic–hypophyseal portal circulation. Gonadal hormones inhibit the release of the hypothalamic and pituitary hormones by negative feedback. In the adult female a positive feedback mechanism is also involved.

GONADOTROPHIN SECRETION AND THE ONSET OF PUBERTY

Until recently it was generally accepted that the secretion of gonadotrophins was not initiated before the onset of puberty. With the introduction of radioimmunological techniques for measurement of small concentrations of plasma gonadotrophins, our knowledge has increased rapidly. We now know that in several species and in man there is a major increase in plasma gonadotrophin levels in earlier periods of life, before puberty starts. Both FSH and LH are synthesised and secreted by the fetal pituitary under the influence of Gn-RH by 10–12 weeks of gestation (for a review see Kaplan *et al.* 1976).

Early in gestation both fetal pituitary gonadotrophins and placental gonadotrophin (chorionic gonadotrophin) stimulate the secretion of testo-

* We prefer to use the initials Gn-RH to LH-RH, because LH-RH takes no account of FSH secretion.

sterone by the Leydig cells of the fetal testes. During a critical period (8–12 weeks of gestation) fetal testosterone will induce development of external genital organs and stimulate differentiation of the genital ducts (Wolffian ducts). The concentrations of fetal serum LH and FSH rise to very high levels (adult castrate levels) at mid-gestation and then decrease. FSH and LH values in umbilical venous blood at term are low (Grumbach 1978). Grumbach and coworkers have postulated that the decrease in serum FSH and LH values toward term is due to the development of steroid receptors in hypothalamus and pituitary gland and maturation of the negative feedback system (Grumbach & Kaplan 1973; Kaplan *et al.* 1976).

During the first months after birth, serum gonadotrophin concentrations in both sexes, as well as testosterone concentrations in boys and oestradiol concentrations in girls, rise to prepubertal values or even higher (Faiman & Winter 1971; Forest *et al.* 1974; Bidlingmaier *et al.* 1974; Winter *et al.* 1975; for review see Sizonenko & Aubert 1978). This indicates that the hypothalamic–pituitary–gonadal system at this time operates at a different level as compared with late gestation. Serum LH values are higher in male infants, serum FSH values higher in female infants. Both serum gonadotrophin and sex steroid levels are decreasing by 6 months of age, but there is much individual variation and it may not be until 2 years of age that serum gonadotrophin values are at the low levels present throughout childhood until the onset of puberty. In particular plasma FSH values in the female may remain at pubertal levels (P2–P3) until 2 years of age. The increased activity of the hypothalamic–pituitary–gonadal axis in infancy is also demonstrated by the high pituitary responsiveness to Gn-RH in both male and female infants during the first six months of life (Job *et al.* 1977).

From about the age of 2 years until the onset of puberty plasma gonadotrophin levels are low. In agonadal patients (XO-syndrome) plasma gonadotrophin concentrations are elevated in infancy and early childhood, indicating that the negative feedback mechanism is operative in normal children at this age (Conte *et al.* 1975, 1980). However in these agonadal patients gonadotrophin levels decrease at about 4 years of age and remain low until the age of about 11 years, when they increase to very high levels as compared with normal children during puberty. These findings suggest that in normal children, as well as in agonadal patients, from early childhood until puberty the secretion of gonadotrophins is restrained. In 1970 Knobil and coworkers were the first to observe a rhythmic, pulsatile pattern of gonadotrophin secretion in ovariectomised monkeys (Dierschke *et al.* 1970). Recent experimental work by the same group has shown that the intermittent, pulsatile release of gonadotrophins is regulated by pulsatile release of Gn-RH. The central control system is located in the arcuate nucleus in the medial basal part of the hypothalamus. This 'oscillator' regulates frequency and amplitude of gonadotrophin release. Knobil *et al.* were able to induce ovulatory menstrual cycles in prepubertal rhesus monkeys by pulsatile administration of Gn-RH. These findings strongly suggest that puberty is normally initiated by reactivation of the hypothalamic 'oscillator' which regulates gonadotrophin release through Gn-RH (for review see Knobil 1980). Observations in the human as well as animal experiments support these findings. In the human, precocious puberty often is caused by lesions in the posterior part of the hypothalamus. Donovan and van der Werff ten Bosch (1956, 1965) and other experimental endocrinologists have shown that in the rat destructive lesions in the anterior and posterior part of the hypothalamus cause precocious puberty.

Boyar *et al.* (1972) were the first to describe increased, pulsatile plasma LH-levels during sleep, in late prepubertal and pubertal children. In late puberty also, pulsatile LH-secretion is observed during the day. Both frequency and amplitude increase during puberty until the adult type of pattern is achieved (about one pulse per hour).

In boys in whom puberty is commencing, increased LH secretion during sleep leads to increased plasma concentrations of testosterone (Boyar *et al.* 1974). The suggested reactivation of the hypothalamic 'oscillator' can also be observed in agonadal patients: during the early pubertal period there is increased secretion of LH during sleep (Boyar *et al.* 1973a). Sleep related gonadotrophin release has also been demonstrated in children with idiopathic precocious puberty (Boyar *et al.* 1973b).

Grumbach and co-workers postulated that the onset of puberty could be explained by a change in sensitivity of the negative feedback mechanism (for review see Grumbach 1978). However, most if not all data available from studies in the human can now be explained by the animal experiments from Knobil and co-workers. In the late prepubertal stage the 'oscillator' in the hypothalamus, which controls the pulsatile release of Gn-RH, is reactivated: from very low activity to nocturnal activation and gradually to adult frequency. What causes this reactivation, after a long period of low activity during childhood, is unknown. It is very likely that there is a release of inhibition by central nervous system 'input', which in some way is related to the maturation of the organism as a whole (for review see Visser 1973). Genetic and environmental factors affect body growth, skeletal maturation and the onset of puberty. Puberty usually does not start until there is a certain degree of skeletal maturation. Frisch and Revelle (1969, 1970) have suggested that there is a relation between the onset of puberty and a critical body weight. Apparently there is a correlation of brain maturation and maturation of the body as a whole. At a certain critical point pubertal development starts.

During the past century in the industrial countries a constant trend to earlier maturation and onset of puberty has been observed. The onset of puberty has become earlier at a rate of about 1 year per 25 calender years (for a review see van Wieringen 1978). The cause of the trend toward earlier maturity is not clear, but may be explained by constant improvement in environmental conditions at an early age.

Animal experiments support these findings in the human. There is evidence that acceleration of growth and earlier sexual maturity is connected with earlier maturation of the brain. The postponement of puberty after birth is a distinctive evolutionary trait in primates and man, and it is an intriguing question why this is so. The hypothalamic–pituitary–gonadal axis is fully active in infancy but it is inactive during childhood until puberty starts. The syndrome of true precocious puberty in young children shows that maturation of gonads and secondary sex characteristics can start at a much earlier age than the usual time of puberty.

From a biological point of view a long maturation period of the brain is advantageous for learning; the maturation of the reproductive system has to wait for the maturation of the brain.

HORMONAL CHANGES IN PUBERTY

Many studies reporting plasma gonadotrophin and sex-steroid concentrations during childhood and pubertal development are now available (for a review see Winter 1978; Sizonenko 1978a). Cross-sectional studies are not suitable for the demonstration of rapid incremental changes in hormone levels during puberty, because of individual differences in timing of puberty. Longitudinal studies indicate changes in individual hormonal concentrations in relation to pubertal stages.

FSH and LH levels rise in parallel in both sexes during puberty. Rapid increments in sex-steroids (testosterone in boys, oestradiol in girls) can be seen during pubertal stages 3–4 (Table 13.1). Of great practical importance is the finding that there is a difference in pituitary responsiveness to Gn-RH between prepubertal and pubertal children. The increase of plasma gonadotrophin levels (particularly LH) after intravenous injection of Gn-RH can be observed in late prepubertal children, before the appearance of secondary sex characteristics and before elevation of basal gonadotrophin levels in blood. This demonstrates that the release of gonadotrophins is conditioned by previous stimulation with endogenous Gn-RH, the first indication of the onset of puberty (Grumbach *et al.* 1974; Job *et al.* 1977).

We perform the Gn-RH test after an overnight fast using 0.1 mg of Gn-RH administered intravenously as a bolus. Blood samples are taken for measurement of LH and FSH at 0, 20, 30, 60, 90 and 120 minutes after the injection. No side-effects have been observed.

There is a wide variation in individual responses, depending on age and sex. Moreover, results from various laboratories differ and cannot be compared. Therefore each laboratory has to establish its own reference ranges but for obvious reasons these are difficult to obtain. Interpretation of results in individual patients is often difficult and longitudinal studies in the same individual may be necessary to confirm or exclude a suspected diagnosis. There is good evidence that LH responsiveness increases in boys

Pubertal stages	FSH (IU/l)	LH (IU/l)	Testosterone (nmol/l)	Androstenedione (nmol/l)	DHA (nmol/l)	DHAS (μg/dl)	Oestradiol (nmol/l)	Oestrone (nmol/l)	17α-OH-progesterone (nmol/l)
Males									
P1	2.4 (1.0–6.0)	3.5 (1.3–9.1)	0.27 (0.14–0.54)	0.67 ± 0.51	2.8 (0.8–9.9)	41.5 ± 31.7	0.03 ± 0.01	0.04 ± 0.01	0.93 ± 0.18
P2	3.0 (1.7–5.4)	5.8 (2.9–11.7)	0.81 (0.28–2.3)	1.5 ± 0.79	6.1 (1.7–19.4)	62.7 ± 35.6	0.04 ± 0.01	0.05 ± 0.01	1.23 ± 0.27
P3	3.1 (2.2–4.5)	6.7 (3.8–11.6)	5.3 (0.9–29.9)	1.8 ± 0.70	9.2 (4.1–20.5)	73.4 ± 48.8	0.06 ± 0.02	0.07 ± 0.01	2.34 ± 0.30
P4	3.2 (1.5–6.7)	6.6 (4.3–9.3)	9.4 (2.9–28.2)	3.6 ± 2.0	11.7 (6.2–22.3)	103.0 ± 55.7	0.08 ± 0.03	0.11 ± 0.02	3.84 ± 0.33
P5						123.4 ± 48.2	0.08 ± 0.02	0.11 ± 0.02	
Females									
P1	2.2 (0.9–5.1)	4.1 (1.8–9.2)	0.26 (0.10–0.64)	1.3 ± 0.8	2.6 (0.6–10.5)	43.1 ± 18.8	0.03 (< 0.03–0.07)	0.04 (< 0.03–0.10)	0.87 ± 0.18
P2	3.2 (1.4–7.0)	5.9 (2.0–16.6)	0.60 (0.22–1.6)	2.1 ± 1.5	10.2 (1.6–66.1)	48.9 ± 21.9	0.05 (< 0.03–0.13)	0.05 (< 0.03–0.13)	1.38 ± 0.21
P3	4.3 (2.4–7.7)	8.7 (5.6–13.6)	1.0 (0.59–1.8)	2.4 ± 1.5	16.1 (4.3–60.0)	71.3 ± 53.7	0.11 (0.03–0.22)	0.10 (0.03–0.18)	2.31 ± 0.45
P4	4.1 (1.5–11.2)	10.0 (7.0–14.4)	1.2 (0.68–2.3)	4.8 ± 1.9	15.6 (5.3–45.8)	86.4 ± 48.3	0.17 (0.04–0.34)	0.13 (0.03–0.26)	3.57 ± 0.54
P5	4.2 (1.9–9.3)	8.8 (4.4–17.5)	0.9 (0.47–1.7)		17.7 (5.6–56.2)	120.4 ± 42.3	0.23 (0.04–0.93)	0.18 (0.04–0.48)	

FSH, LH, testosterone, DHA mean values and 95% confidence limits (Sizonenko & Paunier 1975). Androstenedione mean ± 1 SD (Korth-Schutz *et al.* 1978). DHAS mean ± 1 SD (de Peretti & Forest 1978). Oestradiol and oestrone (males) mean ± 1 SD (Bidlingmaier *et al.* 1973). Oestradiol and oestrone (females) median and range (Bidlingmaier *et al.* 1977). 17α-OH-progesterone mean ± SE (Sippell *et al.* 1980). P1–P5: pubertal stages (Tanner 1962) with minor modifications by the authors.

Table 13.2 Basal and peak plasma LH and FSH levels in prepubertal and pubertal children during an intravenous Gn-RH test.

	Pubertal stage	LH (IU/l) Basal	LH (IU/l) Peak	FSH (IU/l) Basal	FSH (IU/l) Peak	Ref.
Boys						
Prepubertal	testicular vol. < 4 ml	0.9 (0.5–3.0)	4.1 (0.5–12.2)	1.3 (0.5–2.6)	4.0 (1.7–11.5)	[1]
	P1	1.9	8.8	2.5	6.6	[2]
	P1	0.5 (0.2–1.0)	1.3 (0.6–1.8)	0.6 (0.5–0.8)	1.5 (1.0–3.0)	[3]
Pubertal	testicular vol. ⩾ 4 ml	0.9 (0.5–3.5)	8.3 (3.5–14.0)	1.2 (0.5–4.0)	3.6 (1.5–7.0)	[1]
	P2–P5	3.5	23.0	5.3	9.0	[2]
	P2	0.8 (0.6–1.2)	5.9 (4.5–7.5)	1.0 (0.5–1.8)	2.4 (1.3–4.7)	[3]
	P4	1.3 (0.6–2.4)	14.9 (7.2–24.0)	2.1 (1.0–3.4)	3.0 (2.3–4.3)	[3]
Girls						
Prepubertal	P1	1.3	4.6	1.7	11.7	[2]
	P1	0.4 (0.2–1.0)	3.0 (2.1–3.8)	0.6 (0.5–0.9)	2.0 (1.0–3.2)	[3]
Pubertal	P2–P5	3.3	21.7	5.6	12.0	[2]
	P2	0.8 (0.4–1.4)	7.0 (3.1–13.5)	2.2 (0.7–3.6)	6.2 (1.5–13.0)	[3]
	P4	1.6 (0.6–2.6)	13.5 (7.1–25.0)	2.0 (1.2–2.8)	4.5 (2.6–8.0)	[3]

[1] Savage *et al.* (1981); prepubertal: testicular volume < 4 ml; pubertal: ⩾ 4 ml; Gn-RH 0.1 mg i.v.
[2] Job *et al.* (1972); pubertal all stages; Gn-RH 0.1 mg i.v.
[3] Dickermann *et al.* (1976); prepubertal P1, pubertal P2, P4 (Tanner 1962); Gn-RH 0.05 mg i.v. All values are mean values, with range between parentheses.

with a bone age greater than 11 years. Such data in girls are not yet reported, but corresponding bone age in girls would be about 10 years.

Basal and peak plasma LH and FSH levels in prepubertal and pubertal boys and girls during an intravenous Gn-RH test are presented in Table 13.2. In the prepubertal stage the basal levels of LH and FSH are low. Peak levels are 2 to 5 times basal levels. When puberty begins basal LH levels increase and peak levels are 5 to 10 times basal levels in both boys and girls. Basal FSH levels increase as well, but peak levels are 2 to 5 times the basal levels found before puberty.

Adrenarche

During puberty there is a considerable increase of the plasma concentration of the major adrenal androgens dehydroepiandrosterone (DHA), its sulphate (DHAS) and androstenedione. Cross-sectional and longitudinal studies have indicated that this increase begins at the chronological age of 7–8 years, before any change in plasma concentrations of gonadotrophins and gonadal steroids can be observed (Table 13.3; Sizonenko & Paunier 1975; Korth-Schultz *et al.* 1976; Ducharme *et al.* 1976). This first sign of increased adrenal androgens production has been called adrenarche (compared to gonadarche: the increase of gonadal steroid production at the onset of puberty).

The increase of plasma DHAS concentration is related to the development of the zona reticularis of the adrenal gland. The physiological significance of adrenarche is unknown and has been described only in man and primates. Adrenal androgens have a weak androgenic effect; they contribute to the growth of pubic and axillary hair, particularly in the female.

Several hypotheses for the control of adrenal androgen secretion in man have been advanced (for review see Grumbach *et al.* 1978). Available data suggest that the increased production of adrenal androgens as puberty approaches might be the result of stimulation by a specific pituitary hormone (adrenal androgens stimulating hormone), but so far this substance has not been isolated. There is no adolescent spurt in the production of cortisol and aldosterone (calculated per m^2 body surface). In children pubic and axillary hair may develop at an early age without the appearance of the other symptoms of sexual maturation. Bone maturation is advanced but rarely more than one year. This syndrome, premature adrenarche (or pubarche), is mostly seen in girls. Plasma concentrations of DHEAS and

Table 13.3 Plasma DHA and DHAS concentrations in normal boys and girls in relation to bone age. Mixed longitudinal data (Sizonenko *et al.* 1976). Note the increase of DHAS (and in lesser degree DHA) between bone age 6 and 8 years (adrenarche).

Bone age (y)	6	8	10 (boys) 9 (girls)	11	13
Boys					
DHA nmol/l	1.5 ± 0.3	2.6 ± 0.3		3.6 ± 0.4	9.3 ± 1.7
DHAS nmol/l	194 ± 31	403 ± 74	581 ± 128		1443 ± 317
μg/dl	7.6 ± 1.2	15.7 ± 2.9	22.7 ± 5.0		56.3 ± 12.4
Girls					
DHA nmol/l	1.7 ± 0.3	3.3 ± 0.6	4.8 ± 1.4	11.4 ± 2.7	18.9 ± 4.4
DHAS nmol/l	131 ± 23	307 ± 49	340 ± 72	988 ± 156	
μg/dl	5.1 ± 0.9	11.9 ± 1.9	13.3 ± 2.8	38.6 ± 6.1	

androstenedione are elevated (Korth-Schutz *et al.* 1976; Sizonenko & Paunier 1975). Most patients with premature adrenarche enter puberty within the normal age range, but in these patients early puberty is sometimes observed. The fact that in some patients with premature adrenarche disorders of the central nervous system (such as hydrocephalus) are found, as in some patients with sexual precocity, is in favour of the existence of a specific pituitary adrenal androgen stimulating hormone.

Normal levels of plasma DHA have been found in patients with Turner's syndrome or isolated gonadotrophin insufficiency. In patients with idiopathic delayed puberty low plasma DHA concentrations have been reported. In patients with Addison's disease both normal and very low plasma DHEA values have been found. The association of Addison's disease with hypogonadotrophic hypogonadism has been observed, and obviously in such patients puberty will not begin (for a review see Sizonenko 1978b).

Excessive adrenal androgen production in patients with congenital adrenal hyperplasia is associated with accelerated growth and advanced bone age. In such patients there is advanced maturation of the hyothalamic–pituitary–gonadal axis, and if such children are treated with glucocorticoids after attaining a bone age of 10 years in girls or 12 years in boys precocious puberty will start.

In conclusion, there is no evidence that adrenal androgens play a role in the onset of puberty in normal children. The hypothalamic–pituitary–gonadal axis and hypothalamic–pituitary–adrenal androgen axis operate independently. In some clinical situations both regulatory systems may be involved, but it is not clear how they interact.

Premature thelarche

Isolated premature development of the breasts (premature thelarche) is usually found in young girls (1–3 years of age). There are no other signs of sexual maturation, except for minimal oestrinisation of the vaginal smear in some patients. Growth and bone maturation are normal for age.

Slightly raised plasma concentrations of LH, FSH and oestrogens have been reported in some patients with premature thelarche (Kenny *et al.* 1969; Bidlingmaier *et al.* 1973; Radfar *et al.* 1976; Escobar *et al.* 1976). It is tempting to speculate that in these girls the hypothalamic–pituitary–gonadal axis which is normally highly active in infancy, is still active at a somewhat older age. After stimulation with Gn-RH, release of LH and particularly FSH is usually similar to the response seen in normal girls during the first six months of life and sometimes similar to that of normal late prepubertal girls (Job *et al.* 1977). Usually the breast enlargement is transient, but it may persist until the onset of puberty at a normal age. Rarely in girls with early premature thelarche at a later age, premature adrenarche and/or precocious puberty is observed.

NORMAL VARIATIONS IN THE ONSET OF PUBERTY

Many somatic changes occur during puberty. Boys and girls are different before puberty, but during puberty these differences become much greater. There is growth of the gonads and genitalia, appearance of secondary sex characteristics, a remarkable increase in body growth—both in size and shape (the adolescent growth spurt), and there are marked changes in body composition. These somatic changes during puberty have been reviewed in great detail by Tanner (1962), and more recently by Marshall (1978). Table 13.4 summarises mean ages (± 2 SD, as percentiles) for girls and boys, in The Netherlands, Switzerland and England when different stages of pubertal development are reached (data from van Wieringen *et al.* 1968; Marshall & Tanner 1969, 1970).

In girls (50th percentile) pubertal development starts at the age of 10.4–11.7 years in these countries, with the simultaneous development of breasts and pubic hair. The peak of the adolescent growth spurt in height is reached at 12 years of age (stage B3, PH3). Average age at menarche is 13.4 years.

In boys (50th percentile) pubertal development starts with enlargement of the testes at age 11.0–11.6 years. Pubic hair appears about 1 year later (11.7–13.4 years). Peak height velocity is at 14.0 years of age (stage G4, PH4).

It is apparent that there are great individual variations in the age at which puberty begins, as well as in the temporal relationships of the different pubertal events such as development of breasts and pubic hair in girls.

In Switzerland in 95% of boys pubertal development (G2) starts between 9.1 and 13.9 years of age (11.5 ± 2.4). For girls (B2) This is 9.0–13.4 years (11.2 ± 2.2). The term 'precocious' puberty is usually defined as the appearance of pubertal signs at an age less than 2.0 to 3.0 standard deviations of the mean age of onset of puberty. For most western countries this is the appearance of breasts and/or pubic hair before the age of 8 years, or menarche before 10 years in girls, and in boys the enlargement of testes before the age of 9 years, and appearance of pubic hair before 10 years of age. Obviously these 'definitions' change with time because of the phenomenon of continuous earlier maturation during childhood.

'Delayed' puberty may be defined as the lack of sexual maturation at an age more than 2.0 to 3.0 standard deviations of the mean age of onset of puberty. By definition this includes 2.5% of all children.

In the average girl skeletal maturation at the onset of puberty (PH2, B2) is about 11 years and in the average boy (PH2, G2) about 12.5 years. There is great individual variation. In children with early puberty skeletal age is advanced for chronological age; in delayed puberty skeletal age is retarded for chronological age.

Table 13.4 Mean ages (±2 SD, as percentiles) at which different stages of puberty are reached: girls and boys, in three different countries.

	Netherlands [1]			Zurich, Switzerland [2]			England [3]		
	10	50	90	5	50	95	5	50	95
Girls									
Pubic hair									
PH2	9.5	11.3	13.0	8.0	10.4	12.8	9.3	11.7	14.1
PH3	10.2	12.2	13.9	9.8	12.2	14.6	10.2	12.4	14.6
PH4	11.4	13.3	14.7	10.8	13.0	15.2	10.9	13.0	15.1
PH5	13.1	14.9	17.7	11.4	14.0	16.6	12.2	14.4	16.6
Breast									
B2	9.0	11.0	12.7	8.5	10.9	13.3	8.9	11.1	13.3
B3	10.6	12.1	13.7	10.0	12.2	14.6	9.9	12.1	14.3
B4	11.2	13.4	15.3	11.4	13.2	15.0	11.1	13.1	15.1
B5	12.7	15.2	—	11.6	14.0	16.4	11.9	15.3	18.7
Menarche	11.8	13.4	14.9	11.2	13.4	15.6	11.5	13.5	15.5
PHV	—	—	—	10.0	12.2	14.4	10.3	12.1	13.9
Boys									
Pubic hair									
PH2	—	11.7	14.1	9.6	12.2	15.2	11.2	13.4	15.6
PH3	11.9	13.4	15.0	11.1	13.5	15.9	11.9	13.9	15.9
PH4	13.2	14.4	15.9	12.0	14.2	16.4	12.2	14.4	16.6
PH5	14.3	16.0	17.7	12.9	14.9	16.9	13.0	15.2	17.4
Genitalia									
G2	—	11.0	13.8	8.2	11.2	14.2	9.4	11.6	13.8
G3	11.2	13.2	14.7	10.5	12.9	15.3	10.8	12.8	14.8
G4	12.9	14.1	16.0	11.6	13.8	16.0	11.8	13.8	15.8
G5	14.1	15.8	18.0	12.5	14.7	16.9	12.7	14.9	17.1
PHV	—	—	—	12.3	13.9	15.5	12.0	14.0	16.0

[1] van Wieringen *et al.* (1968)
[2] Largo and Prader (1981)
[3] Marshall and Tanner (1969, 1970)
For abbreviations see p. 351.

PRECOCIOUS PUBERTY (Table 13.5)

Precocious puberty is usually classified into two groups: true precocious puberty and pseudo-precocious puberty. In true or hypothalamic precocious puberty, the (very) early pubertal development results from a premature release of gonadotrophins, stimulated by hypothalamic Gn-RH. Theoretically true precocious puberty could result from overproduction of gonadotrophins by pituitary tumours, but this has never been reported (for an extensive discussion on precocious puberty the reader is referred to some

Table 13.5 Differential diagnosis of precocious puberty: 'true' precocious puberty (A) and 'pseudo' precocious puberty (B, C, D, E, F).

		Serum FSH/LH (basal values)	Remarks
A	True (hypothalamic) precocious puberty		
	Idiopathic	+[1]	Occasionally familiar type in boys (autosomal dominant)
	Pathology of central nervous system	+	Tumours: congenital abnormalities
	Hormonal 'overlap'	+	In hypothyroidism (rare)
B	Gonadotrophin–producing tumours		Very rare
	Chorion epothelioma; teratoma	+++	
	Hepatoma	+++	Only in boys
C	Congenital adrenal hyperplasia	0	Usually 21-hydroxylase deficiency (plasma 17α-OH-progesterone elevated)
	Adrenal tumours	0	Rare
D	Tumours of ovaries and testes		
	Granulosa cell tumours; theca cell tumours	0	Usually unilateral
	Leydig cell tumours	0	Unilateral testis tumour
E	Cysts in ovaries	0	Rare; only in girls
	McCune–Albright syndrome		
F	Drugs		
	Oestrogens	0	
	Androgens	0	

\+ Elevated basal serum values.
0 Normal basal serum values.
1 In true precocious puberty basal LH/FSH serum concentrations may be normal for age. However after stimulation with Gn-RH, serum LH/FSH concentrations will increase as in normal puberty. This will not be seen in patients with 'pseudo' precocious puberty.

recent review papers: Bierich 1975; Rayner 1976; Styne & Kaplan 1979). In pseudo-precocious puberty (very) early pubertal changes may result from increased production of gonadal hormones (androgens and/or oestrogens) without hypothalamic stimulation of pituitary gonadotrophins. Usually tumours of ovaries and testes are involved. Other causes of pseudo-precocious puberty are congenital adrenal hyperplasia and adrenal tumours. Gonadotrophin-producing tumours such as chorionepithelioma and teratoma are very rare. Ingestion of androgen- or oestrogen-containing medications may be the cause of (usually transient) pseudo-precocious puberty. Young girls are very sensitive to adsorption of oestrogens from oestrogen-containing ointments or cosmetics. A rare syndrome associated with precocious puberty in girls is the McCune–Albright syndrome, with fibrous dysplasia of long bones (especially femur), café-au-lait pigmentation and other symptoms. Functioning ovarian cysts have been reported, but ultimately true precocious puberty may be observed.

True precocious puberty has been described in patients with untreated hypothyroidism; it is postulated that this is due to 'overlap feedback' stimulation of not only thyroid-stimulating hormone (TSH) but also LH and FSH.

True precocious puberty may be caused by central nervous system disorders (congenital anomalies such as hydrocephalus; tumours such as craniopharyngioma). When no definable causes are found the term idiopathic precocious puberty is used. Idiopathic precocious puberty is more common in girls than in boys (in girls 60–80%, in boys 20–40%; Bierich 1975).

Diagnosis

Clinical symptoms are those of normal puberty: development of breasts and pubic hair in girls, enlargement of testes and penis and development of pubic hair in boys. Vaginal bleeding may rarely be the first sign of precocious puberty in girls. An important symptom is accelerated growth.

During the period 1969–80 we observed 67 children (62 girls, 5 boys) with precocious puberty, premature thelarche and adrenarche (Tables 13.6, and 13.7). In two thirds of the patients with precocious puberty acceleration of growth in the preceding year was recognised (Table 13.8).

Careful neurological investigations (including CT-scan) and ophthalmological investigations (including perimetry) are indicated in all patients with precocious puberty in order to exclude underlying cerebral abnormalities. These investigations should be repeated during the follow-up of

Table 13.6 Diagnosis in 38 children with precocious puberty observed in the Sophia Children's Hospital in Rotterdam 1969–80 (Sabbe-Claus *et al.* submitted for publication).

	Girls	Boys	Total
Precocious puberty			
Idiopathic type	30	3	33
CNS lesions	5	1	6
Pseudo-precocious puberty	3	—	3
Primary hypothyroidism	—	1	1
Premature thelarche	13	—	13
Premature adrenarche	11	—	11
Total	62	5	67

Table 13.7 Clinical symptoms in children with precocious puberty, premature thelarche and premature adrenarche (see Table 13.6). B/G = breast development in girls, genital development in boys. PH = development of pubic hair.

Diagnosis	B/G and PH	B/G	PH	Menses
Precocious puberty				
Idiopathic type	18	12	1	9
CNS lesions	6	0	0	1
Pseudo-precocious puberty	3	0	0	2
Premature thelarche	0	13	0	0
Premature adrenarche	0	0	11	0

patients with idiopathic precocious puberty when neurological signs become present. Skeletal maturation (bone age) is advanced in patients with precocious puberty (Table 13.9) and will result in early epiphyseal fusion. Adult height of most (untreated) patients with precocious puberty is below normal.

Hormonal investigations include serum gonadotrophins, oestrogens and androgens. Basal LH and FSH serum levels may be normal in patients with true precocious puberty; however in such patients the Gn-RH test will enable differentiation from pseudo-precocious puberty. Children with true precocious puberty have releasable LH and FSH stores within or above the normal range, corresponding to their clinical pubertal stage (Job *et al.* 1977).

Study of sleep-associated pulsatile release of LH and FSH, which has been observed in children with true precocious puberty, has no practical diagnostic value at this moment.

Single plasma estimations of oestrogens (oestradiol) and androgens (testosterone and Δ4-androstenedione) are not always reliable for the diagnosis of precocious puberty. In some studies of patients with idiopathic precocious puberty, and this is also our own experience, only about 50%

Table 13.8 Acceleration of growth in the year preceding diagnosis of precocious puberty, premature thelarche and premature adrenarche (see Table 13.6).

Diagnosis	Present	Absent	Unknown
Precocious puberty			
Idiopathic type and CNS lesions	29	4	6
Pseudo-precocious puberty	2	1	0
Premature thelarche and adrenarche	0	18	6

Table 13.9 Advanced skeletal maturation in children with precocious puberty versus children with premature thelarche and premature adrenarche (see Table 13.6).

	Bone age minus chronological age Mean (y)	n
Precocious puberty		
Idiopathic type	3	33
CNS lesions	2	6
Pseudo-precocious puberty	1.7	3
Premature thelarche	0.6	13
Premature adrenarche	0.7	11

of the individual values were greater than the upper limit of the normal range (Escobar *et al.* 1976; Bidlingmaier *et al.* 1977). It is our experience that in patients with precocious puberty due to lesions of the central nervous system, and in patients with pseudo-precocious puberty, individual determinations of plasma oestrogens and androgens are somewhat more reliable. Increased plasma levels of 17α-hydroxyprogesterone indicate congenital adrenal hyperplasia (21-hydroxylase defect). In patients with adrenal tumours high plasma levels of oestrogens and DHEA(S) are reported (Drop *et al.* 1981).

Estimation of urinary excretion of gonadotrophins, oestrogens and androgens (17-oxosteroids) is of no practical value anymore, as radio-immunoassay measurements of gonadotrophins and steroids in serum are now generally available.

In summary, precocious puberty is primarily a clinical diagnosis. Laboratory tests include serum measurements of gonadotrophins (basal and after stimulation with Gn-RH), oestradiol, testosterone and 17α-hydroxyprogesterone.

DELAYED PUBERTY (Table 13.10)

Delay of sexual maturation and the growth spurt of puberty can be a matter of great concern for the affected children and their parents. The most frequent cause is the so-called constitutional delay of growth and adolescence (puberty). We are dealing here with one end of the normal variation in the onset of puberty, which affects about 2.5% of all boys and girls (Table 13.10). The prognosis of this variant of late, but normal, pubertal development is excellent. It is interesting that in medical practice more boys than girls with delayed puberty are seen. This is probably due to the fact that the psychosocial consequences of delayed growth and puberty are felt

Table 13.10 Differential diagnosis of delayed puberty.

Constitutional delay of growth and puberty
Systemic chronic diseases
Undernutrition (anorexia nervosa)
Chronic administration of glucocorticosteroids
Hypothalamic–pituitary insufficiency
Isolated gonadotrophin deficiency (with anosmia: Kallmann syndrome)
Tumours, encephalitis, pituitary insufficiency (with growth hormone deficiency)
Primary gonadal failure
Congenital (XO, Turner syndrome; XXY, Klinefelter syndrome)
After surgery, trauma, infection
After radiation, cytostatic and immunosuppressive drugs

more by boys than by girls. Obviously the syndrome must be present equally in both sexes.

Apart from systemic chronic diseases (such as Crohn's disease, severe asthma), chronic undernutrition (e.g. anorexia nervosa) and chronic administration of glucocorticosteroids, the differential diagnosis includes two groups of disorders: hypothalamic–pituitary insufficiency and primary gonadal failure. For a more detailed discussion of the different aspects of delayed puberty the reader is referred to some recent review articles (Prader 1975; Rayner 1976; Sizonenko 1978b; Styne & Kaplan 1979; Prader *et al.* 1980).

Hypothalamic–pituitary insufficiency may rarely be caused by lesions from tumours and encephalitis, but the two main causes are isolated gonadotrophin deficiency (with and without anosmia) and pituitary insufficiency (with growth hormone deficiency). Primary gonadal failure can be seen after surgery (for cryptorchidism), trauma, infection, radiation and the use of cytostatic and immunosuppressive drugs. It also includes chromosomal disorders such as Turner's syndrome and Klinefelter's syndrome.

Diagnosis

The clinical diagnosis of constitutional delay of growth and puberty generally presents no problems. The growth rate in such children often decreases slowly during childhood and the growth curve may drop to a lower percentile. Skeletal maturation is retarded; usually bone age is 13 years or less in boys, 11 years or less in girls. Very often there is a positive family history. There is good evidence that many, if not all, children with constitutional delay of growth and puberty have partial (transient) growth hormone deficiency. Short term administration of androgens or oestrogens to boys with short stature, delayed bone age and puberty can improve the impaired plasma growth hormone response to various stimuli (Martin *et al.* 1968; Illig & Prader 1970) (see p. 307). A prospective study has shown that such children have normal growth spurts as sexual maturation progresses at puberty (Martin *et al.* 1979). Not only in these children but also in normal children, GH secretion and peak GH response to standard stimuli are increased during puberty. Priming with sex steroids is unlikely to alter the outcome of GH testing if marked growth deceleration is evident at prepubertal age and such patients will have true persistent GH deficiency (Drop *et al.* 1982).

Recent studies by Prader *et al.* (1980) have indicated that children with constitutional delay of growth and puberty also may have transitory reduced responsiveness of the hypothalamic–pituitary–adrenal axis. Urinary tetrahydro-S values after administration of metyrapone 500 mg/m^2 surface area were subnormal. The results became normal with the onset of puberty.

Plasma concentrations of the adrenal androgens DHA and DHAS in children with constitutional delay of growth and puberty are low for chronological age, but normal for bone age (Copeland *et al.* 1977). There is evidence that in patients with hypogonadotrophic hypogonadism adrenal androgens are normal for chronological age and bone age, but this has to be confirmed in more studies. The differential diagnosis between children (particularly boys) with constitutional delay of growth and puberty and patients with hypogonadotrophic hypogonadism due to isolated gonadotrophin deficiency may present great problems. The latter syndrome is a relatively rare, sometimes familial disorder, but has to be considered. Patients with isolated gonadotrophin deficiency usually grow normally during childhood into the pubertal age period. Boys with constitutional delay of growth and puberty, when seen at the age of 15–16 years, usually have a bone age of no more than 13.5 years. When their testicular volume is 4 ml or more, one may be more confident that pubertal development will begin in the next year. When a 16-year-old boy at physical examination has very small testes ($\leqslant$ 2 ml), a bone age of 14 years and no pubertal signs, the diagnosis of isolated gonadotrophin deficiency may be suspected. It is important to investigate the ability to smell. In some patients there is anosmia or hyposmia with or without associated abnormalities such as cleft lip, cleft palate, colour blindness and nerve deafness (Kallmann syndrome).

Several investigators have used the Gn-RH test to differentiate between constitutional delay of growth and puberty and isolated gonadotrophin deficiency. Single standard Gn-RH tests cannot reliably differentiate individual children in either group (Job *et al.* 1977). About 30% of patients with isolated gonadotrophin deficiency will have normal LH and FSH responses (Marshall *et al.* 1972), suggesting that at least in some patients the syndrome is due to lack of Gn-RH itself, rather than to a pituitary deficiency.

Kelch and co-workers have shown that a standard Gn-RH test may give reassuring results in boys with a bone age greater than 12 years (Kelch *et al.* 1976). Savage *et al.* (1981) were able to confirm these findings. On the other hand, a blunted response or failure to respond to Gn-RH is not diagnostic of permanent gonadotrophin deficiency (Kelch *et al.* 1980). This can be seen especially in obese boys with delayed puberty.

Several modifications of the standard Gn-RH test have been proposed in order to distinguish between the two groups of patients, e.g. Gn-RH stimulation before and after administration of clomiphene-citrate and a 4-hour continuous infusion of Gn-RH. Clomiphene has oestrogen-like properties and stimulates LH and FSH release in normal adult men. Premedication with clomiphene may rise Gn-RH-induced peak LH in patients with hypogonadotrophic hypogonadism and depress peak LH in constitutional delayed puberty (Snoep *et al.* 1977). A 4-hour infusion of Gn-RH may give continuously rising levels of plasma LH in constitutionally delayed puberty, while in hypogonadotrophic hypogonadism the pituitary release of LH is decreasing (de Lange *et al.* 1978). It is clear that although in many cases the diagnosis of constitutional delay of growth and puberty can be made relatively easily, in other cases the situation may be more difficult and require the expertise of a pediatric endocrinologist. Longitudinal studies with repeated Gn-RH tests may be necessary to establish a definite diagnosis of isolated gonadotrophin deficiency. A short course of therapy with testosterone or an anabolic steroid may advance pubertal maturation in boys with constitutional delay of puberty and in this way can be helpful in the differential diagnosis with isolated gonadotrophin deficiency (and see Chapter 11).

In children with primary gonadal failure (anorchia, testicular atrophy, Turner's syndrome) the LH and FSH response to Gn-RH is markedly increased, at least after the age of about 11 years (Job *et al.* 1974; Illig *et al.* 1975). In all girls with small stature, decreasing growth rate and delay of puberty, chromosomal analysis should be done. When the classical physical symptoms of Turner's syndrome are absent, mosaic variants may be found.

Basal plasma levels of testosterone and oestradiol in children with primary gonadal failure will not differentiate these patients from normal prepubertal children. Stimulation tests may be very useful. In boys human chorionic gonadotrophin (hCG) can be used. Several protocols with hCG administration of different duration have been proposed. We use one injection of hCG (Pregnyl) 1500 IU intramuscularly and measure plasma testosterone before and 72 hours after infection. Basal values will normally increase by two to three times (Smals *et al.* 1979). This test will not differentiate between constitutional delay of puberty and gonadotrophin deficiency (and see Chapter 11).

In the female the ovaries can be stimulated with human menopausal gonadotrophin (hMG); 150 IU is given daily for 5 days, intramuscularly. Plasma oestradiol and progesterone are measured. Studies giving normal data have not been reported.

In summary, the most frequent cause of delay in sexual maturation is the syndrome of constitutional delay of growth and puberty. The diagnosis is first of all a clinical one. Laboratory tests include serum measurement of growth hormone to differentiate from true growth hormone deficiency, gonadotrophins (basal and after stimulation with Gn-RH) and adrenal androgens to differentiate from isolated gonadotrophin deficiency. Longitudinal studies may be necessary to establish the latter diagnosis with certainty. Serum measurements of testosterone and oestradiol (basal and after gonadal stimulation) are needed to diagnose primary gonadal insufficiency.

REFERENCES

BIDLINGMAIER F., BUTENANDT O. & KNORR D. (1977) Plasma gonadotropins and estrogens in girls with idiopathic precocious puberty. *Pediatr. Res.* **11**, 91–4.

BIDLINGMAIER F., VERSMOLD H. & KNORR D. (1974) Plasma estrogens in newborns and infants. In Forest M. G. and Bertrand J. (eds.) *International symposium on sexual endocrinology of the perinatal period*, Lyon, Inserm 32, pp. 299–314.

BIDLINGMAIER F., WAGNER-BARNACK M., BUTENANDT O. & KNORR D. (1973) Plasma estrogens in childhood and puberty under physiologic and pathologic conditions. *Pediatr. Res.* **7**, 901–7.

BIERICH J. R. (1975) Sexual precocity. *Clin. Endocrinol. Metab.* **4**, 107–42.

BOYAR R. M., FINKELSTEIN J. W., DAVID R., ROFFWARG H. P., KAPEN S., WEITZMAN E. D. & HELLMAN L. (1973b) Twenty-four hour patterns of plasma luteinizing hormone and follicle-stimulating hormone in sexual precocity. *New Engl. J. Med.* **289**, 282–6.

BOYAR R. M., FINKELSTEIN J. W., ROFFWARG H. P., KAPEN S., WEITZMAN E. D. & HELLMAN L. (1972) Synchronization of augmented luteinizing hormone secretion with sleep during puberty. *New Engl. J. Med.* **287**, 582–6.

BOYAR R. M., FINKELSTEIN J. W., ROFFWARG H. P., KAPEN S., WEITZMAN E. D. & HELLMAN L. (1973a) Twenty-four-hour luteinizing hormone and follicle-stimulating hormone secretory patterns in gonadal dysgenesis. *J. Clin. Endocrinol.* **37**, 521–5.

BOYAR R. M., ROSENFELD R. S., KAPEN S., FINKELSTEIN J. W., ROFFWARG H. P., WEITZMAN E. D. & HELLMAN L. (1974) Human puberty: simultaneous augmented secretion of luteinizing hormone and testosterone during sleep. *J. Clin. Invest.* **54**, 609–18.

CONTE F. A., GRUMBACH M. M. & KAPLAN S. L. (1975) A diphasic pattern of gonadotropin secretion in patients with the syndrome of gonadal dysgenesis. *J. Clin. Endocrinol.* **40**, 670–4.

CONTE F. A., GRUMBACH M. M., KAPLAN S. L. & REITER E. O. (1980) Correlation of luteinizing hormone-releasing factor-induced luteinizing hormone and follicle-stimulating hormone release from infancy to 19 years with the changing pattern of gonadotropin secretion in agonadal patients: relation to the restraint of puberty. *J. Clin. Endocrinol.* **50**, 163–8.

COPELAND K. C., PAUNIER L. & SIZONENKO P. C. (1977) The secretion of adrenal androgens and growth patterns of patients with hypogonadotropic hypogonadism and idiopathic delayed puberty. *J. Pediatr.* **91**, 985–90.

DICKERMANN Z., PRAGER-LEWIN R. & LARON Z. (1976) Response of plasma LH and FSH to synthetic LH-RH in children at various pubertal states. *Am. J. Dis. Child.* **130**, 634–8.

DIERSCHKE D. J., BHATTACHARYA A. N., ATKINSON L. E. & KNOBIL E. (1970) Circhoral oscillations of plasma LH levels in the ovariectomized rhesus monkey. *Endocrinology* **87**, 850–3.

DONOVAN B. T. & VAN DER WERFF TEN BOSCH J. J. (1956) Precocious puberty in rats and hypothalamic lesions. *Nature* **178**, 745.

DONOVAN B. T. & VAN DER WERFF TEN BOSCH J. J. (1965) *Physiology of puberty.* Arnold, London.

DROP S. L. S., BRUINING G. J., VISSER H. K. A. & SIPPELL W. G. (1981) Prolonged galactorrhoea in a 6-year-old girl with isosexual precocious puberty due to a feminizing adrenal tumour. *Clin. Endocrinol.* **15**, 37–43.

DROP S. L. S., SABBE-CLAUS L. & VISSER H. K. A. (1982) The effect of puberty and short-term oral administration of testosterone undecanoate on GH tests and sex-steroid related plasma components in GH deficient patients. *Clin. Endocrinol.* **16**, 375–81.

DUCHARME J. R., FOREST M. G., PERETTI E. DE, SEMPÉ M., COLLU R. & BERTRAND J. (1976) Plasma adrenal and gonadal sex steroids in human pubertal development. *J. Clin. Endocrinol.* **42**, 468–76.

ESCOBAR M. E., RIVAROLA M. A. & BERGADA C. (1976) Plasma concentrations of estradiol-17 beta in premature thelarche and in different types of sexual precocity. *Acta Endocrinol.* **81**, 351–61.

FAIMAN C. & WINTER J. S. D. (1971) Sex differences in gonadotrophin concentrations in infancy. *Nature* **232**, 130–1.

FOREST M. G., SIZONENKO P. C., CATHIARD A. M. & BERTRAND J. (1974) Hypophysogonadal functions in humans during the first year of life. I. Evidence for testicular activity in early infancy. *J. Clin. Invest.* **53**, 819–28.

FRISCH R. E. & REVELLE R. (1969) The height and weight of adolescent boys and girls at the time of peak velocity of growth in height and weight: Longitudinal data. *Hum. Biol.* **41**, 536–59.

FRISCH R. E. & REVELLE R. (1970) Height and weight at menarche and a hypothesis of critical body weights and adolescent events. *Science* **169**, 397–9.

GRUMBACH M. M. (1978) The central nervous system and the onset of puberty. In Falkner F. and Tanner J. M. (eds.) *Human Growth* II, pp. 215–38. Baillière Tindall, London.

GRUMBACH M. M. & KAPLAN S. L. (1973) Ontogenesis of growth hormone, insulin, prolactin and gonadotropin secretion in the human foetus. In Comline K. S., Cross K. W., Dawes G. S. & Nathanielsz P. W. (eds.) *Foetal and neonatal physiology*; Proceedings of the Sir Joseph Barcroft Centenary Symposium 1972, pp. 462–87. Cambridge University Press, Cambridge.

GRUMBACH M. M., RICHARDS G. E., CONTE F. A. & KAPLAN S. L. (1978) Clinical disorders of adrenal function and puberty: An assessment of the role of the adrenal cortex in normal and abnormal puberty in man and evidence for an ACTH-like pituitary androgen stimulating hormone. In James V. H. T., Serio

GRUMBACH M. M., ROTH J. C., KAPLAN S. L. & KELCH R. P. (1974) Hypothalamic–pituitary regulation of puberty in man: Evidence and concepts derived from clinical research. In Grumbach M. M., Grave G. D. and Mayer F. E. (eds.) *Control of the onset of puberty*, pp. 115–66. Wiley, New York.

D., Guisti G. & Martini L. (eds.) *The endocrine function of the human adrenal complex*; Proceedings of the Serono symposia 18, pp. 583–612. Academic Press, New York.

Illig R. & Prader A. (1970) Effect of testosterone on growth hormone secretion in patients with anorchia and delayed puberty. *J. Clin. Endocrinol.* **30**, 615–18.

Illig R., Tolksdorf M., Mürset G. & Prader A. (1975) LH and FSH response to synthetic LH-RH in children and adolescents with Turner's and Klinefelter's syndrome. *Helv. Paediatr. Acta* **30**, 221–31.

Job J. C., Chaussain J. L. & Garnier P. E. (1977) The use of luteinizing hormone-releasing hormone in pediatric patients. *Horm. Res.* **8**, 171–87.

Job J. C., Garnier P. E., Chaussain J. L., Binet E., Rivaille P. & Milhaud G. (1972) Effects of synthetic luteinizing hormone-releasing hormone (LH-RH) on serum gonadotropins (LH and FSH) in normal children and adults. *Rev. Europ. Etudes Clin. Biol.* **17**, 411–14.

Job J. C., Garnier P. E., Chaussain J. L., Scholler R., Toublanc J. E. & Canlorbe P. (1974) Effect of synthetic luteinizing hormone-releasing hormone (LH-RH) on the release of gonadotropins in hypophyso-gonadal disorders of children and adolescents. V. Agonadism. *J. Clin. Endocrinol. Metabolism* **38**, 1109–14.

Kaplan S. L., Grumbach M. M. & Aubert M. L. (1976) The ontogenesis of pituitary hormones and hypothalamic factors in the human fetus: Maturation of central nervous system regulation of anterior pituitary function. *Recent Progr. Horm. Res.* **32**, 161–243.

Kelch R. P., Hopwood N. J. & Marshall J. C. (1980) Diagnosis of gonadotropin deficiency in adolescents: limited usefulness of a standard gonadotropin-releasing hormone test in obese boys. *J. Pediatr.* **97**, 820–4.

Kelch R. P., Markovs M. & Huss J. (1976) LH and FSH responsiveness to intravenous gonadotropin-releasing hormone (GnRH) in children with hypothalamic or pituitary disorders: lack of effects of replacement therapy with human growth hormone. *J. Clin. Endocrinol.* **42**, 1104–13.

Kenny F. M., Midgley A. R., Jaffe R. B., Garces L. Y., Vazquez A. & Taylor F. H. (1969) Radioimmunoassayable serum LH and FSH in girls with sexual precocity, premature thelarche and adrenarche. *J. Clin. Endocrinol.* **29**, 1272–5.

Knobil E. (1980) The neuroendocrine control of the menstrual cycle. *Recent Prog. Horm. Res.* **36**, 53–88.

Korth-Schutz S., Levine L. S. & New M. I. (1976) Serum androgens in normal prepubertal and pubertal children and in children with precocious adrenarche. *J. Clin. Endocrinol.* **42**, 117–24.

Korth-Schutz S., Virdis R., Saenger P., Chow D. M., Levine L. S. & New M. I. (1978) Serum androgens as a continuing index of adequacy of treatment of congenital adrenal hyperplasia. *J. Clin. Endocrinol.* **46**, 452–8.

Lange W. E. de, Snoep M. C. & Doorenbos H. (1978) The effect of LH-RH infusion on serum LH, FSH and testosterone in boys with advanced puberty, delayed puberty and hypogonadotrophic hypogonadism. *Acta Endocrinol.* **89**, 209–16.

Marshall J. C., Harsoulis P., Anderson D. C., McNeilly A. S., Besser G. M. & Hall R. (1972) Isolated pituitary gonadotrophin deficiency: gonadotrophin secretion after synthetic luteinizing hormone and follicle stimulating hormone-releasing hormone. *Br. Med. J.* **4**, 643–5.

Marshall W. A. (1978) Puberty. In Falkner F. & Tanner J. M. (eds.) *Human Growth* II, pp. 141–81. Baillière Tindall, London.

Marshall W. A. & Tanner J. M. (1969) Variations in pattern of pubertal changes in girls. *Arch. Dis. Child.* **44**, 291–303.

Marshall W. A. & Tanner J. M. (1970) Variations in the pattern of pubertal changes in boys. *Arch. Dis. Child.* **45**, 13–23.

MARTIN L. G., CLARK J. W. & CONNOR T. B. (1968) Growth hormone secretion enhanced by androgens. *J. Clin. Endocrinol.* **28**, 425–8.

MARTIN L. G., GROSSMAN M. S., CONNOR T. B., LEVITSKY L. L., CLARK J. W. & CAMITTAE F. D. (1979) Effect of androgen on growth hormone secretion and growth in boys with short stature. *Acta Endocrinol.* **91**, 201–12.

PERETTI E. DE & FOREST M. G. (1978) Pattern of plasma dehydroepiandrosterone sulfate levels in humans from birth to adulthood: Evidence for testicular production. *J. Clin. Endocrinol.* **47**, 572–7.

PRADER A. (1975) Delayed adolescence. *Clin. Endocrinol. Metabolism* **4**, 143–55.

PRADER A., ZACHMANN M. & BUCHER H. (1980) Constitutional delay of growth and puberty: auxological and endocrine characteristics. In Cacciari E. & Prader A. (eds.) *Pathophysiology of puberty*, pp. 123–35. Academic Press, London.

RADFAR N., ANSUSINGHA K. & KENNY F. M. (1976) Circulating bound and free estradiol and estrone during normal growth and development and in premature thelarche and isosexual precocity. *J. Pediatr.* **89**, 719–23.

RAYNER P. H. W. (1976) Puberty: precocious and delayed. *Br. Med. J.* **I**, 1385–7.

SAVAGE M. O., PREECE M. A., CAMERON N., JONES J., THEINTZ G., PENFOLD J. L. & TANNER J. M. (1981) Gonadotrophin response to LH-RH in boys with delayed growth and adolescence. *Arch. Dis. Child.* **56**, 552–6.

SIPPELL W. G., DÖRR H. G., BIDLINGMAIER F. & KNORR D. (1980) Plasma levels of aldosterone, corticosterone, 11-deoxycorticosterone, progesterone, 17-hydroxyprogesterone, cortisol and cortisone during infancy and childhood. *Pediatr. Res.* **14**, 39–46.

SIZONENKO P. C. (1978a) Endocrinology in preadolescents and adolescents I: Hormonal changes during normal puberty. *Am. J. Dis. Child.* **132**, 704–12.

SIZONENKO P. C. (1978b) Preadolescent and adolescent endocrinology: Physiology and physiopathology II: Hormonal changes during abnormal pubertal development. *Am. J. Dis. Child.* **132**, 797–805.

SIZONENKO P. C. & AUBERT M. L. (1978) Pre- and perinatal endocrinology. In Falkner F. and Tanner J. M. (eds.) *Human Growth* I, pp. 549–592. Baillière Tindall, London.

SIZONENKO P. C. & PAUNIER L. (1975) Hormonal changes in puberty III: Correlation of plasma dehydroepiandrosterone, testosterone, FSH, and LH with stages of puberty and bone age in normal boys and girls and in patients with Addison's disease or hypogonadism or with premature or late adrenarche. *J. Clin. Endocrinol.* **41**, 894–904.

SIZONENKO P. C., PAUNIER L. & CARMIGNAC D. (1976) Hormonal changes during puberty. V Longitudinal study of adrenal androgen secretions. *Horm. Res.* **7**, 288–302.

SMALS A. G. H., PIETERS G. F. F. M., DRAYER J. I. M., BENRAAD T. J. & KLOPPENBORG P. W. C. (1979) Leydig cell responsiveness to single and repeated human chorionic gonadotropin administration. *J. Clin. Endocrinol.* **49**, 12–14.

SNOEP M. C., LANGE W. E. DE, SLUITER W. J. & DOORENBOS H. (1977) Differential response of serum LH in hypogonadotropic hypogonadism and delayed puberty to LH-RH stimulation before and after clomiphene citrate administration. *J. Clin. Endocrinol.* **44**, 603–6.

STYNE D. M. & KAPLAN S. L. (1979) Normal and abnormal puberty in the female. *Pediatr. Clin. N. Am.* **26**, 123–48.

TANNER J. M. (1962) *Growth at adolescence*, 2nd edn. Blackwell Scientific Publications, Oxford.

VISSER H. K. A. (1973) Some physiological and clinical aspects of puberty. *Arch. Dis. Child.* **48**, 169–82.

WIERINGEN J. C. VAN (1978) Secular growth changes. In Falkner F. and Tanner J. M. (eds.) *Human Growth* II, pp. 445–73. Baillière Tindall, London.

Wieringen J. C. van, Wafelbakker F., Verbrugge H. P. & Haas J. H. de (1968) *Groeidiagrammen Nederland* 1965; Nederlands Instituut voor Praeventieve Geneeskunde. Groningen, Wolters-Noordhoff N.V.

Winter J. S. D. (1978) Prepubertal and pubertal endocrinology. In Falkner F. and Tanner J. M. (eds.) *Human Growth* II, pp. 183–213. Baillière Tindall, London.

Winter J. S. D., Faiman C., Hobson W. C., Prasad A. V. & Reyes F. I. (1975) Pituitary-gonadal relations in infancy. I. Patterns of serum gonadotropin concentrations from birth to four years of age in man and chimpanzee. *J. Clin. Endocrinol.* **40**, 545–51.

Chapter 14

Disorders of Bone and Mineral Metabolism

JOSEPH M. GERTNER

Calcium and phosphorus combine in the mineralized portion of the skeleton and for this reason their metabolic roles are usually discussed together. Magnesium is not as important a constituent of bone mineral but since it resembles calcium in circulating as a divalent cation and its metabolism impinges on calcium homoeostasis, some aspects of its turnover will be discussed in the present chapter.

THE PHYSIOLOGY OF CALCIUM HOMOEOSTASIS

Total calcium is commonly measured by atomic absorption spectrophotometry. Where measurement of ionised calcium is desired the most convenient methods available make use of ion-sensitive electrodes. These instruments are reliable, but expensive to purchase and maintain so that, for most purposes, it is still most convenient to calculate the ionised calcium from measurements of total calcium, serum albumen, and blood pH.

Every individual requires calcium in the diet, at least to the extent that excretory losses are balanced. An adult requires a daily intake of 7–10 mg calcium per kg body weight to avoid a net loss of calcium from the body. In the growing child these calculations are complicated by the increased calcium requirements of the growing skeleton, partly offset by the improved calcium absorption which characterises young children.

Parathyroid hormone

The extracellular concentration of ionised calcium is tightly regulated, the principal sensor for this regulation being the calcium-sensitive parathyroid chief cell. As changes in ionised calcium levels are detected, the secretion of parathyroid hormone (PTH) is modulated, high levels of calcium causing an almost complete cessation of hormone secretion while low levels stimulate production of the hormone.

Parathyroid hormone, released in response to the stimulus of hypocalcaemia increases the resorption of calcium from bone and is thus able to effect rapid increases in the level of ionised calcium in the serum. At the renal level the actions of parathyroid hormone are complex, but all tend to increase the extracellular fluid calcium. First, the distal tubular reabsorption of filtered calcium is enhanced, with a consequent reduction in the fractional excretion of calcium. Secondly, proximal tubular phosphate reabsorption is inhibited, leading to phosphaturia and, ultimately, to hypophosphataemia. This, in turn, will act to inhibit bone mineralisation and shift the equilibrium of calcium flux between bone and the extracellular fluid towards the latter. Lastly, parathyroid hormone serves as a crucial stimulus to the activation of vitamin D, a role to be considered in more detail in the next section. By enhancing the 1α-hydroxylation of the vitamin, parathyroid hormone acts to increase the fraction of dietary calcium absorbed into the body, again tending to increase the calcium concentration in the extracellular fluid.

Parathyroid hormone is an 84 amino acid peptide (Niall *et al.* 1970) and was one of the first of the peptide hormones shown to act through the activation of adenyl cyclase (Chase & Aurbach 1967). The formation of cyclic adenosine monophosphate (cAMP) is a first step in the antiphosphaturic and bone resorbing (Chase & Aurbach 1970) actions of PTH, but the role of this compound in PTH-mediated renal tubular calcium reabsorption and 1α-hydroxylation of 25-hydroxyvitamin D is less clear. Since cAMP arising from the renal action of parathyroid hormone is largely excreted in the urine, the renal excretion of this compound can be used as an index of parathyroid hormone activity (Broadus *et al.* 1977).

Vitamin D

Vitamin D, the other essential mediator of calcium homoeostasis, can be considered both as a vitamin and as a hormone. The form of the vitamin taken in unfortified foods and synthesised in the skin by photo-catalysis is cholecalciferol, a sterol whose structure resembles the cyclopentanophenanthrene framework which characterises the adrenal and gonadal steroids. Certain foodstuffs, notably milk in the USA and margarine in the UK are fortified with vitamin D as a public health measure (Working Party 1980). The form of the vitamin used in such fortification is ergocalciferol, which is produced by the irradiation of yeast. Ergocalciferol and cholecalciferol are probably handled identically and their metabolites are, in mammals, considered to exert equipotent effects. Accordingly the term 'vitamin D', when unqualified in this chapter, will apply equally to either compound.

Although the work of Mellanby (1919) and Huldchitnsky (1919) after the First World War showed that vitamin D could be obtained both from natural foodstuffs and from dermal irradiation, the mode of action of this substance was ill-understood until the development of sophisticated techniques for the examination of vitamin D metabolism *in vivo*. These techniques arose from the explosive increase in vitamin D research which followed the synthesis of radiolabelled vitamin D analogues of high specific activity and the development of reliable chromatographic technique for the resolution of vitamin D metabolites dissolved in organic solvents (Lund &

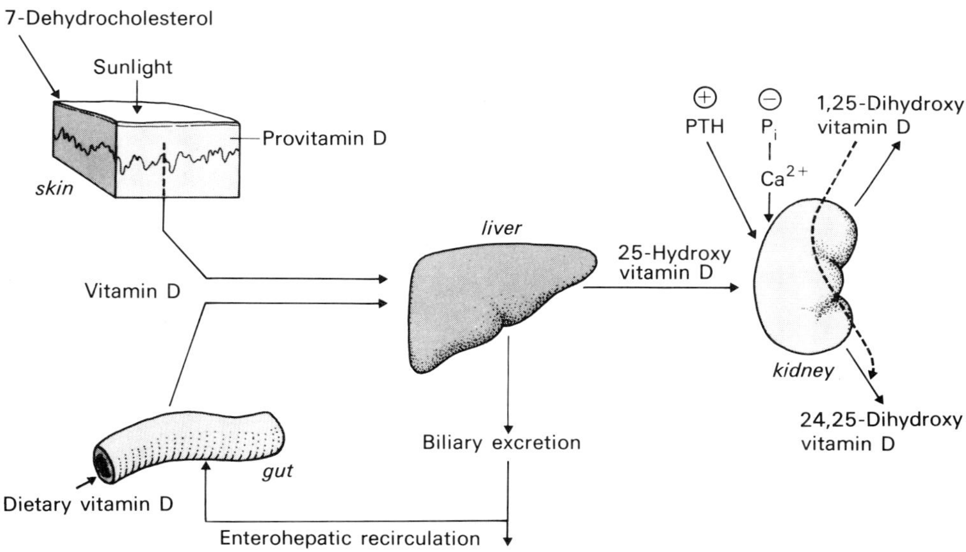

Fig. 14.1 Schematic representation of the sources and metabolic fate of vitamin D in man. Note the enterohepatic recirculation. Factors promoting or antagonising the renal 1α-hydroxylation of 25-OHD are indicated by + and − respectively. 24-hydroxylation is stimulated by 1,25 $(OH)_2D$ as indicated by the dotted line.

DeLuca 1966). Using these new methods it was possible to trace the process whereby vitamin D is synthesised and then hydroxylated before it can exert its biological activity. In addition assays were designed which permitted the determination of circulating levels of nutritionally (Haddad & Chyu 1971) and functionally (Brumbaugh *et al.* 1974) important hydroxylated vitamin D metabolites. As a result of these efforts, it is possible to present a fairly complete picture of vitamin D production and activation (Fig. 14.1). 7-dehydrocholesterol is converted by solar (or artificial ultraviolet) irradiation of the skin to form cholecalciferol. Recent studies (Holick & Clark 1978) have shown that the release of cholecalciferol from the skin follows the gradual transformation of the initial radiation product, previtamin D3, to the vitamin. This is considered a 'smoothing' mechanism, to provide a constant supply of vitamin in the face of fluctuating synthetic rates. Cholecalciferol formed in the skin, and chole- or ergocalciferol absorbed from the diet are transported, via the plasma vitamin D-binding proteins, to the liver for further metabolism, or to adipose tissue and muscle for storage in time of abundant supply. Although the unchanged vitamin forms only a small portion of the circulating vitamin D it forms the bulk of tissue stores of these compounds (Mawer *et al.* 1972).

In the liver vitamin D is rapidly hydroxylated to 25-hydroxyvitamin D (25-OHD), an NADPH-dependent microsomal reaction. This reaction is inhibited by high concentrations of vitamin D (Bhatacharyya & DeLuca 1973), thus serving as a site for feedback control, which again might serve to 'smooth' the supply of metabolically active material in the face of the episodic synthesis or ingestion of its precursor. A proportion of the 25-OHD is excreted into the bile and it has been suggested that this material can be re-used upon subsequent reabsorption from the gut. Interference with this 'entero-hepatic recirculation' of vitamin D (or 25-OHD) could form the basis of some of the disorders of vitamin D nutrition which accompany gastrointestinal disease (Arnaud *et al.* 1975).

25-OHD is tightly bound to vitamin D-binding protein and forms the bulk of the vitamin D-derived material circulating in the plasma. It is the easiest of the vitamin D derivatives to assay in the plasma, its concentration closely following the supply of vitamin D so that 25-OHD measurements provide an index of an individual's vitamin D nutritional status. On account of its protein binding, total 25-OHD plasma levels are closely dependent on plasma protein concentrations, reduction in 25-OHD concentration being noted in, for example, the nephrotic syndrome (Haddad & Walgate 1976).

While 25-OHD, unlike native vitamin D, has a biological potency of its own, it is recognised that, for most of the functions of vitamin D, the most active calciferol metabolite is 1,25-dihydroxyvitamin D (1,25$(OH)_2$D).

This compound was until recently thought to be produced exclusively in the kidney (Fraser & Kodicek 1970). Whereas it now seems possible that other human tissues, for example placenta and perhaps sarcoid granulomata, can synthesize 1,25$(OH)_2$D, it remains true that in healthy non-pregnant man 1,25$(OH)_2$D synthesis is the province of the kidney. Since 1,25$(OH)_2$D is much more potent than 25-OHD (Haussler *et al.* 1971; Reynolds *et al.* 1974), it is from the functional point of view the most significant of the vitamin D metabolites and, not surprisingly, its synthesis is subject to much tighter physiological regulation than is the case for its metabolic precursors.

The renal hydroxylation of 25-hydroxyvitamin D is believed to take place in the mitochondria of cells lining the proximal renal tubule, the enzyme concerned being a cytochrome P450-dependent mixed-function oxidase (Ghazarian *et al.* 1974). The principal factors controlling the activity of 25-hydroxyvitamin D 1α-hydroxylase are represented in Fig. 14.1. The reaction is stimulated by PTH (Garabedian *et al.* 1972) and low concentrations of calcium (Hughes *et al.* 1975) or phosphate (Tanaka & DeLuca 1973). Conversely, the reaction rate is slowed in the absence of PTH or when calcium or phosphate concentration are high. An intriguing feature of the activation of 25-OHD to 1,25$(OH)_2$D is its reciprocal relationship with the formation of another dihydroxylated vitamin D metabolite, 24,25$(OH)_2$D. Under circumstances in which 1,25$(OH)_2$D production is suppressed 24,25$(OH)_2$D production increases (DeLuca & Schnoes 1976). This also occurs when exogenous 1,25$(OH)_2$D is administered to an animal (Tanaka & DeLuca 1974) or man (Rasmussen *et al.* 1980).

The sites of action of vitamin D and the relative importance of the various metabolites in exerting each action are still not fully understood. It is generally accepted that the stimulation of intestinal calcium absorption is a primary function of vitamin D and that, under physiological circumstances, $1,25(OH)_2D$ is the metabolite which governs the rate of calcium absorption (DeLuca & Schnoes 1976). It has been considered that $1,25(OH)_2D$ achieves this result primarily by inducing new protein synthesis following hormone-nucleus interaction (Brumbaugh & Haussler 1974). Recent evidence, however, indicates that $1,25(OH)_2D$ can influence the structure and function of intestinal cell membranes through a mechanism which does not depend on new protein synthesis (Rasmussen *et al.* 1981).

The next best understood action of vitamin D is the stimulation of bone resorption, an action which seems paradoxical in a substance known primarily for its antirachitic action. It is possible that the bone resorbing action of vitamin D is involved in its antirachitic action and that the vitamin is here involved in causing a translocation of calcium from mineralised to hitherto unmineralised regions of bone. Another possibility is that the bone resorbing and osteoid mineralising actions of vitamin D are independent of one another and indeed that different metabolites might be primarily responsible for each action. One view, admittedly speculative, is that $24,25(OH)_2D$ may play a direct part in bone mineralisation while $1,25(OH)_2D$ is concerned with the mobilisation of skeletal calcium.

Calcitonin

Calcitonin is a 32 amino acid polypeptide hormone which is elaborated in the C-cells of the thyroid. It has convincingly been shown that calcitonin has an action on bone that is independent of those of PTH and the vitamin D sterols (Gudmunsson *et al.* 1966). Calcitonin acts rapidly on osteoclasts, reducing their bone resorptive function. The role of the hormone in health, is obscure and its actions will be discussed only insofar as they have clinically important actions on bone metabolism and mineral homoeostasis.

PHOSPHATE HOMEOSTASIS

While extracellular fluid calcium ion concentrations are tightly regulated, such tight regulation does not apply to phosphorus or its compounds. From a teleological viewpoint this can be considered as highlighting the contrast between the direct metabolic role played by calcium ion activity (Ca^{2+}) and the indirect role of phosphorus, expressed through a very large number of organophosphorus compounds (Robertson 1976). In practical terms, this difference results in a predominance of mechanisms controlling total body phosphorus content over factors which affect short-term fluctuations in serum phosphorus concentration. The principal control of phosphorus homeostasis resides at the renal level, the system responsible for this control being much less well understood than the corresponding mechanism for calcium.

Plasma phosphorus

Most of the phosphorus in plasma circulates as organically bound phosphorus, largely as a component of phospholipids. Although these compounds are ultimately in equilibrium with circulating inorganic phosphate through red cell inorganic phosphates, this concentration is not routinely measured and need not be taken into account in a discussion of the circulating inorganic phosphates. Conventionally, phosphate concentrations are expressed in terms of the concentration of elemental phosphorus, thus eliminating the need to calculate equivalents between the orthophosphate ion and its protonated forms. In the absence of pH changes, knowledge of total plasma inorganic phosphate ('plasma phosphorus') provides adequate physiological information.

While normal plasma phosphorus levels (as defined above) vary quite widely, certain underlying trends are clearly visible. The most obvious of these is the finding that values are higher in children than in adults. In post-menopausal women however, the value rises once more, a phenomenon not observed in older men (Morgan 1973).

Renal and hormonal control

The fraction of ingested phosphorus absorbed from the diet is relatively constant and the rate of proximal renal tubular phosphate reabsorption is the main factor governing fasting serum phosphorus levels (Robertson 1976). Reabsorption proceeds via a sodium-dependent mechanism, clearly related to the carrier system responsible for the reabsorption of glucose It is known that the rate of proximal tubular phosphate reabsorption is under hormonal control but a major component of the control mechanism, the manner whereby renal phosphorus rejection is tied to dietary phosphorus intake, is not well understood.

Parathyroid hormone plays an important role in renal phosphate handling and hence in phosphorus homoeostasis in general. The signal to which PTH secretion responds, however, is (Ca^{2+}) and not phosphorus concentration, so that this system does not act primarily in defence of plasma phosphorus. Dietary phosphorus intake has a much more pervasive effect than PTH on renal phosphorus handling, phosphate loading causing a decrease in tubular phosphate reabsorption (Goldman & Basset 1958), while phosphorus restriction can rapidly induce a sharp reduction in urinary phosphorus excretion. This change occurs before a significant fall in fasting serum phosphorus or total body phosphorus is noted, suggesting some other means whereby phosphate intake is 'sensed'. The conservation of phosphorus during phosphorus deprivation is not PTH-dependent since it occurs in parathyroidectomised animals (Troehler *et al.* 1976) and overrides the phosphaturic response to exogenous PTH (Steele *et al.* 1976). The renal capacity to conserve phosphorus appears to depend on bodily needs for the element as well as upon supply. Thus, total starvation will not lead to an increase in phosphate reabsorption; this only occurs if enough energy and nitrogen for protein synthesis are provided during phosphorus restriction. Growing children (Corvilain & Abramow 1972), children receiving growth hormone (Henneman *et al.* 1960), and adults with acromegaly (Catteneo *et al.* 1964) all have higher renal thresholds for phosphate than healthy adults.

Humoral factors other than those depending on dietary intake and PTH secretion affect renal phosphate handling, but their physiological role is unclear and probably not important. These factors include growth hormone (see above), insulin (DeFronzo *et al.* 1976), possibly the insulin-like growth factors/somatomedins (Gertner *et al.* 1981), oestrogens and calcitonin. Considering their profound effect on calcium handling and skeletal metabolism, the vitamin D metabolites seem to play a remarkably small effector role in phosphate homoeostasis. Intra- and probably extracellular fluid phosphate plays an important part in regulating vitamin D activation but there is much less evidence for the converse (Puschett *et al.* 1972; Madsen & Olgaard 1977) (i.e. an effect of 25-OHD or $1,25(OH)_2D$ on renal phosphate handling).

MAGNESIUM AND ITS ROLE IN CALCIUM HOMOEOSTASIS

Body magnesium is largely extraskeletal and intracellular.

Hypomagnesaemia occurring due to dietary deficiency is very rare, but it is more common in the malabsorption syndromes, particularly after small bowel resection (Swenson *et al.* 1974). A specific syndrome of impaired intestinal magnesium absorption with otherwise normal bowel function has also been described in children (Lombeck *et al.* 1975). Hypomagnesaemia and generalised magnesium deficiency can also be caused by inappropriate renal losses of magnesium. This phenomenon is seen as a consequence of the excessive loss of magnesium that is observed in diabetes mellitus, particularly when control is poor (Butler 1950).

The normal range for plasma magnesium in British children is 0.74–1.0 mmol/l (Clayton *et al.* 1980). Hypocalcaemia often follows hypomagnesaemia. This has been attributed both to an inadequacy of PTH production (Rude *et al.* 1978) and to a peripheral unresponsiveness to PTH (Muldowney *et al.* 1970) during states of hypomagnesaemia. The association of hypomagnesaemia with hypocalcaemia has been well described in neonates (Cockburn *et al.* 1973) and in contradistinction to giving calcium, the administration of magnesium to infants with hypocalcaemia and hypomagnesaemia is likely to correct both conditions.

ALKALINE PHOSPHATASE IN BONE MINERALISATION

Alkaline phosphatase is a widely distributed enzyme, recognised and assayed by its ability to hydrolyse organic phosphate esters at alkaline pH. The importance of alkaline phosphatase in a consideration of bone mineral metabolism lies in the crucial role played by the enzyme in the mineralisation process. Whereas the production and physiological function of the enzyme are local, enough spills over into the circulation for measurements of its concentration in serum to serve as a useful guide as to the activity of cells which produce alkaline phosphatase.

Heterogeneity of the enzyme

Several varieties of alkaline phosphatase are recognised. While the phosphatases originating in bone, liver, small intestine and placenta can be distinguished by physical means (such as gel electrophoresis or heat inactivation), some of the isoenzymes bear a close structural resemblance to one another. In particular it appears that bone and liver alkaline phosphatases are made in response to the same gene, modified by post-translational events, while distinct genes are responsible for the synthesis of the intestinal and placental enzymes (Seargent & Stinson 1979).

Role of alkaline phosphatase in skeletal mineralisation

Calcifying cartilage is rich in alkaline phosphatase. Over 30% of the alkaline phosphatase in a crude preparation of such cartilage is associated with extracellular membrane vesicles which appear to derive from the plasma membranes of osteoblastic cells (Rasmussen 1983). The precise function of the alkaline phosphatase in these vesicles is unknown but one hypothesis is that its role is the hydrolysis of inorganic pyrophosphate and the initiation of hydroxyapatite crystal formation. Whatever the precise part played by the enzyme in the calcification of cartilage it is clear that a proportion reaches the circulation so that circulating alkaline phosphatase levels reflect the net activity of bone mineralising cells throughout the skeleton. It is important to bear in mind that alkaline phosphatase levels may be elevated when attempts at mineralisation are ineffectual (as in rickets) or unevenly distributed over the skeleton (as in Paget's disease) (Table 14.1). Two conditions in which childhood alkaline phosphate levels are disturbed will be discussed in detail here: first, transient elevations of bony alkaline phosphatase, and secondly the disease known as hypophosphatasia.

Table 14.1 Alkaline phosphatase in childhood.

Elevated
Alkaline phosphatase originating in bone
Mineralisation defects
Nutritional and metabolic rickets
Increased bone turnover
Generalised hyperparathyroidism
uraemic bone disease
primary hyperphosphatasia
Localised fibrous dysplasia (McCune–Albright)
neurofibromatosis
Englemann's disease
fractures
Transient (cause unknown)
Alkaline phosphatase not originating in bone
Obstructive hepatobilary disease (liver alkaline phosphatase)
Small intestinal mucosal disease (intestinal alkaline phosphatase)
Reduced
Growth arrest (e.g. severe illness or hormonal deficiency)
Hypophosphatasia

Reference ranges for alkaline phosphatase activities during childhood are given in Clayton *et al.* (1980) pp. 27, 28.

Transient hyperphosphatasia

For some years it has been recognised that occasionally apparently healthy children may show transient elevations in alkaline phosphatase levels. This has become more noticeable in recent years, with the tendency to include alkaline phosphatase as one of a battery of automated biochemical tests performed on large numbers of people: while some infants with elevated alkaline phosphatase levels may in addition have apparently unrelated pathological findings (Posen *et al.* 1977), others may be completely healthy in every respect. While most authors consider that the phosphatase in these children is entirely of skeletal origin, Wieme (1978) has presented evidence that both the skeletal and hepatic isoenzymes are involved. No cause for these transient elevations has been found. Peak levels up to 10 times higher than the upper limit of normal for age have been recorded. In most cases values return to normal within 4–6 weeks (Posen *et al.* 1977). Where this does not happen an underlying skeletal or hepatobiliary cause should be pursued, although a few cases of permanent, often familial, hyperphosphatasia are recorded in otherwise healthy children (Wilson 1979).

Familial hypophosphataemia

This is a rare, crippling skeletal disorder, probably inherited as an autosomal recessive trait. Affected children suffer from rickets-like skeletal deformities as well as from craniosynostosis and dental abnormalities. Several variants of the disorder are recognised, including a relatively mild syndrome which has its onset in adult life (Whyte *et al.* 1979). More commonly, however, children are affected. In the most severe cases the onset is in infancy and there is a high incidence of cerebral damage due to the craniosynostosis. Mortality may exceed 50% in these cases (Rasmussen 1983).

While hypercalcaemia may be present, particularly in infants, the principal biochemical abnormalities are the hypophosphatasia and an increased urinary excretion of phosphoethanolamine (PEA). In general serum alkaline phosphatase levels are very low, although occasional patients with normal levels have been described (Scriver & Cameron 1969). It is assumed that functional abnormalities of the enzyme are responsible for this minority of cases. Studies on the fractionation of alkaline phosphatase have shown the reduction in circulating enzyme level to be due not only to a deficiency in the skeletal isoenzyme, as might be expected, but also to a variable reduction in levels of the hepatobiliary isoenzyme. Levels of intestinal alkaline phosphatase are normal but enzyme activity is reduced in the polymorphonuclear leukocytes.

Phosphoethanolamine is a substrate for alkaline phosphatase, albeit only as one of many hydrolysable phosphate esters. It is not known why this particular substance tends to accumulate in the plasma (with excess spill-over into the urine) in hypophosphatasia. It is possible that a block in the reaction

$$NH_2CH_2CH_2{-}O{-}P(OH)_2{=}O + H_2O \xrightarrow{\textit{Alkaline phosphatase}} H_3PO_4 + NH_2CH_2CH_2OH$$

is of major importance in the pathophysiology of hypophosphatasia. Another possibility, however, is that PEA accumulates in preference to other phosphate esters not because of any unique metabolic role, but because there are no alternative pathways for its disposal in the absence of alkaline phosphatase. Inorganic pyrophosphate is also excreted in excess in the urine of patients with hypophosphatasia. This observation supports, but does not prove, the hypothesis that pyrophosphate is a 'true' substrate for alkaline phosphatase in the bone mineralisation process.

HYDROXYPROLINE

This amino acid is unusual in mammalian metabolism since it is formed from its precursor (proline) only after the latter has been incorporated into growing polypeptide chains. Moreover collagen, procollagen, and elastin are the only proteins which contain hydroxyproline. Collagen is by far the most abundant of these so that, effectively, endogenous hydroxyproline derives entirely from the breakdown of collagen or from partly-formed fragments of collagen precursors. In addition to this endogenous collagen-derived hydroxyproline, a component of circulating and urinary hydroxyproline in carnivores, including man, is derived from the ingestion of dietary collagen. Since hydroxyproline cannot be directly incorporated into new collagen, all hydroxyproline liberated by the breakdown of collagen must be either excreted or metabolically degraded. The work which relates hydroxyproline metabolism to collagen turnover has focussed on measurements of urinary hydroxyproline excretion, a topic which has been extensively reviewed by Kivirikko (1970).

Based upon studies of the excretion of hydroxyproline from radiolabelled collagen, and on observations in individuals with an inherited enzymatic block which prevents hydroxyproline breakdown, it has been concluded that only about 10% of the hydroxyproline from degraded collagen appears in the urine, while the remainder is further catabolised. Nevertheless it is felt that the proportion of collagen-derived hydroxyproline appearing in the urine does not show much change under various physiological and pathological conditions, so that total urinary hydroxyproline excretion does provide a useful reflection of collagen breakdown. Free hydroxyproline is normally almost completely reabsorbed in the proximal renal tubule and nearly all the hydroxyproline in the urine is in the form of hydroxyproline-containing peptides. More than 75% of these are small peptides of molecular weight 700–900 (Hanson & Wiederanders 1969).

Harris and Krane (Krane *et al.* 1970) have suggested that the small proportion of high molecular weight (non-dialysable) hydroxyproline-containing peptides that do appear in the urine are derived from collagen precursors while the major (dialysable) portion derives from the breakdown of mature collagen. In this way measurement of non-dialysable hydroxyproline-containing peptides gives an indication of collagen formation rate, while total urinary hydroxyproline serves as an indicator of collagen breakdown and bone resorption.

Measurements of total urinary hydroxyproline are usually made while the patients are on a hydroxyproline-restricted diet. According to Sjoerdsma *et al.* (1965) total urinary hydroxyproline falls significantly even on changing from a restricted diet (no gelatin, jelly, ice cream, etc.) to one which is completely collagen-free. In several epidemiological studies, however, useful data have been obtained from patients who are not on a hydroxyproline-restricted diet. Total urinary hydroxyproline excretion is measured after acid hydrolysis of the urine, to convert peptide bound hydroxyproline to the free form. Values are higher in children, especially during rapid growth (Wharton *et al.* 1972; Round 1980) than in healthy adults. Values are also higher in diseases involving increased bone turnover such as rickets, fibrous dysplasia and hyperparathyroidism. In these children both bone formation and resorption are increased. In the hypercalciuria and hypercalcaemia which follow immobilisation, however, bone resorption is increased without any increase in bone formation. These patients also show increased total urinary hydroxyproline excretion (Minaire *et al.* 1974), strongly suggesting that bone resorption rather than bone formation is the source of the excess hydroxyproline appearing in the urine.

REFERENCE RANGES AND AGE-RELATED CHANGES

The demands placed upon the regulatory systems for calcium, phosphorus and magnesium are subject to change during infancy, childhood and adolescence. The sudden passage from an intrauterine to an independent life, and the subsequent evolution of patterns of nutrition and somatic growth result in changes with age in the observed 'normal' values for concentration of many of the substances involved in mineral homeostasis.

Plasma calcium

The placenta is engaged in the active translocation of calcium from maternal to fetal circulation, with the effect that the concentrations of both total calcium and ionised calcium are higher in cord blood than in adults. Levels then fall to values lower than those of adults and return by 72 hours or so to 'childhood' values that are very close to those found in adults. It has, however, been suggested recently that rapidly growing children have slightly higher plasma calcium levels than adults (Krabbe *et al.* 1982). Representative normal ranges for total calcium for neonates, children and adolescents are given in Table 14.2.

Plasma phosphorus

Plasma phosphorus levels tend to be high in babies (Clayton *et al.* 1980) they then fall during early childhood but not down to adult values. In prepubertal and pubertal children levels rise once more, in a manner closely reflecting the graph of height velocity against time (Round *et al.* 1979), falling rapidly to adult values once the period of peak height velocity is passed. The variation in serum phosphorus with age is effected through age-related changes in renal phosphorus handling. Variation is even seen

Table 14.2 Reference ranges for plasma calcium.

	Total calcium range (mmol/l)
Neonatal (1–3 weeks fasting, capillary) [1]	1.9–2.85
Childhood (fasting capillary) [2]	2.13–2.62
Boys 9–16 yrs (venous, non-fasting) [3]	2.28–2.60
Girls 9–16 yrs (venous, non-fasting) [3]	2.27–2.59

[1] Chapter 2
[2] Clayton *et al.* (1980)
[3] Corrected for specific gravity

in disease states such as X-linked hypophosphataemia, indicating that the mechanism responsible for these temporal changes may still operate when other components of the phosphate-regulating system are absent or malfunctioning.

Parathyroid hormone

The difficulties inherent in the radioimmunoassay of parathyroid hormone (PTH), and the consequent lack of comparability of data from one laboratory to another are beyond the scope of this Chapter. Suffice it to say that as a result of these difficulties there is no agreement on what constitutes a 'normal' PTH level at any given age. It has, however, been possible for single laboratories to carry out prospective and longitudinal measurements of immunoreactive PTH concentrations so that we have some idea of how PTH concentrations change first in the neonatal period and then in childhood.

The 'normal' pattern of change in PTH concentration after birth has been the subject of dispute. In one study in full-term infants (David & Anast 1974) serum PTH rose during the first 24 hours of life while serum calcium fell. In another study, however, the postnatal rise in PTH occurred later, coinciding with the rise in plasma calcium from its nadir at 24–36 hours (Tsang *et al.* 1973). In premature infants some authors have failed to detect any significant rise in PTH in the first day or two of life (Anast & Dirksen 1978), while others have found elevated levels in premature babies less than 48 hours old. This controversy will be discussed further on p. 382. This subject will clearly remain disputed territory until biologically valid assays for PTH (Goltzmann *et al.* 1980) are more widely available.

There has been less interest in developing a picture of changes in parathyroid function with age in older children and adolescents but reference ranges have been given by Clayton *et al.* (1980).

Vitamin D metabolites

Under most circumstances vitamin D is rapidly metabolised to 25-OHD. Assay of the vitamin is, moreover, technically difficult and there is relatively little modern information available as to the normal plasma levels of chole- and ergocalciferol. The major circulating metabolite of vitamin D is 25-hydroxyvitamin D (25-OHD) (Holick *et al.* 1972). Levels of 25-OHD reflect the body's nutritional vitamin D status. This depends largely on dietary intake and dermal synthesis of the vitamin, but may be affected by pathological conditions of the gastrointestinal tract and by the ingestion of drugs which speed the hepatic elimination of vitamin D metabolites (e.g. phenobarbitone). Both in Britain (Preece *et al.* 1974) and in the United States (Haddad & Hahn 1973) surveys have shown that circulating 25-OHD3 predominates over 25-OHD2, suggesting that dermal synthesis (of cholecalciferol–vitamin D3) is a more important source of the vitamin than artificially fortified foods (which contain ergocalciferol–vitamin D2). As might be expected if dermal synthesis of vitamin D were indeed an important source, levels of 25-OHD vary with the seasons of the year, being highest in late summer and lowest in early spring (McLaughlin *et al.* 1974; Stamp & Round 1974).

Table 14.3 gives published normal values for 25-OHD concentration for various countries and population subgroups. While geography, race, dietary habit, and time of year may all influence 25-OHD levels there has been no suggestion that the normal range for concentrations of this metabolite are age-related beyond the neonatal period. Levels of 25-OHD in cord blood are lower than, and correlated with, maternal levels in both the term and premature infant. Levels in newborn infants do not appear to change rapidly with age, continuing at first to reflect the mothers'

Table 14.3 25,hydroxyvitamin D (25-OHD) concentration in normal populations from USA or UK.

Population	25-OHD (ng/ml) range
St Louis USA, general	12–27
St Louis USA, newborn	11–27
Bronx (NY) USA, newborn	15–31
England (summer)	10–43
England (winter)	7–25

Compiled from: Haddad and Chyu (1971); Hillman and Haddad (1975); Rosen *et al.* (1974); Stamp and Round (1974).

Table 14.4 1,25-dihydroxyvitamin D ($1,25(OH)_2D$) concentrations in normal children

Population	Age	$1,25(OH)_2D$ (pg/ml ± 1 S.D.)
Ohio [1]	Birth	19 ± 16
Connecticut [2]	Birth	58 ± 36
Denmark [3]	3 m–2 yr	136 ± 55
New York/Wisconsin [4]	1–2 yr	35 ± 6

[1] Steichen *et al.* (1980)
[2] Gertner *et al.* (1980)
[3] Lund *et al.* (1980)
[4] Chesney *et al.* (1980)

vitamin D nutritional status. Low levels have been associated with prematurity but the relationship is unlikely to be causal (Rosen *et al.* 1974).

While 25-OHD levels reflect nutritional status $1,25(OH)_2D$ levels depend on the forces regulating the production of an active hormone. The serum concentrations of this compound, which is largely responsible for controlling intestinal calcium absorption, are, therefore, tightly regulated according to need. As might be expected $1,25(OH)_2D$ levels vary with age, just as the need to absorb dietary calcium varies. Cord blood $1,25(OH)_2D$ levels are rather higher than non-pregnant adult values and are correlated with maternal levels (Gertner *et al.* 1980). $1,25(OH)_2D$ correlations are said to rise slightly in the first few days of life (Steichen *et al.* 1980) and then to rise to very high levels during the remainder of the first 1–2 years (Lund *et al.* 1980). The high values in infancy observed in the Danish review were not confirmed in an American study (Chesney *et al.* 1980). Both studies agree, however, that values in late childhood are higher than in adults and that they parallel growth rates, reaching a peak in adolescence. $1,25(OH)_2D$ concentrations from normal children are summarised in Table 14.4.

Alkaline phosphatase

Serum levels of alkaline phosphatase, like those of most enzymes, are usually expressed in terms of activity rather than true concentration. Since a variety of assay methods is in current use, comparisons between one series and another are sometimes difficult. Even when expressed in international units the reference range can differ from one method to another, according to the reaction conditions applied. Efforts to standardise the measurement and expression of alkaline phosphatase activity (International Union 1974) have found wide but not universal acceptance.

Posen and Doherty (1981), to facilitate conversion from one system of measurement to another, have expressed various authors' published enzyme levels for children in terms of those same authors' upper reference limit for adults. Expressed in this way the mean alkaline phosphatase level for neonates was slightly higher than the adult upper reference limit in 13 of 17 published series.

Mean serum alkaline phosphatase levels rise during the first year of life and then remain fairly steady until puberty. Mean reported values for pre-pubertal children range from $1\frac{1}{2}$–$2\frac{1}{2}$ times the upper reference limit for adults. When data for both sexes are combined there is little change in alkaline phosphatase during adolescence. When the results are grouped by sex and growth velocity, however, a marked temporal assocation between a rise in alkaline phosphatase and the pubertal growth spurt becomes apparent (Round *et al.* 1979). In childhood, alkaline phosphatase activities may fall outside the reference ranges in a variety of conditions which are summarised in Table 14.1.

Urinary total hydroxyproline

Collagen turnover, and hence urinary hydroxyproline excretion, is increased in infancy and childhood by comparison with adult values. Part of the increase in infancy is due to the large amount of the free amino acid excreted by the very young child (Morrow *et al.* 1967). This increase is due

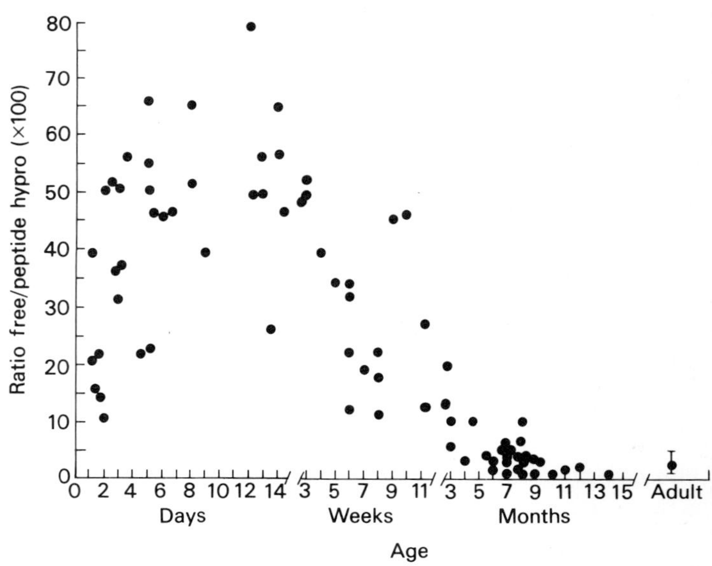

Fig. 14.2 The proportion of excreted hydroxyproline which is in the form of free amino acid falls sharply during the first few weeks of life. From Morrow *et al.* (1967).

to alterations in the catabolism of free hydroxyproline rather than to any change in renal function. Fig. 14.2 shows the fall in the ratio of free to peptide hydroxyproline as measured in the urines of a group of term infants.

When measuring the total urinary hydroxyproline excretion of children beyond infancy, and of adults, a considerable reduction in the scatter of normal values is obtained when the values are expressed in terms of body surface area. This correction should be made wherever possible.

Where measurements of body surface area are not available the data in Fig. 14.3 can be used. The figure shows urinary hydroxyproline excretion data from a large group of children of various ages. The subjects were on a low gelatin diet during the study (Kivirikko & Laitinen 1965).

Measurements made on children on an unrestricted diet are less specific for endogenous collagen breakdown. Nevertheless such data have proved valuable in field studies conducted for epidemiological and nutritional purposes. Such surveys have made use of the hydroxyproline/creatinine

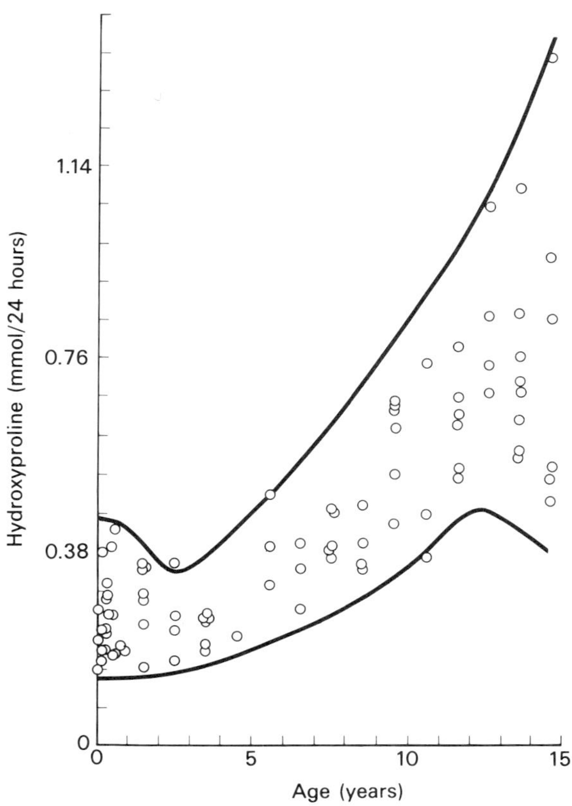

Fig. 14.3. The normal range for 24-hour total hydroxyproline excretion in children on a gelatine-restricted diet. After Kivirikko and Laitinen (1965).

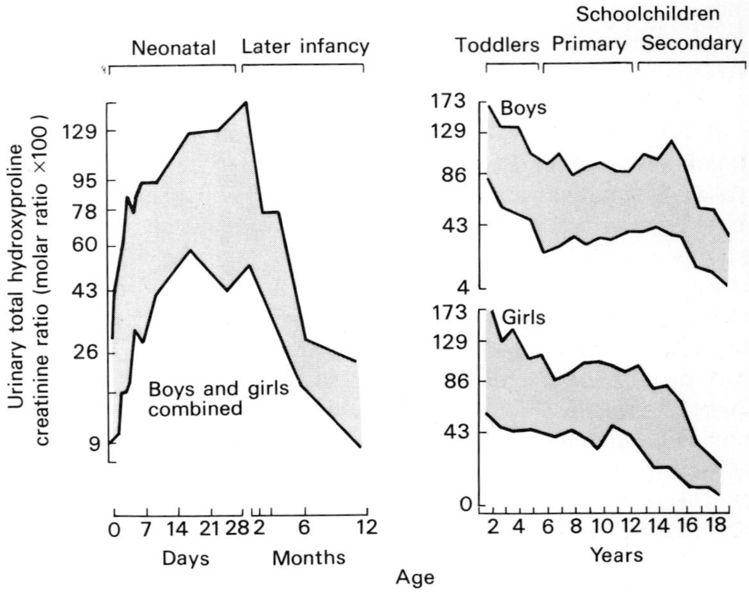

Fig. 14.4. Urinary total hydroxyproline: creatinine ratio measured in spot urine samples obtained from infants and children on an unrestricted diet. From Wharton *et al.* (1972).

ratio in random urine samples (Coward *et al.* 1971). Data on a series of English children studied in this way (Wharton *et al.* 1972) are shown in Fig. 14.3.

DISORDERS OF MINERAL METABOLISM

Neonatal hypocalcaemia

This is the most common disorder of plasma clacium homoeostasis. Estimates of its frequency vary with the way in which the syndrome is defined and with the nature of the patients referred to the clinician (David & Anast 1974; Tsang *et al.* 1973). The most useful definitions of neonatal hypocalcaemia are those that specify a concentration of (ionised) calcium below which symptoms are frequent (Sorrel & Rosen 1975). Thus defined hypocalcaemia means an ionised calcium concentration of less than 0.75 mmol/l, without reference to normal values for age or gestational age and without the necessity to demonstrate the presence of symptoms in affected babies.

Clinical manifestations

Neonatal hypocalcaemia may usefully be divided into 'early' and 'late' forms according to the age at onset. The early form is generally manifest by 24, or at most 48 hours and generally remits within a further day or two. Affected infants are often premature or the offspring of poorly-controlled diabetic mothers.

The overt symptoms of neonatal hypocalcaemia relate to the role of the Ca^{2+} ion in the functioning of excitatory tissue. Jitteriness and partial or complete seizures are frequently observed. The muscular tetany which characterises hypocalcaemia in older children and adults seems to be infrequent in the neonate, possibly because it is often unrecognised or misinterpreted. Cardiac failure has been attributed to neonatal hypocalcaemia (Troughton & Singh 1972) and a low ionised calcium concentration, when present, may well worsen the condition of infants with severe congenital heart disease. Very small babies are liable both to be hypocalcaemic and to have apnoeic attacks, but the frequency with which hypocalcaemia may contribute to such attacks is not known.

Aetiology

Early neonatal hypocalcaemia can be viewed as an accentuation of the drop in ionised calcium concentration that follows the cessation of the flow of calcium across the placenta. The principal defences against such a sudden loss of calcium input are all weak in the neonatal period, and especially so in the sick premature infant. Parathyroid hormone concentrations are low at birth (David & Anast 1974) and take some time to recover. The peripheral response to the hormone may be impaired in the neonate (Linarelli *et al.* 1972). In particular bone resorption may be inhibited by hormones present in high concentration in the small sick neonate, particularly calcitonin and cortisol. Levels of $1,25(OH)_2D$ have not been studied in the context of neonatal hypocalcaemia, largely because of the large volume of plasma needed for the assays. The available evidence does not indicate that levels are depressed in the newborn. However, as the main function of $1,25(OH)_2D$ is to promote intestinal calcium absorption, dietary calcium intake might be at least as relevant as hormonal assays in the assessment of an infant's calcium balance. The average 48-hour-old premature infant may have a weight-adjusted calcium intake only 20% that of a week-old term baby. If the infant is sick his dietary calcium intake may well be zero, leaving him to depend on bone resorption alone.

It has been suggested that marginal vitamin D nutrition is associated with early neonatal hypocalcaemia (Fleischman *et al.* 1980) and this is doubtless true in populations where maternal nutrition is inadequate. Rosen *et al.* (1974), however found that low 25-OHD levels in premature infants were of no prognostic significance as regards the development of hypocalcaemia.

Early neonatal hypocalcaemia tends to remit after a few days, as parathyroid function recovers and as dietary calcium intake rises. Where the condition persists the possibility of phosphate-induced hypocalcaemia or of significant hypoparathyroidism has to be considered.

Apart from the low total and ionised calcium, patients with early neonatal hypocalcaemia show no consistent biochemical abnormalities. Hypomagnesaemia may occur but is frequently absent. Serum phosphorus levels do not usually exceed the already high levels seen in normal babies. As mentioned there is still ignorance of the role of PTH and the vitamin D metabolites in this disorder and there is no consensus as to the values to be expected when these quantities are measured in affected babies. Calcitonin levels, elevated in all neonates (Cannarozzi *et al.* 1976), are not known to be especially high in early hypocalcaemia of the newborn.

The term 'late neonatal hypocalcaemia' describes a condition whose onset affects newborn infants at 5–10 days, in contrast to the early form of the disorder. Symptoms may occur in healthy full-term babies as well as in the premature. Besides hypocalcaemia, most of these infants manifest an important degree of hyperphosphataemia. Hyperphosphataemia reduces bone resorption and diminishes the production of $1{,}25(OH)_2D$ so that one might expect the development of hypocalcaemia in a hyperphosphataemic infant. Thirty years ago a major cause of the condition was the use of poorly modified cow's milk formula in infant feeding (Gardner 1952). Cow's milk has a phosphorus content 8-fold higher than that of breast milk and the ingestion of nutritionally adequate volumes leads to the accumulation of a phosphate load beyond the excretory capacity of the neonatal kidneys. Since the introduction of low phosphate feeds, late hypocalcaemia has become distinctly uncommon. Other causes of late neonatal hypocalcaemia include renal failure, in which phosphate cannot be excreted adequately, and varieties of hypoparathyroidism. Hypoparathyroidism of infancy and childhood is discussed below. Further details on management of hypocalcaemia are in Chapter 2.

Hypoparathyroidism

The clinical significance of hypoparathyroidism lies in the inability of affected patients to maintain a normal plasma calcium. The disorder is due to absence or destruction of the parathyroid glands while a related disturbance, pseudohypoparathyroidism is due to peripheral resistance to parathyroid hormone. Table 14.5 summarises the principal causes of hypoparathyroidism in childhood together with some of the clinical features associated with each disorder.

Clinically, most of these disorders present with the symptoms of hypocalcaemia including tetany, epileptic seizures, and persistent laryngeal stridor. Occasional cases may come to light when serum calcium and phosphorus are measured because the patients show clinical or biochemical

Table 14.5 Clinical features of childhood hypoparathyroidism.

Type	Parathyroid pathology	Age of onset	Associated features
Temporary hypoparathyroidism	Fetal parathyroid suppression	Birth	Maternal hypercalcaemia
DiGeorge Syndrome	Congenital absence of parathyroids	Birth	T-cell dysfunction, thymic aplasia, congenital heart disease
Sporadic congenital hypoparathyroidism	Congenital absence of parathyroids	0–5 years	Occasionally familial (dominant or x-linked)
Candida endocrinopathy	Autoimmune destruction	5–15 years	Mucocutaneous candidiasis, hypoadrenalism
Haemosiderosis	Iron deposition	15–25 years	Transfusion-treated thalassaemia major
Post-surgical	Ischaemic necrosis	10+	Surgery for thyrotoxicosis or thyroid cancer
Magnesium depletion	Reduced PTH secretion	Any	Gut fistulae, steatorrhoea
Pseudohypoparathyroidism	Peripheral resistance to PTH	5–10 years	Moon facies, brachydactyly, short stature

evidence of a condition associated with hypoparathyroidism. In any event, delay and errors in diagnosis are common.

Laboratory investigations reveal hypocalcaemia which may vary from the extreme (total Ca < 1.0 mmol/l) to the very mild, the latter being a feature of some cases of hypoparathyroidism (Parfitt 1972) and many cases of pseudohypoparathyroidism (Nusynowitz *et al.* 1976). Hypophosphataemia due to renal phosphate retention is a constant feature. Serum phosphorus levels, which are higher in normal children than in adults, are further elevated so that very high concentrations may occur. Other routine laboratory tests usually give normal results in hypoparathyroidism, but it should be remembered that hypomagnesaemia can reduce PTH secretion and cause resistance to the effects of PTH. 25-OHD levels, reflecting vitamin D nutrition, are normal but $1,25(OH)_2D$, whose synthesis is, at least in part, PTH-dependent, is deficient in hypoparathyroidism (Rosen *et al.* 1977) and PHP (Drezner *et al.* 1976). Skeletal X-rays are usually normal except for the brachydactyly seen in PHP, but this is part of the abnormal phenotype of the disease and is not a consequence of the metabolic abnormality. Skull X-rays often show basal ganglia calcification which, in mild cases, may be apparent only using computer-assisted tomography.

Patients with pseudohypoparathyroidism often show the phenotype of brachydactyly, short stature, round facies and mental retardation, but these features may all be absent in patients with all the biochemical features of the disorder. The hallmark of the condition is resistance to the renal effects of PTH as evidenced by high immunoreactive PTH levels despite functional hypoparathyroidism and absence of the phosphaturic response to exogenous PTH. The finding that hypocalcaemia may be milder than in hypoparathyroidism and the fact that bony changes suggestive of hypoparathyroidism are not uncommon (Nusynowitz *et al.* 1976) suggest that the resistance to PTH is often incomplete, particularly at the skeletal level. As first demonstrated by Chase *et al.* (1969) the primary cause of pseudohypoparathyroidism appears to be a failure of target cells to release cyclic AMP when they come into contact with PTH. Some patients have recently been shown to lack a regulatory coupling protein involved in cyclic AMP generation, a deficiency not confined to cells sensitive to PTH (Levine *et al.* 1980). This defect was not, however, found in all patients with pseudohypoparathyroidism which, in any case, appears to be a heterogeneous group of diseases rather than a single entity.

The diagnosis of hypoparathyroidism and pseudohypoparathyroidism is generally not difficult since there are few other causes of hypocalcaemia and hyperphosphataemia. Renal failure should be excluded by measurement of serum creatinine and urea. In vitamin D deficiency rickets serum phosphorus is usually low, but occasionally high values are seen in patients assumed to be functionally resistant to PTH. These patients can be distinguished from those with hypoparathyroidism because of their high plasma alkaline phosphatase, low 25-OHD levels, high PTH levels, and radiological rickets. Measurement of immunoreactive PTH is usually too insensitive to distinguish between 'normal' and hypoparathyroid levels. Since, however, almost all patients in which hypoparathyroidism is suspected will be hypocalcaemic the lack of correspondingly elevated PTH levels can be taken as supporting evidence for the diagnosis. Hypo and pseudohypoparathyroidism should be considered in individuals with biochemical hypoparathyroidism unless other features of the clinical state (e.g. history of neck surgery) make it quite clear that the primary lesion is a lack of parathyroid hormone. To distinguish between hypoparathyroidism and pseudohypoparathyroidism the test of Ellsworth and Howard (1934) examines the phosphaturic response to exogenous PTH (200 u given over 15 minutes). Patients with hypoparathyroidism are expected to show a greater than four-fold rise in urinary phosphorus to creatinine ratio while in pseudohypoparathyroidism the ratio rises less than two-fold. Interpretation of the Ellsworth–Howard test suffers from the wide variation in normal phosphaturic response to PTH. Measurement of urinary cyclic AMP/creatinine ratio (Chase *et al.* 1969) after a similar dose of PTH is much more reliable. Patients with hypoparathyroidism should show a 15–20-fold increase in this ratio after PTH, while those with pseudohypoparathyroidism increase the urinary cyclic AMP/creatinine ratio by more than two-fold. Rare variants such as pseudohypoparathyroidism type II (normal cyclic AMP but absent phosphaturic response to PTH) and pseudopseudohypoparathyroidism (morphological phenotype of pseudohypoparathyroidism without any of the biochemical features) will not be further discussed here.

Since human parathyroid hormone is not available for therapy and the bovine material is antigenic, specific hormone replacement therapy cannot be used in hypoparathyroidism. The standard treatment is to give vitamin D or one of its metabolites together with oral calcium supplements. When native vitamin D is used, doses much larger than nutritional maintenance requirements have to be used to sustain normocalcaemia. In patients treated with $1,25(OH)_2D$, however, 'normal' physiological replacement doses suffice. This suggests that lack of the PTH-mediated generation of $1,25(OH)_2D$ may play a contributory role in the development of hypocalcaemia in hypoparathyroidism.

Rickets

Rickets is a disorder in which there is a failure of normal mineralisation of the osteoid of the epiphyseal growth plate. Thus, by definition, rickets can only affect those young enough to have non-fused epiphyses, i.e. growing children. Rickets is distinct from, although frequently associated with, osteomalacia, the latter being a disorder of bone matrix that can occur at any age.

While rickets can have many causes, each of which is clinically distinct, it is useful to review some of the clinical and biochemical features shared by all forms of rickets. Children with rickets tend to show poor growth which is often disproportionate because growth of the lower limbs is more badly affected than that of the trunk. Some degree of deformity of the lower (weight-bearing) limbs is common, producing bow-leg or knock-knee

Table 14.6 Biochemical changes in rickets.

Type of rickets	Ca	P	Alkaline phos-phatase	HCO_3	25-OHD	$1,25(OH)_2D$
Nutritional (D-deficiency)	↓	↓ or N	↑	N	↓	↓↑ or N
Pseudodeficiency type I	↓	↓ or N	↑	N	N	↓
Pseudodeficiency type II	↓	↓ or N	↑	N	N	↑
Familial hypophosphataemic (FHR)	N	↓	↑	N	N	N or ↓
Fanconi's syndrome (cystinosis, etc.)	N or ↓	↓	↑	↓	N	N or ↓
Uraemia	N, ↑ or ↓	↑	↑	↓	N or ↓	↓
Renal tubular acidosis	N	↓	↑	↓	N	?

(Normal N, increased ↑ and decreased ↓).

deformities. The epiphyseal growth plates are expanded with excess unmineralised osteoid, giving rise to the characteristic swellings of the peri-articular regions and costochondral junctions. The bones are often painful and tender on pressure. The most important causes of rickets are their biochemical characteristics, summarised in Table 14.6, which also mentions some of the distinguishing features of each disorder.

Nutritional (vitamin D deficiency) rickets

Historically, vitamin D deficiency was the main cause of rickets all over the world. Fortunately such deficiency is now distinctly uncommon in the developed world, although vitamin D deficiency remains a major public health problem in many less developed countries. Cases of rickets seen in Europe, North America and Japan are usually due either to unusual factors predisposing to vitamin D deficiency or to abnormalities of phosphorus and/or vitamin D metabolism. Thus, outside certain social groups whose vitamin D nutrition is marginal (Rudolph *et al.* 1980) vitamin D deficiency is usually seen in association with malabsorption (Thompson *et al.* 1966), anticonvulsant therapy (Richens & Rowe 1971), or extreme prematurity (Hoff *et al.* 1979).

The clinical features specific to vitamin D deficiency rickets reflect these various associated conditions. In addition, patients are usually hypocalcaemic (see Table 14.7) and may suffer tetany or convulsions. Myopathy commonly accompanies vitamin D deficiency and may be the presenting complaint, regression in walking skills being noted in toddlers.

While a pattern of hypocalcaemia and hypophosphataemia is most typical of nutritional rickets, deviations from this pattern are not uncommon and have been used as the basis for a classification of nutritional rickets in stages (Fraser *et al.* 1967). By this scheme normal or high plasma phosphorus levels typify early disease. Progressive worsening of the vitamin D deficiency leads first to secondary hypoparathyroidism which maintains normocalcaemia while phosphorus levels fall, and finally to a state of bone resistance to PTH in which both calcium and phosphorus levels are low.

Measurements of 25-OHD levels in vitamin D deficiency rickets give the expected low values since the metabolite's concentration is known to reflect nutritional vitamin D status (see above). $1,25(OH)_2D$ levels may be low, and that is certainly the case in animals experimentally deprived of vitamin D. In man, however, there are now reports of cases of nutritional rickets in which $1,25(OH)_2D$ levels were normal (Eastwood *et al.* 1979; Garabedian *et al.* 1972). This carries, of course, the important implication that $1,25(OH)_2D$ is not the only vitamin D metabolite responsible for skeletal mineralisation and normal calcium homoeostasis. Plasma $24,25(OH)_2D$ levels usually tend to reflect the 25-OHD concentration so that low levels of the metabolite are found in vitamin D deficiency states.

Defects of vitamin D metabolism

Rickets due to primary disorders of vitamin D metabolism, or to abnormalities in the peripheral response to the active metabolites of vitamin D is very rare. The best described condition is the so-called pseudodeficiency rickets of Prader *et al.* (1961), and Balsan and Garabedian (1972) in which it has been considered that a simple, inherited, metabolic block to the 1α-hydroxylation of 25-OHD is responsible for all the findings. Low $1,25(OH)_2D$ levels have been found in this condition, but the precise pathogenesis remains uncertain. Clinically the disease is suspected when a patient with all the clinical and biochemical features of nutritional rickets is found not to respond to conventional replacement doses of vitamin D or to have normal circulating levels of 25-OHD. Fraser *et al.* (1973) have reported that $1,25(OH)_2D$ therapy, in doses close to that compound's natural production rate, can cure the disease.

Some patients with skeletal and biochemical features similar to those described above have been found to have high levels of $1,25(OH)_2D$ before treatment and to be resistant to treatment with physiological doses of that compound. Some of these children have had alopecia, the relationship of this finding to the vitamin D $1,25(OH)_2D$ resistance being obscure. Elegant studies of nuclear binding of $1,25(OH)_2D$ have shown that these patients lack the nuclear receptor to $1,25(OH)_2D$, at least in the skin fibroblast lines examined (Eil *et al.* 1981). The condition, which is also familial, has been classified 'pseudodeficiency rickets type II'.

Hypophosphataemic rickets

Although adult osteomalacia (Dent & Winter 1974) and infantile rickets (Rowe *et al.* 1979) can occur as a result of dietary phosphorus deprivation, use of the term 'hypophosphataemic rickets' is confined to rickets primarily due to excessive renal phosphate loss. The defect in renal tubular phosphate reabsorption may be isolated or combined with other manifestations of proximal tubular failure.

This distinction divides patients with hypophosphataemic rickets into two groups which differ markedly from one another and which will now be considered separately. The salient biochemical features of the two groups are presented in Table 14.6.

Familial hypophosphataemic rickets (FHR). This disease, associated with isolated renal phosphate wasting (although renal glycosuria is sometimes seen in long-standing cases), has a rather confusing nomenclature. The phenotype was originally described by Albright *et al.* (1937) who realised that some cases of rickets were not due to vitamin D deficiency and called the syndrome 'vitamin D resistant rickets'. Since then other types of non-nutritional (i.e. vitamin D resistant) forms of rickets have been identified and the term is too imprecise to justify its continued use. While Albright thought that the phosphate wasting was due to secondary hypoparathyroidism, Dent (1970) considered the renal tubular lesion to be primary and classified the syndrome as 'type I renal tubular rickets'. His view of the aetiology has prevailed but the classification into types is not widely used. Rickets due to isolated renal phosphate loss is nearly always congenital and usually familial. Although autosomal recessive and autosomal dominant cases are described, the most common mode of transmission is an X-linked dominant. The term 'familial hypophosphataemic rickets' (FHR) adequately describes all the congenital cases who resemble each other closely even though the different patterns of transmission indicate that a variety of genetic lesions can give rise to the phenotype.

Affected children generally present with rickets and deformity at 1–3 years of age. Myopathy and the consequences of hypocalcaemia are not seen in FHR. Severe growth failure with disproportionate shortening of the lower limbs usually accompanies the rickets. In milder cases, short stature may be the only feature while some heterozygous females with the trait may show only biochemical abnormalities without physical stigmata.

While a defect in renal phosphate reabsorption is now accepted as the primary lesion in FHR, other aspects of the pathophysiology remain obscure. Levels of 25-OHD (Haddad & Chyu 1973) and $1{,}25(OH)_2D$ are generally reported to be normal in untreated cases. Since hypophosphataemia is a potent stimulus to $1{,}25(OH)_2D$ formation, however, the normal levels in FHR can be considered inappropriate and there is evidence that the phosphopenic stimulus to $1{,}25(OH)_2D$ generation is inoperative in FHR (Insogna *et al.* 1983). Parathyroid hormone levels have been reported as normal or slightly elevated and we have found nephrogenous cyclic AMP excretion (reflecting renal responsiveness to endogenous PTH) to be normal or slightly elevated. Studies on the phosphataemic responsiveness to PTH in FHR have variously shown no response (Glorieux & Scriver 1972), a normal response, and an exaggerated response (Short *et al.* 1976). Careful consideration of all the experimental and clinical evidence suggests that PTH and the renal response to that hormone are not very important in the pathogenesis of FHR.

There is evidence that intestinal mucosa also shows defective phosphate transport in FHR. The fact that the bony lesions are not completely cured with supplements of phosphate and $1{,}25(OH)_2D$ (Glorieux *et al.* 1980) and that the degree of bony involvement is not closely related to the degree of hypophosphataemia, have led to the suggestion that an intrinsic defect in skeletal metabolism, possibly related to phosphate transport in bone cells, may exist in the syndrome. The existence of an animal model for FHR (the *Hyp*/Y mouse, in which the phosphate-losing trait is also transmitted on the X chromosome) has permitted an experimental approach to the nature of the renal lesion. In studies on phosphate uptake by renal cortical slices and by vesicles prepared from tubular brush border membranes, the lesions have been localised to the brush border membrane (Tenenhouse *et al.* 1978) but many questions regarding the biochemical lesion remain unanswered.

Acquired isolated phosphaturia leading to osteomalacia and rickets was first described in adults (Dent & Stamp 1971). It is now recognised, however, that children can be affected by this rare syndrome which, in the paediatric age group, is known as 'oncogenous (or "tumour") rickets'. Affected children have a fibrous overgrowth, usually of bone, which seems to secrete a phosphaturic agent. The syndrome has been described in association with bone fibromata, the linear sebaceous naevus syndrome, neurofibromatosis and polyostotic fibrous dysplasia of bone. Nothing is known about the nature of the abnormal phosphaturic substance secreted in these conditions except that it is unlikely to be PTH.

The Fanconi syndrome. A classification of the causes of this syndrome of proximal renal tubular dysfunction is beyond the scope of this chapter. Although any severely affected child with proximal renal tubular failure may, in theory, develop rickets in fact, nearly all cases of the Fanconi syndrome with rickets are suffering from cystinosis. Children with this lysosomal storage disease show progressive renal tubular dysfunction, eventually terminating fatally in uraemia. The biochemical findings seen in a pre-uraemic case are summarised in Table 14.6. Patients with cystinosis usually develop rickets in the second year of life, by which time they are markedly small for age and suffering the effects of polyuria and acidosis. The rickets is due not only to hypophosphataemia but also to acidosis and probably to a reduced capacity to synthesise $1,25(OH)_2D$ in the cystine-damaged cells. There is little direct information on $1,25(OH)_2D$ formation in cystinosis, but it is known that $1,25(OH)_2D$ production is impaired in experimental Fanconi syndrome of animals (Brewer *et al.* 1977). Such a reduction in the capacity to activate vitamin D might explain the hypocalcaemia seen sometimes in cystinosis, but never in FHR.

Renal tubular acidosis

Renal tubular acidosis (RTA) develops when the kidney fails to excrete acid at the rate at which it is being generated systemically and when such failure is due to tubular (as distinct from renal glomerular) pathology. Proximal RTA, as part of a pattern of general proximal tubular disease, has been discussed above. When RTA is due to distal tubular dysfunction it is frequently an isolated phenomenon (i.e. other aspects of renal function are intact). Distal (or type I) RTA is described here because it frequently causes hypercalciuria and nephrocalcinosis and may be associated with rickets.

While type I RTA may be secondary to many factors affecting the distal nephron, cases in infants and young children are usually primary, often as part of a familial trait (Morris *et al.* 1972). Affected children usually present with failure to gain weight, or short stature, before the age of 5 years. Vomiting, polyuria and polydipsia, the passage of renal stones, and rickets may all be presenting features. There are no physical abnormalities apart from small size and, occasionally, rickets. Biochemical testing shows a low blood pH and bicarbonate content. Serum potassium is also often reduced. Renal phosphate wasting usually accompanies systemic acidosis (Lemann *et al.* 1966) and serum phosphate levels are, therefore, reduced. Serum calcium is generally low–normal while urinary calcium excretion is very high. The high urine calcium is found in the fasting as well as the fed state. Intestinal calcium absorption is usually impaired in type I RTA. There is not, as yet, a clear consensus as to whether the hypercalciuria of type I RTA is a response to systemic acidosis in general or whether it is a specific consequence of the distal tubular lesion. Although systemic acidosis can impair the conversion of 25-OHD to $1,25(OH)_2D$ it is not known whether low levels of $1,25(OH)_2D$ are a feature of type I RTA, either with or without rickets.

The diagnosis of type I RTA depends on the finding of systemic acidosis with a urine pH that exceeds 6.5 despite the acidosis. Urinary excretion of bicarbonate is constant even in the face of severe acidaemia. The proportion of filtered bicarbonate that is excreted, however, seldom exceeds 10%, higher percentages being a feature of proximal renal tubular acidosis (McSherry 1981). The hypercalciuria, nephrocalcinosis, rickets (or osteomalacia), and growth failure of type I RTA are all reversible with the administration of alkali without the addition of vitamin D metabolites (McSherry & Morris 1978).

Hypercalcaemia (see also Chapter 15)

Persistent hypercalcaemia in infancy and childhood can lead to gastrointestinal and renal dysfunction, failure to thrive and, possibly, to mental retardation. Four important causes of infantile and childhood hypercalcaemia will be discussed. These topics include: the infantile hypercalcaemia (Williams') syndrome; familial hypocalciuric hypercalcaemia; primary hyperparathyroidism; hypercalcaemia of immobilisation.

Infantile hypercalcaemia (Williams') syndrome

The existence of the infantile hypercalcaemia, peculiar facies, aortic stenosis syndrome was first defined by Garcia *et al.* in 1964. These authors recognised that a large overlap existed between two syndromes described earlier: Lightwood's syndrome (Lightwood & Stapleton 1953) of infantile hypercalcaemia and Williams' syndrome (Williams *et al.* 1961) of supravalvular aortic stenosis and peculiar facies. Since no constant specific physical or biochemical markers exist for the syndrome, its boundaries remain rather vague, nonetheless typical cases resemble each other to an impressive degree and there clearly *is* a syndrome. The disorder is usually referred to as idiopathic infantile hypercalcaemia in Great Britain and by the eponym, Williams' syndrome, in the United States.

The syndrome is usually sporadic rather than familial, and presents in the first 3 years of life. Patients may come to medical attention because of failure to thrive, symptoms of hypercalcaemia (vomiting, constipation, thirst), developmental delay, a systolic murmur, or peculiar facies. Affected children often have a low birthweight. A high vitamin D intake in pregnancy or early infancy has been suggested as predisposing to the condition but such a history is not usually obtained.

Physical examination reveals a small child with the characteristic facies (epicanthic folds, flattening of the nasal bridge, overhanging lower lip and undeveloped mandible). A murmur of supravalvular aortic stenosis or peripheral pulmonary stenosis is heard. Developmental milestones and intellectual development may be delayed. Biochemical tests are often all normal but hypercalcaemia may be a feature, particularly in more severe cases. The hypercalcaemia is usually transient and it is, of course, impossible to tell whether normocalcaemic cases had previously had elevated serum calcium levels. Cases in which the hypercalcaemia is severe may have impaired renal function with elevated serum creatinine and urea and hyperphosphataemia. Other children, however, have been noted to have low phosphate levels in the hypercalcaemic phase (Forfar *et al.* 1956).

The hypercalcaemia is readily reversed by the administration of corticosteroids (Morgan *et al.* 1956) and this, and other evidence has led to the view that IHS is caused by a 'hypersensitivity' to vitamin D. This view certainly gains support from the fact that the syndrome was much more common in Great Britain during the 1950s, when the fortification of infant foods led to substantial vitamin D intakes, and then incidence fell when the amount of vitamin D supplied to infants was decreased. Since very few cases have been studied during the hypercalcalemic phase in recent years the role of possible abnormalities in vitamin D metabolism in pathogenesis remains to be defined. There do not appear to be any abnormalities of the vitamin D-binding proteins (Daiger *et al.* 1978). There is a suggestion that the conversion of vitamin D to 25-OHD is enhanced even in normocalcaemic individuals with the IHS phenotype (Taylor *et al.* 1982). On the other hand in a single patient studied whilst he was hypercalcaemic, Gertner (1982) found normal 25-OHD levels, while $1,25(OH)_2D$ was suppressed, suggesting that another hypercalcaemic factor was being elaborated, or that there was true end-organ hypersensitivity to the active hormone. There is no good information on parathyroid function during the hypercalcaemic phase of infantile hypercalcaemic syndrome.

Familial hypocalciuric hypercalcaemia (FHH) (also called familial benign hypercalcaemia)

With the increasing frequency of routine serum calcium measurements it has become evident that a few apparently healthy persons have elevated serum calcium levels and that this trait for hypercalcaemia is often familial. Originally called 'familial benign hypercalcaemia' by the Harrisons and their colleagues (Foley *et al.* 1972) it has since become clear that affected persons may suffer some of the physical and mental consequences of hypercalcaemia. The disease is clinically and genetically related to the severe form of infantile hyperparathyroidism, but enough differences exist for us to be able to classify FHH as a separate syndrome.

Our knowledge of the clinical and biochemical features of the syndrome is largely due to Marx and his colleagues (Marx *et al.* 1981) on whose work the following account is based. Although occasional children from families with the FHH trait present with severe hypercalcaemia and skeletal demineralisation in association with hyperparathyroidism, hypercalcaemic children with the syndrome are typically asymptomatic or have mild constipation or poor weight gain as their only sign of disease. There is no evidence that affected children or adults suffer from peptic ulcer or renal stones or die prematurely as a result of their hypercalcaemia. Indeed an excess of medical investigations might be their principal complaint.

Children (and adults) with FHH were compared with patients with an equivalent degree of hypercalcaemia, due to primary hyperparathyroidism. The FHH patients had higher creatinine clearances (corrected for age and surface area) than those with hyperparathyroidism. Serum magnesium values were high while they tended to be below normal in hyperparathyroidism. Immunoreactive PTH content, urinary cyclic AMP excretion, and renal phosphate excretion threshold were measured as indices of parathyroid function. In all cases, for the equivalent degree of hypercalcaemia there was less evidence of parathyroid activity in FHH than in primary HPT. Parathyroid function was not, however, fully suppressed in the FHH patients and the suspicion persists that some abnormality of parathyroid function may contribute to the syndrome. The most striking biochemical feature of FHH is the relative hypocalciuria. Urinary calcium excretion may not be low in absolute terms but is distinctly lower than expected for the degree of hypercalcaemia.

The functional pathology of FHH is not yet known. Currently it seems most likely that several tissues, including parathyroid and renal tubule, share a defect in divalent cation (Ca^{2+} and Mg^{2+}) transport. Such a defect might reduce the renal excretion of calcium. The evidence for a relationship to severe infantile hyperparathyroidism is clear, but its significance remains a mystery.

Hyperparathyroidism

Hyperparathyroidism is very rare in childhood. Primary hyperparathyroidism denotes inappropriate hypersecretion of parathyroid hormone while in secondary hyperparathyroidism overactivity of the parathyroids represents an appropriate response to hypocalcaemia. Secondary hyperparathyroidism of childhood is seen in vitamin D deficiency rickets and in uraemia and is discussed under those headings. The development of inappropriate, autonomous hyperparathyroidism following prolonged secondary hyperparathyroidism (so-called 'tertiary hyperparathyroidism') is very rare in children and will not be discussed further.

Primary hyperparathyroidism may arise from hyperplasia of all four of the parathyroid glands or from development of functioning tumours in one or more of the glands. While in adults the relative ratio of hyperplasia to tumour is 1:6 (Broadus 1981), in childhood cases the ratio is much more heavily biased towards hyperplasia. In adults hyperparathyroidism is usually sporadic, but childhood cases are often familial and, in particular, associated with one of three syndromes: severe neonatal hyperparathyroidism (Marx *et al.* 1982) and multiple endocrine neoplasia, types I and II (MEN I and II) (Yamaguchi *et al.* 1980; Sizemore *et al.* 1980). All of these syndromes tend to produce parathyroid hyperplasia, rather than adenomas, thus accounting for the high ratio of hyperplasia to adenoma in childhood.

The clinical effects of excessive secretion of parathyroid hormone are, to an extent, predictable from the earlier discussion of the actions of parathyroid hormone. Patients are hypercalcaemic because of increased bone resorption and increased intestinal calcium absorption, the latter being a consequence of PTH-induced overproduction of $1{,}25(OH)_2D$ from 25-OHD. Serum phosphorus is subnormal because of a reduction in the renal tubular threshold for phosphate reabsorption. Although PTH increases the distal tubular reabsorption of calcium this effect is outweighed by the increased filtered load of calcium so that there is a net increase in urinary calcium excretion. Levels of parathyroid hormone and $1{,}25(OH)_2D$ are generally above normal as is the quantity of cyclic AMP excreted in the urine. There are differences in the relative importance of these pathophysiological changes from one type of hyperparathyroidism to another, accounting for the variable clinical patterns that are observed.

Neonatal primary hyperparathyroidism is a rare, severe disease with a high mortality. Osteopenia and fractures are present at birth and a restricted thoracic capacity has been noted. Patients are markedly hypercalcaemic and most reported cases have died in infancy, early surgical correction offering the only chance for cure. The syndrome may be sporadic or familial. A dominant mode of inheritance is described in familial cases and there is an intriguing relationship with familial hypocalciuric hyperparathyroidism (Marx *et al.* 1982).

Children with either of the familial syndromes of multiple endocrine neoplasia may develop hyperparathyroidism. In these cases the disease is much later in onset and much more benign than in the severe neonatal form described above. Up to 88% of patients with MEN I (adenomas of the pituitary and pancreatic islets, with hyperparathyroidism) may develop hyperparathyroidism. Hyperparathyroidism is the most common endocrine dysfunction in MEN I and often has its onset in childhood. In contrast hyperparathyroidism is third in frequency of the endocrine gland disorders in MEN II (after medullary carcinoma of the thyroid and phaeochromocytoma). Hyperparathyroidism in MEN II is frequently asymptomatic and detected only by screening procedures. Even in affected persons an onset during childhood is uncommon.

Clinically the picture of hyperparathyroidism seen in these children is much closer to the pattern familiar from adult cases than it is to the devastating disease seen in the newborn. The diagnosis rests in the finding of hypercalcaemia and (age-adjusted) hypophosphataemia. Parathyroid hormone levels are often elevated but this is not always a reliable measure of parathyroid overactivity. The excretion of nephrogeneous cyclic AMP (Broadus *et al.* 1977) has proved a sensitive indicator of parathyroid function in adults but has not been applied to the differential diagnosis of hypercalcaemia in children.

Hypercalcaemia of immobilisation

Although severe bone loss, with hypercalcaemia and hypercalciuria can occur in either sex at any age, the typical patient is an adolescent or young adult male. This is due, in part, to the higher bone mass and more rapid bone turnover characteristic of young men but it is also true that prolonged, sudden immobilisation results most commonly from a type of severe trauma most commonly seen in active young men. While most common following traumatic section of the cervical or thoracic spinal cord immobilisation, trauma sufficiently severe to give rise to bone loss can arise from severe burns (Berliner *et al.* 1972), the application of plaster cases for orthopaedic correction and fractures of weight-bearing bones (Rosen *et al.* 1978). Significant loss of skeletal calcium also occurs during the weightlessness of space travel (Pace 1977) although this has not yet become a paediatric problem.

Immobilisation may be complicated by symptomatic hypercalcaemia or by urolithiasis due to massive hypercalciuria. The majority of immobilised patients, however, have no symptomatic evidence of altered calcium homoeostasis although biochemical and radiological changes are very common. The urinary calcium/creatinine ratio rises within a week or two of immobilisation, exceeding normal by the 4th week and reaching a maximum at 16 weeks (Bergman *et al.* 1977). Skeletal radiodensity decreases until 6–12 months after immobilisation, bone mineral content has diminished by about 30%, after which time both bone loss and hypercalciuria diminish.

A recent study of calcium metabolism after immobilisation (Stewart *et al.* 1982) has clarified the mechanisms involved. Fasting calcium excretion was much greater than normal even in normocalcaemic subjects. The suggestion that this was due to suppression of parathyroid activity was supported by the low levels of immunoreactive parathyroid hormone and nephrogenous cyclic AMP and by the high renal tubular threshold for phosphorus reabsorption. There was no calciuric response to an increase in dietary calcium, indirect evidence for a reduction in calcium absorption from the intestine. The poor calcium absorption was associated with a sharply reduced concentration of $1,25(OH)_2D$. All these findings typify the phenomenon of resorptive, or bone derived hypercalciuria. It is presumed that mechanical factors determine the high rate of bone resorption and it is of interest that osteoclast numbers rise after immobilisation despite low levels of parathyroid hormone. The usual coupling between bone resorption and formation breaks down after immobilisation with a marked reduction in bone formation rates (Minaire *et al.* 1974).

REFERENCES

ALBRIGHT F., BUTLER A. M. & BLOOMBERG E. (1937) Rickets resistant to vitamin D therapy. *Am. J. Dis. Child.* **54**, 529–47.

ANAST C. S. & DIRKSEN H. (1978) In: Copp D. H. and Talmage R. V. (eds.) *Endocrinology of calcium metabolism*, p. 12. Excerpta Medica, Amsterdam.

ARNAUD S. B., GOLDSMITH R. S., LAMBERT P. W. & GO V. L. W. (1975) 25-hydroxyvitamin D3; evidence of an enterohepatic circulation in man. *Proc. Soc. Exp. Biol. Med.* **149**, 570–2.

BALSAN S. & GARABEDIAN M. (1972) 25-hydroxycholecalciferol: a comparative study in deficiency rickets and different types of resistant rickets. *J. Clin. Invest.* **51**, 749–59.

BERGMANN P., HEILPORN A., SCHOUTENS A., PATERNOT J. & TRICOT A. (1977) Longitudinal study of calcium and bone metabolism in paraplegic patients. *Paraplegia* **15**, 147–59.

BERLINER B. C., SHENKER I. R. & WEINSTOCK M. S. (1972) Hypercalcemia associated with hypertension due to prolonged immobilization: an unusual complication of extensive burns. *Pediatrics* **49**, 92–6.

BHATTACHARYYA M. H. & DE LUCA H. F. (1973) The regulation of rat liver calciferol-25-hydroxylase. *J. Biol. Chem.* **248**, 2969–73.

BREWER E. D., TSAI H. C., SZETO K. S. & MORRIS R. C. (1977) Maleic acid-induced impaired conversion of 25-(OH)D3 to $1,25\text{-}(OH)_2D3$: implications for Fanconi's syndrome. *Kidney Int.* **12**, 244–52.

BROADUS A. E. (1981) Mineral Metabolism. In Felig P., Baxter J. D., Broadus A. E. & Frohmam L. A. (eds.) *Endocrinology and Metabolism*, pp. 1019–43. McGraw-Hill, New York.

BROADUS A. E., MAHAFFEY J. E., BARTTER F. C. & NEER R. M. (1977) Nephrogenous cyclic AMP as a parathyroid function test. *J. Clin. Invest.* **60**, 771–83.

BRUMBAUGH P. F., HAUSSLER D. H., BRESSLER R. & HAUSSLER M. R. (1974) Radio-receptor assay for 1a,-25-dihydroxyvitamin D3. *Science* **183**, 1089–91.

BRUMBAUGH P. F. & HAUSSLER M. R. (1974) 1α,25-dihydroxychoecalciferol receptors in the intestine. II. Temperature dependent transfer of the hormone to chromatin via a specific cytosolic receptor. *J. Biol. Chem.* **249**, 1258–62.

BUTLER A. M. (1950) Parenteral fluid therapy in diabetic coma. *Acta Paed.* **38**, 59–70.

CANNAROZZI D. B., CANALE D. D. & DONABEDIAN R. K. (1976) Hypercalcitoninemia in infancy. *Clin. Chim. Acta* **66**, 387–92.

CATTANEO C., MARTINI P. F. & MODICA A. (1964) The maximum tubular reabsorption of phosphate in acromegaly. *Acta Endocrinol.* **45**, 203–10.

CHASE L. R. & AURBACH G. D. (1967) Parathyroid function and the renal excretion of 3′,5′ adenylic acid. *Proc. Nat. Acad. Sci. (USA)* **58**, 518–25.

CHASE L. R. & AURBACH G. D. (1970) The effect of parathyroid hormone on the concentration of adenosine 3′5′ monophosphate in skeletal tissue *in vitro*. *J. Biol. Chem.* **245**, 1520–6.

CHASE L. R., MELSON G. L. & AURBACH G. D. (1969) Pseudohypoparathyroidism: defective excretion of 3′5′-AMP in response to parathyroid hormone. *J. Clin. Invest.* **48**, 1832–44.

CHESNEY R. W., ROSEN J. F., HAMSTRA A. J. & DE LUCA H. F. (1980) Serum 1,25-dihydroxyvitamin D levels in normal children and in vitamin D disorders. *Am. J. Dis. Child.* **134**, 135–9.

CLAYTON B. E., JENKINS P. & ROUND J. M. (eds.) (1980) *Paediatric Chemical Pathology*, p. 107 Blackwell Scientific Publications, Oxford.

COCKBURN F., BROWN J. K., BELTON N. R. & FORFAR J. O. (1973) Neonatal convulsions associated with primary disturbance of calcium, phosphorus, and magnesium metabolism. *Arch Dis. Child.* **48**, 99–108.

CORVILAIN J. & ABRAMOW M. (1972) Growth and renal control of plasma phosphate. *J. Clin. Endoc. Metab.* **34**, 452–9.

COWARD D. G., SAWYER M. B. & WHITEHEAD R. G. (1971) Microtechniques for the automated analysis of serum total protein and albumin, urinary urea, creatinine, and hydroxyproline for nutrition surveys in developing countries. *Am. J. Clin. Nutr.* **24**, 940–6.

DAIGER S. P., MILLER M., ROMEO G., PARSONS M. & CAVALLI-SFORZA L. L. (1978) Vitamin D binding protein in the Williams' syndrome and idiopathic hypercalcemia. *New Engl. J. Med.* **298**, 687–8.

DAVID L. & ANAST C. S. (1974) Calcium metabolism in newborn infants. The interrelationships of parathyroid function and calcium, magnesium and phosphorus metabolism in normal, 'sick', and hypocalcemic newborns. *J. Clin. Invest.* **54**, 287–96.

DEFRONZO R. A., GOLDBERG M. & AGUS Z. S. (1976) The effects of glucose and insulin on renal electrolyte transport. *J. Clin. Invest.* **58**, 83–90.

DELUCA H. F. & SCHNOES H. K. (1976) Metabolism and mechanism of action of vitamin D. *Ann. Rev. Biochem.* **45**, 631–66.

DENT C. E. (1970) Rickets (and osteomalacia), nutritional and metabolic (1919–69). *Proc. Roy. Soc. Med.* **63**, 401–8.

DENT C. E. & STAMP T. C. B. (1971) Hypophosphataemic osteomalacia presenting in adults. *Quart. J. Med.* **40**, 303–29.

DENT C. E. & WINTER C. (1974) Osteomalacia due to phosphate depletion from excessive aluminium. Hydroxide ingestion. *Br. Med. J.* **1**, 551–2.

DREZNER M. K., NEELON F. A., HAUSSLER M. R., MCPHERSON H. T. & LEBOWITZ H. E. (1976) 1,25-dihydroxycholecalciferol deficiency: the probable cause of hypocalcemia and metabolic bone disease in pseudohypoparathyroidism. *J. Clin. Endocrinol.* **42**, 621–28.

EASTWOOD J. B., DE WARDENER H. G., GRAY R. W. & LEMANN J. L. (1979) Normal plasma 1,25(OH)2-vitamin D concentrations in nutritional osteomalacia. *Lancet* **1**, 1377–8.

EIL C., LIBERMAN U. A., ROSEN J. F. & MARX S. J. (1981) A cellular defect in hereditary vitamin-D-dependent rickets type II: defective nuclear uptake of

1,25-dihydroxyvitamin D in cultured skin fibroblasts. *New Engl. J. Med.* **304**, 1588–91.

ELLSWORTH R. & HOWARD J. E. (1934) Studies on the physiology of the parathyroid gland. Some responses of normal human kidneys and blood to IV parathyroid extract. *Bull. Johns Hopkins Hosp.* **55**, 296–308.

FLEISCHMAN A. R., ROSEN J. F., NATHENSON G. & FINBERG L. (1980) Prevention and treatment of early neonatal hypocalcemia in premature neonates. In: De Luca H. F. and Anst C. S. (eds.) *Pediatric Diseases Related to Calcium*, pp. 345–54. Elsevier, Amsterdam.

FOLEY T. P., HARRISON H. C., ARNAUD C. D. & HARRISON H. E. (1972) Familial benign hypercalcemia. *J. Pediat.* **81**, 1060–7.

FORFAR J. O., BALF C. L., MAXWELL G. M. & TOMPSETT S. L. (1956) Idiopathic hypercalcemia of infancy: clinical and metabolic studies with special reference to the aetiological role of vitamin D. *Lancet* **1**, 191–8.

FRASER D. R., KOOH S. W., KIND H. P., HOLICK M. F., TANAKA Y. & DELUCA H. F. (1973) Pathogenesis of hereditary vitamin-D-dependent rickets: an inborn error of vitamin D metabolism involving defective conversion of 25-hydroxyvitamin D to 1a,25-dihydroxyvitamin D. *New Engl. J. Med.* **289**, 817–22.

FRASER D. R., KOOH S. W. & SCRIVER C. R. (1967) Hyperparathyroidism as the cause *Ped. Res.* **1**, 425–35.

FRASER D. R. & KODICEK E. (1970) Unique biosynthesis by kidney of a biologically active vitamin D metabolite. *Nature* **228**, 764–6.

GARABEDIAN M., HOLICK M. F., DELUCA H. F. & BOYLE I. T. (1972) Control of 25 hydroxycholecalciferol metabolism by the parathyroid glands. *Proc. Nat. Acad. Sci. (USA)* **69**, 1673–76.

GARCIA R. E., FRIEDMAN W. F., KABACK M. M. & ROWE R. D. (1964) Idiopathic hypercalcemia and supravalvular aortic stenosis. *New Engl. J. Med.* **271**, 117–20.

GARDNER L. I. (1952) Tetany and parathyroid hyperplasia in the newborn infant: influence of dietary phosphate load. *Pediatrics* **9**, 534–6.

GERTNER J. M. (1982) Calcium hyperabsorption with reduced 1,25(OH)2D in the infantile hypercalcemia (Williams') syndrome. *Am. Soc. Bone Min. Res.* Abst. S-22.

GERTNER J. M., GLASSMAN M. S., COUSTAN D. R. & GOODMAN D. B. P. (1980) Feto-maternal vitamin D relationships at term. *J. Pediat.* **97**, 637–40.

GERTNER J. M., TAMBORLANE W. V., HINTZ R. L., HORST R. L. & GENEL M. (1981) The effects on mineral metabolism of overnight growth hormone infusion in growth hormone deficiency. *J. Clin. Endocrinol Metabolism* **53**, 818–22.

GHAZARIAN J. G., JEFCOATE C. R., KNUTSON J. C., ORME-JOHNSON W. H. & DELUCA H. F. (1974) Mitochondrial cytochrome P450. A component of chick kidney 25-hydroxycholecalciferol 1-α hydroxylase. *J. Biol. Chem.* **249**, 3026.

GLORIEUX F. H., MARIE P. J., PETTIFOR J. M. & DELVIN E. E. (1980) Bone response to phosphate salts, ergocalciferol, and calcitriol in hypophosphatemic vitamin D resistant rickets. *New Engl. J. Med.* **303**, 1023–31.

GLORIEUX F. H. & SCRIVER C. R. (1972) Loss of a parathyroid hormone-sensitive component of phosphate transport in X-linked hypophosphatemia. *Science* **175**, 997–1000.

GOLDMAN R. & BASSET S. H. (1958) Renal regulation of phosphorus excretion. *J. Clin. Endocrinol.* **18**, 981–90.

GOLTZMANN D., HENDERSON B. & LOVERIDGE N. (1980) Cytochemical bioassay of parathyroid hormone: characteristics of the assay and analysis of circulating hormonal forms. *J. Clin. Invest.* **65**, 1309–17.

GUDMUNSSON T. V., MCINTYRE I. & SOLIMAN H. A. (1966) The isolation of thyrocalcitonin and a study of its effects in the rat. *Proc. Roy. Soc.* (*B*) **164**, 460–77.

HADDAD J. G. & CHYU K. J. (1971) Competitive protein-binding radioassay for 25-hydroxycholecalciferol. *J. Clin. Endocrinol.* **33**, 992–5.

HADDAD J. G., CHYU J. K., HAHN T. J. & STAMP T. C. B. (1973) Serum concentrations of 25-hydroxyvitamin D in sex-linked hypophosphatemic vitamin-D resistant rickets. *J. Lab. Clin. Med.* **81**, 22–7.

HADDAD J. G. & HAHN T. J. (1973) Natural and synthetic sources of circulating 25-hydroxyvitamin D in man. *Nature* **244**, 515–16.

HADDAD J. G. & WALGATE J. (1976) Radioimmunoassay of the binding protein for vitamin D and its metabolites in human serum. Concentrations in normal subjects and in patients with disorders of mineral homeostasis. *J. Clin. Invest.* **58**, 1217–22.

HANSON H. & WIEDERANDERS B. (1969) Fractionierung Hydroxyprolin-haltiger Harnpeptide mittels Sephadex-gelfiltration. *Clin. Chem. Acta* **25**, 255–62.

HAUSSLER M. R., BOYCE D. W., LITTLEDIKE E. T. & RASMUSSEN H. (1971) A rapidly acting metabolite of vitamin D3. *Proc. Nat. Acad. Sci.* (*USA*) **68**, 177–81.

HENNEMAN P. H., FORBES A. P., MOLDAWAR M., DEMPSEY E. F. & CARROL E. F. (1960) Effects of human growth hormone in man. *J. Clin. Invest.* **39**, 1223–38.

HILLMAN L. S. & HADDAD J. G. (1975) Perinatal vitamin D metabolism. II. Serial 25-hydroxyvitamin D concentrations in sera of term and premature infants. *J. Pediat.* **86**, 928–35.

HOFF N., HADDAD J., TEITELBAUM S., MCALISTER W. & HILLMAN L. S. (1979) Serum concentrations of 25-hydroxyvitamin D in rickets of extremely premature infants. *J. Pediat.* **94**, 460–6.

HOLICK M. F., DELUCA H. F. & AVIOLI L. V. (1972) Isolation and identification of 25-hydroxycholecalciferol from human plasma. *Arch Int. Med.* **129**, 56–61.

HOLICK M. F. & CLARK M. B. (1978) The photobiogenesis and metabolism of vitamin D. *Fed. Proc.* **37**, 2567–74.

HUGHES M. R., BRUMBAUGH P. F., HAUSSLER M. R., *et al.* (1975) Regulation of 1-α25 dihydroxyvitamin D3 by calcium and phosphate in the rat. *Science* **190**, 578–80.

HULDSCHINSKY K. (1919) Heilung von Rachitis durch kunstliche Hohensonne. *Dtsch. Med. Wochenschr.* **45**, 712–13.

INSOGNA K. L., BROADUS A. E. & GERTNER J. M. (1983) Impaired phosphorus conservation and 1,25 dihydroxyvitamin D generation during phosphorus deprivation in familial hypophosphatemic rickets. *J. Clin. Invest.* **71**, 1562–9.

INTERNATIONAL UNION OF PURE AND APPLIED CHEMISTRY AND INTERNATIONAL FEDERATION OF CLINICAL CHEMISTRY (1974) Quantities and units in clinical chemistry. Recommendations. *Pure Appl. Chem.* **37**, 519–46.

KIVIRIKKO K. I. (1970) Urinary excretion of hydroxyproline in health and disease. *Int. Rev. Conn. Tiss. Res.* **5**, 93–163.

KIVIRIKKO K. I. & LAITINEN O. (1965) Clinical significance of urinary hydroxyproline excretion in children. *Ann. Paed. Fenn* **11**, 148–53.

KRABBE S., TRANSBOL I. & CHRISIANSEN C. (1982) Bone mineral homeostasis, bone growth, and mineralisation during years of pubertal growth: a unifying concept. *Arch. Dis. Childh.* **57**, 359–63.

KRANE S. M., MUNOZ A. J. & HARRIS E. D. (1970) Urinary polypeptides related to collagen synthesis. *J. Clin. Invest.* **49**, 716–29.

LEMANN J., LITZEN J. R. & LENNON E. J. (1966) The effects of chronic acid loads in normal man: further evidence for the participation of bone mineral in the defence against chronic metabolic acidosis. *J. Clin. Invest.* **45**, 1608–14.

LEVINE M. A., DOWNS R. W., SINGER M., MARX S. J., AURBACH G. D. & SPIEGEL A. M. (1980) Deficient activity of guanine nucleotide regulatory protein in

erythrocytes from patients with pseudohypoparathyroidism. *Biochem. biophys. Res. Comm.* **94**, 1319–24.
LIGHTWOOD R. & STAPLETON T. (1953) Idiopathic hypercalcaemia in infants. *Lancet* **2**, 255–6.
LINARELLI L. G., BOBIK J. & BOBIK C. (1972) Newborn urinary cyclic AMP and developmental responsiveness to parathyroid hormone. *Pediatrics* **50**, 14–23.
LOMBECK I., RITZL F., SCNIPPERING H. G., MICHAEL H., BREMER H. J., FEINENDEGEN L. E. & KOSENOW W. (1975) Primary hypermagnesemia. Absorption studies. Z. *Kinderheilkd.* **118**, 249–58.
LUND B., CLAUSEN N., LUND B., ANDERSEN E. & SORENSEN O. H. (1980) Age-dependent variations in serum 1,25 dihydroxyvitamin D in childhood. *Acta Endocrinol.* **94**, 426–9.
LUND J. & DELUCA H. F. (1966) Biologically active metabolite of vitamin D from bone, liver, and blood serum. *J. Lipid Res.* **7**, 739–44.
MADSEN S. & OLGAARD K. (1977) Has vitamin D a direct renal effect on the tubular reabsorption of phosphate? In Massry S. G., Ritz E., Rapado A. (eds.) *Homeostasis of Phosphate and Other Minerals*, pp. 111–23. Plenum, New York.
MARX S. J., ATTIE M. F., LEVINE M. A., SPEIGEL A. M., DOWNS R. W. & LASKER R. D. (1981) The hypocalciuric or benign variant of familial hypercalcemia: clinical and biochemical features in 15 kindreds. *Medicine* **60**, 397–412.
MARX S. J., ATTIE M. F., SPIEGEL A. M., LEVINE M. A., LASKER R. D. & FOX M. (1982) An association between neonatal severe primary hyperparathyroidism and familial hypocalciuric hypercalcemia in three kindreds. *New Engl. J. Med.* **306**, 257–64.
MAWER E. B., BACKHOUSE J., HOLMAN C. A., LUMB G. A. & STANBURY S. W. (1972) The distribution and storage of vitamin D and its metabolites in human tissues. *Clin. Sci.* **43**, 413–31.
MCLAUGHLIN M., RAGGAT P. R., FAIRNEY A., BROWN A., BROWN D. J., LESTER E. & WILLS M. R. (1974) Seasonal variations in serum 25 hydroxycholecalciferol in healthy people. *Lancet* **1**, 536–8.
MCSHERRY E. (1981) Renal tubular acidosis in childhood. *Kidney Int.* **20**, 799–809.
MCSHERRY E. & MORRIS R. C. (1978) Attainment and maintenance of normal stature with alkali therapy in infants and children with classic renal tubular acidosis. *J. Clin. Invest.* **61**, 509–27.
MELLANBY E. (1919) An experimental investigation on rickets. *Lancet* **1**, 407–12.
MINAIRE P., MEUNIER P., EDOUARD C., BERNARD J., COUPRON P. & BOURRET J. (1974) Quantitative histological data on disuse osteporosis. *Calc. Tiss. Res.* **17**, 57–73.
MORGAN B. (1973) Ageing and osteoporosis, particularly spinal osteoporosis. *Clin. Endoc. Metab.* **2**, 187–201.
MORGAN H. G., MITCHELL A. G., STROWERS J. M. & THOMSON J. (1956) Metabolic studies on two infants with idiopathic hypercalcaemia. *Lancet* **1**, 925–31.
MORRIS R. C., SEBASTIAN A. & MCSHERRY E. (1972) Renal acidosis. *Kidney Int.* **1**, 322–40.
MORROW G., KIVIRIKKO K. I. & PROKOP D. J. (1967) Catabolism and excretion of free hydroxyproline in infancy. *J. Clin Endocrin.* **26**, 1012–4.
MULDOWNEY F. P., MCKENNA T. J., KYLE L. H., FREANY R. & SWAN M. (1970) Parathyroid-like effect of magnesium replenishment in steatorhea. *New Engl. J. Med.* **282**, 61–8.
NIALL H. D., KEUTMANN H. T., SAUER R., HOGAN M., DAWSON B., AURBACH G. D. & POTTS J. T. (1970) The amino-acid sequence of bovine parathyroid hormone I. *Hoppe-Zeyler's Z. Physiol. Chem.* **351**, 1586–8.
NUSYNOWITZ M. L., FRAME B. & KOLB F. O. (1976) The spectrum of the hypoparathyroid states. *Medicine* **55**, 105–19.
PACE N. (1977) Weightlessness: a matter of gravity. *New Engl. J. Med.* **297**, 32–7.

PARFITT A. M. (1972) The spectrum of hypoparathyroidism. *J. Clin. Endocrinol Metabolism* **34**, 152–8.

POSEN S. & DOHERTY E. (1981) The measurement of serum alkaline phosphatase in clinical medicine. *Adv. Clin. Chem.* **22**, 163–245.

POSEN S., LEE C., VINES R., KILHAM H., LATHAM S. & KEEFE J. F. (1977) Transient hyperphosphatasemia of infancy. *Clin. Chem.* **23**, 292–4.

PRADER A., ILLIG R. & HEIRLI E. (1961) Eine besondere Form der primaren vitamin-D-resistenten Rachitis mit Hypocalcamie und autosomal-dominanten Erbgang: die heriditare Pseudo-Mangelrachitis. *Helv. Paediat. Acta* **16**, 452–68.

PREECE M. A., O'RIORDAN J. L. H., LAWSON D. E. M. & KODECEK E. (1974) A competitive protein-binding assay for 25 hydroxycholecalciferol and 25-hydroxyergocalciferol in serum. *Clin. Chim. Acta* **54**, 235–42.

PUSCHETT J. B., MORANZ J. & KURNICK W. S. (1972) Evidence for a direct action of cholecalciferol and 25 hydroxycholecalciferol on the renal transport of phosphate, sodium and calcium. *J. Clin. Invest.* **51**, 373–85.

RASMUSSEN H. (1983) Hypophosphatasia. In Stanbury J. B., Wyngaarden J. B., Fredrickson D. S., Goldstein J. L., Brown M. S. (eds.) *The Metabolic basis of Inherited Disease*, pp. 1497–505. McGraw-Hill, New York.

RASMUSSEN H., BARON R., BROADUS A. E., DE FRONZO R. A., LANG R. & HORST R. L. (1980) 1,25$(OH)_2$D3 is not the only D metabolite involved in the pathogenesis of osteomalacia. *Am. J. Med.* **69**, 360–8.

RASMUSSEN H., FONTAINE O. & MATSUMOTO T. (1981) Liponomic regulation of calcium transport by 1,25$(OH)_2$D. *Ann. N.Y. Acad. Sci.* **372**, 518–29.

REYNOLDS J. J., HOLICK M. F. & DELUCA H. F. (1974) The effects of vitamin D analogs on bone resorption. *Calc. Tissue Res.* **15**, 333–9.

RICHENS A. & ROWE D. J. (1971) Anticonvulsant osteomalacia. *Br. Med. J.* **4**, 684.

ROBERTSON W. G. (1976) Plasma phosphate homeostasis. In Nordin B. E. C. (ed.) *Calcium, Phosphate, and Magnesium Metabolism*, pp. 219–21. Churchill Livingstone, Edinburgh.

ROSEN J. F., FLEISCHMAN A. R., FINBERG L., EISMAN L. & DELUCA H. F. (1977) 1,25-dihydroxycholecalciferol: its use in the long-term management of idiopathic hypoparathyroidism in children. *J. Clin. Endocrinol Metabolism* **45**, 457–68.

ROSEN J. F., ROGINSKY M., NATHENSON G. & FINBERG L. (1974) 25-hydroxyvitamin D: plasma levels in mothers and their premature infants with neonatal hypocalcemia. *Am. J. Dis. Child.* **127**, 220–3.

ROSEN J. F., WOLIN D. A. & FINBERG L. (1978) Immobilization hypercalcemia after single limb fractures in children and adolescents. *Am. J. Dis. Child.* **132**, 560–4.

ROUND J. M. (1980) Changes in plasma urate creatinine alkaline phosphatase and the 24 h excretion of hydroxyprotein during sexual maturation in adolescents. *Ann. Hum. Biol.* **7**, 83–8.

ROUND J. M., BUTCHER S. & STEELE R. (1979) Changes in plasma inorganic phosphorus and alkaline phosphatase activity during the adolescent growth spurt. *Ann. Hum. Biol.* **6**, 129–36.

ROWE J. C., WOOD D. H., ROWE D. W. & RAISZ L. G. (1979) Nutritional hypophosphatemic rickets in a premature infant fed breast milk. *New Engl. J. Med.* **300**, 293–6.

RUDE R. K., OLDHAM S. B., SHARP C. F. & SINGER F. R. (1978) Parathyroid hormone secretion in magnesium deficiency. *J. Clin. Endocrinol Metabolism* **47**, 800–6.

RUDOLPH M., ARULANANTHAM K. & GREENSTEIN R. M. (1980) Unsuspected nutritional rickets. *Pediatrics* **66**, 72–6.

SCRIVER C. R. & CAMERON D. (1969) Pseudohypophosphatasia. *New Engl. J. Med.* **281**, 604–6.

SEARGENT L. E. & STINSON R. A. (1979) Evidence that three structural genes code for human alkaline phosphatases. *Nature* **281**, 152–4.

SHORT E., MORRIS R. C., SEBASTIAN A. & SPENCER M. (1976) Exaggerated phosphaturic response to circulating parathyroid hormone in patients with familial X-linked hypophosphatemic rickets. *J. Clin. Invest.* **58**, 152–63.

SIZEMORE G. W., HEATH H. & CARNEY J. A. (1980) Multiple Endocrine Neoplasia Type 2. *Clinics in Endocrinology and Metabolism* **9**, 299–315.

SJOERDSMA A., UDENFRIEND S., KEISER H. & LEROY E. C. (1965) Hydroxyproline and collagen metabolism. *Ann. Int. Med.* **63**, 672–94.

SORREL M. & ROSEN J. F. (1975) Ionized calcium: serum levels during symptomatic hypocalcemia. *J. Pediat.* **87**, 67–70.

STAMP T. C. B. & ROUND J. M. (1974) Seasonal changes in human plasma levels of 25-hydroxyvitamin D. *Nature* **247**, 563–5.

STARK H., BARNETT H. L. & EDELMANN C. M. (1965) Renal effects of hypercalciuria in immobilized children. *Proc. Soc. Exp. Biol. Med.* **118**, 870–2.

STEELE T. H., UNDERWOOD J. L., STROMBERG B. A. & LARMORE C. A. (1976) Renal resistance to parathyroid hormone during phosphorus deprivation. *J. Clin. Invest.* **58**, 1461–4.

STEICHEN J. J., TSANG R. C., GRATTON T. L., HAMSTRA A. J. & DELUCA H. F. (1980) Vitamin D homeostasis in the perinatal period: 1,25 dihydroxyvitamin D in maternal, cord, and neonatal blood. *New Engl. J. Med.* **302**, 315–9.

STEWART A. F., ADLER M., BYERS C. M., SEGRE G. V. & BROADUS A. E. (1982) Calcium homeostasis in immobilization: an example of resorptive hypercalciuria. *New Engl. J. Med.* **306**, 1136–40.

SWENSON S. A., LEWIS J. W. & SEBBY K. R. (1974) Magnesium metabolism in man with special reference to jejunoileal bypass for obesity. *Am. J. Surg.* **127**, 250–5.

TANAKA Y. & DELUCA H. F. (1973) The control of 25 hydroxyvitamin D metabolism by inorganic phosphorus. *Arch. Biochem. Biophys.* **154**, 566–74.

TANAKA Y. & DELUCA H. F. (1974) Stimulation of 24,25 dihydroxyvitamin D3 production by 1,25 dihydroxyvitamin D3. *Science* **183**, 1198–200.

TAYLOR A. B., STERN P. H. & BELL N. H. (1982) Abnormal regulation of circulating 25-hydroxyvitamin D in the Williams syndrome. *New Engl. J. Med.* **306**, 972–5.

TENENHOUSE H. S., SCRIVER C. R., MCINNES R. R. & GLORIEUX F. H. (1978) Renal handling of phosphate in vivo and in vitro by the X-linked hypophosphatemic male mouse: evidence for a defect in the brush border membrane. *Kidney Int.* **14**, 236–44.

THOMPSON G. R., LEWIS B. & BOOTH C. C. (1966) Absorption of vitamin D-3H in control subjects and patients with intestinal malabsorption. *J. Clin. Invest.* **45**, 94–102.

TROEHLER U., BONJOUR J.-P. & FLEISCH H. (1976) Inorganic phosphate homeostasis; renal adaptation to the dietary intake in intact and thyroparathyroidectomized rats. *J. Clin. Invest.* **57**, 264.

TROUGHTON O. & SINGH S. P. (1972) Heart failure and neonatal hypocalcemia. *Br. Med. J.* **4**, 76–9.

TSANG R. C., CHEN I.-W., FRIEDMAN M. F. & CHEN I. (1973) Neonatal parathyroid function, role of gestational and postnatal age. *J. Pediat.* **83**, 728–38.

WHARTON B. A., GOUGH G., WILLIAMS A., KITTS S. & PENNOCK C. A. (1972) Urinary total hydroxyproline: creatinine ratio. Range of normal and clinical application in British children. *Arch. Dis. Child.* **47**, 74–9.

WHYTE M. P., TEITELBAUM S. L., MURPHY W. A., BERGFELD M. A. & AVIOLI L. V. (1979) Adult hypophosphatasia. *Medicine* **58**, 329–47.

WIEME R. J. (1978) More on transient hyperphosphatasemia in infancy – an insufficiently recognised syndrome. *Clin. Chem.* **24**, 520–2.

Williams J. C. P., Barratt-Boyes B. G. & Lowe J. B. (1961) Supravalvular aortic stenosis. *Circulation* **24**, 1311–18.
Wilson J. W. (1979) Inherited elevation of alkaline phosphatase activity in the absence of disease. *New Engl. J. Med.* **301**, 983–4.
Working Party on the Fortification of Food with Vitamin D (1980) *DHSS Rep. Health Soc. Subj.* **19**, pp. i–xii, 1–66.
Yamaguchi K., Kameya T. & Abe K. (1980) Multiple Endocrine Neoplasia Type 1. *Clin. Endocrinol. Metabol.* **9**, 261–84.

Chapter 15
The Mentally Subnormal Child

JOSEPHINE E. HAMMOND & JAN STERN

Is he all right? This question is probably the first asked by many a mother about her newborn child. She is justified in seeking assurance, as amongst liveborn infants some 2% have major abnormalities, a further 3% have minor abnormalities. The central nervous system is affected in about 15–20% of these cases and frequently these infants will be mentally retarded.

Mental retardation (synonyms: mental handicap, mental deficiency, subnormality) has been defined as arrested or incomplete development of the mind, originating early in life, and which is associated with impaired maturation, learning and social adjustment. The mildly retarded, (IQ 50–70; 'ESN (M)') make up approximately 1.8%, and the severely retarded (IQ < 50; 'ESN (S)') 0.35–0.4% of the population. These figures refer to adult or school-age populations (the 'prevalence'). At birth the incidence of many disorders associated with severe mental subnormality is considerably higher, perhaps up to 1%, but many of these infants die during the first year of life, often before their mental defect has been ascertained.

Aetiology

The diagnosis of mental handicap in a young infant may be very difficult. Some infants are stillborn or die soon after birth and lesions of the nervous system may then be found at autopsy. In survivors, neuropsychiatric deficits usually become apparent only at a later age. Sadly, some infants may appear completely normal at birth, but as they grow older it becomes obvious that they are mentally subnormal and must have been affected from birth. Sometimes the presence of mental handicap can be inferred from associated physical signs, as, for example, in Down's syndrome.

The results of a careful review of the causes of severe mental retardation in patients born in Northern Ireland in the years 1900–79 are summarised in Table 15.1. Findings in other surveys and in our own group of hospitals have been similar (Bernsen 1981; Gustavson *et al.* 1977; Kirman & Bicknell 1975). It should be noted that in only about half of the severely subnormal patients can the aetiology of the mental defect be established unequivocally, if patients with Downs syndrome are excluded, although some claim that this estimate is unduly pessimistic (Mackay 1982). In the mildly retarded or ESN (M) patients this proportion is usually even lower (Hagberg *et al.* 1981). This presents a continuing challenge to all those who deal with mentally retarded children and their families.

Table 15.1 Causes of severe mental handicap in patients born in Northern Ireland (1955–1976) (Based on records of The Special Care Service and abridged from Elwood & Darragh 1981).

Causes	Number of patients
Genetic disorders	
Chromosomal	
Down's syndrome	688
Other chromosomal disorders	5
Autosomal recessive disorders	
Phenylketonuria	15
Mucopolysaccharide disorders	3
Homocystinuria	2
Other recessive disorders	8
Sex-linked recessive disorders	
Lowe's syndrome, muscular dystrophy, etc.	3
Multifactorial conditions	
Spina bifida, hydrocephalus	36
Other disorders	
Cretinism, Rh/ABO incompatibility, etc.	20
Environmental disorders	
Perinatal	
Prematurity	91
Birth injury	33
Perinatal hypoxia	67
Others	21
Antenatal infections	
Rubella, cytomegalavirus, toxoplasma	9
Postnatal infections	
Meningitis, gastroenteritis	11
Encephalitis and related disorders	
Encephalitis, pertussis immunisation, measles	39
Miscellaneous	
Head injury, poisoning, non-accidental injury	12
Total	1066

Approximately 40% of patients could not be classified.

The aetiological pattern of severe mental defect is not static. For example, the contribution to the prevalence of mental handicap by phenylketonuria, congenital hypothyroidism and rubella embryopathy has been or is being eliminated. On the other hand, with improved techniques in cytogenetics and enzymology new disorders are being discovered at an increasing rate, diminishing the number of unclassified patients. Sadly, industrial pollution and iatrogenic disasters have occasionally added to the causes of mental handicap.

Severe mental handicap may result from the effects of a single gene or a well-defined environmental hazard such as meningitis or head injury. More often than not, the aetiology of the mental defect is multifactorial. A typical case history of a patient in a unit for mentally retarded children might record a small-for-dates infant born after inadequate antenatal supervision to an elderly, multiparous mother living in poverty. Perinatal problems which might include jaundice, hypoglycaemia and anoxic episodes are followed by a deprived childhood. It is then often extremely difficult to aportion the blame amongst the many adverse factors operating in this type of situation.

In general, the more severely retarded a child, the earlier he is likely to present. Even then, parental concern is often not expressed until the child fails to achieve developmental milestones, such as walking and talking. The late R. C. Mackeith often stressed that handicapped children need early recognition, comprehensive assessment, immediate treatment and support for the family, thereafter continuing treatment and support for the family, and periodic reassessment. These needs can only be met by extensive screening of all children for developmental delay and, when relevant, for metabolic disorders. Continued alertness is an essential requirement in all those dealing with children, in whatever capacity. Comprehensive assessment will include a full medical, social and psychological evaluation. It will be up to the physician in charge to decide, in consultation with the laboratory, which biochemical investigations are indicated.

This chapter does not aim to provide an exhaustive account of the many inborn errors which may effect the nervous system. Their detailed investigation is, in most instances, beyond the scope of the routine hospital laboratory. An excellent reference volume is that of Stanbury *et al.* (1983).

CLINICAL CHEMISTRY INVESTIGATIONS FOR MENTALLY RETARDED CHILDREN

Some general comments

The success of biochemical investigations in mentally retarded children depends in no small measure on cooperation between laboratory staff and those looking after the patients. The mentally retarded now often live in hostels or foster homes in the care of houseparents or social workers, rather than nurses or doctors. The former feel more at home with behaviour therapy and pastoral care than with metabolic investigations. Knowledge of simple ward procedures like urine or blood collections can no longer be taken for granted, and detailed instructions and explanations given person to person are essential. Investigations arranged by phone or memorandum often go wrong. Normally the laboratory will be some miles from the patients and problems may arise in the prompt transmission of samples because of lack of flexibility in the time tables of hospital transport, often geared to other tasks. The laboratory must stress to all staff involved the importance of adhering to the recommendations of the Howie Report on the prevention of infection, particularly as blood samples from units for the mentally retarded carry an above average risk of being infected with hepatitis B virus.

For all routine investigations physicians caring for mentally handicapped children have, of course, access to the pathology service of the district in which the patients reside. A case can be made for regular screening of patients on long-term medications with anticonvulsants or psychoactive drugs in order to provide early warning of anticonvulsant rickets or liver damage by drugs. A haematological screen may be carried out at the same time. Thyroid function is abnormal in a considerable proportion of adult mentally retarded patients, particularly those with Down's syndrome, but screening (not to be confused with neonatal screening for congenital hypothyroidism) is not nearly so productive of abnormal results in mentally retarded children. Monitoring of thyroid status and renal function is essential in patients on lithium therapy. Abnormal levels of the sex hormones are seen in conditions associated with hypogonadism, such as Klinefelter's syndrome (Ratcliffe *et al.* 1982).

The problem of how to deal with HBsAg (Australia antigen) positive patients poses many problems in units for the mentally retarded, where the frequency of carriers may be much above that in the community. Some clinicians are averse to screening, since a positive diagnosis may result in severe curtailment of the social life and education of a diagnosed carrier, because of fear of infection amongst staff. Some hospitals confine screening to patients with Down's syndrome, who are particularly vulnerable to hepatitis B, to patients with jaundice, and new admissions. In one hospital simple precautions coupled with education of staff brought about a more than four-fold drop in the frequency of carriers in a group of patients with Down's syndrome (Rundle 1979). The problem should decrease as current tests for the detection of HBs antigen are extended to determine the degree of infectivity of antigen-positive patients. It must be remembered that the major risk is to those handling blood samples from mentally retarded patients and that includes laboratory staff.

Mention must be made of the many non-specific biochemical abnormalities which have been reported in mentally retarded children, particularly in those living in institutions (Crome & Stern 1972). Some were the consequences of the frequent, and often chronic, infections to which these patients are liable. Malnutrition was common, often due to inability or unwillingness to eat, or due to inefficient intestinal absorption. Severely retarded children are often unable to chew. To prevent malnutrition the diet offered has to be of the right consistency, as well as adequate in composition. In many cases hormonal and biochemical assays produce a wider scatter of values than in normal children. Inappropriate response to stress and inadequate homoeostasis are common. More often than not such abnormalities are the consequence of the way retarded children live and are not of significance in the aetiology of the mental defect. They are found less and less frequently as conditions in institutions improve and the great majority of retarded children live in the community.

PHYSICAL STIGMATA AND NEUROLOGICAL SIGNS IN THE MENTALLY RETARDED

Mentally retarded children often look 'different', but genuine physical signs must be differentiated from the impression of oddity sometimes conveyed by delayed growth, immaturity, inappropriate clothing and occasionally unflattering haircuts. A number of retarded children have definite somatic abnormalities. These may be obvious from birth in some chromosomal disorders such as Down's syndrome, or form a recognisable pattern of clinical signs as, for example, in the deLange syndrome ('Amsterdam dwarfs') or Apert syndrome (acrocephaly syndactyly). An excellent monograph for the diagnostic use of physical signs has been published by Smith (1976) who also notes useful laboratory findings.

Sometimes symptoms of one system will suggest a more generalised disease also affecting the central nervous system. For example, the heart lesion found following intrauterine exposure to rubella virus will suggest the possibility that the brain too may have been damaged. Other infants have abnormalities such as neural tube defects which will often be associated with, or lead to, mental retardation.

Ataxia

Ataxia must be distinguished from the immature coordination commonly found in retarded children. In ataxia telangiectasia (Hosking 1982) the ataxia is often of late onset, and progressive. It is associated with a defect of the immune mechanism, defective repair of radiation damage to DNA, and a raised blood level of α-fetoprotein.

Recently Blass (1980) has noted that a number of inborn errors of metabolism may be associated with spinocerebellar disorders and in some instances present with ataxia and neuropathy (Table 15.2). It is important to remember that ataxia may be a sign of toxicity of certain drugs such as phenytoin, often prescribed for retarded children.

Abnormal movements (dyskinesia) and abnormal tone (dystonia) are seen in many of the neurometabolic disorders to which reference is made in this chapter (Table 15.3). Of special interest is the infant with hypotonia.

Table 15.2 Some inborn errors which have presented as spinocerebellar disorders (adapted from Blass 1980).

Disorder	Initial tests
Pyruvate dehydrogenase complex defects	Plasma pyruvate, lactate
Pyruvate carboxylase deficiency	Plasma pyruvate, lactate
Bassen–Kornzweig syndrome (abetalipoproteinaemia)	Plasma cholesterol
Hartnup disease	Urine amino acid screen
Intermittent maple syrup urine disease	Urine amino acid and keto acid screen[1]
Lesch–Nyhan syndrome (HGPRT deficiency variant)	Urine uric acid: creatinine ratio
γ-Glutamyl cysteine synthetase deficiency	Screen for haemolytic anaemia
Juvenile metachromatic leucodystrophy	Metachromatic granules in urine
Juvenile G_{M1} gangliosidosis	Lysosomal β-galactosidase[2]
Tay–Sachs disease variant	Lysosomal hexosaminidase-A[2]

1 During attacks
2 Reference laboratory necessary

Table 15.3 Dyskinesia and dystonia in some neurometabolic disorders (adapted from Hagberg *et al.* 1979).

Disorder	Comment
Most lipidoses and leucodystrophies	see Chapter 16
Most mucolipidoses	see Chapter 16
Most organic acidaemias	see Tables 15.15 and 15.18
Malignant phenylketonuria	see p. 427
Homocystinuria	see p. 428
Urea cycle disorders	see Chapter 4
Pyruvate dehydrogenase deficiency	see p. 434
Lesch–Nyhan syndrome	see p. 416
Pseudohypoparathyroidism	see Chapters 2 and 14
Hypoparathyroidism	see Chapter 14
Wilson's disease	see p. 438
Crigler–Najjar syndrome	Congenital familial non-haemolytic jaundice due to glucuronyl transferase deficiency. Some cases respond to barbiturate therapy
Huntington's chorea	May present in childhood. Biochemical tests are not helpful during lifetime of patient
Neuroaxonal dystrophy	Biochemical tests are not of diagnostic value during lifetime of patient

Muscle weakness and hypotonia

Hypotonia and weakness may result from a disorder of the motor unit (the anterior horn cell, peripheral nerves, neuromuscular junction of muscle fibres), but may also be secondary to disorders affecting levels above the motor neuron. In the newborn in particular, hypotonia is often seen following trauma, hypoxia-ischaemia, intracranial infection or haemorrhage, in numerous metabolic disorders and in some degenerative disorders, giving way at a later date to cerebral palsy.

Investigations may include brain scan, X-rays of spine and skull, and myelography, nerve conduction studies and electromyography, and examination of the CSF and assay of serum creatine phosphokinase (CPK). While the aminotransferases and aldolase are elevated in muscular dystrophy, CPK has proved to be most organ-specific (see Chapter 17). In a child with muscle weakness a normal CPK does not exclude a myopathy, *conversely* a very high CPK does not *necessarily* imply a severe myopathy (Dubowitz 1978).

Examination of the cerebrospinal fluid may be diagnostically useful in the polyneuropathies. Elevation of protein without an increase in cells, 'albumino cytological dissociation', is found, for example, in the Guillain–Barré syndrome (infectious polyneuropathy) and in Refsum's disease.

Nerve conduction velocity depends on the diameter and degree of myelination of the nerve. In the demyelinating neuropathies the nerve conduction velocity is markedly reduced. Results of nerve conduction studies sometimes help to narrow the choice of biochemical investigations, particularly when the peripheral neuropathy is associated with involvement of the central nervous system as, for example, in the leucodystrophies. The relationship between muscle weakness, hypotonia and mental retardation is summarised in Table 15.4.

Sensory organs

Ocular findings are common in severely retarded children. Among 285 such children admitted to our hospital 20% were blind or partially sighted (Kirman & Bicknell 1975). Amongst common signs nystagmus, strabismus, optic atrophy and microphthalmia do not call for specific biochemical investigations, nor does retrolental fibroplasia, nowadays seen only rarely since greater care is taken in the administration of oxygen to premature infants. Laboratory investigations do have a diagnostic contribution to make in cases presenting the signs listed in Table 15.5. Some of the suggested tests are offered by most laboratories, whilst others will probably be referred to specialised centres.

Deafness is probably as prevalent as blindness in the severely mentally retarded. It may be acquired prenatally, as in maternal rubella and other infections, perinatally when it may be associated with cerebral palsy and

Table 15.4 Hypotonia and muscle weakness in mentally handicapped children.

Disorder	Comment
Hypotonia without significant weakness	
Non-specific mental deficiency	Very common in unclassified mental retardation. Often variable and tends to get less severe with age
Down's syndrome	Treatment of the hypotonia with 5-hydroxy-tryptophan produced no lasting benefits. Other trisomies may show increased tone
Amino acidurias	Hypotonia is a striking feature of hyperlysinaemia and non-ketotic hyperglycinaemia
Organic acidurias	Hypotonia is seen in a number of organic acidurias resulting from defects in the metabolism of pyruvate or of the derivatives of the branched chain amino acids
Prader–Willi syndrome	Adiposity, short stature, hypogenitalism with severe hypotonia, endocrine abnormalities (Jeffcoate *et al.* 1980), high risk of diabetes
Zellweger's syndrome	Cerebrohepatorenal syndrome; severe hypotonia, convulsions, hepatic cirrhosis; serum pipecolate level raised
Other metabolic disorders	Hypotonia is found in hypercalcaemia of infancy, untreated hypothyroidism
Muscle weakness with incidental hypotonia	
Congenital myopathies	
Mitochondrial cytopathy	'Ragged red fibre' syndromes; may affect many systems in addition to muscle (Egger *et al.* 1981)
Pompe's disease (glycogenosis type II)	cf. p. 472. Serum CPK may be elevated; other glycogenoses not associated with mental retardation if carefully managed
Congenital myotonic dystrophy	Mental retardation common but non-progressive; serum CPK and CSF protein normal; electro-myogram shows 'dive bomber' effect
Congenital muscular dystrophy	Often associated with mental retardation; serum CPK raised in early stages, later normal; condition non-progressive
Duchenne muscular dystrophy	Non-progressive mental retardation found in some patients, serum CPK grossly elevated in presymptomatic and early stages of the disease
Peripheral neuropathies	Lower motor neuron affected in the leucodystrophies, in some other lysomal disorders (cf. Chapter 16 and in abetalipoproteinaemia, familial dysautonomia, Leigh's encephalopathy and in infectious polyneuropathies

Note that Werdnig–Hoffmann disease, myasthenia gravis, Refsum's disease and most congenital myopathies are not normally associated with mental retardation.

Table 15.5 Investigation of mentally retarded patients with ocular findings.

Sign	Disorder and initial laboratory investigations
Cataracts	Galactosaemia: urine galactose; galactose-1-phosphate; uridyl transferase activity in red cells Pseudohypoparathyroidism: serum calcium, phosphate, alkaline phosphatase Lowe's syndrome: urinary amino acids Rubella syndrome: rubella antibody titre Dystrophia myotonica: electromyography, IgG turnover Wilson's disease: serum copper and caeruloplasmin Galactokinase deficiency: urinary galactose
Corneal clouding	Some mucopolysaccharidosis: cf. Chapter 16 Fabry's disease: cf. Chapter 16
Retinal degeneration	Some lipidoses: cf. Chapter 16 Abetalipoproteinaemia: plasma cholesterol, blood film for 'acanthocytes' Refsum's disease: plasma phytanic acid Laurence–Moon–Biedl syndrome: urinary gonadotrophins
Dislocation of the lens	Homocystinuria: urine nitroprusside test and amino acids, dislocation of the lens also seen in Marfan's syndrome and sulphite oxidase deficiency
Kayser–Fleischer rings	Wilson's disease: serum copper and caeruloplasmin
Telangiectasia of conjunctivae	Ataxia telangiectasia: IgA, α-fetoprotein

Table 15.6 Disorders associated with peculiar odour.

Disorder	Smell	Compound	Comment
Phenylketonuria	Musty	Phenylacetic acid	Noticeable in some older untreated patients only
Maple syrup urine disease	Maple syrup	Branched chain keto-acid derivatives	Many variants; in its classical form the disease presents within a few days of birth
Oasthouse disease	Musty	α-Hydroxybutyric acid	Biochemical abnormalities are the result of methionine malabsorption
Isovaleric acidaemia	Sweaty feet	Isovaleric acid	Smell useful diagnostically in sick acidotic infant
Beta methylcrotonyl glycinuria	Cat's urine	β-Hydroxyisovaleric acid	Disorder clinically variable
Trimethylaminuria	Fishy	Trimethylamine	Not associated with mental retardation
Glutaric aciduria II	Sweaty feet	Organic acids	Associated with hypoglycaemia and metabolic acidosis

Table 15.7 Abnormalities of skin and hair in mentally retarded children.

Disorder	Abnormality	Comments
Kinky-hair disease (Menkes' disease)	Twisted and brittle hair (pili torti and trichorrhexis nodosa)	Low serum copper and caeruloplasmin: death in early infancy usual, but milder mentally retarded patients reported
Argininosuccinic aciduria	Brittle hair	Urinary amino acid chromatogram diagnostic; wide spectrum of severity
Homocystinuria	Hair sparse and brittle, malar flush	See p. 428
Hartnup disease	Pellagra-like rash	Urinary amino acid chromatogam; shows transport defect of neutral amino acids; many patients intellectually normal
Phenylketonuria	Rough dry skin sometimes with eczema; dilution of hair colour	Seen only in untreated patients with classical phenylketonuria

other evidence of an insult to the nervous system, and postnatally, for example, as sequelae to meningitis or the action of ototoxic drugs. Deafness is also a feature of a number of distinct hereditary syndromes. In the majority, biochemical abnormalities have not been identified. Exceptions include Hunter's syndrome (mucopolysaccharidosis II) cf. Chapter 16, and the Pendred syndrome in which nerve deafness and an iodination defect in the synthesis of thyroxine is frequently associated with mental retardation.

Disorders associated with peculiar odours are shown in Table 15.6, those with abnormalities of skin and hair in Table 15.7.

Growth deficiency and endocrine disorders

Smith (1976) has pointed out that many patterns of malformation associated with growth retardation are the result of congenital hypoplasia in the skeletal system and in other organs which will often include the brain. The disorders may be caused by teratogens, chromosomal abnormalities or mutant genes (Table 15.8). In many instances there is good correlation between the degrees of linear growth retardation and brain growth deficiency and mental defect. Postnatally there is usually a lack of catch-up growth, due to irreversible antenatal impairment of brain and skeletal development. In mentally retarded children secondary growth retardation is also common (Table 15.8), but in contrast to primary growth deficiency can sometimes be prevented or reversed by appropriate treatment.

Tall stature is mostly genetically determined, but is a feature of a few mental retardation syndromes (Table 15.8).

Table 15.8 Growth disorders and mental handicap (for details see Smith 1976).

Disorder	Comment
Primary growth deficiency	
Caused by environmental agents	
Fetal alcohol syndrome	A major cause of congenital malformations and mental handicap
Fetal phenytoin syndrome	Risk of withholding anticonvulsant greater than risk of teratogenicity
Fetal thalidomide syndrome	Usually not associated with mental retardation
Fetal aminopterin syndrome	Similar effects produced by methotrexate, a methyl derivative of aminopterin
Fetal rubella syndrome	Incidence falling rapidly with introduction of immunisation; growth retardation also seen in other early intrauterine infections
Caused by chromosomal inbalance	Growth retardation generally seen in autosomal aneuploidy
Produced by mutant genes	The primary defect affects brain and skeletal growth in parallel as in the mucopolysaccharidoses
Pathogenetic mechanism unknown	Growth retardation seen in a large number of syndromes with congenital malformations and mental retardation of unknown aetiology, for example, the deLange and Rubinstein–Taybi syndromes
Secondary growth deficiency	
Environmental	
Malnutrition	May be due to inappropriate diet
Psychosocial deprivation	Formerly common, e.g. in large institutions
Metabolic	
Rickets	May be exacerbated by anticonvulsants
Renal disease	May form part of a syndrome or be the result of repeated urinary infections
Heart disease	Common complication, for example, in Down's syndrome
Respiratory disease	Common in retarded patients, frequently exacerbated by deficient cough reflex
Endocrine dysfunction	Hypothyroidism, see Chapter 12; pseudohypoparathyroidism, see Chapter 14; other endocrine dysfunction is usually non-specific, secondary to suboptimal function of the nervous system
Cerebral palsy	Common, mechanism of growth retardation unknown
Chronic severe infection	Mechanism of action unknown
Tall stature	
XXY and XYY karyotypes	The alleged aggresive behaviour of patients with the XYY karyotype, has not been confirmed
Cerebral gigantism (Soto's syndrome)	No consistent biochemical or endocrinological abnormality has been demonstrated
Homocystinuria	See p. 428; patients with clinically similar Marfan's syndrome are usually not mentally retarded
Beckwith–Wiedemann syndrome	Mild to moderate mental retardation; hypoglycaemia caused by pancreatic hyperplasia often seen

In the mentally retarded child endocrine abnormalities are more often than not the consequence of impairment of the nervous system rather than its cause. Exceptions are hypothyroidism (see p. 316) and hypoparathyroidism (see p. 384). Mental retardation has been found in several of the rare inborn errors of thyroid dyshormogenesis, particularly the iodide peroxidase defect, but also in the dehalogenase defect and in some cases involving deficits in thyroglobulin synthesis or breakdown. Numerically most important as a cause of mental handicap is congenital non-goitrous cretinism associated with thyroid dysgenesis. With an incidence of approximately 1 in 4000 births this is probably the most common endocrine disorder in infants. In the past few years neonatal screening has been introduced in many countries, and this should reduce the prevalence of mental handicap by this cause. It should be noted that the condition can only be diagnosed by clinical signs in the neonate in a small minority of affected infants with congenital hypothyroidism, even by an experienced paediatrician.

In hypoparathyroidism and pseudohypoparathyroidism mental retardation of varying severity is quite common. The precise relationship of the hypocalcaemia to the mental defect is not clear. The clinical syndrome of *pseudohypoparathyroidism* may occur without abnormalities of serum calcium and phosphate (*pseudopseudohypoparathyroidism*). The diagnosis of these conditions is described in Chapter 14. Another disorder of calcium metabolism associated with mental retardation is hypercalcaemia of infancy. This is discussed more fully in Chapter 14.

Self-injurious behaviour, behaviour disturbance, autism

Mild stereotyped self-injurious behaviour occurs in over 10% of mentally retarded patients; severe self-injurious behaviour resulting in irreversible mutilation, blindness or brain injury occurs in about 1 in 10000 of the total mentally handicapped population (Corbett & Campbell 1981). Biochemical factors are probably involved as certain agents known to influence neurotransmitters, such as the 5-hydroxytryptamine precursor 5-hydroxytroptophan, and the gamma-aminobutyric acid (GABA) analogue Baclofen (lioresal) affect this behaviour at least temporarily. The rare *Lesch–Nyhan syndrome* (hypoxanthine–guanine phosphoribosyl transferase deficiency) is the best known syndrome with severe self-mutilation. The diagnosis will be suggested by a raised urinary uric acid/creatinine ratio and can be confirmed by enzyme assay on blood collected on filter paper, as for infant screening tests. Severe self-mutilation is by no means confined to the *Lesch–Nyhan syndrome*; we have seen very similar behaviour in a child with the *Cornelia deLange syndrome*.

Behaviour problems are common in the retarded, but often they are only behaviour appropriate for the mental age of the patient. Severe behaviour problems have been found in two out of three patients with *glutathionuria* (γ-glutamyltranspeptidase deficiency; Wright *et al.* 1979). This condition should be detectable by the urinary nitroprusside test. Phenylethylamine overproduction has been reported in aggressive psychopaths (Sandler *et al.* 1978). Finally mention must be made of *infantile autism*, a syndrome described more than 40 years ago. Affected children fail to develop social relationships, exhibit ritualistic and repetitive behaviour and suffer from disturbance of language and perception. About half these children are mentally retarded. Autistic features have been incidental findings in some children with inborn errors, for example untreated phenylketonuria, homocystinuria and some lipidoses. On the whole, however, biochemical investigations and indeed biochemical research have not been rewarding in autism, which probably comprises an aetiologically heterogeneous group of disorders.

DEGENERATIVE DISEASES OF THE NERVOUS SYSTEM WITH MENTAL HANDICAP

A significant proportion of degenerative disorders are genetically determined and presumably therefore have a biochemical basis. Investigations of suspected cases is by no means straightforward. Neurological examination of severely retarded children is notoriously difficult (Foley 1975). Observations over weeks or months and psychological and electroencephalographic investigations may be needed to establish first slowing, then arrest of

Table 15.9 Some disorders leading to intellectual deterioration[1].

Tumours
Infections (subacute sclerosing panencephalitis)
Chronic poisoning (lead, organic mercury)
Autoimmune and post-infectious disorders (Schilder's disease)
Heredodegenerative disorders
Neurocutaneous disorders (tuberous sclerosis)
Spinal and spinocerebellar degenerations (Friedreich's ataxia)
Lysosomal storage disorders (cf. Chapter 16)
Leucodystrophies (Krabbe's, metachromatic, adrenoleucodystrophy)
Huntington's chorea
Wilson's disease
Childhood autism

1 Examples are shown in brackets. Conditions presenting in neonates have been excluded (cf. Chapter 2).

Table 15.10 Investigations of children with suspected degenerative disease of the nervous system.

Investigation	Comments
Blood calcium, phosphate, alkaline phosphatase	Onset of hypocalcaemia may be insidious in pseudo-hypoparathyroidism
Blood glucose (fasting)	
Blood lactate and pyruvate	One hour after a meal; good screening test for disorders of carbohydrate metabolism
Liver function tests	Commonly abnormal in children with Wilson's disease before appearance of neurological signs
Blood copper and caeruloplasmin	Caeruloplasmin is not invariably low in Wilson's disease
Blood lead	Regression only very rarely attributable to chronic lead poisoning
Blood and urine amino acids	Useful in confirmation of diagnosis of homocystinuria, Lowe's syndrome
Urine uric acid/creatinine ratio	Screen for Lesch–Nyhan disease [1]
Urine mucopolysaccharides	Screen for mucopolysaccharidoses [2]
Urine oligosaccharides	Screen for mucolipidoses [3]
Urine intracellular metachromasia	Screen for metachromatic leucodystrophy
Urinary porphyrins	Screen for acute intermittent porphyria
Vacuoles in lymphocytes	Used in investigations of lysosomal disorders
Metachromatic granules in lymphocytes	Used in investigation of mucopolysaccharidoses
Bone marrow	For the investigation of storage disorders
Measles antibody titre	For the diagnosis of subacute sclerosing panencephalitis (SSPE)
EEG	Diagnostic in SSPE and some late onset lipidoses; Note that over 10% of normal children have abnormal EEGs
ERG (electroretinography)	Useful for diagnosis of disorder with retinal degeneration
VER (visual evoked response)	Tests integrity of visual pathways
Nerve conduction study	Useful when peripheral neuropathy present
X-ray of skull and spine	Characteristic in the mucopolysaccharidoses
Enzyme assays on blood or fibroblasts	Definitive diagnosis of inborn errors of metabolism (reference laboratories usually)

[1] Clayton *et al.* (1980)
[2] Pennock (1976)
[3] Tsai and Marshall (1979)

acquisition of skills, and finally regression. Regression is not the same as a drop in IQ. When slow progress is related to age it appears as a lowering of the IQ.

Sometimes clinical signs, such as the retinal cherry-red spot in some sphingolipidoses or the characteristic appearance of children with mucopolysaccharidoses or mucolipidoses, will narrow the diagnostic possibilities.

While regression is the hallmark of many hereditary metabolic disorders, it is also seen in many encephalopathies due to other causes (Table 15.9). Conversely, in some lysosomal disorders (see Chapter 16) pathological processes may be active before birth. In these patients perinatal complications are common and their effects may mask the underlying disease process.

Definitive diagnosis of most heredodegenerative disorders involves assay of the enzyme or isoenzyme affected. These assays are nearly always done in reference centres. The centres would soon be swamped with specimens unless some selection is exercised in the submission of specimens. Table 15.10 lists some exploratory investigations, many of which biochemical tests should be within the scope of the average clinical chemistry laboratory. If used judiciously they should lessen the load on reference laboratories.

CONVULSIONS AND MENTAL RETARDATION

Convulsions are relatively common in childhood: 50 per 1000 in toddlers (Miller *et al.* 1960) and 7 per 1000 in school children (Rutter *et al.* 1970). They occur more frequently in the mentally retarded. Epileptics as a group are somewhat less intelligent than non-epileptics (Cooper 1965), although some are not only normal but intellectually brilliant. In the mentally retarded, seizures may be the primary cause of the handicap; seizures and retardation may both be the result of an underlying disease process, or they may be iatrogenic, related to therapy.

Seizures causing retardation

The view that convulsions harm the brain is deep-rooted. Many neurologists and neuropathologists have maintained that fits themselves, or their consequences such as cerebral hypoxia, asphyxia or vascular disturbances and associated biochemical changes, may cause permanent injury. This may occur in a previously well or retarded child. Sometimes, a vicious circle will be set up, the lesions caused by fits, then producing further fits. The acute illness responsible for fits may, for example, be an infection or a toxic encephalopathy. Acute metabolic disturbances are not uncommonly associated with the condition also causing the fits. For example, we have often seen electrolyte and acid–base disturbances in meningitis. It must not be assumed, without careful consideration of all the evidence, that such disturbances are causally related to any pre-existing or ensuing retardation. Transient metabolic changes under stress are more common in the mentally retarded than in normal children.

Seizures and retardation arising from a common cause

A disease process, perhaps hitherto unrecognised, may give rise to both fits and mental retardation. Examples are some storage diseases and some inherited metabolic disorders in which the biochemical mechanism may be known or unknown. While characteristically seen in infants and toddlers these disorders may present at any age.

Iatrogenic seizures

The sudden withdrawal of anticonvulsants or administration of other drugs which interfere with the metabolism of anticonvulsants can precipitate fits. The risk is increased in the mentally retarded child in whom both compliance with therapy and absorption of oral drugs can be unreliable.

Common investigations of children with fits are listed in Table 15.11. Depending on the severity and duration and nature of the fits, the majority, but probably not all of these investigations will be carried out. EMI scan, ultrasound and specialised biochemical investigations may then be requested, according to the findings. Clinical aspects of epilepsy and mental retardation have been reviewed by Laidlaw and Richens (1976) and Pond (1979), and pathological aspects by Crome and Stern (1972).

Table 15.11 Investigation of children with mental retardation and convulsions.

Investigation	Comments
Acute infection screen	Includes haematology and microbiology tests
TORCH (under age 3 years)	Tests for toxoplasma (T); rubella (R); cytomegalovirus (C); herpes (H). Commonly positive in well children over age 3 years
EEG	Diagnostic in a small number of degenerative disorders
Electrolytes and urea (plasma)	Often abnormal due to vomiting or failure to drink
Calcium (plasma)	Alkaline phosphatase and inorganic phosphate should also be measured
Magnesium (plasma)	Low priority after 3 months of age
Glucose (plasma)	Often around 2.2 mmol/l but not identifiable as the cause of fits
Blood gases	May be affected by breathholding and vomiting
Blood ammonia	Preferably measured 1 hour after a meal
Blood lactate and pyruvate	Preferably measured 1 hour after a meal
Liver function tests	Often abnormal due to long-term effects of drugs, infections
Lead (blood)	Often moderately raised, but not the cause of the mental defect or fits
Copper (serum)	Low in Menkes' disease. Low priority after age 3 months
Drug screen	Poisoning may be non-accidental

Calcium and folate metabolism

Subclinical hypocalcaemia is common in children taking some anticonvulsants long-term (particularly phenytoin, phenobarbitone, primidone and pheneturide), although clinical symptoms are rare. The anticonvulsants induce hepatic microsomal drug-metabolising enzymes which accelerate the metabolism of cholecalciferol and thereby increase the requirement for vitamin D (Laidlaw & Richens 1976). Silver *et al.* (1974) demonstrated mild rickets in a significant proportion of mentally retarded patients in our hospital. Moderate supplements of vitamin D (75 μg/week) resulted in normalisation of the calcium, phosphate and alkaline phosphatase concentrations in blood.

It is well established that there is an increased prevalence of folate deficiency in severely retarded treated epileptic patients (Reynolds 1973). This deficiency can be simply and cheaply prevented by regular yeast supplements (Eastham *et al.* 1975).

Therapeutic monitoring

The indications for monitoring plasma drug levels have been outlined by Richens and Warrington (1979) and are summarised in Tables 15.12 and 15.13. These authors conclude that routine monitoring is particularly valuable for phenytoin and often helpful for patients receiving carbamazapine and ethosuximide, all drugs widely used for the management of epilepsy in the mentally retarded. Monitoring is less useful for phenobarbitone and primidone; the active metabolite of primidone is also phenobarbitone. With this drug there is progressive change in the responsiveness to its central action and the plasma levels above which toxicity becomes common are not well defined. In the case of sodium valproate there is apparently no direct relationship between plasma concentration and clinical response. The therapeutic effect appears to correlate better with

Table 15.12 Reasons for monitoring plasma drug levels in mental retardation (adapted from Richens & Warrington 1979).

Wide individual variation in the rate of metabolism of a drug
Steep relationship between dose and plasma level
Therapeutic doses are close to toxic doses
Signs of toxicity are difficult to recognise clinically
The patient suffers from gastrointestinal, renal or hepatic diseases
Drug interactions are suspected
Non-compliance is suspected

Table 15.13 Some drugs commonly monitored in the management of epilepsy in mentally retarded children (adapted from Richens & Warrington 1979; Richens & Marks 1981).

Drug	Therapeutic range (μmol/l)	Comments
Phenytoin	40–80	Drug difficult to use without plasma level monitoring
Phenobarbitone	45–110	Wide intra- and interpersonal variation in response to increased plasma levels. Toxic levels not well defined
Primidone	15–16	Most anticonvulsant activity is due to phenobarbitone, a major metabolite of primidone
Carbamazepine	16–50	Monitoring helpful in optimising dose
Sodium valproate	350–?700	Drug action not accurately reflected by serum level. Routine monitoring at present not justified
Ethosuximide	300–700	Monitoring helpful in optimising dose

dose per kg body weight than actual plasma level. Routine monitoring of plasma levels is of little value.

Fits are common in the retarded, particularly in the severely retarded, of whom about a third may be affected compared to 0.5–1.0% in the normal population. Monitoring of anticonvulsant levels is thus an essential contribution of the clinical laboratory to the care of the severely retarded, more so as eliciting of neurological signs of intoxication is difficult in these patients.

THE CHILD WITH MILD RETARDATION

The aetiology of mild mental retardation has been the subject of some controversy. For a long time the so-called high-grade defectives were thought to constitute a 'subcultural' group. Their mental handicap was ascribed to 'normal biological variation', largely conditioned by the operation of polygenes. This approach does not commend itself to the pathologist. Where the brains of mildly retarded patients have been examined they have mostly shown changes similar to, but less severe than, those of severely retarded patients (Crome & Stern 1972). The handicap associated with monogenic disorders such as tuberous sclerosis, some chromosomal disorders, or that caused by environmental agents such as rubella virus, may span the full range from normal to the grossest forms of mental subnormality.

It is a daunting task to identify aetiologically significant biochemical factors in the mildly retarded. Sometimes it is like looking for a needle in a haystack—in the dark and with no guarantee that the needle is actually

Table 15.14 Factors pointing to a biochemical contribution to the aetiology of mild mental retardation.

Intelligence while not grossly impaired is significantly lower than that of parents and rest of family
Consanguinity in parents
More than one slow child in normal family
Behaviour disorder or psychiatric disturbance
Neurological or physical signs
Deteriorating performance at school
Backwardness not explained by birth history or environmental deprivation
Possibility of exposure to drugs or poisons

there! Nevertheless there are some circumstances in which there is a somewhat higher chance that biochemical factors are involved and these are outlined in Table 15.14.

When biochemical abnormalities are detected it is usually by no means easy to establish a causal relationship to the mental retardation. For example, at our hospital we have seen a girl with hyperprolinaemia type II who has epilepsy and near normal intelligence. However, her parents, who are first cousins, are both graduates, as are her three unaffected sisters. Often, patients with this disorder are intellectually normal. An adolescent, untreated phenylketonuric presented with mild mental retardation and an anxiety state, a mildly retarded girl with γ-glutamyltransferase deficiency had severe behaviour problems. Clearly, some patients with inborn errors will be only mildly retarded. Some late-onset degenerative disorders will also be initially so classified (cf. Chapter 16), and minor intellectual deficits may be produced by exposure to harmful substances in the environment.

In general the pattern of biochemical investigations will be the same as that for the severely retarded. It should be mentioned that abnormalities of the sex chromosomes are often associated with mild mental retardation. Sometimes, these abnormalities are associated with recognisable clinical features (Smith 1976). Recently it has been reported that a minor structural defect of the X-chromosome, a 'fragile site', may be associated with a wide spectrum of mental handicap, but also with normal intelligence (Berry 1981). It has been suggested that there are several X-linked genes which can cause non-specific mental retardation. These genes are said to be of variable penetrance, and in many cases the resulting mental defect is therefore not severe. At present it is not known if the 'fragile site' is also the site of the mental retardation genes (Annotation 1981).

THE MENTALLY RETARDED CHILD WITHOUT PHYSICAL SIGNS

A problem which recurs in mental deficiency practice is what biochemical tests to request for a child who presents with mental retardation but no obvious physical signs. There is no simple answer to this question but the following points may be made.

Damage to the brain may occur at any stage of its development and affect any of its functions. Cerebral palsy and marked hydrocephalus may coexist in some children with normal intelligence. Conversely, severely retarded patients may exhibit no abnormal neurological signs and no biochemical abnormalities may be detectable. After death, most of these patients have abnormal brains, but in a few no lesions can be demonstrated with routine neuropathological techniques (Crome & Stern 1972). Damage sustained by the brain during early development may not become apparent until months or even years later, as demands on the nervous system increase and ability for abstract reasoning is tested.

The agent or agents responsible for the handicap may have been active over a limited period only and be no longer detectable when the child is seen. Again, some metabolic defects may only be unmasked during infections or other environmental stress. When biochemical abnormalities are recorded in the history caution is necessary before assuming causal relationships, more so in relating a single abnormal blood finding such as hypoglycaemia, hypoxia, hyperbilirubinaemia, acidosis, hypernatraemia or minor elevation of the blood lead level to later poor development (Lucey 1982). Irreversible brain damage often results from multiple causes, effective in combination though tolerated singly (Eisenberg 1977), and should not be too readily attributed to whichever abnormality happens to have been recorded.

Some investigations which may detect mild and atypical cases are listed in Table 15.15. It is unlikely that a physician will request all tests for every patient. Ultimately, clinical experience and intuition will be the main factors determining the choice of tests.

INDIVIDUAL DISORDERS

Phenylketonuria

Neonatal screening and effective dietary treatment have dramatically reduced the number of phenylketonuric children who are mentally subnormal (Bickel 1980; Bickel *et al.* 1980; Berry *et al.* 1982; MRC Steering Committee 1981). Unsolved problems include the nature of the pathogenetic process or processes, the role of factors other than dietary control which may affect the results of treatment, the risks of relaxing the

Table 15.15 Biochemical investigation in patients with unexplained mental retardation and negative physical findings.

Investigation	Comments
Blood and urine amino acids and in selected cases blood ammonia	Perform whilst child is receiving protein (1 g protein/kg)[1]; also look at urinary pyrimidines and ammonia creatinine ratio (Palmer & Oberholzer 1977)
Urinary mucopolysaccharides and oligosaccharides	Some patients particularly with the Sanfilippo syndrome only develop characteristic appearances in late childhood; in some mucolipidoses (e.g. mannosidosis) the disease may develop very slowly
Blood lead	To allay parental anxiety; blood levels below 1.5 μmol/l are unlikely to be of aetiological significance
Liver function tests	Elevation of ALT and difficulties at school may be earliest manifestation of Wilson's disease; cf. p. 438
Blood glucose, calcium	Low values are nearly always a consequence of the damage to the nervous system rather than its cause
Urine reducing substances, acetone, pH	Also see Table 15.19
Nitroprusside screening test	To detect homocystinuria in cases who do not yet show characteristic appearance, cf. p. 428
Karyotype	Minor chromosomal abnormalities, (deletions, fragile X, etc.) and now more often found in normal-looking mentally retarded patients
TORCH	Blood toxoplasma, rubella, cytomegalovirus and herpes titres in the younger child
Thyroid function tests	Characteristic signs of hypothyroidism may be absent particularly in mild cases
Serum CPK (in selected cases only)	In the early stages of some disorders of muscle, symptoms may not be obvious; great care must be taken in collecting samples

1 A load may be given if the child is not too ill.

The following non-biochemical investigations may also give useful information: EEG, skull X-ray, bone age, CT scan, ultrasound.

diet in older patients, and the chances of normal development for the fetuses of phenylketonuric mothers.

The success of treatment has usually been assessed by monitoring changes in blood phenylalanine levels and IQ measurements. Unfortunately, the use of IQ scales has certain disadvantages. Some well treated patients do not maintain at school the progress expected from their psychometric assessments and problems of behaviour and language development are common. It should be emphasised that even in normal children IQ is by no means constant; in a recent study 25% showed IQ changes of 22 points or more between the ages of 3 and 17 years (Hindley & Owen 1978). Clearly great care is needed in interpreting small changes in IQ. The effects of diet and genotype may be enhanced by environmental factors. Often, careful dietary control and a stimulating environment go hand in hand, as do dietary failure and unsatisfactory home conditions (Williamson *et al.* 1981).

On balance there is now strong, though perhaps not conclusive evidence, that uncontrolled blood phenylalanine levels are harmful in later childhood, although less so than during the period of rapid development of the brain just before and for the first 1–2 years after birth. Children with phenylketonuria should therefore continue dietary treatment beyond 8 years of age, although a measure of dietary relaxation can probably be tolerated (Smith *et al.* 1978; Koch *et al.* 1982). Some centres will allow blood phenylalanine levels up to 1.2 mmol/l in adolescents and 2.0 mmol/l in adults.

The devastating effects of untreated maternal phenylketonuria on the fetus are well documented (Lenke & Levy 1980). It used to be held that provided the maternal blood phenylalanine level did not exceed 1–1.2 mmol/l the risk to the fetus was tolerable. Recently it has become clear that blood levels in phenylketonuric mothers should be controlled at or below 0.5 mmol/l, starting before conception, to avoid the teratogenic effects of phenylalanine early in pregnancy (Komrower *et al.* 1979; Smith *et al.* 1979).

Mothers with phenylketonuria are of two kinds. Those diagnosed as neonates by mass screening are likely to be used to the low phenylalanine diet, and will in general be of normal intelligence. For some of them treatment may be effective with respect to fetal outcome. There are also in the community a small number of undiagnosed phenylketonuric women who grew up before treatment was available, but unlike 95% of untreated phenylketonuric women they do not suffer from severe mental handicap. These women often are of dull normal intelligence or mildly retarded, and frequently have additional psychiatric problems. They are not used to the diet and when diagnosed before or during pregnancy they are often unable or unwilling to cooperate in treatment. Only a very small proportion of untreated phenylketonuric women are potential mothers and mass screening of all pregnant women for PKU has not found favour (American College of Obstetricians and Gynecologists 1973). However a maternal blood phenylalanine assay and urine test for phenylpyruvate should be requested for any woman of child-bearing age who (a) is herself mentally retarded, (b) has a family history of phenylketonuria, mental retardation or microcephaly or (c) has one or more children with clinical signs which may be caused by maternal phenylketonuria, i.e. intrauterine growth retardation, microcephaly, mental retardation and congenital malformations.

The conversion of phenylalanine to tyrosine requires the liver enzyme phenylalanine hydroxylase, dihydropteridine reductase and the cofactor tetrahydrobiopterin (BH4). Reduction in the activity of phenylalanine hydroxylase gives rise to hyperphenylalaninaemia. Patients with 'classical' phenylketonuria present few diagnostic problems. They have less than 1%

of normal phenylalanine hydroxylase activity and on a free diet blood have phenylalanine levels well above 1.2 mmol/l. The situation is different in milder variants of phenylketonuria collectively referred to as 'atypical phenylketonuria' and characterised by 1–5% of normal hydroxylase activity and on a free diet by blood phenylalanine levels of 0.24–1.2 mmol/l. Whilst breast fed the blood phenylalanine of some of these infants may be well below 0.6 mmol/l but on weaning and with an increased protein intake the blood phenylalanine may rise to 1.0 mmol/l or even higher. The decision whether to treat or not to treat with a low phenylalanine diet may then be difficult, particularly if the infant is developing normally. The management of various types of phenylketonuria has been reviewed by Walker *et al.* (1981), and details of the dietary treatment will be found in Francis (1983).

Blood phenylalanine levels may be assayed and treatment monitored by the Guthrie inhibition technique, fluorimetry, spectrophotometry or ion-exchange chromatography. There are some pitfalls in testing for the characteristic urinary metabolites. Phenylpyruvate is unstable, particularly in alkaline urine and may not be formed by transamination at blood phenylalanine levels around 1 mmol/l. Similarly, urinary *o*-hydroxy phenylacetic acid, which has been used in screening for phenylketonuria, may not be detected by some of the less sensitive thin layer chromatographic methods in patients with moderately raised blood phenylalanine levels.

About 1–3% of infants with phenylketonuria have a deficiency of BH4 (Kaufman 1980; Berlow 1980). This cofactor is also required for the hydroxylation of tyrosine to dopa, and tryptophan to 5-hydroxytryptophan. Deficiency of BH4 may be due to impaired synthesis of dihydrobiopterin

Table 15.16 Screening phenylketonuric patients for tetrahydrobiopterin deficiency.

Defect	Plasma biopterin[1]
Hydroxylase defect	
Phe ↑	↑
Phe normal (with diet)	↓
BH4 deficiency	
Due to impaired biosynthesis of BH2	
Phe ↑	↓
Phe normal (with diet)	↓
Due to DHPR deficiency	
Phe ↑	↑
Phe normal (with diet)	↑

1 Crithidia fasiculata assay (BH4 + BH2 + sepiapterin).
↑ increased, ↓ decreased.

(BH2) or to impaired activity of dihydropteridine reductase, leading to a failure of conversion of BH2 to BH4. If untreated, such patients usually develop a severe progressive neurological illness, malignant phenylketonuria. Treatment with a low phenylalanine diet does not alter the course of the disease (Berlow 1980; Kaufman 1980). Since the response to replacement therapy with neurotransmitters is very encouraging it is important to identify these patients early in life, before irreversible neurological damage has occurred. Screening for BH4 deficiency has been discussed in the MRC/DHSS Phenylketonuria Register Newsletter 1982. One approach to screening is the measurement of total biopterin activity in plasma using the Crithidia fasciculata assay (Leeming *et al.* 1976a and b). This should be performed prior to the introduction of diet and again after satisfactory levels of plasma phenylalanine have been attained (see Table 15.16 for details of findings).

Homocystinuria

Excessive excretion of homocystine in urine is found in several inborn errors of metabolism (Bondy & Rosenberg 1980). Of these cystathionine synthase deficiency is the most important. Patients usually have fine fair hair, a malar flush, long thin limbs and skeletal abnormalities reminiscent of Marfan's syndrome. However they also have osteoporosis, which is not seen in Marfan's syndrome. The most important clinical sign is dislocation of the lenses. The characteristic appearance develops only gradually, dislocation of the lenses is usually present by 2–3 years of age.

Deficiency of cystathionine synthase results in a block of the trans-sulphuration pathway from methionine to cystine with accumulation of methionine and homocystine in the body fluids, and excessive excretion in the urine of homocystine and other sulphur amino acids, some formed by transamination and disulphide exchange reactions. Most patients are diagnosed clinically as screening presents a number of problems as biochemical abnormalities are often minimal (Whiteman *et al.* 1979). The role of the laboratory is thus often limited to confirmation of the clinician's suspicion which has been aroused by the patient's appearance or eye abnormalities. In a patient with the clinical features of homocystinuria a negative screening test does not exclude the diagnosis. The blood methionine level remains elevated in homocystinuric patients 24 hours after a dose of 100 mg/kg body weight of L-methionine, and a load test will therefore usually enable a diagnosis to be made. A methionine load test is probably not justifiable in a child who is mentally retarded but has none of the other signs of homocystinuria.

Only about half the patients with homocystinuria are severely retarded but some of the others have psychiatric disorders. Of four patients seen in this hospital two children were mildly retarded, with behavioural problems; one adult, of normal intelligence, had a schizophrenia-like illness; another suffered from depressions and a personality disorder. Homocystinuria carries a high risk of life-threatening thromboembolic crises, which often occur when the patients undergo minor surgery. Cerebral thromboses probably contribute significantly to the neuropsychiatric symptomatology.

The disorder is genetically heterogeneous. In nearly half the patients the biochemical abnormalities can be corrected by massive doses of pyridoxine (100–500 mg/day). Patients who do not respond to pyridoxine may be treated with a low methionine diet. Good biochemical control and satisfactory physical and intellectual development has been achieved by this diet or by pyridoxine in individual patients but further long-term follow-up is required in order to assess the role of treatment.

Monitoring of the diet is not straightforward. Homcystine is unstable in blood even at low temperatures, so that samples must be deproteinised promptly. The urinary excretion pattern of the sulphur amino acids may be affected by disulphide exchange, oxidation and bacterial contamination and is often difficult to interpret. It is advisable to seek the advice of specialised centres whenever homocystinuric patients are to be treated and particularly when surgery is contemplated.

Histidinaemia

Biochemical findings in histidinaemia parallel those in phenylketonuria. Instead of the hydroxylation of phenylalanine to tyrosine the transformation of histidine to urocanic acid is blocked, with consequent accumulation of histidine in body fluids and excessive excretion of imidazolepyruvic acid and other metabolites in urine. The effect of histidinaemia on intellectual development is still in dispute. This problem is of more than theoretical importance since the condition is readily detectable in the neonate by the Guthrie bacterial inhibition technique or by one dimensional chromatography of the blood amino acids. The diagnosis can be confirmed by assay of histidase in skin homogenates or of urocanic acid in sweat. Biochemical control of blood histidine levels by a diet low in histidine is readily achieved (Snyderman *et al.* 1979).

Detection of cases of histidinaemia in psychiatric units is by itself insufficient to establish a causal link between mental retardation and biochemical deficit. Popkin *et al.* (1974) attempted to correct for bias in selection by looking for neuropsychiatric handicap in siblings of affected probands and concluded that at least 40% of untreated histidinaemics would be handicapped. On the other hand, follow-up of over 40 untreated histidinaemic infants diagnosed by mass screening, provided no evidence of adverse effects on intelligence (Clayton 1974; Levy *et al.* 1974). It is therefore difficult to visualise how a clinical trial could demonstrate any benefits of a low-histidine diet. However, the data of Popkin *et al.* (1974) and Ghadimi (1981) do suggest an enhanced vulnerability to neuropsychiatric handicap or epilepsy in histidinaemia, even though the great majority of histidinaemic children are likely to develop entirely normally. We have seen a patient with histidinaemia and an IQ of 100 who had her first grand mal fit at 15 years of age, after entirely normal development. Her histidinaemic brother, who has an IQ of 80, is also an epileptic. Another girl with histidinaemia who was diagnosed by chance at age $2\frac{1}{2}$ years has an IQ of 120 which she has maintained beyond adolescence. She is free from fits and neurological symptoms but her EEG has intermittently shown paroxysmal features.

It may be difficult to decide what to do about a chance finding of histidinaemia in a healthy child. Some paediatricians believe that the parents should not be told, as the anxiety caused in the family may do harm while dietary treatment is not without risk and offers no proven benefits. Others however, advocate treatment at least during the first few years of life, and failure to offer such treatment might be construed as negligence should the child subsequently develop neuropsychiatric symptoms rightly or wrongly attributed to the histidinaemia.

Other aminoacidurias

Selected aminoacidurias related to mental retardation are listed in Table 15.17. Hyperammonaemia is considered in Chapter 4. For a full review see Bremer *et al.* (1981).

Organic acidurias in the mentally retarded child

Organic acidurias are characterised by excessive excretion in the urine of aliphatic or aromatic carboxylic acids. They are often hereditary and may or may not be associated with metabolic acidosis and abnormalities of carbohydrate or amino acid metabolism. Mostly these disorders present as emergencies soon after birth (cf. Chapters 2 and 3). However, some patients with milder variants may present at a later age, sometimes with mental retardation. Organic acidurias do not lend themselves readily to investigation by paper or thin layer chromatography and gas chromatography/mass spectrometry may be necessary to elucidate the often very complex excretion patterns.

Using these techniques Watts *et al.* (1980) examined urine from 2000 severely retarded patients living in institutions. They found that nearly all the abnormalities observed in urinary organic acids could be attributed to drugs, diet or other environmental factors, or were artefacts produced, for example, by bacteria. Thus, organic acidurias do not contribute significantly to the prevalence of severe mental handicap. The question arises how far organic acidurias should be considered in the differential diagnosis of the mentally retarded child, as opposed to the sick neonate. In the latter, despite the impressive biochemical heterogeneity, the clinical findings are surprisingly similar: failure to feed, vomiting, lethargy, fits, hypotonia and coma. Even in late-presenting cases there have been major clinical manifestations: ataxia in disorders of pyruvate dehydrogenase, progressive muscular atrophy (without acidosis) in a child with β-methylcrotonyl-CoA

Table 15.17 Aminoacidurias and mental handicap.

Disorder	Site of defect	Comments
Generally associated with mental defect		
Phenylketonuria	Phenylalanine hydroxylase complex	see p. 424
Homocystinuria	Cystathionine synthase	see p. 428
Hyperammonaemias	Urea cycle	see Chapter 4
Non-ketotic hyperglycinaemia	Glycine cleavage system	see p. 66
Tyrosinosis	Cytoplasmic tyrosine aminotransferase	Mental retardation in Richner–Hanhart syndrome
Hyperlysinaemia	Saccharopine dehydrogenase	$NADP^+$ linked lysine-forming occasionally normal IQ
Saccharopinuria	? Saccharopine dehydrogenase	$NADP^+$ linked glutamate-forming
α-Aminoadipic aciduria	? α-Ketoadipate dehydrogenase	May be of normal intelligence
Pipecolic acidaemia	? L-Pipecolate dehydrogenase	Seen in Zellweger syndrome
Glutathionuria	γ-Glutamyl transpeptidase	see p. 417
β-alaninaemia	β-Alanine aminotransferase	Somnolence, fits, excess GABA in body fluids
Carnosinaemia	Carnosinase	Somnolence, fits; carnosine is a neuropeptide
Hypervalinaemia	Valine aminotransferase	Distinct from MSUD
Hyperleucinaemia	Leucine/isoleucine aminotransferase	Distinct from MSUD
Lowe's syndrome	?	Renal aminoaciduria
No causal relationship to mental defect?		
Histidinaemia	Histidase	Usually benign; enhanced vulnerability of CNS
Sarcosinaemia	Sarcosine oxidase	Benign trait
Hyperprolinaemia I	Proline oxidase	Benign trait
Hyperprolinaemia II	o-Pyrroline-5-carboxylate dehydrogenase	Some cases mildly retarded with epilepsy
Hydroxyprolinaemia	Hydroxyproline oxidase	Probably benign trait
Cystathioninaemia	Cystathioninase	Benign trait
Hartnup disease	Transport of monocarboxylic amino acids	Symptoms of pellagra occasionally seen

carboxylase deficiency, a slowly progressive encephalopathy in two siblings with β-methylglutaconic aciduria. Unexplained metabolic acidosis will at any age suggest a screen for organic acidurias. However, affected patients are not always acidotic and neither in the neonate nor in the older child are the clinical signs specific. Ultimately the clinician must decide how much weight to give to the possibility that the patient has an organic aciduria. Selected organic acidurias are shown in Table 15.18, while Table 15.19 lists the investigations which should be done before a patient is referred to a specialised centre. This field has been reviewed recently by Chalmers and Lawson (1982) and Duran and Wadman (1981).

Table 15.18 Some organic acidurias and mental handicap (adapted from Chalmers *et al.* 1980).

Disorder	Enzyme defect	Diagnostically useful urinary metabolites
Generally associated with neuropsychiatric deficits		
Maple syrup urine disease[1]	Branched chain keto acid decarboxylase	Branched chain amino and keto acids
Isovaleric acidaemia[1]	Isovaleryl-CoA dehydrogenase	Isovaleric and 3-hydroxyisovaleric acids, isovaleryl glycine
3-Methylcrotonylglycinuria	3-Methylcrotonyl-CoA carboxylase	3-Methylcrotonic and 3-hydroxyisovaleric acids, 3-methylcrotonyl glycine
3-Methylglutaconic aciduria?	? 3-Methylglutaconyl-CoA hydratase	3-Methylglutaconic, 3-methylglutaric
3-Hydroxy-3-methylglutaric aciduria[1]	3-Hydroxy-3-methylglutaryl-CoA lyase	3-Hydroxy-3-methylglutaric, 3-methylglutaconic, 3-methylglutaric and 3-hydroxyisovaleric acids
Propionic acidaemia[1]	Propionyl-CoA carboxylase	Propionic, 3-hydroxypropionic and methylcitric acids, glycine
Methylmalonic acidurias (6 variants)	Methylmalonyl-CoA mutase	Methylmalonic, methylcitric, propionic and lactic acids
Holocarboxylase synthetase deficiency	Multiple carboxylase deficiency	Propionic, hydroxypropionic, methyl, citric and 3-hydroxyisovaleric acids, 3-methylcrotonyl glycine
Pipecolic acidaemia	Pipecolate catabolism	Piperidine-2-carboxylic acid (pipecolic acid)
α-Ketoadipic aciduria	2-Ketoadipic acid decarboxylase	2-Ketoadipic, 2-hydroxyadipic and 2-aminoadipic acids
Glutaric aciduria[1]	Glutaryl-CoA dehydrogenase	Glutaric, 3-hydroxyglutaric and glutaconic acids
C_6-C_{10}-Dicarboxylic acidurias[1]	Defect in β-oxidation of fatty acids	Adipic, suberic, sebacic and 5-hydroxycaproic acids
Sidbury syndrome	Green acyl dehydrogenase	Butyric and caproic acids
Glyceric aciduria Type I	D-Glyceric acid dehydrogenase	D-glyceric acid, glycine
Glyceric aciduria Type II	?	D-glyceric acid
Pyroglutamic aciduria	Glutathione synthetase	L-2-Pyrrolidine-5-carboxylic acid (pyroglutamic acid)
Generally not associated with neuropsychiatric deficits		
β-Ketothiolase deficiency	β-Ketothiolase	2-Methylacetoacetic and 2-methyl-3-hydroxybutyric acids, glycine
Tildon and Cornblath syndrome	Succinyl-CoA: 3-ketoacid CoA transferase	Acetoacetic and 3-hydroxybutyric acids

1 Disorder has been described in patients with Reye's syndrome.

Table 15.19 Preliminary tests in cases of suspected organic acidurias.

	Investigation	Comments
1	Blood: Electrolytes and urea, blood gases, glucose, creatinine, calcium, magnesium, culture	Look for anion gap and metabolic acidosis (but acidosis may be absent in an affected patient)
	CSF: Protein, glucose, cells, appearance, culture	Save CSF for biochemical investigations, exclude infection and structural disease of the CNS
	Urine: pH, odour, S.G., glucose, ketones, protein, cells and culture, creatine, ammonia	Assess severity and persistence of ketones, if present (and see Table 15.15)
2	Blood: Lactate, pyruvate, ammonia, lead, copper, liver function tests, amino acids	See also Table 15.10
	Urine: Amino acids, phenolic acids, pyrimidines, possibly TLC screen for isovalerylglycine and methylmalonate, urine ammonia: creatinine ratio	See also Table 15.10

Galactosaemia

In most infants with galactosaemia, life-threatening symptoms are seen by the end of the first week of life. The disorder may, however, be milder, patients presenting at a later age with cataracts and mental retardation, while some stay virtually symptom-free and mentally normal. Galactose-1-phosphate uridyl transferase, the enzyme affected in galactosaemia, exhibits polymorphism and this may at least in part explain the wide range of clinical manifestations (Table 15.20).

Galactosaemia may be detected in infant screening using blood collected on filter paper, by the Beutler fluorescence test for galactose-1-phosphate uridyl transferase, or the Paigen modification of the Guthrie microbiological inhibition test for galactose (Bickel *et al.* 1980). However, there is debate about the timing of the blood sampling. These tests are also suitable for older children. While it is straightforward to demonstrate deficiency of the enzyme, identification of the genotype will usually require quantitative assay of the enzyme and investigation of its electrophoretic mobility and stability.

Treatment is by a diet essentially free of galactose and may be monitored by assay of the level of red cell galactose-1-phosphate. Unfortunately, even well-treated patients have intellectual deficits of about 15 IQ points compared with their parents or unaffected siblings. Behaviour problems are common, even when intelligence is in the normal range. The correlation between control of the red cell galactose-1-phosphate level and

Table 15.20 Variants of galactosaemia (after Gitzelmann & Hansen 1980).

Variant	Enzyme activity (% red cells)	Enzyme properties	Symptoms
Classical (Gal)	0–4		Yes
Negro	0	Approx. 10% activity in liver, intestine	Sometimes
Duarte	50	Fast 3-banded on electrophoresis	No
Rennes	7–10	Slow on electrophoresis	Yes
Indiana	0–40	Slow on electrophoresis, reduced stability	No
Los Angeles	100–140	Fast on electrophoresis (3 bands)	No
Munster	30		Yes
Bern	40		?
Chicago fast	25	25% activity in fibroblasts, fast on electrophoresis	?
Chicago labile		Reduced stability	?

Indiana/Gal and Chicago (fast)/Gal compound heterozygotes have shown symptoms; Duarte/Gal asymptomatic in most cases.
? Outcome as yet uncertain.

ultimate intellectual attainment is by no means perfect and it is difficult to give a prognosis in early infancy. There is an excellent review of this topic by Komrower (1982).

Other disorders of carbohydrate metabolism

Pyruvate occupies a key position in metabolism. Even moderate impairment of some of its metabolic pathways may result in neurological symptoms and lead to mental handicap, particularly if neurotransmitter synthesis is involved (Blass 1979). A number of patients with *pyruvate dehydrogenase deficiency* have been described, mutations affecting one or other of the proteins forming this enzyme complex (Blass 1980). The severity of the disorder is determined by how much enzyme activity remains. Cases with less than about 15% of normal activity normally present in early infancy with lactic acidosis, severe mental and motor retardation, hypotonia and fits. Deficiencies with 20–25% of normal enzyme activity are associated with a milder course and later onset, while deficiencies of 35–50% have been reported in some patients with spinocerebellar degeneration (Blass 1980). It is these milder cases who are likely to be seen in assessment centres for the mentally handicapped. Retarded mental and motor development is also found in *pyruvate carboxylase deficiency*. Other signs include hypotonia, fits, lactic acidosis and sometimes fasting hypoglycaemia.

Blood lactate and pyruvate should be measured whenever a disorder of pyruvate metabolism is suspected. Elevation of lactate (> 2.0 mmol/l)

Table 15.21 Some causes of lactic acidosis (Israels *et al.* 1976).

Environmental
Shock, cardiopulmonary disease
Uraemia, liver disease
Acute infections, septicaemia
Diabetic ketoacidosis
Drugs
Hereditary
Von Gierke's disease (glucose-6-phosphatase deficiency)
Fructose 1,6-diphosphatase deficiency
Methylmalonic acidaemia, propionic acidaemia
Disorders of pyruvate dehydrogenase complex
Pyruvate carboxylase deficiency
Mitochondrial myopathies
Idiopathic

and pyruvate ($>$ 0.1 mmol/l) may be only intermittent, but pyruvate is always high ($>$ 0.115 mmol/l) an hour after a glucose load. Urinary lactate may be markedly increased during acute episodes. The upper limit of normal for the molar lactate: creatinine ratio is approximately 0.1. Blood and urine alanine are elevated in some patients, indicating reduced gluconeogenesis. It must be remembered that lactic acidosis is common in infants (Table 15.21). The assay of blood lactate and pyruvate is essentially a screening test. Raised levels indicate that an abnormality exists somewhere in a large area of metabolism. This abnormality is not even specific for a particular pathway or cause. Enzyme deficiencies may be primary or secondary. For example, pyruvate dehydrogenase activity may be reduced in thiamine deficiency and heavy metal poisoning, indeed by any toxic substance or metabolite which causes mitochondrial damage. The assay of the enzyme is capricious and best left to laboratories specialising in this field.

Hypoglycaemia

Hypoglycaemia has long been recognised as a cause of brain damage and mental retardation. Its causes are manifold (Table 15.22). Neonatal hypoglycaemia is discussed in detail in Chapter 3. Here we should like to point out that pre-existing brain damage may exacerbate the tendency to develop hypoglycaemia with neurological signs ('neuroglycopenia'). In the neonate hypoglycaemia is often associated with additional adverse factors such as anoxia, septicaemia, jaundice and hypocalcaemia, particularly in low birth weight infants. The individual contributions of these factors to the damage caused in the brain may be exceedingly difficult to disentangle.

Severe intractable hypoglycaemia during the first year of life is in the

Table 15.22 Some aetiological factors in hypoglycaemia in infants and children (cf. Marks & Rose 1981; see also Chapter 3).

Environmental
Alcohol
Salicylate, paracetamol
Septicaemia
Malnutrition
Reye's syndrome
Endocrine
Hyperinsulinism
Hypopituitarism
Adrenal insufficiency
Maternal diabetes
Hereditary
Defects in gluconeogenesis
Galactosaemia
Hereditary fructose intolerance
Most glycogen storage disorders
Maple syrup urine disease
Propionic acidaemia
Methylmalonic acidaemia
Multiple causes
Transient neonatal hypoglycaemia

majority of cases due to hyperinsulinism caused by nesidioblastosis (Aynsley-Green 1981; and see Chapter 3). In this condition no ketone bodies are present; normally these can at least in part, replace glucose as a fuel for the brain. In their absence the hypoglycaemia is therefore particularly dangerous. While some of the infants respond to diazoxide with chlorthiazide, sub-total or total pancreatectomy is often necessary. With energetic medical or surgical treatment normal intellectual development may be achieved, but mortality is high in this disorder and brain damage in the survivors is common.

In our experience, the fasting blood glucose in older, severely retarded institutionalised mentally retarded patients is often lower than normal but not below 2.2 mmol/l, and is asymptomatic. As pointed out already, in these patients poor homoeostatic control is not confined to glucose.

Lead

Mental retardation as the result of an acute lead encephalopathy is rare in this country. There is however, considerable concern as to whether or not impaired psychological development and disturbed behaviour may result from low-level lead exposure, associated with blood lead levels below 1.8 μmol/l (36 μg/100 ml), which for many years was accepted as the paediatric upper limit of normal.

Blood lead is the most suitable test for detecting mildly increased exposure to lead as seen in the UK. Its disadvantage is that it reflects recent exposure but provides no information about duration or previous intoxications. Assays of lead levels in teeth, used in some epidemiological studies, probably reflect the body burden of lead averaged over a period of time, but there still remains the problem of distinguishing short periods of high level exposure from more moderate exposure over a longer period.

In a careful epidemiological study Needleman *et al.* (1979) compared the intelligence of children with high and low lead exposure, as reflected by the lead content of dentine, and found a difference of 4 IQ points. A drop of this magnitude would be serious for any population. These findings, and those from other studies are not conclusive (Rutter 1980; Bremer *et al.* 1981; Ernhart *et al.* 1981), since a multiplicity of interacting factors make interpretation difficult, but while there is any doubt, it is clearly desirable to reduce lead in the environment as far as possible.

A rather different problem is the extent to which overexposure to lead may play a part in the aetiology and severity of the handicap of patients seen in units for the mentally retarded. Bicknell *et al.* (1968) found that amongst the severely mentally retarded patients in our hospital 70% of those with pica ('dirt-eaters') ingested excessive amounts of lead, as shown by blood lead levels and urinary lead excretion in response to oral chelating agents. Pica is common in disturbed severely retarded patients. It is often impossible to prevent access to lead in paint or soil and these patients are therefore liable to ingest excessive amounts of the metal. There is no evidence that in the vast majority of cases excessive ingestion of lead has any part in the aetiology of the mental defect, results in intellectual deterioration or increases ill health. An increased body burden of lead is nevertheless potentially dangerous in subjects who are prone to infections and who may also suffer from other intercurrent illnesses or the adverse effects of drugs. Regular monitoring of the blood lead level in patients who are at risk and periodic treatment with oral penicillamine where appropriate is therefore recommended practice (Kirman & Bicknell 1975).

Finally we must mention the child with mild retardation, learning difficulties or behaviour disturbance whose parents, perhaps encouraged by well-meaning environmentalists, believe that their child's problems are due to subclinical lead poisoning. Usually, this appears highly unlikely when their environment and economic circumstances are considered but a blood sample should be sent for a lead assay to a reference laboratory producing precise and accurate results so that the paediatrician may be able to calm parental fears. The determination of lead in hair and urine for this purpose is not recommended.

Copper

In *Wilson's disease* patients may present with hepatic or neurological signs. The clinical picture may include ataxia, tremor, intellectual deterioration and evidence of progressive liver damage, caused by accumulation of copper in brain, liver, kidney and cornea, where the copper deposition produces the characteristic Kayser–Fleischer rings. The plasma levels of copper and copper-binding plasma protein caeruloplasmin are nearly always low, but normal levels have been found in a few authentic cases. While caeruloplasmin copper is low, non-caeruloplasmin copper, which is more loosely bound to other proteins, is raised. This loose binding favours tissue deposition, and produces the effects of heavy metal toxicity with abnormal liver function tests and later on impaired renal tubular reabsorption of glucose, phosphate, uric acid and amino acids.

Early diagnosis of Wilson's disease is important as effective treatment with copper-chelating agents is available (Delves 1980). There are few diagnostic problems once neurological or hepatic symptoms have become established. Unexplained failure at school may be an early sign in neurological cases, while the only abnormality in early hepatic cases may be a moderately raised ALT. Most difficult is the differentiation of presymptomatic cases from heterozygotes. Measurement of caeruloplasmin and serum copper is not helpful as low levels are found in up to a quarter of heterozygotes who, it must be remembered, are about a hundred times more numerous in the general population than homozygotes (Gibbs & Walshe 1979). Even studies of the excretion of radiocopper following an intravenous dose are not always conclusive and a liver biopsy may be necessary.

Menkes' syndrome (kinky hair disease) is probably caused by a defect in copper transport. Clinical manifestations are non-specific (failure to thrive and rapid neurological deterioration), except for the characteristic appearance of the hair which shows pili torti and trichorrhexis nodosa (Delves 1980). Serum copper and caeruloplasmin are low, as is the copper content of liver. There is as yet no proven effective treatment. Milder cases have been described in whom biochemical abnormalities were minimal. It was the characteristic appearance of the hair and mental retardation which suggested the diagnosis (also see p. 171).

REFERENCES

AYNSLEY-GREEN A. (1981) Nesidioblastosis of the pancreas in infancy. *Dev. Med. Child Neurol.* **23**, 372–9.

AMERICAN COLLEGE OF OBSTETRICIANS AND GYNECOLOGISTS (1973) *Maternal Phenylketonuria.* Technical Bulletin No. 25.

ANNOTATION (1981) X-linked mental retardation. *Lancet* **i**, 1086–7.

BERLOW S. (1980) Progress in phenylketonuria: defects in the metabolism of biopterin. *Pediatrics* **65**, 837–9.

Bernsen A. H. (1981) Severe mental retardation among children in a Danish urban area. In Mittler P. (ed.) *Frontiers of Knowledge in Mental Retardation*, Vol. 2, pp. 53–61. University Park Press, Baltimore.

Berry C. (1981) X-linked mental retardation. *Arch. Dis. Child.* **56**, 410–11.

Berry H. K., Hsieh M. H., Bofinger M. K. & Schubert W. K. (1982) Diagnosis of phenylalanine hydroxylase deficiency (phenylketonuria). *Am. J. Dis. Child.* **136**, 111–14.

Bickel H. (1980) Phenylketonuria: past, present, future. *J. Inherit. Metab. Dis.* **3**, 123–32.

Bickel H., Guthrie R. & Hammarsen G. (1980) *Neonatal Screening for Inborn Errors of Metabolism*. Springer Verlag, Heidelberg.

Bicknell J., Clayton B. E. M & Delves H. T. (1968) Lead in mentally retarded children. *J. Ment. Def. Res.* **12**, 282–93.

Blass J. P. (1979) Disorders of pyruvate metabolism. *Neurology* **29**, 280–86.

Blass J. P. (1980) Pyruvate dehydrogenase deficiencies. In Burman D., Holton J. B. & Pennock C. A. (eds.) *Inherited Disorders of Carbohydrate Metabolism*, pp. 252–7. MTP Press, Lancaster.

Bondy P. K. & Rosenberg L. E. (1980) *Metabolic Control and Disease*. Saunders, Philadelphia.

Bremer H. J., Duran M., Kamerling J. P., Przyrembel H. & Wadman S. K. (1981) *Disturbances of Amino Acid Metabolism: Clinical Chemistry and Diagnosis*. Urban & Schwarzenberg, Munich.

Chalmers R. A. & Lawson A. M. (1982) *Organic Acids in Man*. Chapman and Hall, Andover.

Chalmers R. A., Purkiss P., Watts R. W. E. & Lawson A. M. (1980) Screening for organic acidurias and aminoacidopathies in newborns and children. *J. Inherit. Metab. Dis.* **3**, 27–43.

Clayton B. E. (1974) Population screening. In Raine D. N. (ed.) *Molecular Variants in Disease*, pp. 145–9. Journal of Clinical Pathology, London.

Clayton B. E., Jenkins P. & Round J. M. (1980) *Paediatric Chemical Pathology*. Blackwell Scientific Publications, Oxford.

Cooper J. E. (1965) Epilepsy in a longitudinal survey of 5000 children. *Br. Med. J.* **i**, 1020–2.

Corbett J. A. & Campbell H. S. (1981) Causes of severe self-injurious behaviour. In Mittler P. (ed.) *Frontiers of Knowledge in Mental Retardation*, Vol. 2, pp. 285–92. University Park Press, Baltimore.

Crome L. & Stern J. (1972) *Pathology of Mental Retardation*, 2nd ed. Churchill Livingstone, Edinburgh.

Delves H. T. (1980) Inborn errors of trace metal metabolism. In Ellis R. (ed.) *Inborn Errors of Metabolism*, pp. 85–94. Croom Helm, London.

Dubowitz V. (1978) *Muscle Disorders in Childhood*. Saunders, Philadelphia.

Duran M. & Wadman S. K. (1981) Organic acidurias. In Alberti K. G. M. M. and Price C. P. (eds.) *Recent Advances in Clinical Chemistry*, Number 2, pp. 103–27. Churchill Livingstone, Edinburgh.

Eastham R. D., Jancar J. & Cameron J. D. (1975) Red cell folate and macrocytosis during long-term anticonvulsant therapy in non-anaemic mentally retarded epileptics. *Br. J. Psych.* **126**, 263–5.

Egger J., Lake B. D. & Wilson J. (1981) Mitochondrial cytopathy. A multisystem disorder with ragged red fibres on muscle biopsy. *Arch. Dis. Child.* **56**, 741–52.

Eisenberg L. (1977) Development as a unifying concept in psychiatry. *Br. J. Psych.* **131**, 225–37.

Elwood J. H. & Darragh P. M. (1981) Severe mental handicap in Northern Ireland. *J. Ment. Def. Res.* **25**, 147–55.

Ernhart C. B., Landa B. & Schell N. B. (1981) Subclinical levels of lead and developmental deficit—a multivariate follow-up reassessment. *Pediatrics* **67**, 911–19.

FOLEY J. (1975) Neurological examination. In Kirman B. & Bicknell J. *Mental Handicap*, pp. 321–35. Churchill Livingstone, Edinburgh.

FRANCIS D. E. M. (1983) *Diets for Sick Children*, 4th ed. Blackwell Scientific Publications, Oxford.

GHADIMI H. K. (1981) Histidinaemia. *Am. J. Dis. Child.* **135**, 210–11.

GIBBS K. & WALSHE J. M. (1979) A study of the caeruloplasmin concentrations found in 75 patients with Wilson's disease, their kinships and various control groups. *Quart. J. Med.* **48**, 447–63.

GITZELMAN N. R. & HANSEN R. G. (1980) Galactose metabolism, hereditary defects and their clinical significance. In Burman D., Holton J. B. & Pennock C. A. (eds.) *Inherited Disorders of Carbohydrate Metabolism*, pp. 61–101. MTP Press, Lancaster.

GUSTAVSON K. H., HAGBERG B., HAGBERG G. & SARS K. (1977) Severe mental retardation in a Swedish county. *Neuropaediatrie* **8**, 293–304.

HAGBERG B., HAGBERG G., LEWERTH A. & LINDBERG V. (1981) Mild mental retardation in Swedish school children. *Acta Paed. Scand.* **70**, 441–52.

HAGBERG B., KYLLERMAN M. & STEEN G. (1979) Dyskinesia and dystonia in neurometabolic disorders. *Neuropaediatrie* **10**, 305–20.

HINDLEY C. B. & OWEN C. F. (1978) The extent of individual changes in I.Q. for ages between 6 months and 17 years, in a British longitudinal sample. *J. Child Psych. and Psychiat.* **19**, 329–50.

HOSKING G. (1982) Ataxia telangiectasia. *Developmental Medicine and Child Neurology* **24**, 77–80.

ISRAELS S., HAWORTH J. C., DUNN H. G. & APPLEGARTH D. A. (1976) Lactic acidosis in childhood. *Adv. Paediatr.* **22**, 267–303.

JEFFCOATE W. J., LAURANCE B. M., EDWARDS C. R. W. & BESSER G. M. (1980) Endocrine function in the Prader–Willi syndrome. *Clin. Endocrinol.* **12**, 81–9.

KAUFMAN S. (1980) Differential diagnosis of variant forms of hyperphenylalaninaemia. *Pediatrics* **65**, 840–2.

KIRMAN B. & BICKNELL J. (1975) *Mental Handicap*. Churchill Livingstone, Edinburgh.

KOCH R., AZEN C. G., FRIEDMAN E. G. & WILLIAMSON M. L. (1982) Preliminary report on the effects of diet discontinuation in PKU. *J. Ped.* **100**, 870–5.

KOMROWER G. M. (1982) Galactosaemia—thirty years on: the experience of a generation. *J. Inherit. Metab. Dis.* **5**, 96–104.

KOMROWER G. M., SARDHARWALLA I. B., COUTTS J. M. J. & INGHAM D. (1979) Management of maternal phenylketonuria: an emerging chemical problem. *Br. Med. J.* **i**, 1383–7.

LAIDLAW J. & RICHENS A. (1976) *A Textbook of Epilepsy*. Churchill Livingstone, Edinburgh.

LENKE R. R. & LEVY H. L. (1980) Maternal phenylketonuria and hyperphenylalaninaemia. *New Engl. J. Med.* **303**, 1202–8.

LEVY H. L., SHIH V. E. & MADIGAN P. M. (1974) Routine newborn screening for histidinaemia. *New Engl. J. Med.* **291**, 1214–19.

LEEMING R. J., BLAIR J. A., GREEN A. & RAINE D. N. (1976a) Biopterin derivatives in normal and phenylketonuric patients after oral loads of phenylalanine, L-tyrosine and L-tryptophan. *Arch. Dis. Child.* **51**, 771–7.

LEEMING R. J., BLAIR J. A. & REY F. (1976b) Biopterin derivatives in atypical phenylketonuria. *Lancet* **i**, 99–100.

LUCEY J. F. (1982) Bilirubin and brain damage—a real mess. *Pediatrics* **69**, 381–2.

MACKAY R. I. (1982) The causes of severe mental handicap. *Dev. Med. Child Neurol.* **24**, 386–8.

MARKS V. & ROSE F. C. (1981) *Hypoglycaemia*, 2nd edn. Blackwell Scientific Publications, Oxford.

MRC STEERING COMMITTEE FOR THE MRC/DHSS PKU REGISTER Routine screening for phenylketonuria in the United Kingdom 1964–78. *Br. Med. J.* **282**, 1680–4.

MILLER F. J. W., COURT S. D. M., WALTON W. S. & KNOX E. G. (1960) *Growing up in Newcastle upon Tyne*. Oxford University Press, Oxford.

MILSTIEN S., KAUFMAN S. & SUMMER G. K. (1980) Hyperphenylalaninaemia due to dihydropteridine reductase deficiency: diagnosis by measurement of oxidised and reduced pterins in urine. *Pediatrics* **65**, 806–10.

MRC/DHSS (1982) *Phenylketonuria Register Newsletter* No. 8. (Institute of Child Health, London.)

NEEDLEMAN H. L., REED M., PARESIE H., MAHER C. & BARRETT P. (1979) Deficits in psychological and classroom performance of children with elevated dentine lead levels. *New Engl. J. Med.* **300**, 689–95.

PALMER T. & OBERHOLZER V. G. (1977) Diagnosis of urea cycle disorders. *Ann. Clin. Biochem.* **14**, 136–8.

PENNOCK C. A. (1976) A review and selection of simple laboratory methods used for the study of glycosaminoglycan excretion and the diagnosis of the mucopolysaccharidoses. *J. Clin. Path.* **29**, 111–23.

POND D. (1979) Epilepsy and mental retardation. In Craft M. (ed.) *Tredgold's Mental Retardation*, 12th edn., p. 33. Baillière Tindall, London.

POPKIN J. S., CLOW C. L., SCRIVER C. R. & GROVE J. (1974) Is hereditary histidinaemia harmful? *Lancet* **i**, 721–2.

RATCLIFFE S. G., BANCROFT J., AXWORTHY D. & MCLAREN W. (1982) Klinefelter's syndrome in adolescence. *Arch. Dis. Child.* **57**, 6–12.

REYNOLDS E. H. (1973) Anticonvulsants, folic acid and epilepsy. *Lancet* **i**, 1376–8.

RICHENS A. & MARKS V. (1981) *Therapeutic Drug Monitoring*. Churchill Livingstone, Edinburgh.

RICHENS A. & WARRINGTON S. (1979) When should plasma drug levels be monitored? *Drugs* **17**, 488–500.

RUNDLE A. T. (1979) The use and abuse of the laboratory. In Craft M. (ed.) *Tredgold's Mental Retardation*, 12th edn., pp. 163–9. Baillière Tindall, London.

RUTTER M. (1980) Raised lead levels and impaired cognitive/behavioural functioning: a review of the evidence. *Dev. Med. Child Neurol.* **22** (suppl. 1), 1–26.

RUTTER M., GRAHAM P. & YULE W. (1970) A neuropsychiatric Study in Childhood (*Clinics in Developmental Medicine* **35/36**). Spastics International Medical Publications, London.

SANDLER M., RUTHVEN C. R. J., GOODWIN B. L., FIELD H. & MATHEWS R. (1978) Phenylethylamine overproduction in aggressive psychopaths. *Lancet* **ii**, 1269–70.

SILVER J., DAVIES J. T., KUPERSMITT E., ORME M., PETRIE A. & VAJDA F. (1974) Prevalence and treatment of vitamin D deficiency in children on anticonvulsant drugs. *Arch. Dis. Child.* **49**, 344–9.

SMITH D. W. (1976) *Recognizable Patterns of Human Malformation*, 2nd edn. Saunders, Philadelphia.

SMITH I., ERDOHAZI M., MACARTNEY F. J., PINCOTT J. R., WOLFF O. H., BRENTON D. P., BIDDLE S. I., FAIRWEATHER D. V. I. & DOBBING J. (1979) Fetal damage despite low-phenylalanine diet after conception in a phenylketonuric woman. *Lancet* **i**, 17–19.

SMITH I., LOBASCHER M. E., STEVENSON J. E., WOLFF O. H., SCHMIDT H., GRUBELL-KAISER S. & BICKELL H. (1978) Effect of stopping low-phenylalanine diet on intellectual progress of children with phenylketonuria. *Br. Med. J.* **ii**, 723–6.

SNYDERMAN S. E., SANSARICQ C. J., NORTON P. M. & MANKA M. (1979) The nutritional therapy of histidinaemia. *J. Ped.* **95**, 712–15.

Stanbury J. B., Wyngarden J. B., Goldstein J. L. & Brown M. S. (Eds) (1983) *The Metabolic Basis of Inherited Disease*, 5th edition. McGraw-Hill, New York.

Tsai M. Y. & Marshall J. G. (1979) Screening for urinary oligosaccharides and simple sugars by thin layer chromatography. *Med. Lab. Sci.* **36**, 85–90.

Volpe J. J. (1981) *Neurology of the Newborn.* Saunders, Philadelphia.

Walker V., Clayton B. E., Ersser R. S., Francis D. E. M., Lilly P., Seakins J. W. T., Smith I. & Whiteman P. D. (1981) Hyperphenylalaninaemia of various types during three-quarters of a million neonates tested in a screening programme. *Arch. Dis. Child.* **56**, 759–64.

Watts R. W. E., Baraitser M., Chalmers R. A. & Purkiss P. (1980) Organic acidurias and aminoacidurias in the aetiology of long-term mental handicap. *J. Ment. Def. Res.* **24**, 257–70.

Whiteman P. D., Clayton B. E., Ersser R. S., Lilly P. & Seakins J. W. T. (1979) Changing incidence of neonatal hypermethioninaemia: implications for the detection of homocystinuria. *Arch. Dis. Child.* **54**, 593–8.

Williamson M. L., Koch R., Azen C. & Chang C. (1981) Correlates of intelligence test results in treated phenylketonuric children. *Pediatrics* **68**, 161–7.

Wright E. C., Stern J., Ersser R. & Patrick A. D. (1979) Glutathionuria: γ-glutamyltranspeptidase deficiency. *J. Inherit. Metabol. Dis.* **2**, 3–7.

Chapter 16
Lysosomal Storage Disorders

PAUL D. WHITEMAN

The subject of this chapter is complex and the detailed biochemistry is the problem of specialised laboratories. Therefore an attempt is made to give an overview of the subject with emphasis on the simpler screening tests.

Although many of the inherited lysosomal storage disorders (LSD) are rare a large number of diseases comprise this group and the overall incidence may approach that of phenylketonuria. The late infantile form of metachromatic leucodystrophy has an incidence in the order of 1 in 40000 live births and the overall incidence for mucopolysaccharidosis is almost certainly greater than this. The incidence of some of these diseases varies considerably in different populations. For instance, Tay–Sachs disease and Type I Gaucher disease occur with a frequency of about 1 in 3000 births among Ashkenazi Jews. Most LSD have an autosomal recessive inheritance but Hunter and Fabry diseases have an X-linked inheritance.

One aspect of the catabolic phase of normal metabolism involves degradation of the many complex molecules, such as proteins, glycoproteins, glycolipids and proteoglycans, which are found to a varying extent in all tissues. The lysosome plays an important role in the degradation of these substances. It is a membrane-bound intracellular organelle which contains numerous degradative enzymes in an acid medium. Material engulfed at the cell surface is isolated in a membrane-bound phagosome which then fuses with a lysosome, allowing degradation to proceed. Lysosomal enzymes are characteristically most active in an acid environment (usually around pH 4–5). Acid phosphatase is perhaps one of the best known lysosomal enzymes and is often used as a histochemical and biochemical marker for the lysosome. A useful account of the structure and functions of lysosomes has been given by Dean and Barrett (1976).

A large number of relatively rare inherited diseases are associated with excessive storage of material within the lysosome. Most of these lysosomal storage disorders are now known to be due to deficiencies of specific lysosomal enzymes. Although the remainder of this chapter will be devoted to these inherited diseases a brief mention should be made of other circumstances in which lysosomal storage may occur. The introduction of foreign substances into the tissues may cause the lysosomes to become engorged with material that cannot be extensively degraded. Lipofuscin granules of uncertain origin are found in association with lysosomes in apparently normal tissues and increase in prevalence with advancing age. An iatrogenic cause of lysosomal storage is continued administration of

Table 16.1 Classification of heritable lysosomal storage diseases.

Disease	Basic enzyme abnormality	Clinical features
Mucopolysaccharidoses (MPS)		
Hurler (MPS IH)	α-L-iduronidase	Mental retardation, bone deformities, hepatosplenomegaly, heart disease, coarse facies, corneal clouding
Scheie[1] (MPS IS)	α-L-iduronidase	Relatively mild, corneal clouding, stiff joints, aortic valve lesions
Hurler–Scheie (MPS IH–S)	α-L-iduronidase	Hurler-like, but without marked mental retardation
Hunter (MPS II)	Iduronate-2-sulphate sulphatase	Hurler-like but no corneal clouding and mental status variable, X-linked recessive, mild to severe forms
Sanfilippo-type A (MPS IIIA)	Heparan sulphate sulphamidase	Mental retardation, relatively mild physical abnormalities
Sanfilippo-type B (MPS IIIB)	N-acetyl-α-D-glucosaminidase	
Sanfilippo-type C (MPS IIIC)	Acetyl-CoA: α-glucosaminide N-acetyl-transferase	
Sanfilippo-type D (MPS IIID)	Heparan sulphate N-acetylglucosamine 6-sulphate sulphatase	
Morquio (MPS IV)	? N-acetylgalactosamine 6-sulphate sulphatase	Dwarfed, bone deformities, corneal clouding
Maroteaux–Lamy (MPS VI)	Aryl sulphatase B (N-acetylgalactosamine 4-sulphate sulphatase)	Dwarfed, bone deformities, corneal clouding, mild to severe forms
MPS VII	Acid β-D-glucuronidase	Mental retardation, bone deformities, variable clinical expression
MPS VIII (?)	? keratan sulphate N-acetylglucosamine 6-sulphate sulphatase	Mental retardation, bone deformities
Lipidoses		
GM_1-gangliosidosis	Acid β-D-galactosidase	Mental and motor deterioration, Hurler-like features and bone deformities at birth, cherry-red spot at macula, milder forms occur
GM_2-gangliosidosis: Tay-Sachs	Hexosaminidase A	Mental and motor deterioration in infancy, startle reflex to sound, doll-like facies, cherry-red spot, later onset form occurs
GM_2-gangliosidosis: Sandhoff	Hexosaminidase A and B	Similar to infantile form of Tay–Sachs
Metachromatic leucodystrophy	Aryl sulphatase A (cerebroside sulphatase)	Mental and motor deterioration with ataxia and paralysis, early and late onset forms
Gaucher	Glucocerebroside β-glucosidase	Rapid neurological degeneration with hepatosplenomegaly, later onset forms can occur without mental retardation

Niemann–Pick	Sphingomyelinase	Rapid neurological degeneration with hepatosplenomegaly, cherry-red spots, less acute forms occur
Fabry	α-Galactosidase	Dermal, scleral and corneal abnormalities, pains in extremities, renal failure: pulmonary, cardiac and cerebrovascular disease; X-linked recessive, female carriers also affected
Krabbe globoid-cell leucodystrophy	Galactocerebroside β-galactosidase	Mental and motor deterioration, optic atrophy
Farber	Acid ceramidase	Hoarseness, swollen joints, skin nodules, mental status variable
Wolman	Acid esterase	Hepatosplenomegaly, anaemia, steatorrhoea, adrenal calcification
Mucolipidoses (MLS)		
I-cell disease (MLS II)	See text	Hurler-like but presents earlier in infancy, marked bone deformities
Pseudo-Hurler polydystrophy (MLS III)	See text	Milder version of I-cell disease
MLS IV	?Lysosomal ganglioside sialidase	Psychomotor retardation, corneal clouding
Austin leucodystrophy	Multiple sulphatase deficiency including aryl sulphatases A, B and C, steroid sulphatase, heparan N-sulphamidase and iduronate sulphatase	Features of metachromatic leucodystrophy and mucopolysaccharidosis
Glycogenosis		
Pompe disease (glycogen storage disease type II)	Acid α-1,4-glucosidase (acid maltase)	Muscle weakness, massive cardiac enlargement
Other diseases		
Mannosidosis	α-Mannosidase	Hurler-like, opacities in capsule of lens
Fucosidosis	α-Fucosidase	Severe cerebral degeneration, cardiomegaly, thick skin
Aspartylglucosaminuria	1-Aspartamido-β-N-acetylglucosamine hydrolase	Hurler-like

1 Scheie disease was originally designated MPS V.
Note: Where clinical variants occur the features of the severe or early onset forms are given.

certain cationic amphiphilic drugs which can enter the lysosome and then apparently form complexes with, or impair, the degradation of material such as phospholipid (drug-induced phospholipidosis) (reviewed by Lüllmann-Rauch 1979). Drugs which fall into this category include diethylaminoethoxyhexestrol and chloroquine.

An outline classification of the heritable LSD and the corresponding enzyme abnormalities are given in Table 16.1. Many of these disorders have been classified according to the nature of the poorly degraded storage material. Thus, acid mucopolysaccharide (now called acid glycosaminoglycan) and sphingolipid are stored in excess in the mucopolysaccharidoses and the sphingolipidoses respectively. The distinction is not always so clear and the so-called mucolipidoses have features resembling mucopolysaccharidosis and lipidosis. The precise biochemical lesion in ophthalmoplegic lipidosis has not yet been determined but histochemical studies suggest that this neurolipidosis might also be a LSD. Cystine accumulates in the lysosomes of patients with cystinosis, but the cause of this disease is not yet known. The last two mentioned conditions will not be considered further in this chapter.

CLINICAL MANIFESTATIONS

Some of the clinical features of LSD are summarised in Table 16.1. The reader is referred elsewhere for detailed accounts (e.g. Stanbury *et al.* 1978; McKusick 1972). Mental retardation or neurological abnormalities are important features in the majority of LSD and are the main presenting features in Sanfilippo disease, Tay–Sachs disease, fucosidosis, acute forms of Gaucher and Niemann–Pick diseases and metachromatic and Krabbe leucodystrophies. Mental retardation is associated with moderate to severe bone abnormalities in some mucopolysaccharidoses, GM_1-gangliosidosis, I-cell disease, Austin leucodystrophy, mannosidosis and aspartylglucosaminuria. Hepatosplenomegaly is also found in these last mentioned conditions and in Gaucher and Niemann–Pick diseases. The heart is affected in fucosidosis, Pompe's disease and in some forms of mucopolysaccharidosis. Skeletal abnormalities occur in the absence of mental retardation in Hurler–Scheie compound, Morquio and Maroteaux–Lamy diseases and in the adult form of Gaucher disease. Coarse facies are found in most of the conditions associated with bone abnormalities and particularly in Hurler disease.

An important point to remember is that particular enzyme deficiencies, as presently defined in the laboratory, may be associated with a variety of clinical phenotypes. Thus, MPS I, which is associated with α-L-iduronidase deficiency, can be subdivided into Hurler disease (MPS IH), which is the most severe form of mucopolysaccharidosis, Scheie disease (MPS IS), which is one of the mildest forms and is compatible with a relatively normal life, and Hurler–Scheie disease (MPS IH–S) which has an intermediary phenotype. Such variants may differ not only in the extent and severity of the disease but also in the age of onset of symptoms. The severe form of metachromatic leucodystrophy becomes apparent around 3 years of age and is associated with progressive neurological degeneration and death in childhood, whereas an adult form, with onset after 21 years, has also been defined, in which dementia, psychosis and progressive neurological degeneration then occurs over several years. The acute neuronopathic form of Gaucher disease appears at about 6 months and death occurs by 2 years, whereas patients with the 'adult' form are mentally normal and sometimes survive to middle age, despite infiltration of various organs with lipid storage cells. Such phenotypic variations could be partly explained in terms of the genetic abnormality. The activity and specificity of an enzyme can be altered in a variety of ways, depending on the particular aberration in its primary structure. Enzymes may display different activities towards the various natural substrates, on which they normally act, and the artificial substrates often used in the laboratory. Thus, Hurler and Scheie diseases are probably due to different aberrations in the gene coding for α-L-iduronidase. Hurler–Scheie disease (MPS IH–S) is thought to be due to the inheritance of a Hurler gene from one parent and a Scheie gene from the other.

RELEVANT BIOCHEMISTRY

A general review of the biochemistry of LSD has been given by Neufeld *et al.* (1975) and accounts of the biochemistry of the sphingolipidoses and the mucopolysaccharidoses have been given by Brady (1978) and Cantz and Gehler (1976) respectively.

The mucopolysaccharidoses

In the mucopolysaccharidoses, due to incomplete degradation, acid glycosaminoglycans are stored and excreted in excess. The terminology and composition of relevant glycosaminoglycans are outlined in Table 16.2. Particular glycosaminoglycans (GAGs) are characterised by the composition of their repeating units, which in most instances contain a uronic acid and a hexosamine residue. In mammalian GAGs uronic acids are either iduronic or glucuronic acid and the hexosamines are either glucosamine or galactosamine. Iduronic acid and hexosamine residues may carry covalently linked ester sulphate groups (—O—sulphate) and hexosamines may also be N-acetylated (—$NHCOCH_3$) or N-sulphated (sulphamide residue on glucosamine in heparan sulphates and heparin). In their natural state most GAGs are linked to protein chains through a few neutral sugar residues. The sugar–protein linkage in most proteoglycans is between a

Table 16.2 Terminology and composition of acid glycosaminoglycans.

Recommended names (after Jeanloz)	Older names	Components of typical disaccharide repeating units: Hexosamine	Uronic acid
Chondroitin-4-sulphate	Chondroitin sulphate A	N-acetyl-β-D-galactosamine (4-sulphated)	Glucuronic
Chondroitin-6-sulphate	Chondroitin sulphate C	N-acetyl-β-D-galactosamine (6-sulphated)	Glucuronic
Dermatan sulphate	Chondroitin sulphate B Dermatoidin sulphate β-heparin	N-acetyl-β-D-galactosamine (4-sulphated; 6-sulphation can also occur)	Iduronic (some sulphated) Some glucuronic
Heparan sulphate	Heparitin sulphate	α-D-glucosamine (N-acetylated, N-sulphated or O-sulphated)	Glucuronic Sulphated iduronic
Hyaluronic acid	Hyaluronic acid	N-acetyl-β-D-glucosamine (non-sulphated)	Glucuronic
Keratan sulphate	Keratosulphate	N-acetyl-β-D-glucosamine (6-sulphated)	None[1]

1 Keratan sulphate contains β-D-galactose (which can be 6-sulphated) instead of a uronic acid. Mannose, sialic acid and fucose are also present and the form in cartilage also contains galactosamine.

xylose residue and a hydroxyamino acid (serine or threonine). The other end of the GAG chain is the non-reducing terminal; lysosomal hydrolases act at this end. Fig. 16.1 shows a simplified version of the non-reducing terminal of heparan sulphate and the sites of action of various lysosomal enzymes which act sequentially from the non-reducing end. A deficiency of one of these enzymes would prevent further degradation of the heparan sulphate chain. The specific sulphatase necessary for the removal of the ester sulphate from a terminal sulphated iduronic acid residue is deficient in Hunter disease. In Hurler disease there is a deficiency of α-L-iduronidase, the enzyme required for the removal of a terminal non-sulphated iduronic acid residue. Sulphated iduronic acid residues occur in both heparan and dermatan sulphates, although they are relatively infrequent in the latter. Dermatan and heparan sulphates are stored and excreted in excess in Hurler and Hunter diseases.

Sanfilippo disease is also associated with impaired degradation and excessive excretion of heparan sulphate. This clinical phenotype can be due to deficiencies of any one of several enzymes (Table 16.1) which are normally required for the degradation of heparan sulphate but not dermatan sulphate (Fig. 16.1). A specific sulphamidase cleaves the sulph-

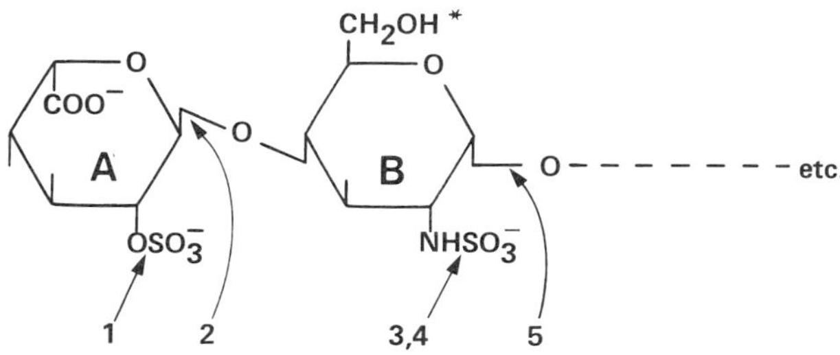

Fig. 16.1 Enzymatic degradation of a non-reducing terminal of heparan sulphate. Carbohydrate residues: A, O-sulphated iduronide; B, N-sulphated α-D-glucosamine. Sequential enzyme attack: 1 Iduronate-2-sulphate sulphatase; 2 α-L-iduronidase; 3 Heparan sulphate sulphamidase; 4 N-acetylation of primary amine group after removal of sulphate by (3); 5 N-acetyl-α-D-glucosaminidase. * This group can be O-sulphated; another sulphatase acts at this site.

amide groups of heparin and heparan sulphates and is deficient in Sanfilippo A disease. N-acetyl-α-D-glucosaminidase splits the α-glucosaminide linkages characteristic of these GAGs and is deficient in Sanfilippo B disease. It has been proposed that glucosamine residues (i.e. non-N-sulphated, as might occur after sulphamidase action) need to be N-acetylated before N-acetyl-α-D-glucosaminidase can act and that the enzyme needed for acetylation is deficient in Sanfilippo C disease. Yet another variant, type D, has recently been demonstrated in which there is a deficiency of a specific heparan sulphate N-acetylglucosamine-6-sulphate sulphatase.

Fig. 16.2 indicates the site of action of some lysosomal enzymes on a representative fragment of dermatan sulphate. Deficiencies of iduronidase and idurono-2-sulphate sulphatase have already been discussed. Maroteaux–Lamy disease appears to be due to a deficiency of a sulphatase which removes the ester sulphate from the N-acetyl-β-D-galactosamine component of dermatan sulphate; this enzyme is usually measured as aryl sulphatase B (the aryl sulphatases are discussed in more detail later). In this condition abnormal amounts of dermatan sulphate, but not heparan sulphate, are excreted in the urine. A few β-D-glucuronic acid residues can occur in dermatan sulphate chains and when they are exposed at the non-reducing terminal they can be removed by lysosomal β-glucuronidase. Some forms of mucopolysaccharidosis have been associated with β-glucuronidase deficiency and have been designated as MPS VII. However, these disorders are heterogenous with respect to reported clinical presentations and patterns of mucopolysacchariduria and further clarification seems desirable before a final classification is adopted.

In the past Hurler disease and some other mucopolysaccharidoses were associated with deficiencies of other enzymes, such as β-galactosidase and hyaluronidase. This was probably a secondary phenomenon due to the

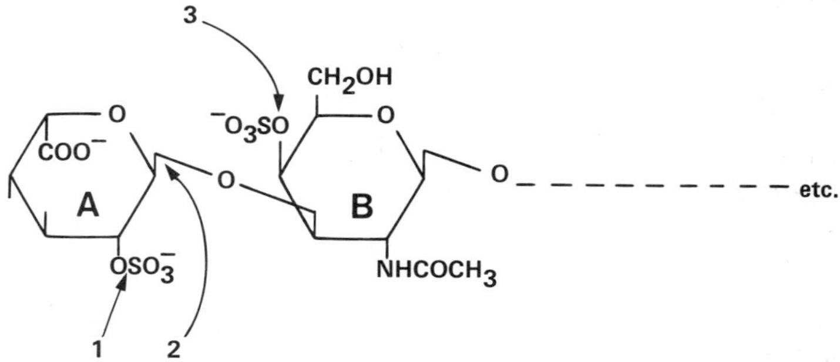

Fig. 16.2 Enzymic degradation of a non-reducing terminal of dermatan sulphate. Carbohydrate residues: A, O-sulphated iduronide; B, N-acetyl-β-D-galactosamine-4-sulphate. Sequential enzyme attack: 1 Iduronate-2-sulphate sulphatase; 2 α-L-iduronidase; 3 N-acetylgalactosamine-4-sulphatase (aryl sulphatase B).

inhibitory effects of large amounts of stored GAGs on the activities of these enzymes in organs such as liver and spleen. A primary deficiency of acid β-galactosidase activity occurs in particular forms of lipidosis, which will be discussed later.

Morquio disease is one of a number of spondyloepiphyseal dysplasias (conditions with particular abnormalities of spine and epiphyses) and is a form of mucopolysaccharidosis. Clinical and radiological criteria need to be strictly applied in order to make an accurate diagnosis. Another specific sulphatase is deficient in some patients with Morquio disease (Table 16.1). Children with this disease have an excessive urinary excretion of keratan sulphate and chondroitin sulphate, which is the main GAG in normal urine. Keratan sulphate is unlike the GAGs mentioned previously in that it does not contain uronic acids. Its repeating unit contains galactose and N-acetyl-glucosamine, either of which can be 6-sulphated. Unfortunately, some of the methods used for the determination of low concentrations of keratan sulphate in urine have not been satisfactory and reports vary as to the prevalence of excessive keratan-sulphaturia in Morquio disease. The problem of interpretation of mucopolysacchariduria will be discussed in a later section but it should be pointed out that this aspect of Morquio disease, together with the problem of clinical diagnosis, has led to some confusion over the question of variant forms of the disease. The term mucopolysaccharidosis is generally restricted to those conditions due to errors of GAG catabolism. However, some clinically similar disorders (e.g. some other forms of spondyloepiphyseal dysplasia) may be due to abnormalities of GAG biosynthesis.

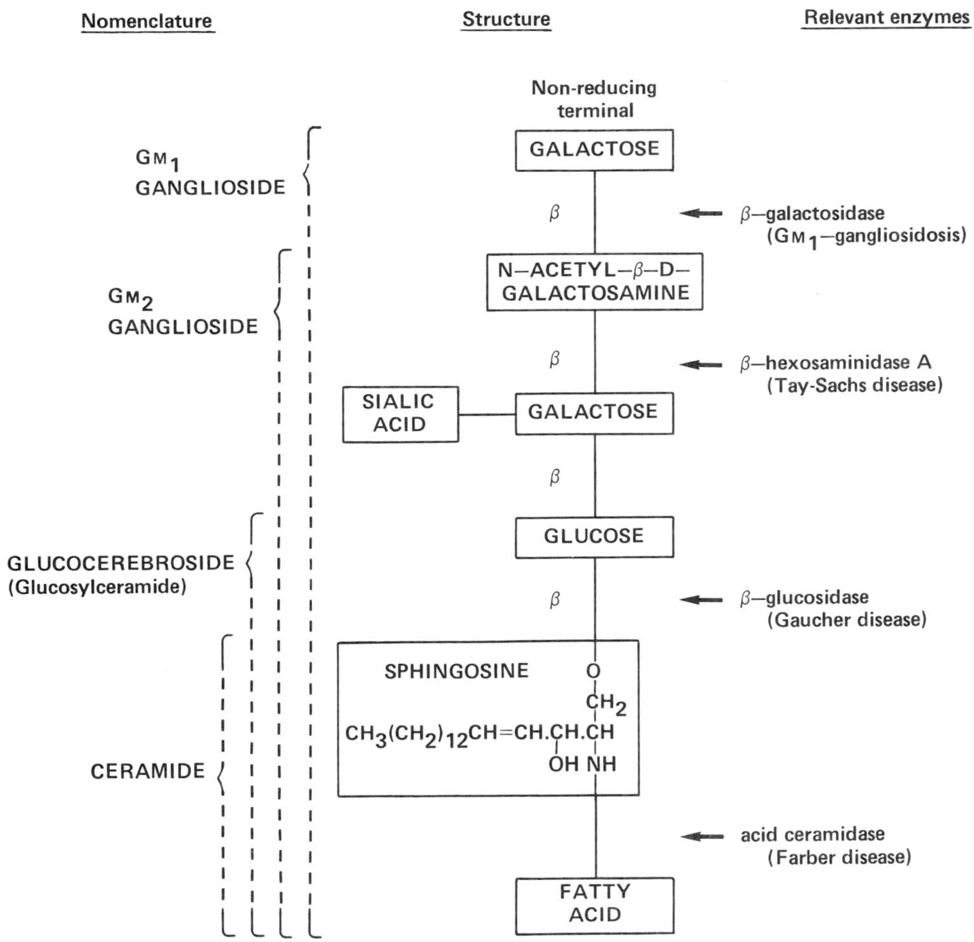

Fig. 16.3 Structure of GM_1 and related glycolipids and sites of action of some relevant lysosomal hydrolases. Lysosomal exohydrolases degrade complex glycolipids by acting sequentially from the non-reducing terminal. Storage diseases resulting from deficiencies of these enzymes are indicated in brackets.

The lipidoses

Some lipidoses are associated with impaired degradation of ganglioside, which is glycosphingolipid containing sialic acid (N-acetyl-neuraminic acid; NANA). GM_1 is a monosialoganglioside and its structure is shown in Fig. 16.3. As is the case of GAG degradation lysosomal enzymes act on this substrate in a stepwise manner. Thus, the terminal β-D-galactose residue must first be removed by acid β-galactosidase before the next enzyme can remove the N-acetyl-β-D-galactosamine residue, and so on. A deficiency of acid β-galactosidase gives rise to G_{M1} gangliosidosis in which G_{M1} accumulates in the brain and viscera. A keratan-sulphate-like material has also been found in the viscera of patients with this disorder (it contains galactose and glucosamine residues). Considerable phenotypic variation occurs in this condition, probably due to residual enzyme activities which behave differently towards the various natural substrates which are synthesised to a different extent in different tissues. G_{M2} is the ganglioside molecule left after the terminal galactose residue has been removed from G_{M1}. In Tay–Sachs disease G_{M2} accumulates in neurons throughout the nervous system, due to a deficiency of hexosaminidase A. Hexosaminidase acts on both N-acetyl-β-D-galactosamine and N-acetyl-β-D-glucosamine residues and in the laboratory it is usually measured as N-acetyl-β-D-glucosaminidase (often referred to as NAG). The enzyme exists in two main isoenzyme forms, hexosaminidase A and B. It is composed of α and β polypeptide sub-units, the A form being composed of α and β sub-units and the B form containing only β sub-units. In Tay–Sachs disease the genetic abnormality affects the α sub-units, so that only the A isoenzyme is deficient. In a related condition called Sandhoff disease there is an abnormality affecting the β sub-units which leads to almost total hexosaminidase deficiency and accumulation of another glycolipid, called globoside, in the viscera, in addition to neuronal storage of G_{M2}. Theoretically, hexosaminidases should play a role in the degradation of glycosaminoglycans; increased storage of glycosaminoglycan has been demonstrated in Sandhoff cells.

Fig. 16.3 also indicates the site of action of β-glucosidase on the structure ceramide-β-glucoside (glucocerebroside). This enzyme is deficient in Gaucher disease in which glucocerebroside accumulates in the brain and lipid-laden histiocytes (Gaucher cells) are found throughout the reticulo-endothelial system. In Farber disease ceramide accumulates in many tissues and a deficiency of acid ceramidase, which splits sphingosine from its attached fatty acid, has been demonstrated in some patients. Excessive storage of sphingomyelin, a normal component of the nerve sheath, occurs in Niemann–Pick disease, due to a deficiency of sphingomyelinase (Fig. 16.4). Another glycosphingolipid, ceramide trihexoside (Fig. 16.5), accumulates in the tissues and plasma of patients with Fabry disease, due to a deficiency of a specific α-galactosidase (isoenzyme A using a synthetic substrate).

Sulphur containing glycolipids are known as sulphatides. Excessive

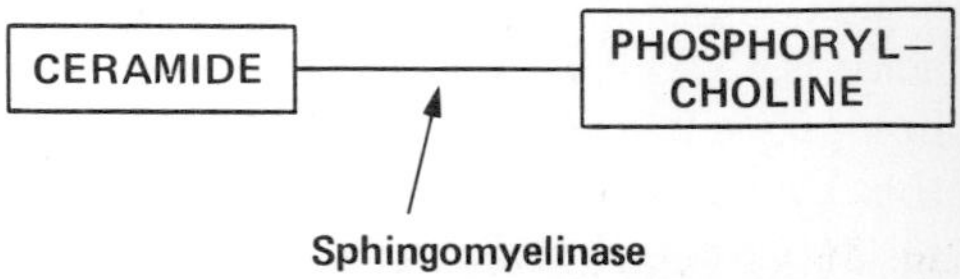

Fig. 16.4 Degradation of sphingomyelin by sphingomyelinase.

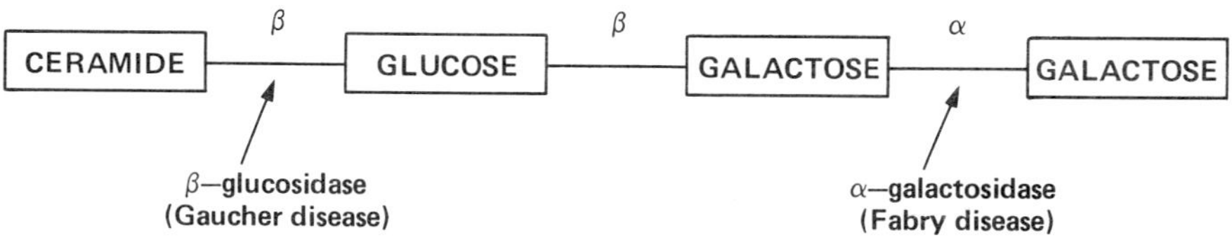

Fig. 16.5 Degradation of ceramide trihexoside. Diseases resulting from deficiencies of these enzymes are indicated in brackets.

amounts of galactosyl sulphatide (cerebroside sulphate) are found in the nervous system, viscera and urinary deposit of patients with metachromatic leucodystrophy (MLD) due to a deficiency of a specific sulphatase (Fig. 16.6). In the laboratory this enzyme is conveniently measured as aryl sulphatase A. Determination of aryl sulphatase activity involves the use of synthetic substrates, such as nitrocatechol sulphate. Since several different enzymes hydrolyse this substrate special methods are used to separate these activities into the so-called aryl sulphatases A, B and C. Aryl sulphatase A is deficient in MLD and as mentioned previously aryl sulphatase B is deficient in Maroteaux–Lamy disease. There is evidence for a common factor in these sulphatase activities. Aryl sulphatases A, B and C and steroid sulphatase are deficient in Austin leucodystrophy (multiple sulphatase deficiency), a condition with features of both MLD and mucopolysaccharidosis. It is interesting that patients with multiple sulphatase deficiency can also develop ichthyosis, and that aryl sulphatase C and steroid sulphatase deficiencies have recently been demonstrated in patients with X-linked ichthyosis. Aryl sulphatase C is not a lysosomal enzyme.

Krabbe disease (globoid cell leucodystrophy) is due to a deficiency of galactocerebroside β-galactosidase, an enzyme which is distinct from the previously mentioned β-galactosidase and which requires a radiolabelled natural substrate for its accurate determination. The site of action of this enzyme is shown in Fig. 16.6. The white matter of the brain is affected in this disease and galactocerebroside accumulates in numerous globoid cells.

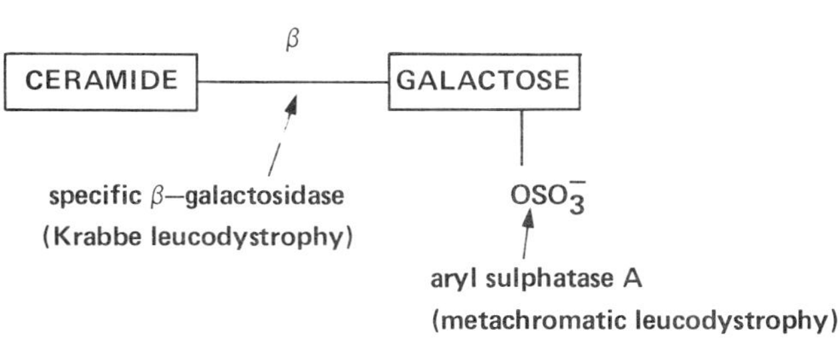

Fig. 16.6 Degradation of cerebroside sulphate (sulphatide). Diseases resulting from deficiencies of these enzymes are indicated in brackets.

Widespread accumulation of cholesteryl esters and triglycerides occur in Wolman disease due to a deficiency of acid esterase (acid lipase; acid cholesteryl ester hydrolase). Unlike the microsomal esterases, which have optimum activity in less acid conditions, acid esterase is resistent to the enzyme inhibitor E600. Acid esterase normally hydrolyses the cholesteryl esters in LDL. Plasma cholesterol and triglycerides are usually normal in Wolman disease. Cholesteryl ester storage disease is also associated with acid esterase deficiency but is a more chronic disorder which can present clinically after childhood. Premature atherosclerosis can occur and the plasma cholesterol is sometimes raised.

The mucolipidoses

Austin leucodystrophy was discussed in the previous section. I-cell disease (mucolipidosis II) presents with many of the clinical features of Hurler disease, but at an earlier stage. Cultured fibroblasts from these patients have characteristic inclusion bodies which are engorged lysosomes (hence 'I-cell'). Some biochemical findings are characteristic of I-cell disease and mucolipidosis III: there is a deficiency of several lysosomal enzymes in cultured fibroblasts, accompanied by an increased activity of these enzymes in the culture medium. However, the activities of acid phosphatase and acid β-glucosidase in I-cell fibroblasts are normal and the enzyme activities in some other tissues are apparently unaffected. The affected enzymes cannot be efficiently taken up by normal fibroblasts. Recent work suggests that there is defective phosphorylation of mannose residues of these glycoproteins (mannose-6-phosphate is thought to be an important 'recognition marker' for selective uptake by fibroblasts). The activities of some lysosomal enzymes in plasma are also increased and the presence of elevated aryl sulphatase A can be used as a screening test for I-cell disease.

Glycogenosis

Of the known glycogen storage disorders only Type 2 (Pompe's disease) is due to deficiency of a lysosomal enzyme. This was the first LSD to be defined biochemically when Hers demonstrated a deficiency of acid α-1,4-glucosidase (acid maltase). This enzyme can split maltose into its monosaccharide components (glucose). In the infantile form tissues such as muscle, liver, heart and tongue are infiltrated with an excess of glycogen of normal structure. Milder forms of the disease occur and in the adult form myopathy may be the main feature. The acid maltase deficiency is less widespread in some of these variants. Pompe's disease does not cause hypoglycaemia.

Other lysosomal storage disorders

Mannosidosis and aspartyl-glucosaminuria are conditions which present with some of the features of Hurler disease, such as coarse facies and bone deformities. They could represent abnormalities of glycoprotein degradation. In mannosidosis there is a deficiency of α-mannosidase and an increased urinary excretion of mannose-containing oligosaccharides with structural similarities to the carbohydrate sequences of glycoproteins. A deficiency of 1-aspartamido-β-N-acetylglucosamine hydrolase has been demonstrated in aspartylglucosaminuria. Aspartamido-β-N-acetylglucosamine forms one of the types of protein-saccharide linkage of glycoproteins and is also found in corneal keratan sulphate.

Fucosidosis is characterised biochemically by excessive storage of fucose-containing glycolipid and a deficiency of α-fucosidase. Profuse sweating with a high sweat sodium is one odd characteristic of this disease.

LABORATORY DIAGNOSIS

Screening tests

Early and accurate clinical and biochemical diagnosis of lysosomal storage disorders is essential now that prenatal diagnosis is possible for a large number of these disorders. This can give rise to a diagnostic dilemma. The clinical features of many lysosomal storage disorders develop slowly, so that the clinician needs to suspect such a disorder, given a minimum of diagnostic clues; laboratory help is desirable at this stage. However, many of the necessary tests can at present only be performed reliably in a few specialist centres, where resources and staff are frequently overstretched. Sometimes relatively simple preliminary tests can narrow down the possible specific diagnoses, thus allowing the specialist laboratory to concentrate on the most relevant specific tests. The ultimate biochemical diagnosis must in general be based on the demonstration of a specific lysosomal enzyme abnormality.

Urine tests

Microscopy of a fresh urine sediment and chemical analysis of the supernatant can provide useful pointers to particular LSD. In Fabry disease the urine sediment of both patients and female carriers usually contains foam cells with birefringent lipid droplets; Maltese-cross configurations are seen when polarised light is used. In MLD and Austin leucodystrophy sulphatide can be demonstrated in the renal epithelial cells of the urine sediment. Sulphatide (like GAG) causes metachromasia (change in light absorption spectrum) with dyes such as toluidine blue and cresyl fast violet.

Fresh urine, but not an early morning specimen, should be used for this test, which is said to be reliable even in the earlier stages of the disease (Lake 1965). Only intracellular metachromatic material counts towards a positive diagnosis. Lake (1981) now recommends the use of toluidine blue for this test rather than cresyl fast violet (Lake 1965).

Excessive urinary excretion of GAGs is an important diagnostic feature of the mucopolysaccharidoses. Various opinions on screening methods can be found in one source (Holton & Ireland 1975) and an excellent review has been written by Pennock (1976). With the possible exception of some patients with Morquio disease the level of total GAG excretion in affected children is generally well above an age-matched control range. Normally, GAG and creatinine excretion display diurnal variation. In practice, the GAG:creatinine ratio provides a useful index of GAG excretion and values for 24 hour urine collections are generally similar to those for random specimens collected between mid-morning and mid-afternoon. The GAG:creatinine ratio is relatively high in babies, falls through infancy and early childhood and falls again after puberty (see for example, Whiteman & Young 1977). Most methods for the determination of total GAG initially involve the formation of precipitable complexes with cationic detergents such as cetylpyridinium chloride (CPC) and cetyltrimethylammonium bromide (CTAB) or cationic dyes such as Alcian Blue 8GX. Optimum formation of these GAG complexes depends on many interacting factors such as: type of GAG and the nature and extent of its anionic groups, molecular weight of the GAG, nature and concentration of the organic cation, pH, temperature and critically on the nature and concentration of inorganic cations. Difficulties arise from the variations found in the molecular weights and charge properties of urinary GAGs which can be taken to represent degraded versions of the larger molecular weight species found in the tissues (purified forms of the latter are generally used as standards). Because of these problems and the large variations which can occur in GAG concentrations at different ages and with renal concentration and dilution, simple spot tests and turbidity tests are not advisable unless special precautions are taken, as in Pennock's CPC-citrate method (Pennock 1969). Classically, analysis of urinary GAG involves the determination of the uronic acid content of the GAG isolated using the precipitation procedures outlined above. The carbazole or modified carbazole reactions are used to determine uronic acids; these involve heating and handling mixtures composed largely of concentrated sulphuric acid. Another method, which involves the measurement of Alcian Blue 8GX complexes (Whiteman 1973), combines simplicity with the discrimination of a quantitative method; for diagnostic purposes it has proved to be as reliable as other quantitative methods. The Alcian Blue method also measures keratan sulphate, which does not contain uronic acid. Laboratories offering GAG analysis should have their own age-related reference ranges based on their methods; such ranges often differ between centres even when apparently similar methods are used.

Table 16.3 Urinary GAG patterns in the mucopolysaccharidoses and related conditions.

Disease	GAG typically excreted in excess[1]
Hurler	DS, HS
Hurler–Scheie	DS, HS
Scheie	mainly DS
Hunter	DS, HS[2]
Sanfilippo	HS[3]
Morquio	KS, CS (6-sulphate)
Maroteaux–Lamy	DS
MPS VII	DS, HS (in some)
MPS VIII	HS, KS
Austin leucodystrophy (multiple sulphatase deficiency)	DS, HS
GM1-gangliosidosis	KS-like substances in some

GAG, glycosaminoglycan; DS, dermatan sulphate; HS, heparan sulphate; CS, chondroitin sulphate; KS, keratan sulphate.

1 CS is the main GAG component in the urine of normal children. In general, for a particular enzyme deficiency severe clinical phenotypes are associated with higher levels of GAG excretion than the milder phenotypes.

2 The HS:DS ratio is usually higher in Hunter disease than in Hurler disease.

3 Some techniques (e.g. two-dimensional electrophoresis) reveal differences in the properties and distribution of heparan sulphate subfractions in some types of Sanfilippo disease and in Hurler and Hunter diseases.

In some patients with Morquio disease and in some adults with mild forms of mucopolysaccharidosis total GAG excretion may not be clearly distinguishable from the reference values. In such cases qualitative analysis of the individual GAG components may be helpful. Qualitative analysis can also provide a reasonably accurate indication of the type of mucopolysaccharidosis. Cellulose-acetate electrophoresis of isolated GAG in 0.1 M-barium acetate (Wessler 1968) allows separation of chondroitin sulphate, dermatan sulphate and heparan sulphate components (in order of decreasing mobility) which can then be stained with Alcian Blue or toluidine blue reagents. One drawback of this system is that keratan sulphate migrates with the chondroitin sulphate component. Keratan sulphate can be demonstrated using thin-layer chromatographic methods or by differential staining of the GAG bands after electrophoresis. Urinary keratan sulphate stains with Alcian Blue 8GX in the presence of 0.5 M magnesium chloride (at pH 5.8) while other urinary GAG, such as chondroitin sulphates, do not form complexes at such high salt concentrations. Greater resolution of urinary GAG components can also be obtained by two-dimensional electrophoresis (Whiteman; in Holton & Ireland 1975). There are of course complicated methods available for GAG analysis, such as fractionation on ion-exchange columns with chemical analysis of the eluate, but in practice the results are rarely more helpful than those obtained by the simple methods outlined above.

Normal children excrete predominantly chondroitin sulphate. Table 16.3 indicates the typical patterns of GAG excretion found in the mucopolysaccharidoses and related conditions. Some patients with the clinical features of Hurler disease have normal or only slightly elevated total GAG excretion and do not have abnormal dermatan or heparan sulphaturia. Diseases such as GM_1-gangliosidosis, mannosidosis, I-cell disease and aspartylglucosaminuria should then be considered. Keratan-sulphate-like material has been demonstrated in the urine of patients with GM_1-gangliosidosis.

Increased GAG excretion has also been reported in other connective tissue disorders such as rheumatoid arthritis, disseminated lupus erythematosus, polymyositis, dermatomyositis, scleroderma, polychondritis, osteogenesis imperfecta and osteopetrosis. In most of these conditions mucopolysacchariduria is of a normal pattern and is usually only slightly elevated above the control range. Accurate interpretation of patterns of urinary GAG excretion requires considerable experience of the methods concerned. 'One-off' attempts with these procedures are generally unhelpful. Bacterial contamination of urine specimens can result in altered GAG content. A preservative such as merthiolate (1 in 10000 w/v) should be added and specimens should be stored deep-frozen until they can be analysed.

Characteristic oligosaccharides and 1-aspartamido-β-N-acetylglucosamine can be demonstrated by special procedures in the urine of patients with mannosidosis and aspartylglucosaminuria respectively but in practice the presence of Hurler-like features together with absence of excessive mucopolysacchariduria suggests the possibility of these conditions.

Other non-specific tests

Plasma aryl sulphatase A is strikingly elevated in I-cell disease and tartrate-stable acid phosphatase is raised in Gaucher disease and sometimes in Niemann–Pick disease. Plasma GAG is increased in mucopolysaccharidosis but the methods used tend to be complicated and so far relatively little is known about the diagnostic value of these tests compared to urinary GAG analysis. Inclusions of various types are found in the white blood cells in: the mucopolysaccharidoses, mucolipidoses, GM_1-gangliosidosis, Niemann–Pick disease, mannosidosis and aspartylglycosaminuria, and these findings can be helpful in differential diagnosis when the services of an experienced histochemist are available. Biopsy and histology of bone marrow, rectum (to demonstrate neuronal storage), peripheral nerves (in metachromatic leucodystrophy) and internal organs should now be largely unnecessary in the diagnosis of those diseases in which the relevant specific enzyme deficiencies can be demonstrated in more accessible material such as blood.

Enzymology

Before services such as prenatal diagnosis can be offered to parents of children affected by lysosomal storage disorders the precise biochemical diagnosis must be established in the index case. This means that the specific enzyme deficiency must be demonstrated whenever possible. In most LSD the enzyme deficiency is not only found in various internal organs but can also usually be demonstrated in easily-accessible materials such as white blood cells and plasma and in cultured skin fibroblasts. Sometimes the abnormality can be demonstrated using urine, tears or hair follicles.

Assay methods used for lysosomal enzymes may involve the use of radiolabelled natural substrates or synthetic substrates which liberate compounds that can be measured by colorimetry or spectrofluorimetry. Colorimetric methods are sometimes too insensitive to determine the low activities of some of these enzymes in relatively small amounts of material. The use of microtechniques and spectrofluorimetry has overcome some of these problems. 4-methylumbelliferone is one particular highly fluorescent compound which can be linked through its phenolic hydroxyl group to substances such as sugars to form glycosides and other derivatives. After incubation of the enzyme with the methylumbelliferyl substrate at the appropriate acid pH the reaction is stopped with an alkaline buffer and the fluorescence of the liberated methylumbelliferone is then measured. The excitation and emission wavelengths used are such that only the fluorescence of the free form is measured. Slow non-specific hydrolysis of the substrate may occur and therefore substrate blanks are required in each assay. In addition, suitable blanks to allow for possible endogenous fluorescence in the sample (e.g. due to drugs) are advisable. Apart from conditions such as substrate concentration (some substrates are poorly soluble), pH and temperature, lysosomal enzyme activity or stability is often influenced by other factors such as the type of buffer used and the presence of inorganic ions and macromolecules (e.g. protein, GAG). Some substrates need to be solubilised with detergents.

Although synthetic substrates are very convenient and can provide great sensitivity they can suffer from the disadvantage of being less specific than the natural substrates. However, the relevant enzyme (or isoenzyme) activity can sometimes be distinguished from other activities. For instance, hexosaminidase A and B (both of which hydrolyse synthetic substrates) are separated by electrophoresis or ion-exchange chromatography and the more labile A form (deficient in Tay–Sachs disease) can be differentially inactivated by heat or altered pH conditions.

Activities of specific lysosomal enzymes in white-cells and cultured skin fibroblasts or amniotic-fluid cells are usually expressed in terms of the protein content of the cells or in relation to a second marker enzyme.

PRENATAL DIAGNOSIS AND CARRIER DETECTION

Prenatal diagnosis of LSD must be carried out by highly trained laboratory personnel; at present only a few centres are providing such services. The disastrous consequences of mistakes in this field are obvious and there are many potential pitfalls for the inexperienced (some of the problems have been discussed by Ellis 1972). Amniocentesis is usually performed at around the fifteenth week of pregnancy and the few live amniotic-fluid cells thus obtained are cultured to provide sufficient material for biochemical studies. This may require 2–4 weeks in culture, although the use of ultramicrotechniques can shorten the time required (see Galjaard 1979). Apart from the time factor other problems sometimes associated with cell culture are infection, failure of cells to grow and overgrowth of the wrong cell type (e.g. maternal rather than fetal). Biochemical tests performed on the cultured cells are usually enzyme assays of the type described in the preceding section, but other methods are sometimes used. For example, pulse-chase methods using ^{35}S-sulphate can indicate abnormal GAG metabolism in the mucopolysaccharidoses. The supernatant amniotic fluid also has an abnormal GAG pattern after the fourteenth week when the fetus is affected by some types of mucopolysaccharidosis; this is probably due to contributions from fetal urine. At this stage the GAG in normal amniotic fluid consists mainly of hyaluronic acid and chondroitin sulphate, the former becoming proportionately less as pregnancy advances. In pregnancies affected by Hurler and Sanfilippo diseases amniotic fluid can contain an excess of dermatan sulphate and heparan sulphate respectively. The reliability of prenatal diagnosis depends very much on the analytical techniques used and the experience of the person interpreting the results.

Carrier detection is possible for a number of LSD. In general, heterozygotes have enzyme activities between those of homozygotes and the control range but ranges for carriers tend to overlap the control ranges in many instances. Discrimination is usually better with cultured fibroblasts than with white-blood cell preparations and is sometimes improved if results are related to the activity of a second reference enzyme or isoenzyme. Large-scale population screening has been undertaken for carriers of Tay–Sachs disease in the Jewish population in some parts of the United States. In the UK carrier-detection has been restricted to affected families.

TREATMENT

Only symptomatic treatment has been generally available to patients with LSD. Recently, there has been considerable interest in the possibility of enzyme replacement therapy for patients with some forms of LSD (e.g.

mucopolysaccharidosis, Gaucher and Fabry diseases). It is known that cultured cells from patients with some LSD can temporarily correct their faulty metabolism by taking up enzymes from a medium in which normal cells were previously grown. Only enzymes containing appropriate recognition residues (probably residing in the carbohydrate portions of the glycoprotein enzymes) are taken up by selective endocytosis. The source of the enzyme is therefore important. Space does not permit a detailed discussion of the various types of specific therapy proposed or tried but a few observations can be made (useful reviews have been provided by Desnick *et al.* 1976; Lloyd & Griffiths 1979). Attempts to supply enzymes through infusions of plasma or white cells have not met with much success and such treatments are impracticable on a long-term basis. Implantation of histocompatible fibroblasts and bone marrow transplantation have recently been tried in some forms of mucopolysaccharidosis and encouraging changes in biochemical, and possibly clinical, parameters have been observed. Clearly, research will continue in this area and will include transplantation and administration of purified enzymes. Most studies to date have involved patients with advanced disease but the effect of therapy started at an earlier stage in the natural history of the disease is now being investigated. Enzyme replacement therapy probably offers the greatest hope to those patients with LSD that do not seriously affect the central nervous system. If enzyme replacement therapy eventually finds a place in the standard management of LSD, then hospital laboratory services can expect to become increasingly involved in the relevant biochemical monitoring.

REFERENCES

BRADY R. O. (1978) Sphingolipidoses. *Ann. Rev. Biochem.* **47**, 687–713.

CANTZ M. & GEHLER J. (1976) The mucopolysaccharidoses: inborn errors of glycosaminoglycan catabolism. *Hum. Genet.* **32**, 233–55.

DEAN R. T. & BARRETT A. J. (1976) Lysosomes. *Essays Biochem.* **12**, 1–40.

DESNICK R. J., THORPE S. R. & FIDDLER M. B. (1976) Toward enzyme therapy for lysosomal storage diseases. *Physiol. Rev.* **56**, 57–99.

ELLIS R. B. (1972) Prenatal screening for inborn errors of metabolism. *Ann. Clin. Biochem.* **9**, 109–11.

GALJAARD H. (1979) Early diagnosis and prevention of genetic disease. *Ann. Clin. Biochem.* **16**, 343–53.

HOLTON J. & IRELAND J. T. (1975) (eds.) Chapters on screening procedures for mucopolysacchariduria by Lewis P. W., Kennedy J. F., Raine D. N. (Ch. 15); Whiteman P. (Ch. 17); Pennock C. A. (Ch. 18). In *Inborn Errors of Skin, Hair and Connective Tissue*. Medical and Technical Publishing, Lancaster.

LAKE B. D. (1965) A reliable rapid screening test for sulphatide lipidosis. *Arch. Dis. Child.* **40**, 284–5.

LAKE B. D. (1981) Metabolic disorders; general considerations. In Berry C. L. (ed.) *Paediatric Pathology*, Chapter 14. Springer-Verlag, Berlin.

LLOYD J. B. & GRIFFITHS P. A. (1979) Enzyme replacement therapy of lysosomal storage disease. *Front. Biol.* **48**, 517–32.

LÜLLMANN-RAUCH R. (1979) Drug-induced lysosomal storage disorders. *Front. Biol.* **48**, 49–130.

MCKUSICK V. A. (1972) *Heritable disorders of connective tissue*, 4th edn. C. V. Mosby Company, St Louis, USA.

NEUFELD E. F., LIM T. W. & SHAPIRO L. J. (1975) Inherited disorders of lysosomal metabolism. *Ann. Rev. Biochem.* **44**, 357–76.

PENNOCK C. A. (1969) A modified screening test for glycosaminoglycan excretion. *J. Clin. Path.* **22**, 379–80.

PENNOCK C. A. (1976) A review and selection of simple laboratory methods used for the study of glycosaminoglycan excretion and the diagnosis of mucopolysaccharidoses. *J. Clin. Path.* **29**, 111–23.

STANBURY J. B., WYNGAARDEN J. B. & FREDRICKSON D. S. (1978) (eds.) Chapters on lysosomal storage diseases written by acknowledged authorities. In *The metabolic basis of inherited disease*, 4th edition. McGraw-Hill, New York.

WESSLER E. (1968) Analytical and preparative separation of acidic glycosaminoglycans by electrophoresis in barium acetate. *Anal. Biochem.* **26**, 439–44.

WHITEMAN P. (1973) The quantitive determination of glycosaminoglycans in urine with Alcian Blue 8GX. *Biochem. J.* **131**, 351–7.

WHITEMAN P. & YOUNG E. (1977) The laboratory diagnosis of Sanfilippo disease. *Clin. Chim. Acta* **76**, 139–47.

Chapter 17

Biochemical Aspects of Muscle Disorders

JOHN Z. HECKMATT & VICTOR DUBOWITZ

The muscle disorders of childhood may be divided into two main groups, the myopathies and the neuropathies. Myopathies include the various progressive muscular dystrophies, the congenital myopathies, congenital dystrophy and metabolic and inflammatory myopathies. The principal neuropathies are spinal muscular atrophy and the motor neuropathies. In the differential diagnosis of these disorders the biochemical measurement of plasma enzymes, electrophysiological studies (EMG) and muscle biopsy are the three main investigations of value. The serum enzymes and electrodiagnostic studies are best regarded as screening investigations, and the muscle biopsy a definitive investigation likely to give a more exact diagnosis.

PLASMA ENZYMES

The release of intracellular enzymes, most notably creatine kinase, from diseased muscle is a well documented phenomenon (Pennington 1980). Measurement of creatine kinase activity in plasma is a useful investigation in the diagnosis of muscle disease in both children and adults. This enzyme, systematic name ATP-creatine N-phosphotransferase (EC.27.3.2), catalyses the following reversible reaction:

$$\text{ATP} + \text{creatine} \rightleftharpoons \text{ADP} + \text{phosphocreatine}$$

It is often referred to as creatine phosphokinase, CPK or CK. Creatine kinase is present in high concentrations in both heart and skeletal muscle and is essential in maintaining ATP levels in muscle cells at times of rapid energy utilisation.

Skeletal muscle disease should be suspected in the presence of chronically elevated plasma CK activity, combined with a suggestive clinical picture. Certain cardiac and cerebral disorders also lead to elevated plasma CK activity, but if the clinical findings are considered this should not cause confusion.

The activities in plasma of lactate dehydrogenase, pyruvate kinase, aldolase and the transaminases may also be raised in muscle disease, but these enzymes are less specific to muscle and, with the possible exception of pyruvate kinase, are a less sensitive index of myopathy than is the level of CK (see also p. 411).

Creatine kinase is present in only very low concentrations in red cells and slight haemolysis before separation of the plasma will not significantly alter the values for CK activity. It is therefore possible, as for example in family studies where blood is collected in the home, to send whole blood samples by post to the clinical laboratory without the risk of haemolysis causing seriously inaccurate results.

There are a number of methods available for the estimation of CK activity in plasma. The enzyme may be measured by the rate of the reaction in either direction but is normally determined by following the rate of ATP generation. Colorimetric, spectrophotometric and fluorimetric methods have all been published and there are also a number of commercial kits available. These kits are preferred by some laboratories because of the complexity, cost, and lack of stability among CK reagents, but not all kits are entirely satisfactory; Pryce *et al.* (1977) reviewed 19 and found only 7 were acceptable.

Because of the wide variation in methodologies, reference ranges vary and each chemical pathology laboratory will derive its own reference values from the method which it employs. The log of plasma CK activity shows a roughly Gaussian distribution (Pennington 1980). Activity is slightly higher in males than in females and in negroes than caucasians; highest values are seen in infancy and activity in cord blood is sometimes grossly elevated (Moss *et al.* 1981; Griffiths 1968). During pregnancy there is an initial fall in values followed by a rise towards term, and this is important when interpreting screening tests on possible Duchenne dystrophy carriers during early pregnancy. Minor injuries and viral illness can cause a short-term rise in CK activity and plasma CK is elevated for several days after strenuous physical exercise. Trained athletes may have levels which are permanently above the upper limit of the reference range.

The most grossly raised plasma CK levels are found in the childhood muscular dystrophies, the most common and severe of these being Duchenne muscular dystrophy.

MUSCULAR DYSTROPHIES

Duchenne dystrophy

Duchenne dystrophy is an X-linked recessive condition which affects only boys, with the onset of symptoms in early childhood. Most children present at 3–6 years because of difficulties in running, jumping and climbing stairs. Though not seriously weak at this early stage, an affected boy tends to fall frequently, and is likely to have some difficulty rising from the floor.

Some boys present earlier, before the development of overt weakness, because of an associated delay in motor or speech development. About half fail to walk until the age of 18 months. This delay is usually associated with intellectual retardation.

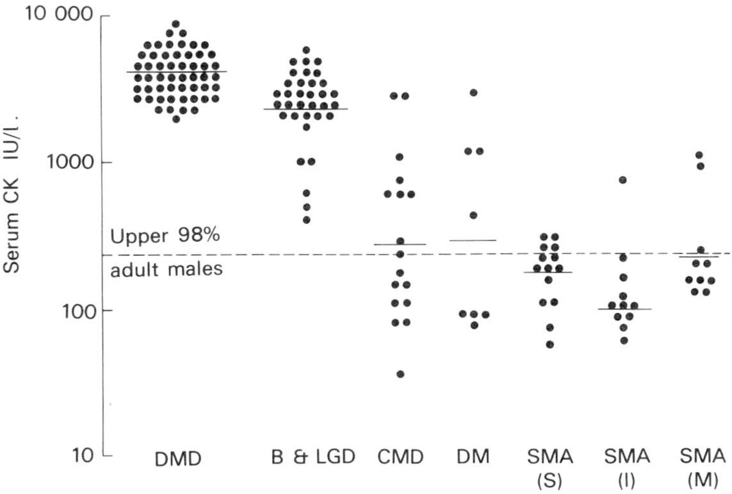

Fig. 17.1 Serum CK activity in neuromuscular disease—personal series. All patients with progressive dystrophy were ambulant; serum CK activity measured at time of diagnostic biopsy; patients with dermatomyositis were those seen by the authors before treatment commenced. CK activity measured by the Scandinavian Reference Method (1979) (*Scand. J. Clin. Lab. Invest.* **39**, 1–5). DMD: Duchenne muscular dystrophy; B&LGD: Becker and limb girdle muscular dystrophy; CMD: Congenital muscular dystrophy; DM: Dermatomyositis; SMA: Spinal muscular atrophy (S: severe, unable to sit; I: intermediate, able to sit, unable to stand; M: mild, able to walk, not to run).

The serum CK activity is grossly elevated and may be used to aid clinical diagnosis (Fig. 17.1). In the early stages of the disease the level should be at least ten times the upper limit of normal. A normal or only moderately elevated level is against the diagnosis. Serum CK is useful as a screening test in excluding dystrophy as a possible cause in boys who present with unexplained delay in motor development and are not walking by 18 months (Gardner-Medwin 1980).

The subsequent clinical picture is of progressive deterioration. Plasma CK tends to fall as the muscle bulk declines, particularly after the child becomes wheel-chair-bound. Most boys die before the age of 20 years, although a few may live until they are about 25 years old.

Other disorders characterised by proximal skeletal muscle weakness include the milder sex-linked Becker dystrophy and the autosomal recessive limb girdle dystrophy. The CK tends to be slightly lower overall in Becker and limb girdle dystrophy than in Duchenne dystrophy, but this is rarely of much practical value in the differential diagnosis of individual cases as the overlap is so great (Fig. 17.1).

Spinal muscular atrophy of mild severity (Kugelberg–Welander syndrome) often presents with a similar clinical picture to the dystrophies, and the plasma CK may be elevated though not so grossly as in Duchenne dystrophy (Fig. 17.1).

The EMG in Duchenne dystrophy shows a myopathic pattern, but a more definite diagnosis is made by examination of a muscle biopsy. We routinely perform needle biopsies of muscle under local anaesthetic (Edwards *et al.* 1980). Unfortunately muscle biopsy is of little help in distinguishing between the Duchenne, Becker and limb girdle types of muscular dystrophy and a final diagnosis must be based on clinical criteria.

Becker dystrophy

This is an X-linked recessive disorder which is genetically distinct from Duchenne dystrophy. The clinical pattern is very similar but the disease follows a milder course. The onset of symptoms is usually later than in Duchenne but some boys who have only a mild degree of weakness may present early with a history of muscle cramps on exertion. These cramps may be accompanied by the passage of dark urine containing myoglobin. Plasma CK at this early stage of the disease is grossly elevated (Fig 17.1). Most boys retain the ability to walk during adolescence and some remain ambulant into adulthood. Examination in the early stages shows mild proximal weakness and usually the presence of calf hypertrophy.

Limb girdle dystrophy

This is an autosomal recessive condition affecting both females and males and is characterised by progressive proximal weakness predominantly of the pelvic girdle. There is considerable variability in the time of presentation, some patients having an onset of weakness in early childhood with a rapid loss of ambulation, while others present later and follow a clinical course similar to that of Becker dystrophy. As in Becker dystrophy muscle cramps on exertion may be a presenting feature.

Congenital muscular dystrophy

Congenital muscular dystrophy is an autosomal recessive condition which is characterised by dystrophic change within the muscle similar to that seen in other dystrophies. The dystrophic process commences *in utero* and clinical signs are already present at birth. There is general muscle hypotonia and weakness with mild facial involvement. There may be some respiratory involvement. Joint contractures can be a serious problem and already be present at birth (arthrogryposis) or commence later. Intellectual impairment is occasionally seen, especially in Japan where the disorder may represent a separate entity. The disorder is non-progressive and there is generally a tendency to functional improvement so long as immobility is avoided and active supportive treatment is given. There is no means of antenatal diagnosis.

Measurement of serum CK activity is of some value in the diagnosis of this disorder, being elevated in about 50% of patients (Fig. 17.1). The severity of the disease is variable but there appears to be no relationship between the clinical severity and the serum CK level.

Carrier detection in Duchenne muscular dystrophy

Although several new techniques recently have been applied to the problem of carrier detection in Duchenne muscular dystrophy (Emery 1980), the measurement of plasma creatine kinase (CK) activity remains the most widely used and is the most thoroughly validated. Unfortunately because the muscle cell is multinucleate and because the Lyonisation of the X chromosome in females is a random event, there is a wide overlap between the CK values in normal women and in Duchenne dystrophy carriers. Creatine kinase is elevated above the 95th centile in about two-thirds of carriers, but not all mothers are carriers as there is a high spontaneous mutation rate, with about one-third of all cases arising as a result of new mutations.

Since the possibility of being a carrier is not eliminated if a woman has CK values within the reference range, techniques for the prediction of carrier status have been developed using the distribution of CK values in normal women and known carriers to calculate probability estimates. Final predictions can then be made by combining risk based on plasma CK with genetic information. One advantage of this technique is that it allows accurate comparisons to be made between plasma CK results from different laboratories (Tippett *et al.* 1982).

As there is a certain amount of day-to-day variability in serum CK activity, it is current practice to do the estimations on three separate occasions. It is important to ensure that there is no reason for spurious elevation, for example heavy exercise within two days before the blood collection.

Serum CK activity decreases with age in both carriers and normal women. There is some evidence to suggest that carriers are more likely to have elevated serum CK activity during childhood and adolescence than when they become adult, though this is still disputed. It must always be remembered that it is difficult to interpret values in children when they are in the borderline adult range. In Becker dystrophy, a similar situation exists with a higher proportion of carriers having elevated serum CK activities in childhood.

Attempts have been made to combine the measurement of other enzymes with plasma CK activity in the hope of enhancing the sensitivity of carrier detection. The measurement of pyruvate kinase and lactate dehydrogenase has been recommended (Percy *et al.* 1979; Roses *et al.* 1977) but this work is controversial (Seay *et al.* 1978; Somer *et al.* 1980). One drawback is that the increased plasma activity of all these various enzymes probably reflects the same abnormality within the muscle, that is the 'leaky membrane'.

A few carriers have overt weakness and calf hypertrophy. Some carrier women have a definite EMG abnormality which is more reliably distinguished using computerised methods to analyse the interference pattern. Approximately 20% of carriers have unequivocal but mild change on muscle biopsy. Detailed quantitation of muscle morphology should help to pick up a further group of carriers (Maunder-Sewry & Dubowitz 1981). A combination of all of these techniques may allow more accurate prediction than will the use of plasma enzyme estimations alone.

Antenatal diagnosis

A woman who is at high risk of having a son with Duchenne dystrophy may opt for antenatal fetal sexing and the abortion of any male fetus. This approach has two obvious drawbacks. First, half the aborted male fetuses will be normal and secondly the number of female carriers will increase as half the females at risk will be carriers.

It was hoped that fetal blood sampling would allow accurate detection of the affected male fetus by enabling the measurement of serum CK activity in fetal serum, but a recent report of false negative results suggests the affected male cannot be detected with confidence in this way (Golbus *et al.* 1979).

Screening for dystrophy

Semi-quantitative micro-assay of plasma CK activity may be carried out by a luciferase method on spots of blood dried on filter paper. The sample can be collected at the same time as the Guthrie test for phenylketonuria (Zellweger & Antonik 1975). Screening has been used to detect Duchenne dystrophy at a preclinical stage and the technique may even be powerful enough to pick up female carriers (Beckman *et al.* 1978). Thus, if the first child turned out to be an affected male or a carrier female, the family could be advised not to have a further son. Advocates argue that this approach will achieve a significant reduction in the incidence of Duchenne dystrophy. Such an approach raises ethical problems whilst no treatment is available. Unfortunately, parents do not always accept the need to avoid further pregnancies (Drummond 1979). At present, screening has not been adopted on any appreciable scale. No doubt if an effective drug treatment could be found, screening and early detection would become a major factor in management.

Malignant hyperthermia (malignant hyperpyrexia)

This disorder occurs as a catastrophic and frequently fatal complication of general anaesthesia. It is particularly tragic in that it affects otherwise healthy children and young people who may be receiving a general anaesthetic for a relatively minor procedure. The first indication may be the onset of jaw rigidity during induction of anaesthesia and this may be so profound as to prevent intubation. The child's temperature rises, sometimes with alarming rapidity, reaching 44 °C or even higher. Generalised muscle rigidity occurs in about 80% of patients.

A profound metabolic disturbance accompanies this acute onset. There is hypoxia, hypercapnia and profound acidosis. Hyperphosphataemia and hypocalcaemia occur. There is active destruction of muscle with gross elevation of muscle enzymes in the plasma. The CK may rise to several thousand units. There is often myoglobinuria, haemoglobinuria and a consumptive coagulopathy. Should an attack develop, intravenous Dantrolene should be administered and strenuous attempts made to limit the rises in body temperature, plasma lactate and potassium.

It is not practical to use serum CK activity to screen every individual about to undergo a general anaesthetic. However the diagnosis of susceptible children can now be made with reasonable confidence using an *in vitro* muscle preparation which, when compared with normal human muscle, shows a greater propensity to contracture in the presence of caffeine (Moulds & Denborough 1974a, b; Ellis *et al.* 1978). Alternatively the measurement of the active form of phosphorylase in the muscle has been used since Willner *et al.* (1980) have shown this to be raised in muscle specimens from susceptible subjects. The use of either of these diagnostic techniques is the province of a specialised laboratory and requires the collection of a few grams of muscle from an open biopsy.

The incidence in the UK is about 1:200000 anaesthetics. Inheritance is autosomal dominant and in about 50% of families the propositus and other susceptible relations have a persistently elevated serum CK. There is also an autosomal recessive variant, usually associated with an overt myopathy and a number of developmental abnormalities including short stature, webbing of the neck, spinal deformity, ptosis and an undescended testicle. Excellent reviews of the condition are available (Denborough 1980; Ellis & Halsall 1980).

Acute rhabdomyolysis: recurrent myoglobinuria

The leakage of myoglobin into the urine reflects acute necrosis of muscle with loss of permeability of the cell membrane. Myoglobin is a relatively small molecule (MW 17000) and is readily cleared by the kidney. In association with muscle necrosis there is muscle contracture, pain, tenderness and pitting oedema. Extensive muscle necrosis is accompanied by respiratory and renal failure. Serum enzymes become grossly elevated.

Myoglobinuria can be classified according to cause although there is overlap between the various groups:

1 Direct trauma: crush syndrome, burns, electric shock and acute ischaemia to a limb.

2 Toxic causes such as poisoning by carbon monoxide, alcohol, barbiturates, Malaya sea snake and hornet venom.

3 Paroxysmal exertional myoglobinuria: associated usually with an underlying myopathy such as dystrophy or a metabolic disorder. The myoglobinuria is accompanied by muscle cramps or stiffness.

4 Exertional rhabdomyolysis: this term implies a more extensive degree of muscle involvement. It may occur in healthy individuals after severe exercise or in an individual with an underlying myopathy.

5 Sporadic and after infection: such as following a viral upper respiratory tract infection. Muscle involvement can be severe. There may be an underlying metabolic myopathy, e.g. carnitine palmityl transferase deficiency.

Myoglobin is usually measured by radioimmunoassay. Kits are available for this technique*.

Dermatomyositis

Childhood dermatomyositis is generally classified along with the collagen disorders. Presentation is with muscle weakness accompanied by general malaise and misery. These signs are very consistent and our rule is: muscle weakness + misery = dermatomyositis, until proved otherwise.

Muscle weakness involves the limbs and is more severe proximally. The muscles of the pharynx and larynx may also be involved. Particularly dangerous is respiratory involvement with the risk of ventilatory failure. At initial presentation many children have a rash. This is frequently seen as a faint bluish–purple discolouration around the eyes. There is tremendous variability in the relative severity of skin and muscle involvement.

The ESR may be moderately elevated. The serum CK is elevated in about half the patients (Fig. 17.1). The EMG is often but not always myopathic. Abnormalities are usually seen on muscle biopsy but a normal result does not exclude the diagnosis.

With correct management the prognosis should be good (Dubowitz 1978). The disease responds to steroids (1mg/kg bodyweight per day) and once the clinical response is sustained the dose can be progressively reduced, with the aim of stopping treatment within a few months. Clinical parameters are much more useful as a guide to progress than is the plasma

* CIS UK Ltd. (Biomedical Products), 354 Ballards Lane, London N12 0ET.

CK level, which may in fact rise as a result of muscle repair. There are risks in using steroids in high dosage for prolonged periods, particularly those of a steroid myopathy. This is illustrated by the case study of Edwards *et al.* (1979), who performed careful nitrogen balance studies in a woman who had acute dermatomyositis and showed that there was net nitrogen loss on high doses of steroids. They postulated that this loss reflected active muscle breakdown and were able to show that their patient went into positive nitrogen balance as the steroids were withdrawn. Their findings, however, should not deter adequate suppression of the disease with steroids. The childhood form of dermatomyositis is particularly sensitive and responds to steroid therapy. With continuing disease activity it may be necessary to employ immunosuppressants to cover steroid withdrawal. We prefer azathioprine as it is relatively free of side-effects.

THE METABOLIC MYOPATHIES

Glycogen storage disorders

Five types of glycogen storage disorders affect muscle, and these have been reviewed by Di Mauro (1980) and Dubowitz (1978, 1980).

Glycogen is a branched-chain polysaccharide which is present in most mammalian cells, the most important sites of storage being liver and muscle. The molecule consists of straight chains of glucose residues which are branched at various points. The residues in the straight chains are joined by $\alpha(1\text{–}4)$ linkages and the branches are initiated by $\alpha(1\text{–}6)$ linkages. Glycogen is broken down in a stepwise process starting with the splitting of the accessible $\alpha(1\text{–}4)$ linkages by the enzyme phosphorylase. Debrancher enzyme is then required to hydrolyse the 1–6 linkages, thus allowing the straight chains that remain to be broken down completely by the further action of phosphorylase.

Indirect tests such as tolerance tests have been largely replaced by the direct assay of enzymes in liver, leucocytes or erythrocytes as appropriate. These must be performed in specialised laboratories and close collaboration between the surgeons and those who will examine a liver or muscle biopsy biochemically or histologically is essential, as the tissue must be dealt with immediately.

Alpha-1,4-glucosidase deficiency (type II glycogenosis; Pompe's disease)

In this disease there is deficiency of the enzyme α-1,4-glucosidase (acid maltase). This enzyme is sited in cell lysosomes and is capable of independently hydrolysing the entire glycogen molecule down to glucose.

Three forms of Pompe's disease are recognised (Di Mauro 1980)—an infantile form, a childhood form and an adult form. In the classic infantile form there is massive accumulation of glycogen in all tissues of the body,

Table 17.1 Glycogenoses affecting muscle.

Type	Names	Enzyme deficiency	Tissues involved besides muscle	Antenatal diagnosis possible
II	Pompe's disease	α-1,4-glucosidase (acid maltase)	All	Yes
III	Limit dextrinosis Forbes' disease Cori's disease	Amylo-1,6-glucosidase (debranching enzyme)	Liver and heart	Yes
IV	Amylopectinosis Andersen's disease	Amylo-1,4-transglucosidase (branching enzyme)	Liver	Yes
V	McArdle's disease	Muscle phosphorylase	—	No
VII	Tarui's disease	Muscle phosphofructokinase	Erythrocytes	No

and this leads to the clinical features which are so characteristic of Pompe's disease: proximal muscle weakness, gross cardiomegaly, hepatomegaly and a large immobile tongue. The infant is often normal at birth and muscle weakness accompanied by heart failure and sometimes respiratory failure develop during the first few months of life; average life expectancy is about five and a half months and death usually occurs before the age of one year. A milder form resembles limb girdle dystrophy.

The electromyogram is a particularly helpful investigation in the diagnosis of Pompe's disease, where it reveals the presence of pseudomyotonic discharges. The electrocardiogram should show a short PR interval and biventricular hypertrophy. On muscle biopsy there is a severe vacuolar myopathy and glycogen can be demonstrated in the vacuoles on PAS staining. Serum CK activity is slightly raised. Deficiency of the enzyme can be demonstrated in leucocytes. Pompe's disease is an autosomal recessive condition. Antenatal diagnosis can be achieved by measuring the enzyme activity in cultured amniotic fluid cells.

Amylo-1,6-glucosidase deficiency (type III glycogenesis)

Type III glycogenosis usually presents with gross hepatomegaly, fasting hypoglycaemia, ketonuria, growth retardation and sometimes mild cirrhosis. In some patients, myopathy is the predominant feature. This may present in infancy as the floppy infant syndrome or later in life with muscle weakness and wasting. There is often exercise intolerance, though muscle cramps are unusual. The EMG is grossly abnormal and the serum CK is frequently elevated. An ischaemic lactate test shows failure of venous lactate to rise. Abnormal glycogen storage can be demonstrated on liver and muscle biopsy. Glycogen storage in muscle occurs even if there is no clinical manifestation.

The diagnosis of this condition is discussed in Chapters 3 and 7.

Amylo-1,6-transglucosidase deficiency (type IV glycogenosis)

This rare disorder is characterised by the storage of excess polysaccharide in most tissues of the body, due to deficiency of brancher enzyme (amylo-1,6-transglucosidase). The polysaccharide that accumulates is abnormal and this stimulates a tissue reaction, in particular the development of hepatic cirrhosis. The infant seems normal at birth but presents in the first year of life with abdominal distension, hepatomegaly and signs of portal hypertension. Muscle is involved but this is usually only of secondary importance clinically. Nevertheless, some infants do have marked hypotonia.

Myophosphorylase deficiency (type V glycogenosis; McArdle's disease)

This condition is characterised by the development of muscle cramps and stiffness on exertion and often the passage of dark urine (myoglobinuria). At rest there is little muscle weakness except sometimes in the fourth and fifth decades of life. Most indexed cases have not been diagnosed until adult life although symptoms could often be traced back to childhood. Because muscle phosphorylase is absent in patients with McArdle's disease the muscle can only exercise while tissue oxygen levels can support oxidative metabolism. Normally after some minutes of heavy exercise glycolytic pathways release glucosyl units in the muscle cells and the anaerobic pathways provide energy for the exercise to continue, this results in an accumulation of lactate which passes out of the muscle cells into the blood.

The diagnosis of myophosphorylase deficiency is suggested by the result of an ischaemic lactate test. In this test a cuff is placed on the arm and blown up to above arterial pressure, the patient is then asked to exercise the forearm by repeatedly clenching and relaxing the fist until limited by pain. Three blood samples are collected, one before, one immediately after and another 25 minutes after the exercise. Lactate produced in the muscle cells during the ischaemic exercise causes the blood lactate to rise. For the correct interpretation of this test it is important to wait about 5 seconds after the cuff is released before collecting the middle blood specimen, so that there is time for any lactate produced to be flushed into the circulation. In a normal individual the middle specimen lactate level should be above the upper limit of the normal reference range. In patients with McArdle's disease, about 20–40 seconds after the start of exercise the muscle goes into a painful contracture which may persist for some minutes after the cuff is released and there is no rise in blood lactate. A drawback to this technique is that the patient may be unwilling or unable to exercise vigorously, so that the test is often proved unsatisfactory by giving a false negative result.

A definitive diagnosis in both children and adults can be obtained by the demonstration of a negative histochemical reaction for phosphorylase on a muscle biopsy sample and the biopsy will also show that the muscle fibres are packed with glycogen.

An exercise test on a cycle ergometer may also be attempted by a patient with myophosphorylase deficiency and although distress is experienced after a few minutes exercise a 'second wind' phenomenon may occur as blood borne glucose and fatty acids are oxidised, and exercise can then be continued.

McArdle's disease usually shows an autosomal recessive inheritance, it is not progressive and most patients are able to live within their limitations.

Phosphofructokinase deficiency (type VII glycogenosis)

This disorder is clinically similar to McArdle's disease, but the 'second wind' phenomenon during an exercise test is very limited because only circulating fatty acids, not glucose, can be utilised. There may be a mild non-spherocytic haemolytic anaemia, suggested initially by a raised reticulocyte count. Phosphofructokinase deficiency in the muscle can be demonstrated histochemically on a needle biopsy sample and confirmed by biochemical assay (see Table 17.1).

Disorders of lipid metabolism

These disorders have been reviewed in detail by Di Mauro *et al.* (1980). Symptoms are those of weakness, exercise intolerance and muscle stiffness and pain, sometimes accompanied by myoglobinuria. They are most evident when free fatty acids are the main substrate for energy metabolism and this is generally during prolonged submaximal exercise, particularly in the fasting state. These disorders may be subdivided as follows:

1 Carnitine deficiency.
2 Carnitine palmityltransferase deficiency.

L-Carnitine (β-hydroxytrimethyl-γ-aminobutyric acid) is responsible for the transport of free fatty acids across the inner mitochondrial wall. L-Carnitine can be obtained from the diet or synthesised from lysine. Its concentration is greater in muscle than in blood and it has been suggested that it is taken up into muscle by an active process (Willner *et al.* 1978).

Carnitine deficiencies

Carnitine deficiencies can be subdivided into myopathic, systemic and secondary.

Myopathic carnitine deficiency. This condition, first described by Engel & Siekert (1972), usually presents with lactic acidosis possibly arising from increased dependence on carbohydrate metabolism for energy. Plasma L-carnitine may be slightly low or normal and muscle carnitine is low. The defect may be due to excess leakage from the muscle or defective uptake into the muscle. The muscle biopsy will show gross lipid storage myopathy. Some patients (Engel & Siekert 1972; Van Dyke *et al.* 1975), but not all (Isaacs *et al.* 1976; Bradley *et al.* 1978) have benefitted from corticosteroid therapy. DL-Carnitine has been used successfully in some (Angelini *et al.* 1970), but not all, patients (Willner *et al.* 1979).

The mechanism of steroid action is not known nor is that of carnitine treatment. In the latter case where improvement was noted, there was a decrease in lipid droplet concentration in the muscle but no demonstrable increase in muscle carnitine (Angelini *et al.* 1978).

Systemic carnitine deficiency. First reported by Karpati and co-workers in 1975, patients have a progressive neuromuscular disorder, nausea and vomiting; hypoglycaemia and dicarboxylicaciduria have also been reported. The condition may progress to coma and death (Karpati *et al.* 1975). Plasma carnitine levels are reduced to as low as 50% of normal and it is suggested that the hepatic synthesis of carnitine may be impaired. Biopsies of liver and muscle show increased lipid droplets.

Oral DL-carnitine has been of benefit to some patients (Karpati *et al.* 1975), the mechanism being unknown. Although blood carnitine was restored to normal, no increase in L-carnitine levels in liver or muscle was found. Not all patients benefit (Cornelio *et al.* 1977).

No enzyme defect has yet been identified in patients with systemic carnitine deficiency, but secondary carnitine deficiency due to renal loss has been postulated (Engel *et al.* 1981). If carnitine deficiency is suspected in a child, specimens of urine and plasma should be collected and stored frozen. Definitive diagnosis will probably require muscle and possibly liver biopsies.

Both systemic and myopathic types of carnitine deficiency are probably genetic and siblings may be affected (Chapoy *et al.* 1980).

Secondary carnitine deficiency. In children low concentrations of carnitine in blood and urine have been reported in: malnutrition (Kahn *et al.* 1977); glutamicaciduria (Dusheiko *et al.* 1979); Spanish oil poisoning syndrome (De La Morena *et al.* 1982); Duchenne muscular dystrophy (Borum *et al.* 1977) and low birth-weight infants receiving total parenteral nutrition (Penn *et al.* 1980).

Carnitine palmitoyl transferase (CPT) deficiency

This disorder was first reported by Engel *et al.* (1970) and the enzyme defect established biochemically by Di Mauro and Di Mauro (1973). Serum and muscle carnitines are normal. Although 2 types of CPT have been described—one situated on the outer (CPT1) and the other (CPT2) on the inner mitochondrial membrane. (Fritz *et al.* 1974; Hoppel *et al.* 1972)—there are indications that these may be the same enzyme (Bergstrom & Reitz 1980).

Clinically there is muscle stiffness and pain on exertion and presentation is similar to that in glycogen storage diseases types V and VII, but there are important differences. Unlike patients with glycogenoses, those with CPT deficiency can undertake short bursts of strenuous activity, but they have problems with prolonged submaximal activity, when free fatty acids would be the main substrates for energy production because CPT deficiency results in an inability to transport long chain fatty acids into the mitochondria.

Patients may have severe attacks of acute rhabdomyolysis with myoglobinuria and acute renal failure.

If a disorder of lipid metabolism is suspected on clinical grounds, a specialist laboratory should be contacted for advice. Specimens of plasma and urine for free and total carnitine should be collected and stored frozen at −30 °C and if the child is moribund fibroblasts should also be collected for tissue culture. Definitive diagnosis at a specialist centre will probably require biopsies of muscle and possibly liver.

Mitochondrial myopathies

The mitochondrial myopathies are a group of disorders in which the muscle mitochondria are abnormal in structure and/or function. The term encompasses a variety of clinical disorders and other systems of the body may also be involved (Table 17.2).

There are often abnormalities of blood biochemistry including lactic-acidaemia, pyruvic acidaemia, and elevated fasting free fatty acids and triglycerides. If the diagnosis is suspected these metabolites should be measured, a specimen of urine and a few millilitres of plasma stored frozen and a specialist laboratory contacted. It is difficult to classify these disorders satisfactorily; an early attempt based on the morphology of the mitochondria was unsuccessful. On muscle biopsy there may be lipid accumulation and a moderate increase in muscle glycogen. Histochemical staining may show the presence of 'ragged red fibres'.

A detailed investigation of mitochondrial function *in vitro* is required in order to fully diagnose these conditions. In the past this has been complicated by the difficulties inherent in obtaining sufficient mitochondria because this entailed the collection of about 5 grams of skeletal muscle from an open biopsy. Methods have recently been developed which enable many

Table 17.2 Mitochondrial myopathy: possible associated clinical features.

Muscle	Progressive proximal weakness Acute relapsing paralysis Distal weakness with foot drop Facioscapulohumeral syndrome Muscle stiffness on exertion with myoglobinuria
Metabolic	Lactic acidaemia Diabetic syndrome Non-thyroidal hypermetabolic state Salt craving
Growth	Short stature Delayed puberty
Central nervous system	Cerebellar ataxia Dementia Seizures
Eyes	External ophthalmoplegia Pigmentary retinopathy Cataract
Respiratory	Hyperventilation Ventilatory failure
Skin	Icthyosis
Facies	Mild facial weakness Ptosis Big tongue
Heart	Heart block Arrhythmias Cardiomegaly
G.I. tract	Hepatomegaly Cirrhosis Malabsorption
Renal	Fanconi syndrome Myoglobinuria Recurrent calculi
Haematological	Sideroblastic anaemia

of these detailed studies to be made using skeletal muscle samples obtained by needle biopsy (Gohil *et al.* 1981).

It is hoped that these microtechniques will aid in the diagnosis and classification of mitochondrial myopathies. At present, it is convenient to divide these disorders into 3 main sub-groups:

1 *Proximal limb girdle weakness and other problems in children and adults.* These various disorders have been reviewed recently (Dubowitz 1978; Di Mauro 1980). Sometimes muscle involvement is only one part of a more

generalised disorder in which other problems predominate, especially ones referrable to the central nervous system and the heart (see Table 17.2). Fatigue out of all proportion to the degree of weakness may be a prominent symptom.

2 *Kearns–Sayre syndrome* (*ophthalmoplegia plus*). This is a severe multi-system disorder. Affected patients have a fairly characteristic facial appearance comprising marked ptosis and external ophthalmoplegia (Dubowitz 1978). The skeletal muscle weakness is usually overshadowed by other problems which include ataxia, pigmentary retinopathy, heart block, deafness, short stature, delayed puberty and mental retardation. The disorder seems to be sporadic.

3 *Excess lipid and glycogen storage with mitochondrial myopathy in infancy*. This probably embraces a number of distinct biochemical defects. Muscle weakness is observed in the first few weeks or months of life. The prognosis is generally very poor because of the presence of respiratory involvement. There is, however, a benign variety which may be recognised on the basis of hepatomegaly and a large tongue (Jerusalem *et al.* 1973).

Two infants have been described with an associated De Toni–Fanconi–Debre type of renal tubular defect. Both had a defect of the cytochrome chain. A patient of ours had a defect of cytochrome oxidase but no renal involvement (Dubowitz 1980).

The inheritance of these disorders is usually autosomal recessive.

PERIODIC PARALYSIS

These disorders have been reviewed by Griggs (1977) and Edwards and Jones (1983). They are characterised by periodic attacks of paralysis with associated flaccidity. Different types are recognised depending on whether the plasma potassium is high or low during an attack.

Hypokalaemic periodic paralysis

This disorder has an autosomal dominant inheritance but is only partially expressed in the female. Attacks usually commence in the second decade and reach peak frequency in early adult life. The attacks are precipitated by heavy exercise followed by a period of rest or sleep. Other precipitating factors include anxiety, alcohol, cold, trauma, and a large meal rich in carbohydrates. Between attacks the patient is usually normal (Gamstorp 1962).

Characteristically the weakness is accompanied or preceded by a fall in plasma potassium which may be reduced to as low as 1.5 mmol/l. The degree of weakness is not however related to the potassium level, which may return to normal before muscle strength is restored. Potassium is not lost from the body during an attack but moves into the muscle cells. During an attack the muscle becomes inexcitable; relief of paralysis by oral potassium is accompanied by repolarisation of the muscle membrane (Reicker & Bolte 1966).

To make the diagnosis, plasma potassium must be measured during an attack. If necessary an attack can be induced by the administration of oral glucose (2 g/kg body weight) and subcutaneous insulin (5–15 units). Careful monitoring is necessary, particularly of the ECG and plasma potassium during the next eight hours. In an affected individual weakness should commence once the plasma potassium has fallen below 3 mmol/l and be accompanied by EMG changes. It is our experience that more children diagnosed elsewhere as having periodic paralysis do not in fact have the condition and this is evident from the history.

Treatment of attacks is by the administration of oral potassium chloride or citrate. Mild exercise may also help to reduce an attack. In the long-term potassium supplementation, restriction of sodium intake and the avoidance of heavy exercise will all help the patient. Acetazolamide has proved valuable in the treatment of this condition.

Hyperkalaemic periodic paralysis (adynamica episodica hereditaria)

Symptomatology is very similar in this condition to that seen in the hypokalaemic form, but the attacks are often more rapid in onset following exercise and of shorter duration. Early on the patient may be able to 'walk off' an attack. There is a close association with paramyotonia (Layzer *et al.* 1967). The inheritance is autosomal dominant with expression in both sexes.

Attacks of weakness generally begin during the first decade of life. Attacks are provoked by potassium and relieved by glucose and insulin. In these respects the condition is the reverse of the hypokalaemic form. If the patient cannot be studied during an attack one can be provoked by the administration of oral potassium. Careful ECG monitoring is needed if this procedure is undertaken. Attacks can be aborted by glucose and insulin. During an attack the plasma potassium usually rises although the exact extent of this rise is variable. Potassium moves out of the muscle cells and possibly out of other tissues too. Muscle sodium increases during an attack as the muscle potassium falls.

High carbohydrate intake, glucose and insulin all help to minimise attacks probably by encouraging potassium uptake by the muscles. As in hypokalaemic periodic paralysis administration of acetazolamide has proved effective although in both these conditions long-term treatment carries a risk. The successful prevention of hyperkalaemic attacks using oral salbutamol has been described (Wang *et al* 1978; Dahl-Jørgensen & Michalsen 1979).

METABOLIC MYOPATHIES IN OTHER MAJOR SYSTEM DISORDERS

Hepatic disease

Problems occur principally in association with the chronic obstructive cholangiopathies which usually present in infancy with jaundice. As a result of modern surgical and medical techniques, many affected children are surviving with a reasonable quality of life. However, reduction in biliary outflow leads to chronic malabsorption of fat and this has serious long-term consequences.

Problems stem mainly from malabsorption of the fat-soluble vitamins. The association of chronic hepatic disease with rickets is well known (Kobayashi *et al.* 1979). It is worth remembering that symptoms and signs of myopathy may predominate over those of the metabolic bone disease, with loss of ambulation as the principle complaint. Vitamin E deficiency can also develop (Rosenblum *et al.* 1981). Some children with chronic hepatic disease develop a neurological syndrome which includes muscle weakness, neuropathy, ataxia and disorders of eye movement. It seems to be associated with virtually undetectable vitamin E levels in the blood. Present experience suggests the clinical changes can be arrested, but not reversed, by administration of vitamin E.

Rickets

Nutritional rickets occurring in infancy is quite commonly associated with hypotonia. Severe rickets secondary to chronic renal tubular acidosis such as occurs in the De Toni–Fanconi–Debre syndrome may be associated with profound hypotonia and motor delay. We have recently seen this association in a 3-year-old girl. Treatment has resulted in a marked reduction in bone pain and progress in her motor development.

Endocrine myopathies

Skeletal muscle may be affected in hypothyroidism and hyperthyroidism although the latter is unusual during childhood. Overt myopathy with gross elevation of serum CK may occur in hypothyroidism, including the neonatal form. In thyrotoxicosis, there may be an acute myopathy, or chronic generalised or mainly proximal weakness. Occasionally thyrotoxicosis is associated with myasthenia gravis.

Muscle weakness has been described in association with Addison's disease, Conn's syndrome, Cushing's syndrome and pituitary gigantism.

INBORN ERRORS OF INTERMEDIARY METABOLISM

Many of these disorders are associated with central nervous system damage. In consequence there may be a significant degree of hypotonia which may indeed be one of the presenting features (see Stanbury *et al.* (1978) for a review and also see Chapter 15).

REFERENCES

ANGELINI C., LUCKE S. & CANTARUTTI F. (1970). Carnitine deficiency of skeletal muscle: Report of a treated case. *Neurology* **26**, 633–37.

BECKMANN R., SAUER M., KETELSEN U-P. & SCHEEUERRBRANDT G. (1978) Early diagnosis of Duchenne muscular dystrophy. *Lancet* **2**, 105 (letter).

BERGSTROM J. D. & REITZ R. C. (1980) Studies on carnitine palmitoyl transferase: The similar nature of CPT_I (inner form) and CPT_O (outer form). *Arch. Biochem. Biophys.* **204**, 71–9.

BORUM P. R., BROQUIST H. P. & ROELOFS R. I. (1977) Muscle carnitine levels in Neuromuscular disease. *J. Neurol. Sci.* **34**, 279–86.

BRADLEY W. G., TOMLINSON B. E. & HARDY M. (1978) Further studies of mitochondrial and lipid storage myopathies. *J. Neurol. Sci.* **35**, 201–10.

CORNELIO F., DI DONATO S., PELUCHETTI D., BIZZI A., BETAGNOLIO B., D'ANGELI A. & WEISMANN U. (1977) Fatal cases of lipid storage myopathy with carnitine deficiency. *J. Neurol. Neurosurg. Psych.* **40**, 170–8.

CHAPOY P. R., ANGELINI C., BROWN W. J., STIFF J. E., SHUG A. L. & CEDERBAUM S. D. (1980) Systemic carnitine deficiency, a treatable inherited lipid storage disease presenting as Reye's syndrome. *New Engl. J. Med.* **303**, 1389–94.

DAHL-JØRGENSEN K. O. & MICHALSEN H. (1979) Adynamia Episodica Hereditaria. *Acta Paed. Scand.* **68**, 583–5.

DE LA MORENA E. & MONTERO C. (1982) Low serum L-carnitine concentrations in toxic oil syndrome. *Lancet* **i**, 904.

DENBOROUGH M. A. (1980) The pathopharmacology of malignant hyperpyrexia. *Pharm. Therapeut.* **9**, 357–65.

DIMAURO S. & DIMAURO P. M. M. (1973) Muscle carnitine palmityl transferase deficiency and myoglobinuria. *Science* **182**, 929–31.

DIMAURO S., TREVISAN C. & HAYES A. (1980) Disorders of lipid metabolism in muscle. *Muscle Nerve* **3**, 369–88.

DRUMMOND L. M. (1979) Creatine phosphokinase tests in the newborn and their use in screening for Duchenne muscular dystrophy. *Arch. Dis. Child.* **54**, 362–6.

DUBOWITZ V. (1978) *Muscle disorders in childhood.* W. B. Saunders and Co., Philadelphia.

DUBOWITZ V. (1980) The Floppy Infant, 2nd Edition. *Clinics in developmental medicine* No. 78, p. 56. Spastics International Medical publications, Heinemann, London.

DUSHEIKO G., KEW M. C., JOFFE B. I., LEWIN J. R., PATH F. F., MANTAGOS S. & TANAKA K. (1979) Recurrent hypoglycaemia associated with glutaric aciduria type II in an adult. *New Engl. J. Med.* **301**, 1405–9.

EDWARDS R. H. T. & JONES D. A. (1983) Muscle diseases. In Adrian R. and Peachey L. D. (eds.) *American Physiology Society Handbook*, Physiology section, Chapter 'Skeletal muscle'. Raven Press, New York.

EDWARDS R. H. T., WILES C. M., ROUND J. M., JACKSON M. J. & YOUNG A. (1979)

Muscle breakdown and repair in polymyosistis: a case study. *Muscle Nerve* **2**, 223–8.

EDWARDS R. H. T., YOUNG A. & WILES C. M. (1980) Needle biopsy of skeletal muscle in the diagnosis of myopathy and in the clinical study of muscle function and repair. *New Engl. J. Med.* **302**, 261–71.

ENGEL A. G., REBOUCHE C. J., WILSON D. M., GLASGOW A. M., ROMSHE C. A. & CRUSE R. P. (1981) Primary systemic carnitine deficiency. II. Renal handling of carnitine. *Neurology* **31**, 819–25.

ENGEL A. G. & SIEKERT R. G. (1972) A lipid storage myopathy responding to prednisone. *Arch. Neurol.* **27**, 174–81.

ENGEL W. K., VICK N. A., GLUECK J. & LEVY R. I. (1970) A skeletal muscle disorder associated with intermittent symptoms and a possible defect of lipid metabolism. *New Engl. J. Med.* **282**, 697–704.

ELLIS F. R. & HALSALL P. J. (1980) Malignant hyperpyrexia. *Br. J. Hosp. Med.* **24**, 318–27.

ELLIS F. R., HARRIMAN D. G. F., CURRIE S. & CAIN S. (1978) Screening for malignant hyperthermia in susceptible patients. In Aldrette J. A. and Britt B. (eds.) *Malignant Hyerpthermia.* Grune and Stratton, New York.

EMERY A. E. H. (1980) Duchenne Muscular Dystrophy, genetic aspects, carrier detection and antenatal diagnosis. *Br. Med. Bull.* **36**, 117–22.

FRITZ T. B., KOPEC B. & BRISNAN J. T. (1974) In Lundquist F. and Tygstrip (eds.) *Regulation of Hepatic Metabolism.* Academic Press, New York.

GAMSTORP I. (1962) A study of transient muscle weakness. Clinical, biochemical and electromyographical findings during attacks of periodic paralysis and adynamia episodica hereditaria. *Acta Neurol. Scand.* **38**, 3–19.

GARDNER-MEDWIN D. (1980) Clinical features and classification of muscular dystrophies. *Br. Med. Bull.* **36**, 109–15.

GOHIL K., JONES D. A. & EDWARDS R. H. T. (1981) Analysis of muscle mitochondrial function with techniques applicable to needle biopsy samples. *Clin. Physiol.* **1**, 195–207.

GOLBUS M. S., STEPHENS J. D., MAHONEY M. J., HOBBINS J. C., HASELTINE F. P., CASKEY C. T. & BANKER B. Q. (1979) Failure of foetal creatine phosphokinase as a diagnostic indicator of Duchenne Muscular dystrophy. *New Engl. J. Med.* **300**, 860–1.

GRIFFITHS P. D. (1968) The activity of ATP: creatine phosphotransferase in umbilical cord blood. *Clin. Chem. Acta* **20**, 465–72.

GRIGGS R. C. (1977) The periodic paralyses. In Griggs R. G., Moxley R. T. III (eds.) *Advances in Neurology* **17**, pp. 143–59. Raven Press, New York.

HOPPEL C. L. & TOMEC R. T. (1972) Carnitine palmityltransferase, location of 2 enzymatic activities in rat liver mitochondria. *J. Biol. Chem.* **247**, 832–41.

ISAACS H., HEFFRON J. J. A., BADENHORST M. & PICKERING A. (1976) Weakness associated with the pathological presence of lipid in skeletal muscle; a detailed study of a patient with carnitine deficiency. *J. Neurol. Neurosurg. Psych.* **39**, 1114–23.

JERUSALEM F., ANGELINI C., ENGEL A. G. & GROOVER R. V. (1973) Mitochondrial lipid–glycogen (MLG) disease of muscle. *Arch. Neurol.* **29**, 162–9.

KAHN L. & BAMJI M. S. (1977) Plasma carnitine levels in children with protein calorie malnutrition before and after rehabilitation. *Clin. Chim. Acta* **75**, 163–6.

KARPATI G., CARPENTER S., ENGEL A. G., WATTERS G., ALLEN J., ROTHMAN S., KLASSEN G. & MAMER O. A. (1975) The syndrome of systemic carnitine deficiency. *Neurology* **25**, 16–24.

KOBAYASHI A., KAWAI S., OHKUBO M. & OHBE Y. (1979) Serum 25-hydroxy-vitamin D in hepatobilary disease in infancy. *Arch. Dis. Child.* **54**, 367–70.

LAYZER R. B., LOVELACE R. E. & ROWLAND L. P. (1967) Hyperkalemic periodic paralysis. *Arch. Neurol.* **16**, 455–72.

MAUNDER-SEWRY C. A. & DUBOWITZ V. (1981) Needle muscle biopsy for carrier detection in Duchenne muscular dystrophy. Part 1. Light microscopy, histology, histochemistry and quantitation. *J. Neurol. Sci.* **49**, 305–24.

MOSS D. W., WHITAKER K. B., PARMAR C., HECKMATT J., WITKOWSKI J., SEWRY C. & DUBOWITZ V. (1981) Activity of creatine kinase in sera from healthy women carriers of Duchenne muscular dystrophy and cord blood, determined by the 'European' recommended method with NAC-EDTA activation. *Clin. Chim Acta* **116**, 209–16.

MOULDS R. F. W. & DENBOROUGH M. A. (1974a) Identification of susceptibility to malignant hyperpyrexia. *Br. Med. J.* **ii**, 245–7.

MOULDS R. F. W. & DENBOROUGH M. A. (1974b) A study of the action of caffeine, halothane, potassium chloride and procaine on normal human skeletal muscle. *Clin. Exp. Pharmacol. Physiol.* **1**, 197–209.

PENN D., SCHIDT-SOMMERFELD E. & WOLF H. (1980) Carnitine deficiency in premature infants receiving total parenteral nutrition. *Early Hum. Dev.* **4**, 23–34.

PENNINGTON R. J. T. (1980) Clinical biochemistry of muscular dystrophy. *Br. Med. Bull.* **36**, 123–6.

PERCY M. E., CHANG L. S., MURPHY E. G., VERELLEN-DUMOULIN C. & THOMPSON M. W. (1979) Serum creatine kinase and pyruvate kinase in Duchenne muscular dystrophy carrier detection. *Muscle Nerve* **2**, 329–39.

PRYCE F. H., KIM E. K. & LOGAN J. E. (1977) An evaluation of commercial kits for creatine phosphokinase assay. *Clin. Biochem.* **10**, 206–9.

REIKER G. & BOLTE H. D. (1966) Membrane Potentiale einzaner skeletalmuskelzellen bei hypokalliamischer periodischer muskle paralyses. *Klin. Wschr.* **44**, 804–7.

ROSENBLUM J. L., KEATING J. P., PRENSKY A. L. & NELSON J. S. (1981) A progressive neurological syndrome in children with chronic liver disease. *New Engl. J. Med.* **304**, 503–8.

ROSES A. D., ROSES M. J., NICHOLSON G. A. & ROE C. R. (1977) Lactate dehydrogenase isoenzyme 5 in detecting carriers of Duchenne muscular dystrophy. *Neurology* **27**, 414–21.

SEAY A. R., ZITER F. A., WU L. H. & WU J. T. (1978) Serum creatine phosphokinase and pyruvate kinase in neuromuscular disorders and Duchenne muscular dystrophy carriers. *Neurology* **28**, 1047–50.

SOMER H., WILLNER J., DECRESCE R. P. & SOMER M. (1980) Duchenne carriers: lactate dehydrogenase isoenzymen in serum and muscle. *Neurology* **30**, 206–9.

STANBURY J. B., WYNGAARDEN J. B. & FREDERICKSON D. S. (1978) *The Metabolic Basis of Inherited Disease*, 4th Edition, Part 3, pp. 336–429. McGraw Hill, New York.

TIPPETT P. A., DENNIS N. R., MACHIN D., PRICE C. P. & CLAYTON B. E. (1982) Creatine kinase activity in the detection of carriers of Duchenne muscular dystrophy: comparison of two methods. *Clin. Chim. Acta* **121**, 345–59.

VAN DYKE D. H., GRIGGS R. C., MARKESBERY W. & DIMAURO S. (1975) Hereditary carnitine deficiency of muscle. *Neurology* **25**, 154–9.

WANG P., CLAUSEN T. & ORSKER H. (1978) Salbutamol inhalations suppress attacks of hyperkalaemia in familial periodic paralysis. *Monog. Hum. Gen.* **10**, 62–5.

WILLNER J. H., GINSBURG S. & DIMAURO S. (1978) Active transport of carnitine into skeletal muscle. *Neurology* **28**, 721–4.

WILLNER J. H., DIMAURO S., EASTWOOD A., HAYS A., ROOHI F. & LOVELACE R. (1979) Muscle carnitine deficiency Genetic Heterogeneity. *J. Neurol. Sci.* **41**, 235–46.

WILLNER J. A., WOODS D. O. S., CERRI C. & BRITT B. (1980) Increased myophosphorylase A in malignant hyperthermia. *New Engl. J. Med.* **303**, 138–40.

ZELLWEGER H. & ANTONIK A. (1975) Newborn screening for Duchenne muscular dystrophy. *Paediatrics* **55**, 30–4.

Chapter 18
Acute Poisoning

ALEXANDER T. PROUDFOOT

Man uses many substances in his attempts to alter internal and external environments to his benefit, however, most of those used to combat disease produce effects in addition to the one for which the substance is primarily taken. Some of the unwanted effects are unpredictable and are usually said to be due to hypersensitivity or idiosyncrasy, while others are predictable and often directly related to the magnitude of the exposure, whether by inhalation, ingestion or skin application. When unwanted effects complicate the correct use of accepted therapeutic amounts of a drug they are usually termed adverse reactions and when they result from excessive doses, incorrect use or exposure to some substance not used therapeutically, the term poisoning is used. It will be obvious that there is no clear dividing line between most adverse reactions and poisoning. Peter Mere Latham (1789–1875) crystallised the situation by saying that 'poison and medicine are often-times the same substance given with different intents'.

CLASSIFICATION OF CHILDHOOD POISONING

The classification of poisoning is unsatisfactory. It is based on at least three factors, including the age at which poisoning occurs, the circumstances preceding it and whether it is self-inflicted or caused by another person. Five classes are commonly recognised.

Neonatal poisoning

This type of poisoning occurs very early during life. It is caused by transfer of drugs from mother to child across the placenta before or during delivery. It is rarely, if ever, due to transmission through breast milk.

Therapeutic poisoning

This term is used when intoxication occurs during the therapeutic use of drugs which are often given in excessive dosage for the age and weight of the patient. Unfortunately the response to drugs often varies widely from one individual to another and cannot be predicted with certainty. At one extreme a 'standard' dose may be sub-therapeutic and ineffective while at the other, features of overdosage rapidly develop. In some cases therapeutic intoxication is due to inherited factors or immaturity or limited capacity of some drug-metabolising enzymes. Clearly this type of poisoning cannot be confined to a specific age group, although it is more common in younger children and infants.

Accidental poisoning

Accidental poisoning is the commonest type of childhood poisoning and involves the child eating some substance which is potentially harmful. It is uncommon before the age of nine months but after that age the incidence of accidental poisoning rises rapidly, reaching a peak in the third and fourth years of life. This is simply because crawling and walking permit curious children, who explore the environment with their mouths as well as their hands and eyes, to get access to potential poisons in an unsupervised moment. In the fifth year of life the incidence declines sharply to continue at a low level until the age of about eight years.

Occasionally accidental poisoning results from a child being given a substance by another person, e.g. drugs proffered as sweets by another youngster who knows no better, or surgical spirit given in mistake for gripe water by a mother (an infrequent but regularly recurring cause of enquiries to poisons information services).

Self-poisoning

This term is used to describe the knowing, deliberate consumption of an overdose of a drug or to the ingestion (very rarely inhalation) of some other substance. Such behaviour is the most common cause of poisoning in adults and is seldom encountered before the age of ten years. However it is well-documented at younger ages and from the age of eight years upwards any drug overdose should be regarded with suspicion as it may not be accidental. Failure to appreciate the true nature of the poisoning may prevent the identification and possible correction of the interpersonal, social and occasionally psychiatric factors which commonly precipitate or contribute to such acts. Self-poisoning is often the response to a disagreement with a close relative and in this context often appears unreasonable or excessive. However to the individual no other solution or escape may be apparent at the time of crisis. Less frequently, self-poisoning is intended to punish the other person or manipulate him into doing something he would otherwise not do.

Non-accidental poisoning

Non-accidental poisoning is an unsatisfactory term used to describe the deliberate, covert poisoning of a child, usually by one of its parents. This form of child abuse has only been recognised in recent years and presents considerable diagnostic challenges. The illness which results from non-accidental poisoning may be used by one spouse to distress the other or, by making the child the focus of attention, the parents may be able to co-operate more harmoniously for a period.

EPIDEMIOLOGY OF CHILDHOOD POISONING

Size of the problem

It is impossible to estimate with any certainty the number of children poisoned in a year. Many incidents probably never come to medical attention. Hospital admissions indicate the minimum size of the problem and analysis of 10% samples for England and Wales showed that the number of admissions annually for poisoning in children under the age of 15 more than doubled between 1964 and 1969, to reach 28 000. Thereafter the number rose to 29 000 in 1976 and in the last year or two there has been a slight fall. The admission rate for children up to 4 years of age is about ten times that for those aged 5–9 years and in both age groups boys predominate.

The poisons involved

Young children seem to be able to eat the most unpalatable and unlikely things. The range of substances taken varies widely but drugs are responsible for about 60% of incidents. Poisoning with simple analgesics such as aspirin and paracetamol accounts for about half of these but antidepressants, tranquillisers and hypnotics make an important if smaller contribution. The non-drug poisons are drawn from a vast array of household products kept in cupboards below kitchen sinks and in bathrooms. Polishes, cosmetics, bleaches, disinfectants and turpentine-substitute, amongst other things, are consumed with monotonous regularity, as are house plants and bulbs. Older children may experiment with 'magic' mushrooms and solvent sniffing.

Morbidity from poisoning

In 1964 in England and Wales 66% of children admitted because of poisoning had a duration of stay in hospital of one day or less. In 1969 the corresponding figure was 77% and in 1976, 81%. These figures suggest that the morbidity from childhood poisoning is usually minimal and that admission is usually a precautionary measure. A detailed analysis of 178 episodes in children under the age of 15 showed that only 22% had symptoms of poisoning while only 1% suffered permanent injury. In a further 13% of cases there was good evidence that the child had taken some substance though symptoms did not develop. The majority however (65%) were regarded as 'scares' either because there was no evidence that the child had absorbed any poison or because it was inherently innocuous or only taken in small amounts. Even with psychotropic drugs serious intoxication is uncommon. Unfortunately the distinction between 'scares' and true poisoning can only be made with certainty in retrospect and 'unnecessary' hospital admissions can never be completely avoided.

Mortality from poisoning

Deaths from poisoning in British children aged less than 10 years have recently been reviewed. 598 deaths were recorded between 1958 and 1977, with the peak incidence in 1964 and 1965. That 81% of all deaths were due to drugs highlights their importance, but the pattern of deaths due to individual drugs or groups of drugs has changed over the years. Whereas salicylates were by far the most important in the early part of the period studied, their contribution has greatly declined while that of tricyclic antidepressants has increased to the extent that they are now the single most important cause of drug deaths. Barbiturates and iron salts also feature less prominently now while others such as digoxin, quinine, potassium salts, orphenadrine and opiates (particularly in anti-diarrhoeal drugs) continue to make small but significant contributions.

Lead was the single most important cause of non-drug deaths but its source was not given sufficiently frequently to permit interpretation of the statistics. Corrosive substances, hydrocarbons and paraquat are other important causes of non-drug deaths. In contrast, plants accounted for only three of the 598 deaths in 20 years.

CLINICAL CHEMISTRY IN ACUTE POISONING

The importance of clinical chemistry in the management of acute poisoning should not be underestimated. Poisons tend to cause multiple organ damage and of the 20% of children who have symptoms as a result of accidental poisoning, only a minority will be suitable for treatment with antidotes or techniques intended to enhance elimination of the poisons. In the majority of cases the clinician has no choice but to support vital functions until the toxin is metabolised and/or excreted and the patient recovers. Biochemical investigations are therefore essential to:

1 Detect acid–base and electrolyte disturbances.
2 Detect depression of function (e.g. respiration).
3 Assess the severity of organ damage (principally renal and hepatic).
4 Monitor the efficacy of treatment intended to correct these abnormalities.

Arterial blood gas analysis

Arterial blood gas analysis is probably the single most important and urgent biochemical investigation in acute poisoning. Elevation of the arterial carbon dioxide tension indicates inadequate ventilation, whether spontaneous or mechanically assisted, and a need for corrective measures. Hyperventilation indicated by a low arterial carbon dioxide tension, is usually due to overdosage of salicylates and rarely to sympathomimetic drugs (mainly bronchodilators and 'slimming' pills) or phenoxyacetate ('hormone') weedkillers.

Low arterial oxygen tensions may also be the result of ventilatory failure, such as may occur in deep coma due to a variety of drugs in patients whose lungs are clinically and radiologically normal and in the absence of significant carbon dioxide retention. Mismatching of ventilation and perfusion is the most likely explanation. In other cases, hypoxia is secondary to pulmonary aspiration of gastric contents or pneumonia.

Arterial blood gas analysis is also essential to detect and assess the severity of any acid–base disturbance. Metabolic acidosis is a common finding in any severe poisoning particularly with the drugs and chemicals listed in Table 18.1. In such cases the acid–base abnormalities are usually associated with clinical evidence of serious toxicity. While profound changes in arterial hydrogen ion concentration may be fatal they may also enhance or diminish the toxicity of some poisons by altering their distribution throughout body fluids. Thus acidaemia (elevated arterial hydrogen ion concentration) facilitates the movement of weakly acidic drugs such as salicylates from the plasma into the brain, thereby increasing toxicity (see below). Conversely the toxic effect on the central nervous system of phencyclidine, a basic drug widely abused in North America, is reduced by metabolic acidosis.

Table 18.1 Poisons which may cause metabolic acidosis.

Benzene hexachloride	Metaldehyde
Carbon monoxide	Methanol
Cyanide	Paraquat
Digoxin	Quinidine
Ethylene glycol	Salicylates
Fluoride	Solvent inhalation
Iron	Theophylline and derivatives
Isoniazid	

Plasma electrolyte concentrations

The possible alterations of plasma electrolyte concentrations which may occur in association with acute poisoning or its treatment are listed in Table 18.2. Only in recent years has it become apparent that the over-enthusiastic use of salt solutions to induce vomiting may produce fatal hypernatraemia in adults as well as children. Syrup of ipecacuanha is now the emetic of choice.

The acute changes in plasma potassium concentrations found in association with poisoning with cardiac glycosides and sympathomimetic drugs are probably due to redistribution of potassium between the intracellular and extracellular fluids. These changes are particularly important in the myocardium and overdosage with these drugs is frequently associated with cardiac arrhythmias. Digoxin, the most widely used cardiac glycoside, inhibits the membrane sodium–potassium pump, thus permitting leakage of potassium from the cells. Hyperkalaemia may result and in adults at least, has been used as an indication of the severity of poisoning. In contrast, ephedrine stimulates cyclic-AMP and theophylline inhibits phosphodiesterase. Both of these drugs therefore increase intracellular potassium at the expense of extracellular concentrations and the resulting hypokalaemia can be severe. Hypokalaemia has also been reported in association with salbutamol and terbutaline overdosage. Clearly acute changes in potassium gradients across the myocardial cell membrane may be important in the genesis of arrhythmias.

Plasma calcium concentrations may be severely reduced after the ingestion of oxalic acid or inorganic fluorides. These compounds combine with calcium to form insoluble calcium oxalate and fluoride respectively

Table 18.2 Alteration of plasma electrolyte concentrations in poisoning.

	Plasma concentration	
Electrolyte	Raised	Reduced
Sodium	Saline emetics	
Potassium	Metabolic acidosis Cardiac glycoside poisoning Overdosage with KCl or potassium diuretic combinations	Sympathomimetic poisoning Forced alkaline diuresis
Calcium	Chronic hypervitaminosis D	Ethylene glycol Oxalic acid Inorganic fluorides Forced alkaline diuresis
Bicarbonate		Metabolic acidosis Respiratory alkalosis

which in turn may be deposited in tissues and excreted in the urine. Ethylene glycol is also important in this respect since it is widely used as an antifreeze and is metabolised by alcohol and aldehyde dehydrogenases to produce oxalic acid. Calcium also stabilises the myocardium and frequent recurring ventricular fibrillation is a well-documented result of severe hypocalcaemia after ingestion of fluorides. Fortunately the fluoride preparations available in the community (mainly for the prophylaxis of dental caries) contain very small quantities and ingestion of even 100 tablets or more is unlikely to produce serious toxicity.

Despite the importance of vitamin D in the control of plasma calcium concentrations, acute overdosage with this vitamin does not cause problems. Hypercalcaemia is only likely with chronic hypervitaminosis D.

Forced alkaline diuresis is commonly used for the treatment of salicylate and phenobarbitone poisoning and may produce hypocalcaemia and hypokalaemia.

Reduction of the plasma bicarbonate concentration may be the result of either metabolic acidosis or respiratory alkalosis and may be the first indication of an acid–base disturbance. Its presence should prompt more detailed assessment by arterial blood gas analysis.

Plasma urea and creatinine

The measurement of plasma urea and creatinine to monitor renal function is too well known to warrant comment. It should be remembered that in the presence of severe hepatic necrosis (as in paracetamol poisoning) the plasma urea is a poor indicator of renal function and that the plasma creatinine is much more reliable.

Liver function tests

Paracetamol, iron salts, phenylbutazone and halogenated hydrocarbons may cause acute hepatocellular necrosis. Serial measurement of the plasma bilirubin and alanine aminotransferase and alkaline phosphatase activities is useful for following the severity and course of the liver damage, but for the clinician the prothrombin time ratio is usually available more quickly and is more relevant. Severe hepatic damage soon causes hypoprothrombinaemia which may lead to potentially lethal haemorrhage. The risk of this complication may be reduced by administration of fresh frozen plasma or clotting factor concentrates. The time course for abnormalities of the prothrombin time tends to parallel the biochemical tests of liver function.

Plasma glucose

Measurement of the plasma glucose concentration as an emergency is valuable in ascertaining whether or not undiagnosed coma is due to hypoglycaemia. It is also mandatory for the assessment of the effects of overdosage with hypoglycaemic drugs such as insulin and the sulphonylureas and for monitoring the effects of treatment. Plasma glucose estimations may have to be made frequently over the course of several days when chlorpropamide has been ingested, since the plasma half-life of this drug is 48 hours or longer and hypoglycaemia may recur despite adequate carbohydrate intake for up to 5 days after acute poisoning. Patients who have been severely hypoglycaemic for protracted periods may not recover consciousness after administration of intravenous dextrose and it may be important for the clinician to confirm that the treatment has achieved adequate plasma concentrations, since failure to respond may indicate underlying cerebral oedema. Regular measurement of the plasma glucose concentration is also indicated when a poison has produced severe hepatocellular necrosis, since hypoglycaemia may be one of the few treatable complications.

Hypoglycaemia may rarely result from ethanol and salicylate overdosage in childhood.

Hyperglycaemia and glycosuria have been reported after poisoning with isoniazid, organophosphates and sympathomimetic drugs.

Methaemoglobinaemia

Methaemoglobinaemia is an uncommon complication of acute poisoning, though it is well-documented after ingestion of chlorates, aniline, nitrates, sulphonamides, phenacetin, dapsone, some local anaesthetics and phenazopyridine. It has rarely been reported after ingestion of paraquat monolinuron formulations and phenol. Severe methaemoglobinaemia is usually complicated by haemolysis with subsequent haemoglobinuria and jaundice due to an increase in unconjugated bilirubin.

Other enzymes

Apart from aspartate and alanine aminotransferases the plasma activity of various enzymes may be increased after acute poisoning. Lactate dehydrogenase and creatine kinase activity is often greatly increased in patients who have been unconscious and may reflect skeletal muscle damage. Myoglobinuria and acute renal tubular necrosis with its usual biochemical features may follow. Myopathy and rhabdomyolysis are uncommon and have usually been reported in adults after poisoning with substances such as carbon monoxide, isopropanol, phencyclidine, benzene hexachloride, colchicine and 2,4-dichlorophenoxyacetic acid.

Cholinesterase activity may be severely reduced after exposure to organophosphate and carbamate insecticides.

Examination of the urine

The colour of the urine may change after ingestion of some poisons. Bilirubinuria secondary to severe hepatocellular necrosis does not require comment. A green or blue discolouration of the urine usually indicates the presence of methylene blue which is contained in some over-the-counter 'liver and kidney' pills or given therapeutically for methaemoglobinaemia. Rifampicin, phenindione and phenazopyridine make the urine turn orange/red in colour while phenols and cresols produce shades varying from grey to black. The presence of the latter is readily confirmed by their characteristic odour. Treatment of iron poisoning with desferrioxamine also produces an orange/red-coloured urine, which is due to the presence of the iron/desferrioxamine complex, ferrioxamine. A brown-coloured urine may also be found in association with myoglobinuria or haemoglobinuria.

Unusual crystaluria occasionally occurs in acute poisoning. Oxalate crystals may be present after ingestion of ethylene glycol, and the anticonvulsant primidone may crystalise out of the urine as it cools after massive overdosage.

PLASMA DRUG CONCENTRATIONS

The majority of poisoned patients recover with supportive care alone and knowledge of the plasma concentration of the poison seldom alters clinical management. However, the demonstration of a poison in plasma and measurement of its concentration is potentially useful in order to:

1 Confirm exposure to and absorption of significant amounts. This alone is usually of little value except in patients in whom non-accidental poisoning or brain death is suspected.

2 Assess the severity of poisoning and determine the need for treatment with specific antidotes (e.g. iron salts, paracetamol, methanol, ethylene glycol) or measures to enhance elimination (most frequently forced alkaline diuresis for salicylates and phenobarbitone and, rarely haemodialysis or haemoperfusion for barbiturate hypnotics, paraquat, ethylene glycol, methanol, theophylline and ethanol).

3 Monitor the efficacy of treatment. This is particularly important when techniques such as forced diuresis, exchange transfusion, peritoneal dialysis, haemodialysis and haemoperfusion are used in attempts to increase elimination of poison.

4 Help determine when to re-start regular drug therapy. The usual clinical situation is an epileptic who has had superimposed acute poisoning with his anticonvulsants.

5 Investigate the pharmacokinetics of serious poisons so that rational treatment can be devised. This indication for measurement of plasma drug concentrations may appear superfluous for a routine hospital laboratory and more a research function, but such poisonings occur only sporadically and maximum information from each episode is essential if progress is to be made. Paraquat poisoning is a recent, useful illustration. In the knowledge that the mortality from paraquat ingestion is high, many patients were treated in blind desperation with protracted peritoneal dialysis, haemodialysis and haemoperfusion and suffered significant complications from the treatment. The development of a satisfactory paraquat assay made assessment of the severity of the poisoning much easier and in retrospect it became clear that many patients who were not seriously poisoned had (with the best of intentions) been exposed unnecessarily to the hazards of a 'treatment' which was probably of no value.

Interpretation of plasma drug concentrations

The plasma concentration of a poison is determined by many factors, including the dose, route of absorption, volume of distribution, rates of distribution, metabolism and excretion and the time since exposure, while age, tolerance, other drugs, biologically active metabolites, and some diseases (particularly hepatic and renal) influence toxicity. With so many variables it is hardly surprising that the correlation between plasma concentrations and toxicity is often poor. Equally it will be clear that statements about 'fatal' plasma concentrations and concentrations above which certain treatments should be implemented must be interpreted with extreme caution. Though the results of laboratory drug analyses are important they can only be interpreted in the context of the overall clinical situation. In the great majority of cases it is important to treat the patient and not the plasma concentration of the poison.

It must be emphasised that at present there is considerable confusion concerning the way in which drug concentrations are expressed. Both molar and mass units are used, even though drugs are prescribed in mass units. This is a dangerous situation and can lead to confusion and delay when reporting results by phone to a clinician.

The time response for drug assays

The degree of urgency of a drug analysis should be determined by how much the result will influence clinical management. In Table 18.3 toxicological analyses have been divided into three groups. The first group requires an immediate response from the laboratory since the results largely determine whether specific antidotes are administered or steps are taken to increase the elimination of the poison. It must be remembered that most antidotes (particularly that for iron poisoning) are potentially toxic and their indiscriminate use cannot be condoned.

Table 18.3 The laboratory response time for measurement of plasma drug concentration.

Immediate response	Salicylate Paracetamol Iron Lithium Paraquat
Response within 6 hours	Phenobarbitone Barbiturate hypnotics Phenytoin
'Leisure' response	New drugs Serious poisoning With 'old' drugs

Since all but a very small number of patients poisoned with central nervous system depressants are managed by supportive measures alone, emergency measurement of the plasma concentrations of these drugs is rarely indicated. However such requests are justified in very severe poisoning if some special measure such as forced diuresis, haemodialysis or haemoperfusion is being contemplated.

Investigation of new or unfamiliar drugs or poisons usually requires the development and validation of new assays and even if urgent analysis was possible it is unlikely that the result would be helpful since lack of data would make interpretation uncertain. Though a 'leisure' response is therefore sufficient in this situation, the importance of collecting as much information as possible should not be underestimated. Similarly many of the drugs which have been in clinical use for many years have not been adequately investigated after overdosage and would bear scrutiny using present-day laboratory techniques.

Unfortunately (and like many laboratory investigations) clinicians frequently fail to ask themselves how the result of a drug analysis will alter treatment, and many requests (particularly salicylates and barbiturates in unconscious patients) seem to originate at a spinal reflex level. The only mitigating factor in this approach is that the responsibility for the immediate management of acute poisoning frequently falls on the most junior and inexperienced members of the medical teams who gain comfort and confidence (which is not always justified) from toxicological investigations.

Analytical methods

The responsibility for choice of methods lies with the laboratory staff and it is essential that they should use specific methods which measure the concentrations of active drugs only. Problems have arisen with, for example, paracetamol poisoning because some laboratories use assays which measured not only paracetamol but also its inactive metabolites, thus producing totally misleading results upon which physicians no doubt act. Other commonly used assays, particularly for sedative drugs, may also be non-specific and of limited value. Highly specific methods now need not be more complex and time consuming.

In some cases the effort spent trying to measure active drug only would not be justified because toxicity correlates poorly with plasma concentrations. In general however, there is a need for laboratories which perform drug analysis only intermittently to review their methods so that valid results can be obtained within the restraint imposed by the need for speed and simplicity.

Some analytical methods use one member of a groups of drugs (e.g. barbiturates) as the standard for measuring plasma concentrations of all members of that group. 'Conversion factors' are then used to modify the result for the particular member of the group involved. This practice is undesirable.

DRUG SCREENING

Clinicians frequently request drug 'screens' when faced with unconscious patients, including children, in whom there is no obvious focal neurological lesion. These screens most frequently involve measuring plasma barbiturate and salicylate concentrations, though, regretfully, few physicians seem to appreciate that salicylates seldom cause impairment of consciousness or that routine barbiturate assays are increasingly irrelevant as poisoning with these substances becomes less common. Drug 'screens' are tedious, expensive and time-consuming and even the best laboratories are unable to identify the agent responsible in every case of confirmed poisoning. Inevitably drug screening has to be based on the expectation that drugs commonly prescribed (mainly psychotropic drugs and analgesics) are likely to be consumed accidentally. A screen should not be undertaken without prior detailed discussion with the consultant looking after the child. It is not only essential to determine what drugs may have been available for poisoning, but also to obtain a list of those which may have been given during the course of resuscitation or subsequent treatment, lest the latter appear in the screen and create confusion. Drug screens may justifiably be requested when:

1 A patient is desperately ill and the clinician feels that it is necessary to know which poison, in case specific treatment or methods to enhance elimination are required.

2 A patient who is ill with a presumed diagnosis of poisoning. In the absence of improvement or in the face of deterioration the clinician may wish to confirm that poison is present and, if so, which, in case treatment should be modified.

3 When non-accidental poisoning of children is suspected.

Drug analysis should never become a substitute for careful clinical and social assessment, or a comforter for insecure physicians. However not all doctors deal with acute poisoning sufficiently frequently to gain the confidence that allows them to dispense with requests for drug analysis except in the circumstances stated above. Careful discussion between the clinicians and the laboratory staff should be able to reduce the number of requests for drug 'screens' to those which are likely to be clinically profitable.

SPECIFIC POISONINGS

Salicylate poisoning

Salicylates are derived from salicylic acid, which was originally obtained from willow bark. Salicylic acid itself is now only used in corn solvents and ointment for removing the scales of skin diseases such as psoriasis. Aspirin (acetylsalicylic acid) has analgesic, anti-inflammatory and antipyretic properties and is commonly kept as a household remedy for a wide variety of common minor ailments. Paediatric preparations, some attractively flavoured and coloured, are also available and ease of access, particularly in bathroom medicine cabinets, may be partly responsible for the high incidence of accidental salicylate poisoning in childhood. Methyl salicylate (oil of wintergreen) is toxic in relatively small amounts and is fortunately seldom encountered now. Newer salicylates such as benorylate and diflunisal have not yet been shown to be important causes of poisoning in childhood.

Mode of action

Salicylates have a direct toxic action on the inner ear and renal tubules and stimulate the respiratory centre in the mid-brain. They also uncouple oxidative phosphorylation and have complex effects on intermediary metabolism.

Pharmacokinetic considerations

Salicylates are readily absorbed in the gastrointestinal tract, but absorption may be delayed if enteric-coated or delayed release formulations have been ingested. Salicylic acid in ointments may be absorbed through the skin. Salicylates are extensively bound to plasma albumin and are metabolised in the liver to salicyluric acid (the major metabolite), salicyl acyl and phenolic glucuronides and gentisic acid. Salicyluric acid formation is readily saturated, even with therapeutic doses, and children are at great risk of therapeutic intoxication. Renal excretion becomes an important route of elimination as plasma concentrations increase. Salicylates are weak acids and their distribution in body fluids is profoundly influenced by changes in hydrogen ion concentration. Acidaemia reduces ionisation and facilitates transfer of salicylates from plasma into tissues, while the reverse occurs with alkalaemia.

Features of poisoning

In overdosage salicylates commonly cause nausea, vomiting, ringing in the ears, deafness and some degree of dehydration due to sweating and hyperventilation. The metabolic rate is increased as a result of uncoupling of oxidative phosphorylation and this in turn causes warm, flushed extremities and bounding pulses. Fever may occur, especially in young children. In severe poisoning consciousness may be impaired with delirium and rarely convulsions, hypoglycaemia, pulmonary oedema or renal failure may develop.

The acid–base disturbance

The acid–base changes in acute salicylate poisoning are complex. Most patients have a combination of respiratory alkalosis secondary to the hyperventilation produced by stimulation of the respiratory centre and metabolic acidosis due to disturbances of intermediary metabolism as a result of uncoupling of oxidative phosphorylation. Increased urinary excretion of lactic acid and keto-acids has been demonstrated in poisoned children. The age of the patient is the most important factor in determining which component dominates the acid–base disturbance. Children under the age of 4 years commonly present with a dominant metabolic acidosis while older children and the vast majority of adults have a dominant respiratory alkalosis. Since the arterial hydrogen ion concentration has such an important influence on the distribution of salicylate throughout body fluids, the interest in the acid–base disturbance is more than academic. Knowledge of the arterial hydrogen ion concentration in an ill patient is as essential to assessment as measurement of the plasma salicylate concentration. The longer the duration of poisoning the more likely the child is to have a dominant metabolic acidosis.

Treatment

Treatment of salicylate poisoning is directed to preventing further absorption by emptying the stomach and replacement of fluids and electrolytes lost by vomiting, sweating and hyperventilation.

The symptoms of salicylate poisoning can be alleviated more rapidly by enhancing elimination of the drug from the body. It is rarely necessary to resort to haemodialysis or haemoperfusion. The majority of patients with moderate or high plasma salicylate concentrations can be satisfactorily managed by enhancing renal elimination. Forced alkaline diuresis has been the standard treatment for many years. Children suitable for this form of treatment are those under 5-years-old who have plasma salicylate concentrations exceeding 350 mg/l, while in older children the critical level can be increased to 500 mg/l as in adults. The reason for differentiating the two groups is that the acid–base disturbance may seriously increase toxicity in younger children.

Forced alkaline diuresis involves giving a large fluid load intravenously together with sufficient sodium bicarbonate to make the urine alkaline (preferably between pH 7 and 8). Though alkalinisation is the most important component of this treatment and diuresis as such probably makes little contribution, a combination of both continues to be used. In practice, alkalinisation of the urine is more difficult to achieve than one might anticipate and may require the administration of very large quantities of sodium bicarbonate leading to, or exacerbating, alkalaemia. As a consequence hypokalaemia and hypocalcaemia may occur and appropriate monitoring and replacement therapy is essential.

Plasma salicylate concentrations

Plasma salicylate concentrations are widely used to assess the severity of salicylate poisoning, though in general they correlate less well with clinical severity than one might wish. This is hardly surprising since the commonly used assays measure free (active) and protein-bound (inactive) drug, the former only becoming the greater at concentrations encountered in severe acute poisoning. Cerebrospinal fluid salicylate concentrations correlate much better, presumably because they reflect the concentration of unbound drug in plasma, but it is inconceivable that CSF salicylate concentrations will ever be accepted as the routine method of assessing the severity of poisoning.

Tricyclic antidepressants

Tricyclic antidepressants are widely available. Toddlers are more likely to eat those prescribed for older children for the treatment of enuresis, though the consensus view is that they are of little value in this condition. Children may also accidentally consume tricyclic antidepressants prescribed for their parents.

Toxicity

The toxicity of tricyclic antidepressants is largely due to their ability to antagonise the effects of acetylcholine and to potentiate the effects of catecholamine at peripheral nerve endings, ganglia and in the brain. They also have a direct depressant effect on the activity of the myocardium.

Features of poisoning

The peripheral features of poisoning with this group of drugs include dilatation of skin blood vessels (producing flushed warm skin), absence of sweating, dilatation of the pupils and paralysis of accommodation, leading to blurred vision. These manifestations are neatly summarised as 'red as a beet, hot as a hare, dry as a bone and blind as a bat'. Urinary retention may also occur. Some patients can also be described as being 'mad as a wet hen' since the central nervous system anticholinergic effects include drowsiness, confusion and delirium with visual hallucinations.

Large overdoses cause coma, convulsions and cardiac effects, including hypotension, tachycardia, bizarre abnormalities of electrical conduction and arrhythmias. The latter cardiac complications are serious and occasionally fatal, though with the exception of sinus tachycardia, they are fortunately uncommon.

Treatment

Treatment of tricyclic antidepressant poisoning is almost entirely a matter of supporting vital functions. A patent airway and adequate ventilation must be maintained and the stomach emptied by aspiration and lavage. Hypotension and hypothermia are seldom serious problems and sinus tachycardia alone does not merit treatment. The management of more serious arrhythmias is unsatisfactory and a matter of clinical debate. However correction of hypoxia and metabolic acidosis is more important than the administration of drugs.

In some countries there has been a vogue for the use of physostigmine salicylate, a tertiary ammonium compound with anticholinesterase properties which is able to cross the blood–brain barrier. There is no doubt that the intravenous administration of this drug to patients with tricyclic antidepressant poisoning produces rapid improvement in the level of consciousness, reduces sinus tachycardia and abolishes delirium and hallucinations. However its effect is very short-lived and it has potentially serious adverse actions which make its routine use inadvisable.

Laboratory investigation

Since the management of tricyclic antidepressant poisoning is supportive the main demand on the laboratory should be for blood–gas analysis and electrolyte estimations.

Plasma concentrations of tricyclic antidepressants can be measured and are likely to be greater than 1 mg/l in patients with evidence of serious toxicity. In the author's opinion there is little merit in measuring such concentrations as an emergency, but it has been claimed that this may be indicated in a child with disturbed behaviour or coma without a history of possible ingestion.

Barbiturate poisoning

Until recently, barbiturate hypnotics were among the most potent and lethal central nervous system depressants available in the community, but they have been largely supplanted by the benzodiazepines, which are much safer in overdosage. However they continue to be prescribed for elderly people who may have been taking them for years and a possibility of accidental poisoning while visiting grandparents is still real. Phenobarbitone is widely used as an anticonvulsant and is likely to be so for the forseeable future.

Toxicity

Barbiturates depress all central nervous system functions. As a result, there is impairment of consciousness, often leading to profound coma which may be prolonged, respiratory depression, circulatory failure and hypothermia.

Pharmacokinetic considerations

Barbiturate hypnotics are lipid-soluble and extensively metabolised in the liver. Very little active drug is excreted in the urine. In contrast, phenobarbitone is more polar and as a consequence only about 75% is metabolised, the remainder being eliminated unchanged in the urine. These properties have important implications for the management of acute poisoning.

Treatment

Once the stomach has been emptied, the treatment of barbiturate poisoning is usually supportive and includes ensuring a patent airway and adequate ventilation, treating shock and prevention of further heat loss in hypothermic patients. These measures are continued until the drug is metabolised and/or excreted and consciousness is regained.

In the case of phenobarbitone poisoning, recovery can be hastened somewhat by increasing renal elimination of the drug by means of forced alkaline diuresis. The rationale for this treatment is that phenobarbitone is weakly acidic and ionised at alkaline pH. Alkalinisation of the urine (achieved by giving intravenous sodium bicarbonate) reduces the reabsorption of phenobarbitone from tubular urine and correspondingly increases elimination. As with forced alkaline diuresis for salicylate poisoning, fluid retention, hypokalaemia and hypocalcaemia are potential hazards, particularly since the procedure usually has to be continued for several days to produce appreciable benefit. This form of treatment is totally inappropriate for poisoning with barbiturate hypnotics.

All barbiturates can be removed from the circulation effectively by perfusing the patient's blood through a column packed with activated charcoal or ion-exhange resin (haemoperfusion), but platelets and other blood constituents are also removed. The use of this technique is reserved for patients who are desperately ill.

Laboratory investigations

Arterial blood gas analysis is usually the most important investigation when managing patients severely poisoned with barbiturates. The results determine the need for assisted ventilation but obtaining reliable samples can be extremely difficult in patients for whom the investigation is most necessary. The body core temperature often has to be taken into account so that gas tensions can be corrected for hypothermia.

Self-ventilating, unconscious adults have carbon dioxide tensions at the upper limit of normal and low oxygen tensions, the latter being progressively reduced with increasing depth of coma. Arterial hydrogen ion concentrations are also often increased, usually due to a mixed metabolic and respiratory acidosis, but less commonly the latter alone.

Arterial blood gas analysis is also mandatory to confirm that assisted ventilation is adequate and may have to be repeated several times during the first few hours when the severity of poisoning and defects of resuscitation are being assessed.

When the diagnosis of poisoning with barbiturate hypnotics is beyond doubt there is no justification for emergency measurement of plasma drug concentrations since management, with rare exceptions, is supportive and unlikely to be altered by the result. However if there is doubt about the nature of the barbiturate involved it is reasonable for the clinician to ask the laboratory to determine whether it is phenobarbitone or one of the others, since forced alkaline diuresis might be appropriate treatment.

Measurement plasma barbiturate concentrations is mandatory before embarking on forced diuresis or any other technique intended to enhance elimination, and also thereafter to monitor the efficacy of treatment.

Poisoning with iron salts

Overdosage with iron salts, particularly ferrous sulphate, has long been one of the most serious forms of childhood poisoning. Despite this the mortality from iron poisoning has probably been exaggerated by biased reporting of serious cases, as with other poisonings.

Features of poisoning

Iron salts have corrosive effects on the gastrointestinal tract and early features of poisoning include vomiting, abdominal pain and diarrhoea. Tablet residues may make the vomitus and stool dark grey or black in colour and blood may also be present if ulceration occurs. Hypotension and impairment of consciousness may develop later and are the best clinical indicators of severe poisoning. Jaundice due to hepatic necrosis may develop after 3–4 days and lead to hepatic encephalopathy and death in rare cases. Renal failure may also occur. Weeks or months after ingestion children may present with vomiting and a metabolic alkalosis secondary to pyloric stricture subsequent to healing of mucosal ulceration by fibrosis.

Treatment

Absorption of iron from the gastrointestinal tract is reduced by emptying the stomach and rendering the iron insoluble by leaving sodium bicarbonate or phosphate solutions in the stomach. Alternatively, desferrioxamine may be used to chelate any remaining iron. Desferrioxamine may also be given by intramuscular or intravenous injection to chelate circulating iron. These measures are supplementary to supportive care including replacement of electrolytes, fluid and blood lost through the gut and, if necessary, conventional treatment for hepatic and renal failure.

Laboratory investigations

Reliable assessment of the severity of iron poisoning is seldom possible from the early features and since parenteral desferrioxamine has potentially serious adverse effects it should not be given indiscriminately. The laboratory must therefore assist with the measurement of the serum iron concentration as an emergency. The limited data available indicate that concentrations peak within 2–3 hours of ingestion, sometimes attaining values such as 90 mg/l (1600 μmol/l) which then decline very rapidly as the iron is distributed to tissues. Chelation with desferrioxamine is most effective if it is given while plasma iron concentrations are high and values in excess of 5 mg/l (90 μmol/l) are usually accepted as indications for desferrioxamine administration. Subsequent serum iron estimations may be necessary to monitor the efficacy of treatment. It should be noted that ferrioxamine (the iron-desferrioxamine complex) is excreted in the urine and colours it orange/red.

The child with serious iron poisoning also requires measurement of plasma electrolytes to guide replacement therapy together with arterial blood gas analysis for correction of acid–base disturbance, urea and creatinine and liver function tests.

Paraquat poisoning

Weedkillers containing paraquat are undoubtedly amongst the most toxic chemicals which are freely available in the community. Many of the early deaths from paraquat, including some children, were due to accidental ingestion of liquid concentrates such as *Gramoxone* (20% paraquat) which had been transferred to soft drink bottles from the original containers. Since then the sale of this formulation has been restricted by law to farmers and professional horticulturists and as a consequence serious accidental poisoning is now less common and most cases are due to deliberate self-poisoning by adults. However children may still get access to paraquat as contained in granular formulations such as *Weedol* and *Pathclear* which are available to the general public. Fortunately the paraquat content of these is low (2.5%) and the risk of poisoning is small unless about 15 g (about $\frac{1}{4}$ of a sachet) or more is ingested, though the lethal dose obviously depends on the age, size and weight of the child.

The ingestion of fruit or vegetables recently sprayed with properly diluted paraquat is exceedingly unlikely to lead to significant poisoning. Inhalation of paraquat aerosols does not cause systemic poisoning.

Toxicity

Solutions of paraquat are strongly alkaline and directly corrosive to the upper gastrointestinal tract and skin. It also causes renal failure, hepatic damage and pulmonary alveolar cell proliferation with fibrosis which results in progressive respiratory failure and death.

Pharmacokinetic considerations

In adults ingested paraquat is very rapidly absorbed, with peak plasma concentrations being attained within 2–3 hours. Plasma concentrations then decline rapidly as paraquat is taken up by tissues, particularly the lungs and kidneys. Most of the absorbed paraquat is excreted in the urine.

Features of poisoning

While paraquat ingestion in childhood is likely to be uneventful the potential for serious harm should not be underestimated. Paraquat causes nausea, vomiting, abdominal pain and diarrhoea. About two days after ingestion burns of the mouth, tongue, pharynx, and oesophagus become apparent, making swallowing, coughing and speaking difficult. These features are a function of the concentration of the paraquat solution swallowed. In severe poisoning acute renal tubular necrosis and renal failure may develop within 24–36 hours and haemodialysis may be necessary. Mild hepatocellular damage also occurs but is less common. Death from paraquat poisoning is usually due to profound hypoxia secondary to progressive pulmonary fibrosis, but with massive ingestion death from cardiorespiratory failure may occur within a few hours.

Treatment

Treatment of paraquat poisoning involves preventing further absorption by emptying the stomach and by giving adsorbents such as Fuller's earth or bentonite. Thereafter, there is probably no effective treatment. Efforts to remove paraquat from the circulation by haemodialysis and haemoperfusion or to mitigate its cellular effects by giving corticosteriods, immunosuppressive drugs and superoxide dismutase have been remarkable only for their lack of success in preventing a fatal outcome.

Laboratory investigation

The toxicity of paraquat is known to many members of the public and, not surprisingly, any possibility of poisoning creates considerable alarm. The laboratory has a vital role in allaying these anxieties at an early stage and protecting patients from potentially hazardous, unnecessary or ineffective treatment.

Absorption of paraquat can be readily confirmed or refuted by adding small quantities of sodium dithionite and sodium bicarbonate to 5 ml of urine. If paraquat is present a colour, varying between pale green and dark blue, develops instantly. A negative test on urine passed within about 4 hours after alleged ingestion indicates that little or no paraquat has been absorbed and that there is no risk. If the urine test is positive, interpretation is more difficult because the severity of systemic poisoning cannot be inferred from the intensity of the blue colour developed on testing the urine. Strongly positive tests are compatible with both insignificant and significant poisoning, depending on the time interval between ingestion and passing the urine tested. Accurate assessment of severity requires measurement of the plasma paraquat concentration, which must then be related to the time interval since ingestion. Concentrations exceeding 2.0, 0.6, 0.2 and 0.1 mg/l at 4, 6, 13 and 24 hours after ingestion, respectively, are likely to be fatal.

Paracetamol

In Britain paracetamol has overtaken aspirin as the standard non-prescription analgesic. As a consequence there has been a considerable increase in the number of adults deliberately taking overdoses of this drug, often with serious or fatal consequences. Surprisingly enough, serious childhood poisoning with paracetamol is uncommon and only a few cases of severe toxicity have been reported. The reasons for this are not clear though it may be related to the early induction of vomiting, failure to ingest toxic doses and qualitative differences in the metabolism of the drug in children which lead to the reduced conversion to the toxic metabolite.

Mode of toxicity

Paracetamol is extensively metabolised in the liver to form sulphate and glucuronide conjugates. However a small proportion is converted to a highly reactive intermediary metabolite by hepatic mixed function oxidases and normally inactivated by conjugation with hepatic glutathione. The glutathione conjugate is eventually excreted in the urine as cysteine and mercapturic acid conjugates. When a large amount has been taken however, the hepatic stores of glutathione are rapidly exhausted and the metabolite is then free to combine with the SH-groups in cell macromolecules, leading to cell death. The same process occurs in the renal tubules but renal failure is an uncommon complication of paracetamol poisoning in adults and has not been reported in children.

Features of poisoning

Most children are unlikely to develop symptoms. If they do, they may, in the early stages, be nothing more than nausea and vomiting. Normally these will clear within 12–24 hours, to be followed by abdominal pain, hepatic tenderness and mild jaundice. These features are the result of hepatocellular necrosis. Proteinuria and haematuria may presage the onset of renal tubular necrosis.

Maximum abnormality of liver function tests occur 2–3 days after ingestion with plasma alanine aminotransferase activities, which must be among the highest recorded. Values of up to 15 000 IU/l are not uncommon in severe poisoning in adults, though unlikely to be attained in children. The plasma bilirubin is only moderately elevated and the alkaline phosphatase is usually normal. Patients with very severe poisoning may go into hepatic encephalopathy and die 3–7 days after ingestion from haemorrhage, infection or cerebral oedema. It must be emphasised that such occurrences are rare in accidental childhood poisoning.

Laboratory investigation

The severity of paracetamol poisoning cannot be assessed from the dose alleged or thought to have been taken, or the symptoms in the first few hours. In adults the likelihood of severe liver damage can be predicted from the plasma paracetamol concentration related to the time interval since ingestion. The limited data available for children suggests that the same relationships pertain and the laboratory must therefore be prepared to measure plasma paracetamol concentrations as an emergency. In doing so it is essential to ensure that a specific assay is used, since some methods also measure inactive metabolites and overestimate plasma concentrations of unchanged drug by over 700%.

Liver function tests should be monitored for 3–4 days if the patient is at risk of liver damage, but not otherwise. Measurement of the plasma urea, creatinine and electrolytes are obviously required if there is any suggestion of renal failure.

Treatment

Most children who ingest excessive quantities of paracetamol will require no more treatment than emptying the stomach to reduce further absorption. However if the plasma concentration, related to the time since ingestion, indicates a risk of hepatic damage it would be appropriate to give specific antidotes such as N-acetylcysteine, or if this is not available, methionine, in doses appropriate to the weight of the patient. In adults treatment with intravenous N-acetylcysteine within eight hours gives complete protection against liver damage. It is generally assumed that children will benefit similarly.

Solvent inhalation (Glue sniffing)

The practice of inhaling the volatile hydrocarbons contained in glues, dry-cleaning agents, paint-thinners and petrol has increased greatly in recent years and is now a cause for considerable concern. A variety of solvents are involved including acetone, carbon tetrachloride, toluene, trichlorethane, trichloroethylene and tetrachloroethylene. Deliberate solvent inhalation is usually a group activity with boys predominating, the peak incidence being between the ages of 13 and 15 years, but with some individuals as young as 8 years. Adhesives are usually squeezed into a polythene or potato-crisp bag before being inhaled, while dry cleaning agents are sniffed after being poured on to a piece of cloth.

Features

The solvents mentioned above are lipid soluble and their major early effects are on the brain, causing features similar to alcohol intoxication. There may be euphoria or mild excitement with dysarthria, ataxia and disturbed behaviour. Some children may experience hallucinations or become suicidal. As exposure continues, consciousness is depressed, leading to coma, convulsions and metabolic acidosis. Ulceration of the lips and nostrils may be present. Rarely, deaths occur most frequently from asphyxia caused by the polythene bags used to contain the adhesive and less frequently from burns and the toxic effects of the solvents.

Recovery from solvent inhalation is usually complete but a variety of short- and long-term complications have been reported, including permanent brain damage (petrol, toluene), reversible renal damage (toluene and halogenated hydrocarbons), Fanconi's syndrome, renal tubular acidosis and peripheral neuropathy (hexane, methyl-isobutyl ketone).

Treatment

Removal from exposure is as much treatment as is required in most cases. Supportive measures are necessary if consciousness is impaired and renal and hepatic damage are managed conventionally.

Laboratory investigations

Laboratory investigations are unnecessary in most cases of solvent inhalation. If exposure has been prolonged or heavy, renal and liver function tests may be required. In severe poisoning arterial blood gas analysis is indicated.

A diagnosis of solvent inhalation is not always possible from the clinical features alone. Toluene is present in most of the abused adhesives and its presence in the blood may be demonstrated by gas chromatography.

Other poisons

Other poisons which occasionally cause childhood deaths (mainly digoxin, quinine, anticholinergic drugs and narcotic analgesics) have few implications for clinical laboratory investigation other than the biochemical investigation required for appropriate supportive care.

BIBLIOGRAPHY

General

PROUDFOOT A. T. (1982) *Diagnosis and Management of Acute Poisoning*. Blackwell Scientific Publications, Oxford.

VALE J. A. & MEREDITH T. J. (1981) *Poisoning: Diagnosis and Treatment*. Update Books, London.

PRESCOTT L. F. (1980) Pharmacological principles in the treatment of drug overdosage and poisoning. In Avery G. S. (ed.) *Drug Treatment*. 2nd edn. ADIS Press, Sidney.

Epidemiology

CALNAN M. W., DALE J. W. & DE FONSEKA C. P. (1976) Suspected poisoning in children. Study of the incidence of true poisoning and poisoning scare in a defined population in North East Bristol. *Arch. Dis. Child.* **51**, 180.

McLEAN W. (1980) Child poisoning in England and Wales: some statistics on admission to hospital, 1964–76. *Health Trends* **12**, 9.

FRASER N. C. (1980) Accidental poisoning deaths in British children, 1958–77. *Br. Med. J.* **1**, 1595.

Laboratory

CIBA FOUNDATION SYMPOSIUM 26 (new series) (1974) *The Poisoned Patient: the role of the laboratory*. Elsevier, North Holland, Amsterdam.

FLANAGAN R. J. *et al.* (1981) Value of toxicological investigation in the diagnosis of acute drug poisoning in children. *Lancet* **ii**, 682.

HELLIWELL M. *et al.* (1979) Value of toxicological investigations in differential diagnosis of coma. *Br. Med. J.* **2**, 819.

INGELFINGER J. A. *et al.* (1981) Reliability of the toxic screen in drug overdose. *Clin. Pharmacol. Ther.* **29**, 570.

PRESCOTT L. F. (1981) Toxicological monitoring. In Richens A. and Marks V. (eds.) *Therapeutic Monitoring*. Churchill Livingstone, Edinburgh.

SIMPSON D. (1980) Units for reporting the results of toxicological measurements. *Ann. Clin. Biochem.* **17**, 328.

STEWART M. J. *et al.* (1979) Inappropriate methods for the emergency determination of plasma paracetamol. *Ann. Clin. Biochem.* **16**, 89.

Non-accidental poisoning

FLANAGAN R. J. *et al.* (1981) Value of toxicological investigation in the diagnosis of acute drug poisoning in children. *Lancet* **ii**, 682.

ROGERS D. *et al.* (1976) Non-accidental poisoning: an extended syndrome of child abuse. *Br. Med. J.* **1**, 793.

Specific poisonings

Salicylate

PROUDFOOT A. T. (1983) Salicylate poisoning. In Haddad L. M. and Winchester J. F. (eds.) *Clinical Poisons and Drug Overdose*. W. B. Saunders, Philadelphia.

Tricyclic antidepressant

BROWN T. C. K. (1975) Overdosage: the rise and fall of tricyclic antidepressants. *Aust. Paediat. J.* **11**, 190.

CROME P. & NEWMAN B. (1979) The problem of tricyclic antidepressant poisoning. *Postgrad. Med. J.* **55**, 528.

CRONIN A. J. *et al.* (1979) Poisoning with tricyclic antidepressants: an avoidable cause of childhood deaths. *Br. Med. J.* **1**, 722.

GOEL K. M. & SHANKS R. A. (1974) Amitriptyline and imipramine poisoning in children. *Br. Med. J.* **1**, 261.

GREENBLATT D. J. *et al.* (1976) Accidental poisoning with psychotropic drugs in children. *Am. J. Dis. Child.* **130**, 507.

LEADING ARTICLE (1979) Tricyclic antidepressant poisoning in children. *Lancet* **2**, 511.

PETIT J. M. & BIGGS J. T. (1977) Tricyclic antidepressant overdoses in adolescent patients. *Pediatrics* **59**, 283.

ROLLIN P. & ANTAKI A. J. (1979) L'intoxication par antidepresseurs tricycliques chez l'enfant et l'adolescent. *Can. Med. Assoc. J.* **120**, 951.

Barbituate

MATTHEW H. (1979) Barbiturates. *Clin. Toxicol.* **8**, 495.

Iron

BARKER G. A. *et al.* (1974) Acute iron poisoning—clinical features and management. *Anaesth. Intens. Care* **4**, 345.

GREENBLATT D. J. *et al.* (1976) Accidental iron poisoning in childhood. *Clin. Pediatr.* **15**, 835.

Haddad L. M. (1976) Iron poisoning. *JACEP* **5**, 691.
Stein M. *et al.* (1976) Acute iron poisoning in children. *West. J. Med.* **125**, 289.

Paraquat

Proudfoot A. T. *et al.* (1979) Paraquat poisoning: significance of plasma paraquat concentrations. *Lancet* **ii**, 330.

Paracetamol

Meredith T. J. *et al.* (1978) Paracetamol poisoning in children. *Br. Med. J.* **2**, 478.
Peterson R. G. & Rumack B. H. (1981) Age as a variable in acetaminophen overdose. *Arch. Intern. Med.* **141**, 390.
Prescott L. F. (1981) Treatment of severe acetaminophen poisoning with intravenous acetylcysteine. *Arch. Intern. Med.* **141**, 386.

Chapter 19
Malignant Disease

JON PRITCHARD & ALAN W. CRAFT

INTRODUCTION TO CHILDHOOD CANCER

Childhood cancer is uncommon, affecting only 1 in 600 children before their fifteenth birthday. Children account for only 1% of all patients with malignant diseases but, taking into account work-years lost, child cancer assumes much importance in economic terms. Not only is cancer the second most common cause of death in children aged 1–14 years (Draper *et al.* 1981) but it is becoming increasingly important in terms of inpatient and outpatient medical care. In 1977, 1.7% of all inpatient days were occupied by children with cancer (Hospital Inpatient Enquiry (HIPE) 1977). About 1400 children in Britain will develop malignant disease each year: accurate registration of all cases occurring in a defined geographical area has been undertaken in the North of England since 1954 and 1968 for the Manchester and Newcastle regions respectively. In these two areas, with a combined total population of almost nine million, there is a remarkably constant incidence rate of 105 cases per million children per year (Table 19.1).

In contrast to adults with cancer, in whom carcinomas (epithelial tumours) predominate, most childhood malignancies are of mesodermal origin (embryonal sarcomas or haemopoietic tumours). The frequency of specific tumour types occurring in the Manchester and Newcastle regions is shown in Table 19.1 (Marsden & Steward 1976). Leukaemia accounts for almost one-third of registered cases and brain tumours one-fifth. International comparisons of the incidence of various types of malignant disease in childhood have shown some similarities, but also some interesting differences. There is, for instance, an inverse relationship between incidences of acute leukaemia and of lymphoma. In several 'third world' countries lymphomas—especially Burkitt-type non-Hodgkin lymphomas—are more common than leukaemia; in the 'developed' world, by contrast, leukaemia is three times more frequent than lymphoma. Ewing's sarcoma of bone is rare in negro children whether they live in Africa or North America, perhaps suggesting an inborn genetic resistance to the development of this tumour. Neuroblastoma is uncommon (2–3% of all tumours) in Africa, whereas retinoblastoma occurs twice as frequently (6–8% incidence) as in the UK or USA. Nephroblastoma (Wilms' tumour) is remarkable for the constancy of its occurrence throughout the world. Wilms' is therefore regarded as the 'index' tumour against which the incidence of other

Table 19.1 Annual incidence of malignant disease per million children in the Northern and North-Western Regions.

Category of tumour	Newcastle	Manchester
Leukaemia	34.7	31
Lymphoma	9.7	10
Glioma	23.2	19
Neuroblastoma	6.2	8
Retinoblastoma	4.8	3
Soft tissue sarcoma	10.1	12
Wilms'	5.8	6
Gonadal	1.3	1
Teratomas	0.4	3
Epithelial	2.9	4
Ewing's	1.6	2
Miscellaneous	4.7	6
Total	105.4	105

malignant diseases, in a particular geographical setting, can be standardised. There are two peak ages of incidence of childhood tumours, the first at 1–4 years and the second during the pubertal growth spurt. Each type of cancer occurs at a more-or-less characteristic age. Acute lymphoblastic leukaemia, Wilms' tumour and neuroblastoma are diagnosed most commonly between 2 and 4 years, whilst Ewing's tumour and non-Hodgkin's lymphoma peak around 7 years of age. Acute myeloid leukaemia, osteosarcoma and Hodgkin's disease occur more commonly in older children. A difference in the sex incidence of childhood cancer is striking only for non-Hodgkin's lymphoma which, for unknown reasons, is 3–4 times more common in boys than girls.

The aetiology of embryonal tumours is considered by Knudsen (1976) to be the result of a 'two-hit' (i.e. two mutation) process. The first mutation may be inherited (somatic) or the result of environmental influence(s) whilst the second 'hit' is always environmental. Knudsen's theory neatly explains the earlier occurrence of inherited compared with sporadic retinoblastoma, but is less readily applicable to those embryonal tumours (Wilms' and neuroblastoma) whose familial incidence is very low.

Large-scale epidemiological studies have been and are being carried out to try to identify factors important in the aetiology of childhood cancer. Genetic influences certainly play a part. Bilateral retinoblastoma, for instance, is transmitted in an autosomal dominant fashion; some patients have a constitutional deletion from the long arm of chromosome 13 which may, for these cases, represent Knudsen's 'first-hit'. Children with certain other chromosomal abnormalities also have a high incidence of cancer. Acute leukaemia, usually lymphoblastic, occurs 15 times more frequently in children with Down's syndrome than in the general population. Patients with congenital immunodeficiency (e.g. Wiskott–Aldrich syndrome) or 'chromosomal instability syndromes' (e.g. ataxia telangiectasia, Bloom's syndrome, Fanconi anaemia) are particularly susceptible to the development of lymphoreticular malignancies or, in the case of xeroderma pigmentosum, skin cancer. Environmental carcinogens are hard to identify. Prenatal exposure to X-ray causes a slight increase in the risk for development of acute leukaemia although, in the past, the danger may have been exaggerated. Now that the prenatal administration, to the mother, of stilboestrol has been identified as the cause of infant adenocarcinoma of the vagina and has been discontinued, this iatrogenic form of childhood cancer does not occur. The association of Epstein–Barr virus with Burkitt lymphoma is well known; whether the virus is oncogenic or merely a 'passenger', is still uncertain. The relationship of a recently-identified C-type virus to the cause of some cases of T-cell lymphoblastic leukaemia is equally uncertain. Thyroid cancer (Refetoff *et al.* 1975) and second malignancies in patients cured of their first tumour by radiation or drugs (Mike *et al.* 1982) are the only clear-cut examples of child cancer that are recognisably induced by environmental factors.

As a result of improvements in social standards and the control of most infectious disease, children aged 1–14 years in the 'developed' world more commonly die from cancer than from any other disease group. This statistic conceals recent major progress in treatment. Whereas, twenty years ago, a diagnosis of malignant disease in a child was an almost inevitable death sentence, over half the patients should now be long-term survivors and probably cured of their disease. Table 19.2 shows the predicted five year survival rates for 10 of the more common tumours. Factors contributing to this encouraging state of affairs are listed in

Table 19.2 Expected 5-year disease-free survival for children with 10 most frequently diagnosed childhood cancers.

	% Survival
Retinoblastoma	95
Hodgkin's disease	85
Wilms' tumour	80
Non-Hodgkin's lymphoma	65
Rhabdomyosarcoma	60
Acute lymphoblastic leukaemia	50
Ewing's sarcoma	45
Medulloblastoma	45
Neuroblastoma	25
Acute myeloid leukaemia	10*

* If HLA–MLC-compatible sibling marrow is available for transplantation and the child is treated with high-dose chemoradiotherapy, the figure is 50+%.

Table 19.3 Factors contributing to improved survival of children with cancer.

Introduction of combination chemotherapy
Introduction of 'new agents' e.g. cis-platinum, epipodophyllotoxins
Use of 'delayed' surgery where appropriate
Refinements of radiotherapy technique e.g. megavoltage equipment, interstitial radiation, 'shrinking fields'
Better 'imaging' investigations
Refinements in histopathological sub-classification
Use of 'tumour markers'
Better supportive care e.g. biochemical support, blood product support, antibiotics, antiviral agents
Development of multi-disciplinary teams within paediatric oncology centres
Establishment of national collaborative groups e.g. United States Children's Cancer Study Group, United Kingdom Children's Cancer Study Group
Early diagnosis (retinoblastoma)

Table 19.3. The introduction and rational use of chemotherapy, along with the continual introduction of new 'active' agents has been of central importance but refinements of surgical and radiotherapy technique have also contributed. Careful screening of families with a known predisposition to retinoblastoma has led to early diagnosis and more effective therapy in this particular group. The establishment of multicentre co-operative groups such as the United Kingdom Children's Cancer Study Group (UKCCSG) means not only that information about these rare diseases is carefully collated and shared, but also that clinical trials can reach statistically significant conclusions much more quickly, because there are more patients available for study than would be the case for single centres. Improved 'support' facilities have also been crucial and co-operation with the clinical chemistry laboratory is of special importance. On the one hand, the use of 'tumour markers' is of extreme importance in monitoring the progress of patients with certain tumours (see p. 517) and, on the other hand, the better understanding of biochemical complications of the diseases and their treatment has enabled more aggressive therapy to be given with greater safety (p. 540).

With improving survival there is heightened awareness of problems resulting from the medium- and long-term effects on the child of both disease and treatment (p. 544 *et seq*). As cancer patients, children are unique because they are growing and therefore particularly susceptible to both emotional and physical handicap. 'Quality of survival' of both child *and* family is now at the forefront of the concerns of the paediatric oncologist: thus, current philosophies of treatment have changed from 'cure-at-any-cost' to 'cure-at-least-cost'.

GENERAL PRINCIPLES OF DIAGNOSIS, STAGING AND MANAGEMENT

Diagnosis

Optimal management depends upon precise pre-treatment diagnosis and staging. Diagnosis of cancer in childhood may be strongly suggested by the history, physical examination and imaging investigations, but confirmatory histopathological tissue diagnosis or supportive biochemical evidence is almost always sought. Exceptions, where reliance is placed upon the clinical diagnosis alone, include brain stem tumours: here, even computerised axial tomographic (CAT) scans may not define the tumour and surgical biopsy is hazardous. Tissue for microscopic diagnosis of solid tumours is usually obtained by open biopsy, but sometimes—for example, where an abdominal mass is considered surgically 'unresectable'—by closed needle biopsy. As well as conventional histological studies, special histochemical stains are used and immunological techniques using fluorescence-labelled hetero- or monoclonal antisera have been introduced for the diagnosis of leukaemias and, very recently, neuroblastoma (Pritchard & Kemshead, 1983). Histopathological sub-classification is of more than academic interest: the tumour sub-type is of major prognostic significance for both leukaemia and solid tumours. The treatment plan for patients with 'favourable histology' Wilms' tumour, for example, is quite different from that for patients with 'unfavourable histology' tumours.

Once diagnosis is secure, staging investigations are carried out. Staging has three main functions: it indicates the extent of disease at diagnosis and thus contributes to treatment planning, it enables an estimate of prognosis to be made and it allows for comparison of the results of treatment in different centres.

At present, there is a great deal of variation between staging systems used in different centres: the adoption of an internationally-accepted system (e.g. the 'TNM' system) would have major advantages. For acute lymphoblastic leukaemia, the height of the white cell count at diagnosis is considered the most important 'staging' investigation (Palmer & Hann 1980) and may determine the intensity of treatment used for individual patients. As yet, it is not clear whether the presence of 'null', 'T' or 'B' markers on leukaemic cells have independent prognostic significance. As

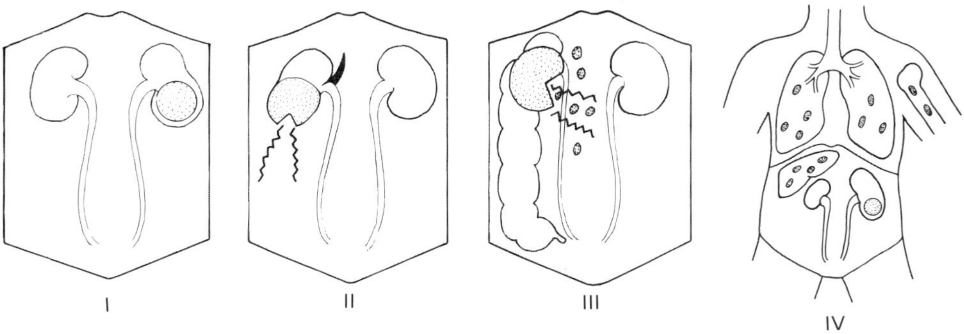

Fig. 19.1 Staging system used in third USA National Wilms' Tumour Study and in first UK Children's Cancer Study Group Wilms' Trial. Stage II includes tumour rupture localised to the flank and IVC involvement. Patients with lymph node involvement and those with measurable residual disease are classified as Stage III.

an example of solid tumour staging, the system used for Wilms' tumour is shown in Fig. 19.1. In general, the 'higher' the stage the more extensive the disease.

In the individual child, the aims of staging investigations are to indicate the extent of the primary and to seek evidence of metastatic spread. The limits of the primary tumour may be defined surgically or by 'imaging' investigations using diagnostic X-rays, ultrasound, isotope scanning and CAT scanning. Most tumours have a characteristic pattern of metastatic spread: investigation can thus be rationalised, saving the child unnecessary trauma and the hospital unnecessary expense. Even where disease appears to be confined to the primary site the presence of micrometastases—too small to be shown by presently available techniques—is usually assumed. There are hopes that new 'immunolocalisation' techniques, using monoclonal antibodies to tumour-associated antigens, may provide a more sensitive and economical way of seeking occult tumour. Although there are some important exceptions, for example stage I and II neuroblastoma, systemic therapy is invariably indicated even for apparently localised disease.

Treatment principles

Generally, attempts at ablation of primary tumour by surgery and radiotherapy are combined with systemic drug treatment to control overt or occult metastatic disease. Complete eradication of some 'sensitive' tumours, especially leukaemias and lymphomas, can often be achieved with chemotherapy alone, but primary excision of less chemosensitive tumours is indicated, providing the surgery is not mutilating. Thus, it is acceptable to perform a nephrectomy for Wilms' tumour, but total pelvic extentoration for rhabdomyosarcoma is now rarely indicated since this tumour can often be eradicated by a combination of chemotherapy and radiotherapy. Presurgical chemotherapy can often make a procedure easier for the surgeon. Similarly, pre-radiation chemotherapy often allows the radiotherapist to use a 'shrinking field' technique, concentrating the highest doses on tumour residue rather than on the whole of the original tumour-bearing volume. Chemotherapy is usually the most appropriate initial treatment for patients with obvious metastatic disease. Unless the secondaries can be controlled, an attack on the primary is clearly illogical. With a precise knowledge of histological type and stage, it is possible to plan for each patient a treatment strategy using surgery, radiotherapy and chemotherapy, carefully combined to give maximum benefit with a minimum of immediate and late toxicity.

'Routine' biochemical studies

Clinicians, often without thinking carefully, frequently ask for regular biochemical studies as part of their 'routine' management of patients. Such an attitude should be discouraged. There is little point, for example, in carrying out repeated liver function tests on children with 'favourable' histology stage I Wilms' tumour whose only post-surgical treatment is vincristine; the situation is quite different for patients with leukaemia receiving regular methotrexate (see p. 545). The choice of investigations should be tailored to the needs of the individual patient or, after discussions with laboratory staff, to the requirements of thoughtfully-designed investigational protocols. These days many patients are entered into national clinical trials and data sent to a central reference point. This often entails the submission of sera for special investigations under standardised conditions. New ideas are constantly evolving and it is valuable practice to store pre-treatment serum, at −70 °C and in aliquots if possible, for future studies.

Clinical trials and organisation of services

Most advances in chemotherapy have come about as a result of pilot studies of single- or combination-drug therapy followed by carefully controlled clinical trials. Although initial studies are promising (Hill & Whelan 1981), *in vitro* sensitivity testing of tumour cells against cytotoxic agents is not yet clinically applicable so that new drugs have to be tested in a series of clinical investigations, termed phase I, II and III studies. In phase I, toxicity is studied whilst in phase II studies an agent is tested in patients with a variety of tumours already resistant to other drugs. Promising results in phase II trials are extended by selecting 'responsive' tumours and studying larger numbers of patients with these tumours in a phase III trial. In this way an accurate 'response rate' can be measured. The dramatic recent improvement in the prognosis for men with testicular cancer (Einhorn & Williams 1980) can be directly attributed to these sorts of studies.

The prognosis for children with cancer is significantly better when diagnosis, staging and treatment planning are carried out in a recognised paediatric oncology centre (Lennox *et al.* 1979). It is therefore no longer appropriate for these children to be treated 'sporadically' in small centres without the personnel and facilities required for optimal medical and psychosocial care. Whilst clinicians should be discriminating in their requests the clinical chemistry laboratory in each paediatric oncology centre should be prepared to offer, on an emergency basis when necessary, facilities for the investigations described in the following sections.

DIAGNOSIS AND MONITORING OF INDIVIDUAL TUMOURS

General aspects of 'tumour markers'

Until recently, there was speculation that cancer-specific antigens might exist and be identifiable. The diagnostic and therapeutic potential of such antigens is not difficult to imagine. Optimism has now given way to a more sober realisation that, with very rare exceptions such as the abundant production of monotypic immunoglobulins by myeloma cells, tumour-specific antigens simply do not exist. For certain specific tumour types, however, it has been established that amplified expression of secreted antigens, expressed in the normal tissue by only rare cells or at low frequency, does occur after malignant transformation. The characteristics of the malignant clone determine the type of protein produced. If transformation has occurred in a 'committed' fetal cell the 'tumour marker' is often a normal fetal product, for example HbF in juvenile chronic myeloid leukaemia and α-fetoprotein (αFP) in liver and yolk sac tumours. Such 'onco-fetal' proteins are easily detected because of their very low normal blood levels after the first few months of life. If, on the other hand, transformation occurs in a 'committed' adult cell a marker will be detectable only if it is readily separable in the laboratory from other normal components of plasma (e.g. 'M-spike' in myeloma) or if it has physiological activity (ACTH from adrenal carcinoma). A third cellular mechanism—aberrent gene activation in tumour cells—leads to 'ectopic' production of tumour markers, for example ACTH by occasional Wilms' tumours.

Clinically useful 'tumour markers', disappointingly only available for a few tumours, are discussed in the relevant parts of the next section (Table 19.4).

Table 19.4 Chemical 'markers' for paediatric tumours.

Brain tumours	
Medulloblastoma	None known
Ependymoma	None known
Germ cell tumour	αFP/βhCG
Glioma	None known
Pinealoma	Melatonin (?)
Carcinoid	5-HIAA*
Endocrine tumours	
Adrenal adenoma/carcinoma	Cortisol/androgens*
Parathyroid tumours	Calcium/PTH*
Phaeochromocytoma	Urinary VMA* Plasma catecholamines*
Thyroid (C-cell) tumour	Calcitonin*
Ewing's sarcoma	LDH (?)
Histiocytosis-X	None known
Leukaemia/lymphoma	LDH (?) Lysozyme (AMMoL only)
Liver tumours	
Hepatoblastoma	αFP*
Hepatoma	αFP*/transcobolamin I*
Neuroblastoma	Urinary HVA/VMA* Ferritin VIP
Osteogenic sarcoma	Alkaline phosphatase
Rhabdomyosarcoma	None known
Wilms' tumour	None known
Yolk sac tumour (malignant teratoma)	αFP*/βhCG*

'Tumour markers' are in serum, unless otherwise stated.
* Indicates markers of *established* clinical value.
Abbreviations defined in the text.

Brain tumours

Primary brain tumours are, as a group, the most common 'solid' tumours of childhood. Computerised tomographic (CT) scanning has made pre-operative assessment of brain tumours much easier for both patient and doctor. Improvements in surgical technique, especially the use of the operating microscope, and in radiotherapy, have recently been reflected in an improving prognosis, especially for patients with medulloblastoma and ependymoma (Bloom 1982). However, although several agents, for example nitrosoureas (CCNU, BCNU), vincristine, procarbazine, VM-26 and methotrexate give measurable response rates (Levin & Wilson 1975; Bloom

1982) when used singly or in combination, there is little evidence, by contrast, that chemotherapy has contributed to the improvement. A high incidence of late sequelae in survivors of radiation therapy (major shortening of adult height, neurological and endocrine dysfunction, second tumours) (Broadbent *et al.* 1981) indicate the urgent need for new strategies in the management of brain tumours.

Attempts to identify possible 'tumour markers' in the cerebrospinal fluid (CSF) of patients with various types of brain tumour were recently summarised and reviewed (Seldenfeld & Marton 1979). Isocitrate dehydrogenase, desmosterol and polyamines are possible candidate compounds, but further study is required before measurements of these substances can be regarded as having clinical value.

The radioimmunoassay of serum melatonin may be of diagnostic help in patients with inoperable pineal masses for whom the clinician would prefer to avoid surgical biopsy. In pinealoma, serum melatonin levels may be significantly raised (Barber *et al.* 1978). So-called 'nyctohemeral' variation in melatonin production by pineal cells, even in tumours, indicates the need for measurement in serial samples over a 24 hour period (Wurtmann & Moskowitz 1977). The value of serial serum melatonin monitoring in patients after treatment for malignant pineal tumours has still to be established. The monitoring of germ-cell tumours in the pineal region, using serial measurements of αFP and/or βhCG (Neuwelt *et al.* 1979) is discussed in the section on malignant teratomas.

Carcinoid tumour

In children a diagnosis of carcinoid tumour is most frequently made incidentally during the histopathological examination of appendicectomy specimens. Origin in other abdominal or extra-abdominal locations is occasionally recorded. The presence of metastases seems to depend less on the histopathological appearance of a given carcinoid than on its size at diagnosis. It is rare to find evidence of secondaries when the primary tumour is < 1 cm in diameter but $> 80\%$ of tumours of 2 cm diameter or more have already spread to liver, lung and sometimes other sites. The 'carcinoid syndrome' is the result of the production of one or more physiologically active agents by the 10% of tumours that are 'secretors'. Serotonin secretion, for example, causes diarrhoea, hypertension, arthropathy and endomyocardial fibrosis, whilst bradykinin and prostaglandins give rise to flushing and overproduction of histamine can lead to bronchospasm. Intermittently elevated levels of some or all of these substances can be documented in carcinoid patients, but measurement of 24-hour secretion of the 5-hydroxytryptamine (5-HT, serotonin) breakdown product 5-hydroxyindole acetic acid (HIAA) is the most useful single diagnostic investigation (Melmon 1975). So long as dietary restrictions (no bananas, nuts or pineapple) are imposed and there is no malabsorption, the 24-hour excretion of > 47 µmol/24 hours (> 9 mg HIAA/24 hours) is considered diagnostic of carcinoid tumour.

If complete resection of small tumours is achieved surgery is almost always curative. Successful control of unresectable or metastatic disease, with or without symptoms of the carcinoid syndrome, has recently been achieved with streptozotocin, 5-fluorouracil and other chemotherapeutic agents (Brennan 1982).

Endocrine tumours

Adrenal carcinoma and adenoma

Of 141 primary adrenal tumours diagnosed at a single UK centre, 130 were neuroblastomas and only 8 were tumours of the adrenal cortex (Stewart *et al.* 1974). The clinical features and histopathological appearance of carcinoma and adenoma are similar and the distinction between them is often made on the basis of tumour size (smaller masses being more commonly benign) and apparent invasiveness. Presentation is most frequently with an endocrine syndrome—Cushing's, virilisation, feminisation (very rarely) or a 'mixed' endocrine picture—but nearly half these tumours are functionally 'silent' and present as an abdominal mass. Some patients with adrenal carcinoma have features of the so-called Fraumeni–Miller syndrome which includes hemi-hypertrophy, skin haemangiomas and brain tumour.

The value of investigations used in the diagnosis of Cushing's syndrome has recently been reviewed (Gold 1979). Primary adrenal pathology is suspected when, in the presence of a fasting plasma cortisol > 276 nmol/l (> 10 µg/dl), plasma ACTH is reduced. The distinction between adrenal hyperplasia and tumour is readily made by abdominal ultrasound, IVU or CAT scanning. The following findings make a diagnosis of carcinoma, in a child, more likely than adenoma: (a) a palpable mass, (b) no suppression of plasma cortisol by high-dose dexamethasone and (c) increased urinary 17-ketosteroids. Abdominal tumours (Wilms' (Pombo *et al.* 1982), hepatoblastoma, or phaeochromocytoma) secreting ACTH 'ectopically' may mimic an adrenal cortical tumour; in such cases ^{131}I-19-iodocholesterol scanning or selective venous sampling may aid tumour localisation.

Whilst surgery is curative for patients with adenoma, the survival of children with adrenal carcinoma is only 20–30% after surgery alone, despite the fact that few patients have overt metastases at diagnosis. Radiation has no known place in management: o,p′-DDD, which acts as a metabolic 'blocker' rather than as a cytocidal agent, is commonly used in treatment of widespread disease (Hogan *et al.* 1978).

Parathyroid tumours

Fasting hypercalcaemia in childhood is more commonly 'secondary' than the consequence of a parathyroid tumour. Children with metastatic neuroblastoma or leukaemia/lymphoma, especially when bony deposits are numerous, may develop hypercalcaemia resulting from the action of soluble bone-resorbing factors released locally by tumour cells. Recently, a rare histological sub-type of non-metastatic Wilms' tumour has been associated with the production of prostaglandins (PGE_2, PGF_2 and prostacyclin) circulating in sufficient concentrations to cause 'distant' bone resorption (Harvey *et al.* 1982) with consequent hypercalcaemia.

As a cause of primary hyperparathyroidism in childhood, parathyroid carcinoma (Toledano 1979) is even less common than adenoma, but the former tumour is more likely to be visible or palpable (Holmes *et al.* 1969). Metastatic spread to lymph nodes or distant organs may be present at diagnosis. Hypercalcaemia resulting from high-circulating levels of parathyroid hormone (PTH) occurs in most instances, but 'non-secreting' tumours, presumably less differentiated than the secreting tumours, have also been described. When 'peripheral' serum PTH levels are borderline, measurement of PTH in 'selective' venous blood samples, obtained by catheterisation of great vessels, may aid tumour localisation (Gluckman *et al.* 1977). The possible diagnosis of a 'MEN' syndrome (Brennan 1982) should be considered when a diagnosis of parathyroid hyperplasia, rather than parathyroid tumour, is made in a child.

Surgery is the treatment of choice; neither radiation therapy or chemotherapy have any proven role in the management of these tumours.

Phaeochromocytoma

In children, phaeochromocytomas are very rare and are invariably benign. Headache, visual blurring and sweating are the most common symptoms. The tumours are rarely palpable but hypertension, either sporadic or persistent, is almost always found (Stringel *et al.* 1980). The presence of multiple tumours (20% of cases) or a family history of phaeochromocytoma or other endocrine tumour may suggest the diagnosis of a multiple endocrine neoplasia (MEN) syndrome (Brennan 1982) in which the incidence of phaeochromocytoma may be as high as 90%. Studies in adults indicate that measurements of catecholamines and their metabolites are more likely to be accurate when plasma rather than urine is tested (Bravo *et al.* 1979). Most phaeochromocytomas secrete noradrenalin; a raised level of vanil-mandelic acid (VMA, HMMA) rather than homovanillic acid (HVA), is therefore more commonly found in 24-hour urine specimens of these patients. In adults, excretion of both HVA and VMA is said to indicate that the tumour is malignant but the same excretion pattern has been recorded in several children with benign phaeochromocytoma (Gitlow *et al.*

1972). Statistically, of course, neuroblastoma is a more likely diagnosis in the child with elevated urinary excretion of both HVA and VMA.

Plasma for catecholamine estimation should, ideally, be taken at rest, but this state of affairs is difficult to achieve without unacceptably heavy sedation in young children in whom 'borderline' values may therefore be difficult to interpret. Inhibition tests, for example with clonidine (Bravo *et al.* 1981), may be helpful, but false negative results have been recorded. Stimulation tests are likely to provoke dangerous hypertension in a child and are contraindicated. There is poor correlation between the degree of hypertension and the amount of circulating total plasma catecholamines and metabolites; in practice, this means that plasma need not be sampled during a hypertensive episode.

Recently, successful attempts at tumour immunolocalisation have been made using ^{131}I-labelled meta-iodobenzylguanidine (MIBG). This relatively non-invasive technique may, in the future, replace angiography and serial venous sampling for accurate tumour localisation (Sisson *et al.* 1981).

In a single instance (Spark *et al.* 1979) a child with phaeochromocytoma presented with Cushing's syndrome due to 'ectopic' ACTH secretion by the tumour. Mild hypercalcaemia occurs in some patients: in most instances, the cause is unknown, but in a few the raised serum calcium may be a manifestation of clinically occult parathyroid hyperplasia or adenoma. Where primary hyperparathyroidism is proven, a MEN syndrome should be strongly suspected.

Successful management of patients with phaeochromocytoma requires excellent collaboration between paediatrician, surgeon and anaesthetist. Adequate pre-operative control of blood pressure is achieved using a combination of α- and β-blocking drugs, usually phenoxybenzamine and propranolol. At operation the surgeon should scrutinise not only both adrenal glands but also other possible sites for multicentric phaeochromocytoma, e.g. pararenal areas, sympathetic chains and the organ of Zuckerkandl.

Thyroid carcinoma

Thyroid cancer in young people is often iatrogenic, for instance the result of ill-advised neck radiation in early childhood (Refetoff *et al.* 1975). There are four histological types of tumour; the follicular and papillary varieties occur more commonly than anaplastic and medullary carcinomas. Thyroid cancer usually presents as a visible neck swelling in an euthyroid patient. Thyroid scanning, using ^{99m}Tc pertechnetate, shows a cold nodule in the gland. Lymph node and lung secondaries may be present. There are no associated biochemical abnormalities. Monitoring of serum thyrocalcitonin is of utmost value in members of families, especially those with a MEN II syndrome (Brennan 1982; Goltzman *et al.* 1974), at risk for development of medullary carcinoma of the thyroid (MCT). Whereas patients with sporadic MCT characteristically develop the tumour during middle age, MEN patients develop medullary carcinoma much earlier, at a median of 19 years for patients with a MEN II syndrome. Calcitonin is a sensitive 'tumour marker' and monitoring of MEN II patients should start during the second decade. When resting serum calcitonin levels are borderline or when there is clinical concern, provocative tests—using intravenous elemental calcium or pentagastrin—can be carried out.

The principal treatment for thyroid cancer is surgical, either full or partial thyroidectomy, with or without neck node dissection. Since the growth of a high proportion of tumours is thyroid stimulating hormone (TSH)-dependent, thyroxin treatment can be effective in controlling recurrent disease. External beam radiation therapy and ^{131}I are less often used in children than in adults because of the risk of induction of a second tumour, especially in the thyroid when only partial thyroidectomy has previously been performed.

Very rare malignant endocrine tumours in childhood

There have been occasional reports of malignant behaviour of tumours that are usually considered benign (Voorhess 1980). Thus occasional examples of malignant aldosteronoma, gastrinoma, glucagonoma, somatostatinoma and VIPoma have been described, principally in adults. Residual or metastatic disease may respond to streptozotocin (Schein 1972), a compound nitrosourea.

Ewing's sarcoma

The histogenesis of Ewing's tumour is still uncertain although recent evidence suggests an origin from a bone marrow cell rather than from vascular endothelium as suggested by Ewing himself. The primary most commonly arises in bone ('long' and 'flat' bones are affected with about equal frequency) but can also arise in soft tissue. Presentation is usually with pain and/or associated soft tissue swelling. Nearly all patients have overt or occult metastases at diagnosis, most commonly in lymph nodes, lung and bones. Investigation of a newly diagnosed patient should therefore include chest X-rays, with or without CT scan of lungs, bone scan and, for lower limb tumours, lymphangiography.

It has been confidently stated (Brereton *et al.* 1975) that serum lactate dehydrogenase (LDH) is a useful prognostic indicator, values of less than 200 u/l at diagnosis predicting a more favourable outcome than higher values. Subsequent studies have not confirmed this finding and the matter is still under investigation. Perhaps systematic studies of LDH iso-enzymes might be helpful.

Pulsed chemotherapy, using vincristine, actinomycin D, cyclophosphamide and sometimes adriamycin, often leads to impressive shrinkage of the primary and control of metastases (Rosen *et al.* 1981). Local control of bulky disease, however, is virtually never achieved without either high-dose radiation (45–60 Gy) or, where the cosmetic result is acceptable, surgical exicision of the bone in which the primary has arisen. Apart from serial LDH estimations, biochemical studies are not indicated as a routine during follow up.

Histiocytosis X

The pathogenesis of this multisystem disorder, characterised histopathologically by a proliferation of Langerhans-type histiocytes, is still poorly understood but possibly has an immunopathological basis (Osband *et al.* 1981). Its clinical behaviour is certainly not typical of a malignant disease. Patients can present with seborrhoeic or purpuric skin rashes, chronic ear discharge, hepatosplenomegaly, oral involvement, respiratory difficulties, diarrhoea, anaemia, diabetes insipidus, bone pain, scalp swellings, proptosis or any combination of these features. Because of extensive clinical and histopathological overlap between them, sub-classification into eosinophilic granuloma, Hand–Schuller–Christian disease and Letterer–Siwe disease is now inappropriate. Although the natural history is characteristically fluctuant (Pritchard 1979), with periods of quiescence or even spontaneous regression, most children with multiple organ involvement eventually require systemic therapy. Prednisone and a vinca alkaloid (either vincristine or vinblastine) are most commonly used; administration of more 'aggressive' combination chemotherapy regimens can lead to complications without an improved response rate (Pritchard 1979). Responses to therapy are generally prompt but may not be long-lasting. Multi-system disease has a significant mortality (up to 20%) and morbidity (up to 30%). Evidence of 'failure' of a vital organ system, especially bone marrow, liver or gut, is ominous.

The most sensitive indices of liver dysfunction in this disease are the level of serum albumen and coagulation studies. A serum albumen concentration of less than 20 g/l is considered to be of unfavourable prognostic significance; prothrombin time, partial thromboplastin time and thrombin time may be prolonged in these circumstances and are not corrected by vitamin K; hyperbilirubinaemia is usually a late but ominous finding. Serum levels of liver enzymes (APT, ALT) are not usually raised in histiocytosis X (H–X), even with extensive infiltration; elevated values would suggest a complicating hepatitis. Liver failure virtually never occurs in patients with normal-sized livers, so liver function tests are unnecessary, although mandatory in patients with hepatomegaly. It should be borne in mind that children receiving steroids may have liver enlargement because of fatty change in the cytoplasm of hepatocytes; such patients usually have normal serum albumen levels. Cirrhosis, with or without portal hypertension, is an occasional late manifestation of liver involvement.

Diabetes insipidus occurs in 20–30% of all patients, about one half at diagnosis and one half at a later stage, sometimes in patients otherwise in complete remission. All children with newly diagnosed H–X should have an early morning urine specimen tested for specific gravity (SG). An SG < 1.030 should be followed up by a formal water deprivation test. There is usually no enlargement of the pituitary fossa in H–X children with diabetes insipidus and CT scan rarely shows a hypothalamic lesion. Replacement therapy with pitressin is effective. The parents of patients without diabetes insipidus at the time of diagnosis should be warned to look out for an altering pattern of thirst or micturition. Some clinicians re-test patients' early morning urine SG only if there is a clinical indication, whilst others test routinely every 4–6 months.

Leukaemia and malignant lymphoma

Acute leukaemia

The acute leukaemias are the most common single type of childhood malignancy. Acute lymphoblastic leukaemia (ALL) occurs about four times more often than acute non-lymphoblastic leukaemia (ANLL). Recently, immunological approaches, including the development of heterologous and monoclonal antisera to leukaemia-associated antigens, have considerably improved understanding of the biology of the acute leukaemias and have led to greater diagnostic accuracy in individual cases (Chessells 1982). Management is with multi-agent chemotherapy and prophylactic cranial irradiation. Disappointingly, however, the prognosis for children with ALL and acute myeloblastic leukaemia (AML) has improved little over the last decade. Bone marrow transplantation following ablative chemoradiotherapy offers a chance for cure but, because only 10–20% of all children with acute leukaemia have a histocompatible sibling marrow donor, this approach has yet to make an impact on overall survival figures. Biochemical investigations are crucial in the management of the early phase of induction chemotherapy, particularly the 'tumour lysis syndrome' (see p. 541), but are only rarely of use in diagnosis.

The mono-myelocytic (AMMoL) and monocytic types of acute ANLL are sometimes associated with hyperlysozymuria (Wiernick & Serpick 1979). Lysozyme (synonym: Muramidase) is synthesised and secreted by mono-myeloblasts and is excreted by renal tubules: its action on renal tubular epithelium can lead to significant wastage of K^+. Thus, although hypokalaemia at diagnosis in a patient with acute leukaemia may reflect poor oral intake because of recent declining general health it may also be a pointer towards a diagnosis of this particular sub-variety of AML. It has been suggested that measurement of cerebrospinal fluid (CSF) beta-2-microglobulin may help in the early diagnosis of CNS infiltration by leukaemia or lymphoma (Maviglit *et al.* 1980): significant elevations of beta-2-microglobulin have, however, also been noted during the post-radiotherapy 'somnolence syndrome' (Demeocq *et al.* 1981), with no evidence of CNS leukaemia so, although further study seems justified, it is unlikely that this investigation will become routine in the surveillance of children receiving treatment for acute leukaemia.

Lymphoma

Lymphomas in children, as in adults, are classified either as Hodgkin's disease or non-Hodgkin's lymphoma (NHL).

Hodgkin's disease. 85% of children with this highly curable form of lymphoma present with painless lymphadenopathy, most commonly in the neck and/or mediastinum (Kaplan 1980). Only a minority of newly diagnosed patients have 'B symptoms', i.e. weight loss, fever or night sweats. Of relevance to the clinical biochemist is the rare patient who presents with a nephrotic syndrome due either to renal venous compression due to massive para-aortic node enlargement or as a result of an immune mechanism (Ghosh & Muehrcke 1970; Routledge *et al.* 1976). After biopsy confirmation of diagnosis, patients are clinically staged, using chest films, intravenous urography, abdominal ultrasound and computerised tomographic (CT) scanning with lymphangiography in older children. Staging laparotomy is not routinely employed, partly because in many centres chemotherapy, which appears to be effective in controlling 'occult' abdominal disease, is now the treatment of first choice for stage II, III and IV patients and also because of the risks of splenectomy in young children. Although serum levels of liver enzymes (ALT, APT and alkaline phosphatase) were at one time considered to be of help in determining whether or not a newly-diagnosed patient had liver involvement, current opinion is that, with the possible exception of the bromsulphthalein test in selected patients, liver function tests are not of value (Abt *et al.* 1974). Serial serum copper and zinc estimations have been regarded by some as useful tumour markers in Hodgkin's disease and other lymphomas, but recent evidence suggests that careful clinical examination, with regular chest X-rays, is just as effective for the early detection of recurrent disease (Kaplan 1980).

Non-Hodgkin's lymphoma (NHL). NHL and acute lymphoblastic leukaemia have many features in common. For instance, approximately 30% of NHL patients present with a mediastinal mass—either in thymus or mediastinal lymph nodes—and immunological studies of biopsy material shows the tumour to consist of T-derived lymphoblasts; the natural history of patients with T-cell NHL is very similar to that of children with T-cell ALL. Another 25% of childhood NHL patients present with abdominal masses; these tumours are usually of B-cell origin, as evidenced by the presence of monotypic cell membrane immunoglobulin. In contrast to adults with multiple myeloma, neither whole immunoglobulin nor free light and/or heavy Ig chains are secreted by the cells, so electrophoresis of serum or urine reveals no evidence of either Bence–Jones protein or of an M-spike. As in Hodgkin's disease and ALL there are no known 'tumour markers'. Although serial LDH measurements have, especially in patients with Burkitt's lymphoma, been considered to be moderately accurate indicators of residual disease, routine use cannot be recommended at present. Serial estimations of beta-2-microglobulin have been used in adults with NHL to monitor disease activity (Cooper & Plesner 1980), but there is no available equivalent data for childhood NHL.

Malignant histiocytosis. This is a true histiocytic lymphoma, previously known as 'histiocytic medullary reticulosis'. As in acute monomyelocytic leukaemia, to which it is closely related, lysozyme is synthesised by tumour cells (Wiernik & Serpick 1979) and often elevated in serum and urine at diagnosis or in relapse (Lampert *et al.* 1978).

Familial haemophagocytic reticulosis. This extremely rare familial malignancy, inherited as an autosomal recessive, presents with pancytopenia, hepatosplenomegaly and sometimes features of CNS involvement. Occasionally, hyperlipidaemia has been noted at diagnosis. Although some have considered this finding to suggest that the disorder also involves a primary defect in fat metabolism, it is likely that the hyperlipidaemia simply indicates deranged fat handling in immature liver grossly infiltrated by tumour (Gross-Kieselstein *et al.* 1981). To date, there have been no long-term survivors and therefore, no opportunity to study fat metabolism in patients in remission.

Liver tumours

Liver tumours are rare in childhood. The most 'common' histological variety—hepatoblastoma—is diagnosed only 10–15 times every year in the UK and other varieties of liver tumour—hepatoma, malignant teratoma and undifferentiated sarcoma—are even rarer. Most children with hepatoblastoma are in the 6-month to 3 year age group, whilst hepatoma occurs in later childhood and adolescence (Exelby *et al.* 1975). Unlike adults, children with liver tumours hardly ever have cirrhosis or other pre-existing liver disease. Children with glycogen storage disorders, especially von Gierke's disease, develop multiple hepatic adenomata (Howell *et al.* 1976) but the progression of these benign nodules may be decellerated by biochemical control of hypoglycaemia and none, to date, have been noted to 'transform' to a malignant phenotype. A combination of intravenous urography, abdominal ultrasound and isotope liver scan usually allows the clinician to differentiate between hepatic and non-hepatic origin of upper abdominal masses. The 30–40% of patients cured of their disease have resectable unifocal primary tumours and no visible evidence of metastases in lungs, which are by far the most common site of tumour spread. Adjuvant chemotherapy, using adriamycin, cis-platinum or a combination of cyclophosphamide, vincristine and 5-fluorouracil has been shown to be effective in destroying micrometastatic disease and, given pre-operatively, can sometimes make excisional surgery easier. Conventional liver function tests are usually normal at diagnosis in patients with uncomplicated hepatoblastoma, although gamma-glutamyl transferase can be elevated in some patients and is probably the product of tumour cells rather than normal hepatocytes.

Serum alpha fetoprotein (αFP) estimation is of especial value in diagnosis and follow up of liver tumours in childhood. αFP can be demonstrated by an immunoperoxidase technique within the cytoplasm of malignant hepatoblastoma, so the bulk of circulating αFP is certainly tumour-derived. In our experience αFP is grossly elevated, at diagnosis, in the serum of more than 90% of children with hepatoblastoma: the range in 10 cases recently studied by us was 480–750000 μg/l (median 70000 μg/l, normal < 5 μg/l) (Pritchard *et al.* 1982). The rarity of hepatoblastoma during the first 6 months of life means that there is seldom difficulty in distinguishing tumour-derived αFP from the declining 'physiological' αFP level that results from normal fetal hepatic synthesis of the protein. The interpretation of αFP levels measured during the first 6 months of life is discussed in the section on malignant teratoma/yolk sac tumours. After partial hepatectomy, serum αFP should be monitored at least weekly for 4 weeks so that a decay curve can be established and compared with the curve predicted if complete excision has been achieved. Fig. 19.2 shows serial αFP measurement in two patients with liver tumours; in patient A αFP declined with a normal half life of 4–5 days whilst patient B had a more shallow curve indicating residual αFP-producing tumour. Patient A was a long-term disease-free survivor whilst patient B, who had an inoperable primary tumour and lung secondaries, had a transient response to chemotherapy but then succumbed to drug-resistant metastatic disease.

The specificity and sensitivity of αFP immunoassay has meant that careful follow up of patients serum levels can yield information about impending recurrence. Fig. 19.3 shows the value of serial monitoring in a child whose αFP rose to 100–200 μg/l 9 months after right hepatic lobectomy for an αFP-synthesising hepatoblastoma. Conventional chest X-rays were negative but CAT lung scanning revealed three 2 mm lung secondaries. The child's chemotherapy was changed and the lung secondaries disappeared. Her long-term survival, despite metastatic spread of this relatively chemoresistant tumour, undoubtedly reflects the early diagnosis of recurrent disease; because tumour 'bulk' was low the child was curable by non-cross-resistant chemotherapy/radiation. 'False-positive' elevation of serum αFP levels during liver regeneration after partial hepatectomy or episodes of inflammatory liver disease presents a

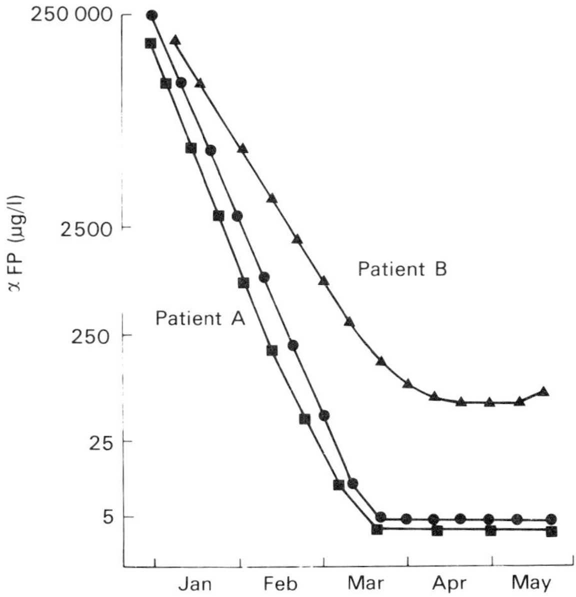

Fig. 19.2 Diagrammatic representation of serum αFP levels (log scale) after resection of αFP-secreting tumours. The 'ideal' fall of αFP levels indicating complete tumour control by therapy is shown (●——●). Patient A (■——■) underwent complete surgical resection of a primary hepatoblastoma and micrometastatic disease was successfully controlled by chemotherapy. Patient B's (▲——▲) tumour was incompletely resected; clinical recurrence was correctly predicted by the 'shallow' αFP decline curve.

possible problem, not encountered by us in practise. Liver regeneration after surgery is usually very rapid and, even if some αFP was secreted by newly-formed hepatocytes, it would almost certainly be 'masked' by the very high tumour-derived αFP burden; if hepatitis is suspected, liver enzyme studies should reveal the diagnosis. 'False-negative' αFP estimations are rarely seen. We have managed one patient with αFP-producing tumour whose recurrence was associated with normal αFP levels. Therefore, regular chest X-rays and abdominal ultrasound should, along with serial αFP measurements, be used to monitor the pre- and post-operative progress of these patients.

In hepatoma, the incidence of elevated αFP levels is higher at diagnosis (around 80%) than in adults (65–70%). Recently, a histologically distinctive form of hepatoma—the so-called 'fibro-lamellar' variant—has been described in association with normal serum αFP levels (Paradinas *et al.* 1982). Serum transcobalamin and vitamin B_{12} levels were, by contrast, grossly elevated and should serve as 'tumour markers' for this rare liver tumour, which has been described in children (usually girls) down to 10 years though not as yet, younger than this. Cystathionine levels are

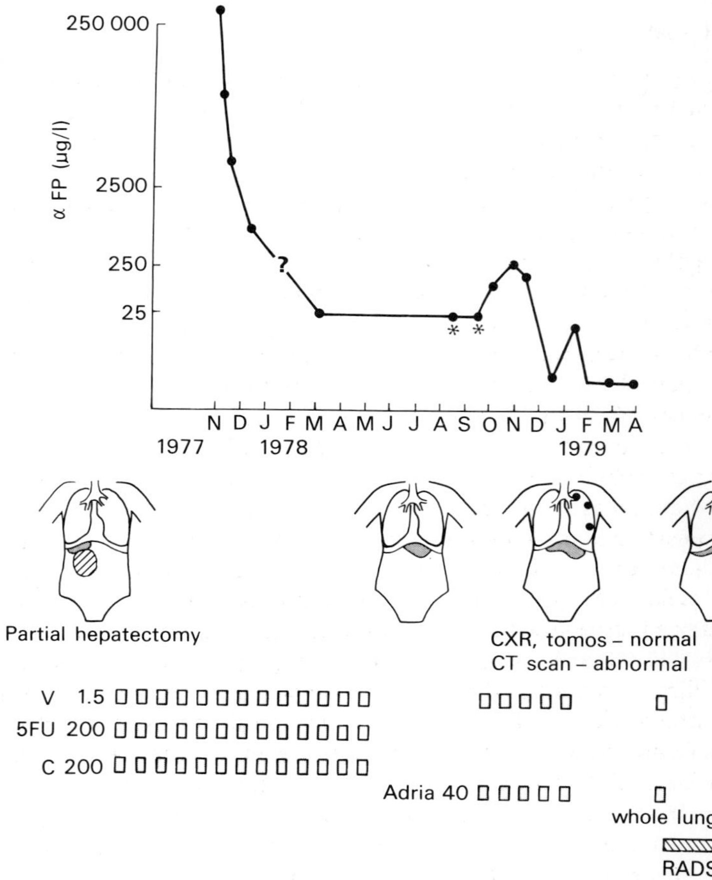

Fig. 19.3 Usefulness of serum αFP monitoring in the detection of recurrence of hepatoblastoma (see text). Asterisks indicate the lcwest limit of detection of αFP by immunoassay at that time (1978). V = vincristine; 5FU = 5-fluorouracil; C = cyclophosphamide; Adria = adriamycin; figures indicate doses/m².

elevated in the urine of some children with liver tumours, but serial monitoring suggests, as in neuroblastoma, that cystathionine is a poor 'tumour marker' (Geiser & Shih 1980).

Other very rare malignant childhood liver tumours include undifferentiated sarcoma, for which no 'tumour marker' is known, and malignant teratoma. The presence of yolk sac (endodermal sinus) or trophoblastic elements in a teratoma is implied if serum αFP or βhCG levels are elevated at presentation. The report (Todani *et al.* 1977) of a 'high' αFP level (25 000 μg/l) in serum at diagnosis of a histopathologically benign teratoma in a 4-month-old baby is a good example of mis-interpretation of a level that was well within the normal range for the infant's age (Wu *et al.* 1981). Fortunately, the child was given no chemotherapy or radiotherapy and, since the tumour contained no malignant elements, survived.

Neuroblastoma

It is over 25 years since increased quantities of catecholamine metabolites were demonstrated in the urine of patients with neuroblastoma (Mason *et al.* 1957; Labrosse *et al.* 1976). Fig. 19.4 shows the principle biochemical pathways of catecholamine metabolism. In 90% of newly-diagnosed cases, increased amounts of vanilmandelic acid (VMA, HMMA—a metabolite of adrenalin and noradrenalin) and/or homovanillic acid (HVA—a dopamine metabolite) can be measured in serum or urine. In some centres, estimation of VMA + vanilglycol (VG) is preferred to HVA estimation alone. Even when urinary VMA/HVA excretion appears normal (Voorhess 1971), electron microscopy may reveal neurosecretory granules in tumour cells; such neuroblastomas are presumably 'non-secretors' rather than 'non-producers' of catecholamines. Occasionally cultured neuroblastoma cells from a child with normal urinary VMA and HVA secrete large amounts of catecholamines, but the low overall cloning efficiency of primary tumours means that this approach has little diagnostic potential. Rare neuroblastomas produce acetylcholine rather than catecholamines, indicating alternative autonomic committment. The Labrosse spot test, which can yield both false-negative and false-positive results (Evans *et al.* 1971), has been superseded as a screening test for estimation of urinary catecholamines. A major advantage of chromatographic methods, usually thin layer chromatography, is that authentic catecholamine metabolites can be separated from dietary contaminants and drugs (Niehaus *et al.* 1979). The Pisano method is most commonly used for quantitation of VMA and VG, whilst HVA can be quantitated fluorimetrically (Sato 1965). Results are most precisely expressed as 'μg VMA or HVA or VG/mg creatinine'. Although VMA/HVA excretion in random specimens is moderately accurate for diagnosis, diurnal variation of catecholamine excretion by tumours means that 24-hour collections of urine—into acid so that the metabolites are chemically stabilised—are much preferred. Quantitation is not simply of academic interest; recent evidence indicates that the VMA:HVA ratio is of prognostic significance. Patients with a ratio of > 1 have a more favourable short-term prognosis than those with the reverse ratio (Laug *et al.* 1978). Whether or not VMA:HVA ratio measurement proves to be of importance as a single variable factor awaits more detailed studies and multi-variate analysis, including other prognostic factors such as age, stage, site of tumour and histological differentiation. To date, no systematic studies of plasma catecholamines have been made but may, in the future be of more value than urinary estimations.

10–15% of children with neuroblastoma have hypertension at diagnosis. There is no consistent pattern of catecholamine secretion that distinguishes these children from those with normal blood pressure.

No child with a solid retroperitoneal mass should undergo anaesthesia without estimation of 24 hour urinary VMA and HVA excretion, because the cardiovascular instability that results from the removal of a mass of

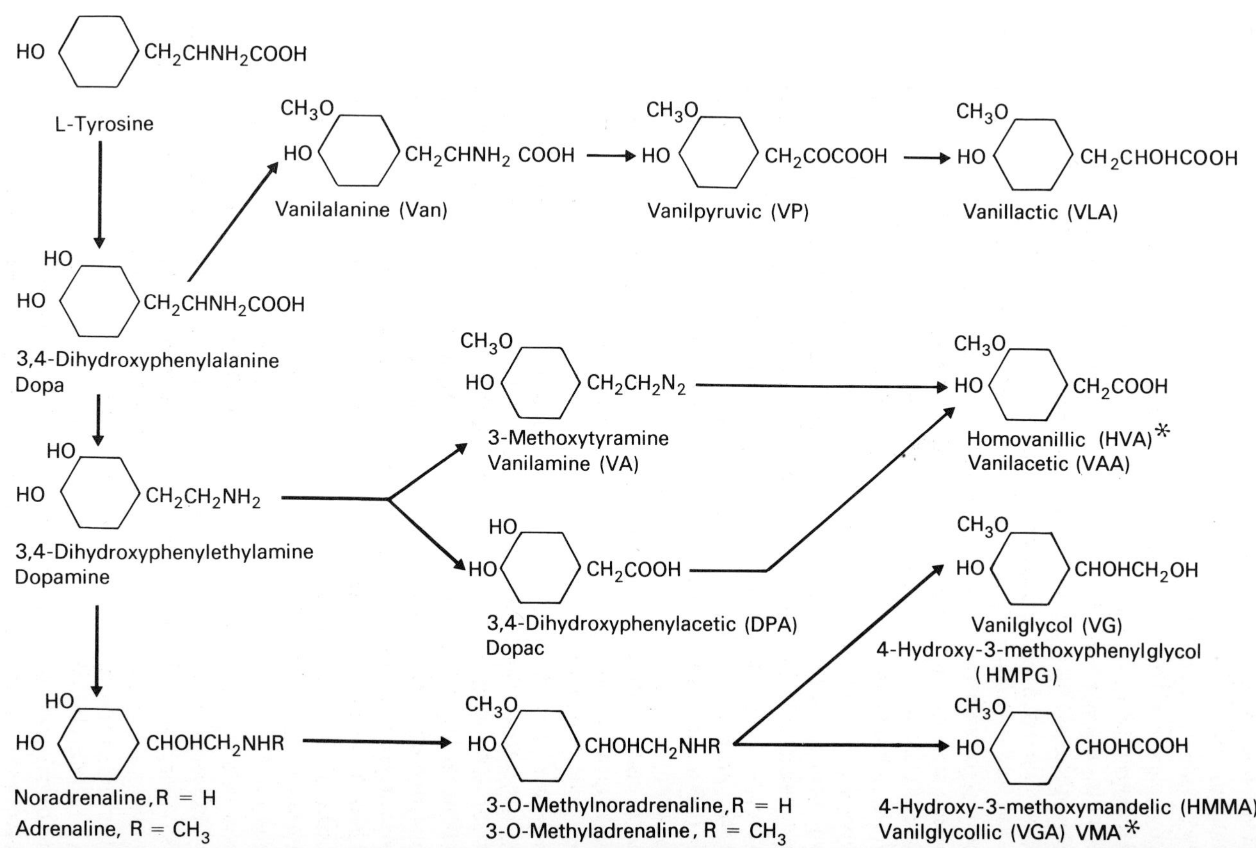

Fig. 19.4 Principle pathways of catecholamine synthesis and degradation. The terms 4-hydroxy-3-methoxymandelic (HMMA), vaniglycollic

catecholamine-secreting tumour may have serious or even fatal consequences. Since occasional neuroblastomas originate from within the kidney, and on intravenous urography and abdominal ultrasound exactly mimic Wilms' tumour, this rule should always be observed. Slight elevations in urinary catecholamines are occasionally seen in patients with abdominal masses other than neuroblastoma, presumably the result of 'stress'. Catecholamine-induced hypertension requires control by both alpha- and beta-blocking agents before surgery. Results of urinary VMA/HVA excretion should therefore be available as soon as possible and certainly within 48 hours of the child's admission to hospital.

On the whole, serial VMA/HVA monitoring after diagnosis has been disappointing. In our experience, it is unusual for rising urinary VMA/HVA to be the sole manifestation of tumour recurrence; usually there is concurrent clinical evidence either of regrowth of tumour at the primary site or of metastatic disease. Occasionally, florid relapse can recur without an elevation of urinary catecholamine levels, presumably because the resistant tumour population is 'non-secretor'. On the other hand, false-positive results hardly ever occur. Most paediatric oncologists follow urinary VMA/HVA excretion in their patients on a regular basis and find, as do the patients' parents, that normal levels are reassuring. However, critical cost-benefit analysis is needed to assess the real value of such a policy.

Occasionally, children with neurogenic tumours present with profuse watery diarrhoea and are shown to have hypokalaemia and gastric achlorhydria (Jansen-Goemans & Engelhardt 1977; Tiedemann *et al.* 1981b). This presentation is similar to that of adults with VIP-secreting pancreatic tumours and other 'VIPomas' and adds to the evidence suggesting that neural crest cells and 'APUD' cells have a common histogenetic origin (Bloom *et al.* 1973). Histological examination of such tumours usually shows a good deal of differentiation (ganglioneuroblastoma or ganglioneuroma) and, using an immunoperoxidase technique, vasoactive intestinal polypeptide (VIP) can be demonstrated in the cytoplasm of the more mature tumour cells, especially ganglion cells (Mendelsohn *et al.* 1979). Plasma VIP levels are mildly elevated, without evidence of diarrhoea, in some children with newly-diagnosed neuroblastoma (Tiedemann *et al.* 1981a). Plasma VIP is thus another useful marker for neural crest origin of tumours and is occasionally of diagnostic help in patients with normal urinary catecholamine levels.

It has recently been claimed that plasma ferritin is also helpful as a tumour marker in neuroblastoma and, in particular, that the measurement of this iron storage protein may help discriminate between patients with good and bad prognosis (Hann *et al.* 1980, 1981). Ferritin is synthesised by neuroblastoma cells in tissue culture but is not a tissue-specific product. Serum ferritin has proved disappointing as a tumour marker in acute leukaemia so further study is required before concluding that serial measurement is valuable in neuroblastoma. Carcino-embryonic antigen (CEA), cystathionine and LDH have been suggested as alternative 'tumour markers' in neuroblastoma, but have proved even less useful than catecholamine measurements (Mann *et al.* 1978). Quantitative assessment of serum immune complexes is held by one group of workers to be helpful in monitoring disease activity but the complexity of the assay means that it is unlikely to be used in routine laboratories.

Children with localised neuroblastoma have an excellent prognosis (100% 2-year survival for stage I patients, 85% for stage II) but unfortunately the majority of patients have stage III or IV disease when diagnosed. If X-rays, including an intravenous urogram, show characteristic features or if tumour cells are present in bone marrow and urinary VMA$\pm$HVA are significantly elevated, patients do not require tissue diagnosis (Pritchard & Kemshead 1983). Well over 50% of patients with advanced neuroblastoma respond, often dramatically, to combination chemotherapy, but disease returns in the majority between 6 and 24 months from diagnosis. The 2-year survival figure for patients with stage III disease is around 30% but less than 10% of stage IV patients survive. Attempts to improve these dismal statistics using high-dose cyclophosphamide, high-dose melphalan and other aggressive combination chemotherapy regimes have been largely unsuccessful, though there are some long-term survivors. Truly 'ablative' therapy using total body irradiation combined with chemotherapy and followed by marrow transplantation (either allogeneic or autologous) deserves further study.

Osteogenic sarcoma

This notorious malignancy is almost unheard of in children under 5 years and is most common in the adolescent age group. Presentation is invariably with pain in the primary site, most commonly in the lower femur, upper tibia or upper humerus, but axial bones can also be affected. A majority of patients have macro- or micrometastases at diagnosis, most commonly in lung, but sometimes in other bones. Prognostic variables include age at diagnosis, sex, duration of symptoms, tumour site, tumour size and histological subtype (Taylor *et al.* 1978). It has been claimed that the serum alkaline phosphatase level at diagnosis is also an important prognostic indicator: serial measurements are said to correlate with disease status, high values indicating recurrence (Thorpe *et al.* 1979). Such studies require confirmation. Multivariate analysis of results should indicate whether or not serum alkaline phosphatase levels are of independent prognostic significance.

Since osteogenic sarcoma is only moderately chemoresponsive and radioresponsive, surgical ablation of the primary is mandatory for cure. The therapist's preference for local excision, with a view to limb preservation or amputation, depends upon the characteristics of the tumour and the age of the child. Limb-salvage is not, for instance, recommended if the tumour is in the lower limb and the child has not yet achieved full growth potential. The value of chemotherapy in this disease is still uncertain. Many regimes under study include high-dose (0.5–15 g/m²) methotrexate (HD MTX) combined with folinic acid (citrovorum factor) rescue; adriamycin and cis-platinum are also considered active agents (Glicksman *et al.* 1979). Patients should receive HD MTX only in centres where facilities for routine monitoring of serum methotrexate levels and serum creatinine are available. Liver function tests—especially AST, ALT and bilirubin estimations—may also be required in an emergency where acute methotrexate toxicity is suspected. Each high-dose methotrexate treatment involves several days in hospital and major disturbances of the family's domestic routine. Although clinicians should schedule therapy so as to avoid public holidays when clinical chemistry laboratories are staffed only for emergency services, families may have a valid preference for starting or completing treatment at weekends, when it may be easier for them to transport the child to or from hospital. Weekend treatment also reduces the time lost from school.

Rhabdomyosarcoma

Rhabdomyosarcoma is the most common histopathological type of soft tissue sarcoma in childhood; liposarcoma, malignant fibrous histiocytoma, synovial cell sarcoma and fibrosarcoma are investigated and treated similarly, though generally they behave in a more 'low-grade' fashion. Of all rhabdomyosarcomas approximately one-third arise in the neck, one-third in the trunk and limbs and one-third in the genito-urinary system. Before the 1970s, these tumours were almost always fatal although some patients were 'salvaged' by very major surgery, e.g. orbital or pelvic extenteration, with or without irradiation. Since the advent of chemotherapy survival rates have greatly improved (60–70% at 5 years from diagnosis), indicating that most children with rhabdomyosarcoma have micrometastases at diagnosis. Chemotherapy regimes are similar to those used for Ewing's tumour. Attention of therapists is now focussed upon improving the quality of survival by avoiding mutilating surgery, especially in children with head and neck and genito-urinary primaries. Efforts are being made to reduce radiation dosage so that retardation of bone growth is lessened, with the aim of a better final cosmetic result (Glicksman *et al.* 1979).

No particular biochemical investigations are indicated in the diagnostic work-up or follow. It is a little curious that, despite the interest in its potential as a tumour marker for Ewing's sarcoma, there seem to have been no serial studies of the value of serum LDH estimations in rhabdomyosarcomas.

Wilms' tumour

The high cure rates achieved in children with this, by far the most common renal tumour of childhood, have been achieved because the tumour cells are highly sensitive to several chemotherapeutic agents (in particular, vincristine, actinomycin D and adriamycin) and radiation therapy (D'Angio *et al.* 1976). Patients most commonly present with abdominal swelling, a discrete mass sometimes having been felt by parents, health visitor or general practitioner. Because overt metastatic disease is uncommon at diagnosis, most children have no systemic upset and specific symptoms such as haematuria (which occurs in 15% of patients) are uncommon. Between 10–20% of Wilms' tumour patients have raised blood pressure at the time of diagnosis. The most common cause of hypertension in Wilms' tumour is renovascular, and is due to displacement or compression by the tumour of major vessels. Usually, this form of hypertension is mild and careful examination may indicate dependency on the child's posture, presumably reflecting varying pressure on or 'kinking' of the renal vessels when the tumour is in different positions. Some children with Wilms' tumour, especially those with bilateral kidney involvement and other congenital abnormalities such as pseudo-hermaphroditism, have a histopathologically characteristic nephropathy (Drash *et al.* 1970); immunofluorescence studies of 'normal' kidney tissue removed from these tumours show evidence of immune complex deposition. Most patients with this nephropathy have associated albuminuria; if the renal lesion is severe, there may be biochemical evidence of renal decompensation with elevated blood urea and creatinine (Drash *et al.* 1970). Definition of the severity of renal disease is critical in such cases since, if there is significant renal dysfunction, radiation to the remaining kidney tissue should be minimised or avoided altogether. There are several reports of renin secretion by Wilms' tumour (Mitchell *et al.* 1970; Ganguly *et al.* 1973) as well as by the much less common mesoblastic nephroma and juxtaglomerular tumour of childhood (Warshaw *et al.* 1979). Because of the anaesthetic and surgical risks of severe hypertension, surgery should be delayed whilst investigations including urinalysis, plasma renin and 24-hour urinary VMA/HVA are carried out and blood pressure subsequently controlled with appropriate drug therapy.

There have been occasional reports (Hopwood & Dorfman 1978) of production of increased amounts of glycosaminoglycans (GAG) by some Wilms' tumours. The mucoid appearance of the cut surface of some Wilms' tumours presumably reflects this process. No systematic studies of GAG excretion by Wilms' tumours have been reported; our own experience has been that urinary GAG estimation and fractionation is unhelpful in the differential diagnosis of the child with a newly-diagnosed abdominal mass.

Multivariate analysis of prognostic features in the USA National Wilms' Tumour Study has shown that (a) histological sub-type and (b) involvement of lymph nodes are the most important prognostic factors at diagnosis (Breslow *et al.* 1978). As a result, it has been possible to refine therapy for children with 'favourable' histological sub-types of Wilms' tumour, especially if the lymph nodes are negative for tumour. For example, stage I 'favourable histology' patients are currently treated in the UK and USA without radiation and with only single or double agent chemotherapy. By contrast, intensive 3- or 4-drug treatment is now recommended in an effort to improve the prognosis for children with 'unfavourable histology' tumours. Overall results of the treatment of Wilms' tumour are extremely encouraging. Recent analysis of patients entered into the second MRC Wilms' Study show higher than 85% survival of all stage I, II and III patients (Morris-Jones, pers. commun.). About half of patients with stage IV and V (bilateral Wilms') tumours can also be 'cured'.

Yolk sac (endodermal sinus) tumour and malignant teratoma

The terminology of this group of tumours is complex. Teratoma is a generic term used to describe all germ cell tumours; many; especially those arising in the sacro-coccygeal area in children are histologically benign, but other, frankly malignant, tumours may either be undifferentiated or contain elements of embryonal/extra-embryonal (yolk sac/trophoblast) origin (Altman & Schwartz 1978). Immunoperoxidase studies show production of αFP by yolk sac (endodermal sinus tumour) elements and the beta subunit of human chorionic gonadotrophin (βhCG) by malignant trophoblast. In adults, 41% of malignant teratomas secrete αFP only, 38% βhCG only, 90% producing one or both 'markers'. In children, by contrast, elevated serum αFP levels are found in $> 90\%$ and βhCG in 10–15% of cases at diagnosis (Mann *et al.* 1978). As these figures suggest, childhood malignant teratomas almost always contain yolk sac components. In the case of infant malignant teratoma of the testis (orchioblastoma) 'pure' yolk sac histology is the rule. Thus, αFP is more commonly a useful 'tumour marker' than βhCG, though both should be measured in each patient at diagnosis. Apart from the gonads, malignant teratoma can arise in the sacrococcygeal region, genito-urinary tract, mediastinum, ovary, liver, extradural space and pineal region (Altman & Schwartz 1978). Lungs, lymph nodes, bones and liver may be invaded by metastatic disease.

Regular monitoring of 'markers' is of established value in the management of malignant germ cell tumours of adults (Javadpour 1980) and, although the rarity of these tumours in childhood has led to slow data accrual, many anecdotal reports indicate that monitoring has similar value in childhood. As for liver tumours, positive levels of αFP and/or βhCG should be followed at weekly intervals for 4–6 weeks after diagnosis, to establish a 'decay curve'. Graphical comparison of the gradient of this curve with a standard curve calculated from the known half life of αFP

(4–5 days) and βhCG (24 hours) (Javadpour 1980) will indicate whether or not all, or almost all, the 'marker'-secreting tumour has been removed. Too 'shallow' a decay curve or a new rise above the normal range indicates either sub-optimal response to therapy or recurrence; reinvestigation of the patient is, in these circumstances, urgently indicated. Early diagnosis of recurrence increases the chance of successful therapy and possible cure using one of several 'second-line' agents (e.g. iphosphamide, VP-16) now available. In adults 20% of recurrences of previously αFP or βhCG-positive tumours are 'marker negative'. This phenomenon has also been recorded in children. This presumably reflects relative sensitivity to treatment of the yolk sac and/or trophoblastic elements of the tumour, whilst the undifferentiated component is resistant. As for liver tumours, slight elevations in αFP should lead to a search for evidence of intercurrent inflammatory liver disease.

Babies with sacrococcygeal teratoma present special problems for the clinician and the clinical chemist. Frankly malignant yolk sac elements can be identified by the histopathologist in only 5–10% of tumours at diagnosis, but in another 10–15% recurrence, sometimes months or years later, reveals the malignant potential of previously undetected residual or micrometastatic disease. Most patients undergo surgical excision of bulk primary disease within the neonatal period. Monitoring αFP is seriously hampered by the extremely high physiological levels of αFP found in infant serum. In a manner entirely analogous to the postnatal decline of fetal haemaglobin (HbF) levels, 'adult' levels of αFP are reached only between 6 and 12 months of age (Wu *et al.* 1981). When there is minimal residual disease after surgery, it seems intrinsically unlikely that small amounts of tumour-derived αFP would 'surface' above these, albeit declining, physiological levels. There is no information at all concerning normal βhCG levels in infant serum; however, the infrequency of trophoblast histology in sacroccygeal teratoma makes it very unlikely that βhCG will be a useful 'tumour marker' in the individual infant. Our own current policy is to monitor only αFP levels in these babies monthly after tumour excision.

3–5% of childhood malignant teratomas arise in the central nervous system, usually in the pineal region. If yolk sac elements are present, raised αFP levels may be detected in cerebrospinal fluid and serum (Neuwelt *et al.* 1979; Norgaard-Pedersen *et al.* 1978). These tumours are hardly ever completely resected because surgery to the pineal region is hazardous. Estimations αFP and βhCG are therefore urgently indicated in children with newly-diagnosed masses in the pineal region.

70% of boys with apparently localised orchioblastoma will be cured by surgery (radical orchidectomy) alone. Chemotherapy is indicated for almost all the remaining patients. There have been major advances in the treatment of malignant teratoma in adults. The use of combination chemotherapy regimes, containing vinblastine, bleomycin and cis-platinum, together with surgical removal of primary and residual secondary tumours where technically feasible has led to dramatically improved survival rates, especially for men with testicular tumours (Einhorn & Williams 1980).

Although in some children malignant teratoma responds well to 'conventional' chemotherapy (e.g. vincristine, actinomycin D and cyclophosphamide) and radiotherapy, the hope is that with the newer regimes there will be a real improvement in prognosis. Pilot studies are promising but also indicate the need for close monitoring of drug toxicity, for example the lung toxicity of bleomycin and the middle ear/renal toxicity of cis-platinum (see p. 546). The importance of the clinical chemistry laboratory in providing an efficient service for monitoring αFP and/or βhCG levels cannot be overstressed. As malignant teratomas are rare in childhood, supra-regional laboratories are often particularly well placed to provide the rapid and reliable service required for the proper management of these patients.

MONITORING OF DRUG LEVELS

Side effects of cytotoxic drugs can be either idiosyncratic, for example, the pulmonary toxicity of bleomycin, or dose-related. Pharmacokinetic data from human studies is available for most drugs but, because current strategy is to treat patients to 'tolerance', monitoring of blood levels is not standard practise. Thus, although methods for the assay of alkylating agents, cytosine arabinoside, 6 MP, methotrexate (MTX) and other drugs are available, only the MTX assay has direct clinical relevance at present. The cytocidal action of MTX is the result of direct inhibition of the intracellular enzyme, dihydrofolate reductase: as the reductase is critical to DNA synthesis, MTX acts in the S phase of the cell cycle. Recently, escalating doses of MTX have been given in the treatment of leukaemia/lymphoma and osteogenic sarcoma. Six to 24-hours after administration of a sub-lethal or lethal dose (0.5–20 g/m^2) the patient is biochemically 'rescued' by administration of folinic acid (citrovorum factor). By competitive inhibition, folinic acid restores the blocked enzyme pathway within normal and tumour cells, hopefully only after the latter have been irreversibly damaged. Therapy must be continued until the serum MTX level falls to non-toxic levels ($< 10^{-8}$ M or 4.5 ng/ml). The excretion of MTX, principally by the kidneys, is enhanced if urine flow is high and urine pH > 7.5. Liberal intravenous fluids and sodium bicarbonate are therefore usually given with folinic acid. At an acid pH in urine MTX may crystallise in renal tubules. Daily monitoring of renal function is therefore necessary. Estimation of serum creatinine 48 hours after MTX is especially worthwhile: if this value is $> 50\%$ higher than the pre-MTX level of creatinine, delayed MTX excretion can be predicted and prolonged folinic acid 'rescue' should be prescribed. Besides renal compromise, other factors delaying excretion of MTX include intercurrent infection, dehydration and the presence of a 'third space' for example, pleural or pericardial fluid. High-dose MTX therapy should only be given in those centres where biochemistry laboratories can offer urgent service for monitoring of renal function and serum MTX levels. Radioimmunoassay is most commonly used to assay MTX (Aherne & Quinton 1981).

Recently, it has been shown that there is considerable variation in the absorption of oral MTX by children with leukaemia (Craft *et al.* 1981). Studies are now in progress to determine whether children who absorb MTX poorly are at increased risk of relapse. If this turns out to be the case, biochemistry laboratories may become involved in drug monitoring in individual children.

TREATMENT COMPLICATIONS—BIOCHEMICAL ASPECTS

Biochemical abnormalities arising as a consequence of childhood cancer can be acute or of a medium- or long-term nature. Disturbances in newly-diagnosed patients, particularly those with lymphoblastic leukaemia/ lymphoma or tumour involving or obstructing the genitourinary tract may be exacerbated by the introduction of therapy and become life-threatening, but mortality can be reduced if appropriate precautions are taken. Once remission has been achieved, patients receive continuing ('maintenance') therapy for periods for 6–36 months depending on the type of malignancy; during this phase, problems can arise from dysfunction of a specific organ or a system as a result of treatment-induced damage. Improving survival rates in childhood cancer have focussed attention on the late-effects of therapy. Whereas treatment effects in adults are usually obvious immediately, manifestations of tissue damage may become evident in children only during later growth, especially the pubertal growth spurt. Only those long-term treatment consequences of direct relevance to the biochemist are discussed here, but it should be borne in mind that other late effects—for example impaired skeletal growth within radiation fields, psychological impairment as a result of cranial radiation and induction of second tumours—are also of major concern to the clinician. Although new and unforeseen long-term complications may arise in the future, problems occurring in some of today's survivors (for example, renal failure resulting from high-dose radiation therapy) will be seen less frequently because of recent refinements of treatment (see p. 550).

Acute problems

In Table 19.5, the major factors contributing to various biochemical disturbances at the time of diagnosis are listed. The exact causation of renal insufficiency is of particular importance to both clinician and biochemist. In some circumstances, such as renal infiltration by leukaemia or uric acid nephropathy, medical treatment may be indicated urgently, whilst in

Table 19.5 Metabolic abnormalities at diagnosis and their causes.

Substance	Increased/present	Decreased/absent
Amylase	'Tumour lysis pancreatitis'	—
Calcium	Extensive bony metastases 'Distant' production of bone resorbing factors	Secondary to hyper-phosphataemia
Creatinine	Urate nephropathy Renal infiltration by leukaemia/ lymphoma Obstructive uropathy due to bladder base or pelvic tumour	Malnutrition
Lactic acid	Hypoxic bulky tumour	—
Magnesium	—	Secondary to hypercalcaemia
Metabolic acidosis	Renal failure Secondary to sepsis Lactic acidosis	—
Phosphate	'Tumour lysis syndrome' Renal failure	—
Potassium	Renal failure 'Tumour lysis syndrome'	Hyperlysozymuria (AMMoL) Diarrhoea
Sodium	Dehydration Inappropriate ADH secretion	Diabetes insipidus
Urea	Dehydration Other causes (as for creatinine)	—
Uric acid	'Tumour lysis syndrome'	—

others, for example bladder neck obstruction by prostatic rhabdomyosarcoma, surgical treatment may be required urgently. Wherever possible, an attempt should be made to correct an existing biochemical disturbance before the start of chemotherapy or radiotherapy.

Tumour-lysis syndrome

Dramatic biochemical changes may follow the treatment of lymphoblastic lymphoma, particularly Burkitt's lymphoma, and ALL. Lymphoblasts seem to be particularly sensitive to treatment and intracellular components (especially uric acid, potassium and phosphate) may be released into the plasma suddenly and in massive quantities. Fig. 19.5 indicates how rapidly such tumours can 'dissolve'; the biochemical consequences are collectively known as the 'tumour lysis syndrome'. The syndrome is almost never seen

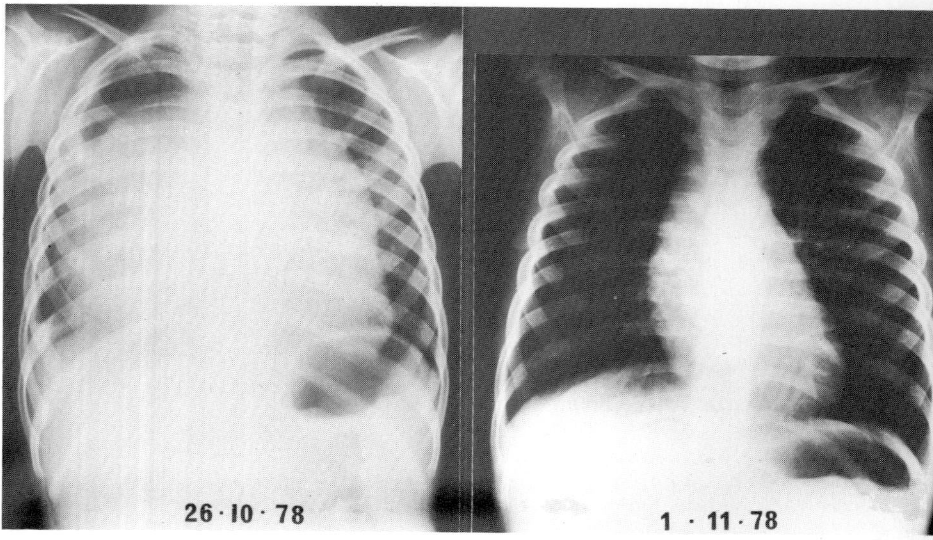

Fig. 19.5 Dramatic response of a T cell non-Hodgkin mediastinal lymphoma (Sternberg type) to combination chemotherapy. A mild urate nephropathy developed but the patient showed no other features of the 'tumour lysis syndrome'.

after therapy of non-lymphoblastic malignancies; the explanation presumably lies in the rather slower lysis rate of such tumours.

Uric acid, the final product of nucleic acid degradation, is only weakly protein-bound in plasma and is filtered at the glomerulus, reabsorbed in the proximal convoluted tubules and resecreted more distally. At high concentrations uric acid is precipitated in the distal tubule and collecting duct and may then cause obstructive neuropathy. At pH < 5.4 uric acid is mostly in the un-ionised, less soluble form, but at high pH its solubility is considerably greater; the solubility of uric acid is inversely related to urine sodium concentration. Urinary excretion of uric acid is markedly elevated before treatment in patients with lymphoblastic leukaemia/lymphoma, especially those with 'bulky' solid masses or a high circulating blast cell count (Hande *et al.* 1981). Introduction of chemotherapy or radiotherapy leads to a sudden increase in the rate of uric acid excretion (Lynch *et al.* 1977). The clinical features of hyperuricaemia, which are anorexia, nausea, vomiting and lethargy, are hard to appreciate in patients who are already systemically unwell because of their tumour; gout and renal colic are rare. Urate nephropathy can usually be avoided by vigorous preventive measures. All children with lymphoblastic tumours should receive the xanthine oxidase inhibitor allopurinol orally for at least 24 hours before anticancer therapy is started. In dire emergencies such as acute airway obstruction, intravenous allopurinol may lead to adequate xanthine oxidase blockade within 12 hours. A fluid intake of 2–3 l/m^2

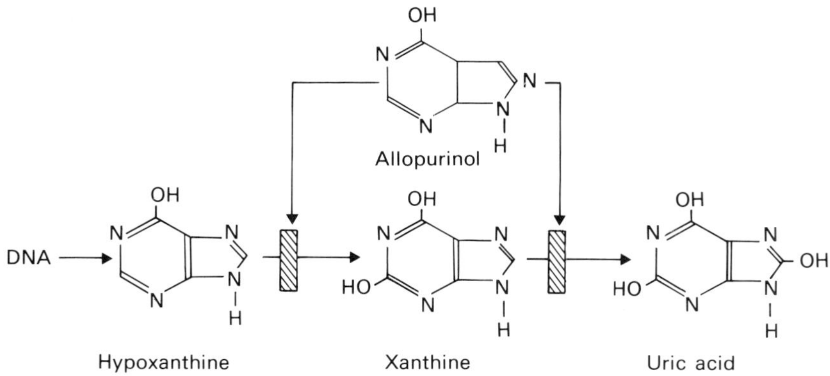

Fig. 19.6 Biochemical inter-relationships of hypoxanthine, xanthine and uric acid; allopurinol inhibits xanthine oxidase activity.

surface area daily should be combined with sufficient intravenous or oral sodium bicarbonate to keep the urine alkaline. Over-alkalinisation can lead to tetany, especially when hyperphosphataemia (see below), a second complication of the 'tumour lysis syndrome', coexists. Allopurinol blockade leads to accumulation of xanthine and its derivatives (Fig. 19.6). Xanthine nephropathy is rare because xanthine and its derivatives are more soluble than uric acid. If, despite maximal 'preventive' therapy, progressive renal decompensation occurs, mannitol diuresis or dialysis must be undertaken.

Hyperkalaemia may be the consequence either of renal failure or the liberation of large amounts of potassium from dying cells as part of the 'tumour lysis syndrome' (O'Regan *et al.* 1977; Lynch *et al.* 1977). As a consequence, life-threatening cardiac arrhythmias can occur, especially after initial treatment of Burkitt's lymphoma. The period of maximum risk for this complication, which occurs in 20–25% of lymphoma patients, is 6–48-hours after the start of therapy.

Lymphoblasts may lyse spontaneously, releasing intracellular cations in anticoagulated blood samples allowed to 'stand' on the bench before separation of plasma. A fictitious diagnosis of hyperkalaemia ('pseudo-hyperkalaemia') may be the consequence, thus a high serum potassium level should be interpreted cautiously in children with ALL, especially in those with high circulating blast counts. In cases of doubt the measurement should be repeated on promptly-separated plasma whilst a search is made for clinical evidence of hyperkalaemia, for example by electrocardiography.

The concentration of phosphate ion in leukaemic blast cells may be greater than in normal cells. Sudden release of intracellular phosphate, another manifestation of the 'tumour lysis syndrome', may lead to hyperphosphataemia and hypocalcaemia; increased renal clearance of phosphate and decreased tubular phosphate reabsorption lead to hyperphosphaturia. Hyperphosphataemia may be exacerbated by renal failure. A high plasma phosphate *per se* probably has few clinical consequences, but the almost inevitably co-existent hypocalcaemia can be responsible for psychological symptoms and, if severe, frank tetany. No specific therapy is required for hyperphosphataemia, but careful intravenous calcium supplementation may be indicated if symptomatic hypocalcaemia occurs. There is a theoretical risk of metastatic calcium phosphate deposition when the serum (ionised calcium × phosphate) product is high, but in practice this complication is exceedingly rare.

Other acute biochemical changes

Hypokalaemia is rare, but in patients with acute monocytic or monomyelocytic leukaemia it may be the consequence of a renal tubular potassium leak induced by high circulating levels of lysozyme (Wiernik & Serpick 1979). Therapy with antibiotics, especially carbenicillin and gentamicin, which are often used in the management of the febrile, neutropenic patient, may cause increased urinary potassium loss.

Hyponatraemia is most commonly due to excessive losses of sodium because of vomiting and diarrhoea or to excessive administration of hypotonic intravenous fluids. The syndrome of inappropriate secretion of anti-diuretic hormone (ADH) is a rare complication of treatment with vinca-alkaloids (Stuart *et al.* 1975) or, exceptionally, other chemotherapeutic agents. Hypernatraemia when it occurs is almost always a consequence of diabetes insipidus due to infiltration of the posterior pituitary gland or hypothalamic–pituitary pathways by tumour, occasionally leukaemia or brain tumour, but most commonly histiocytosis X. An iatrogenic cause of either hypo- or hypernatraemia, especially the delivery of inappropriate intravenous fluids, should be considered in the differential diagnosis of each case.

Disturbances of calcium and magnesium balance are not uncommon. Extensive bony infiltration occurs in up to 20–30% children with ALL and neuroblastoma and rarely with other malignancies. As in metastatic breast cancer, hypercalcaemia in these cases is presumably due to local release by metastatic tumour cells of bone-resorbing factors such as prostaglandins. Occasionally, bone-resorbing activity released from a primary tumour is so potent that 'distant' calcium release from bone is induced (Harvey *et al.*, pers commun.). In the absence of bone pain, hypercalcaemia is most commonly the consequence of a parathyroid tumour (see p. 521). Emergency management of hypercalcaemia is with prednisone or mithramycin. Apart from its occurrence as a consequence of hyperphosphataemia in the 'tumour lysis syndrome' hypocalcaemia occurring in children with cancer has been attributed to malnutrition, antibiotics and therapy with L-asparaginase. Hypocalcaemia may co-exist with hypomagnesaemia, which is an important consequence of therapy with cis-platinum (Blachley & Hill 1981). Various clinical manifestations including anorexia, muscle weakness and disorientation have been attributed to hypomagnesaemia but there is still some controversy about the need for regular magnesium supplementation during cis-platinum therapy, especially since serum levels may fluctuate quite markedly without alteration of dietary magnesium content. Hypophosphataemia has been noted in occasional newly diagnosed children with leukaemia, but is thought to be of little or no clinical consequence.

Lactic acidosis is a rare but well-recognised complication of the production of large amounts of lactic acid by poorly oxygenated tumour cells. It occurs especially where tumour masses are large and when complications, especially septicaemia, further increase tumour hypoxia. This possibility should be entertained in patients in whom a metabolic acidosis cannot be explained by renal failure. Urgent institution of tumour treatment is required (Nadiminti *et al.* 1980). Rapid lysis of intrapancreatic tumour was reported (Spiegel & Magrath 1979) to be the cause of pancreatitis and consequently elevated serum amylase levels in 2 children undergoing treatment for abdominal lymphomas. Mildly elevated blood sugars are often found in children with leukaemia or lymphoma receiving treatment with L-asparaginase. Perhaps because this drug is usually given only for a few weeks, evolution into true diabetes is rare.

Medium-term and long-term effects

Some biochemical side-effects of cancer therapy are listed in Table 19.6.

Nephrotoxicity

Several anti-cancer agents cause renal damage but abnormalities may not be evident for months or years after delivery of treatment. Kidney damage may be manifest by decreasing glomerular function with consequent azotaemia or, less commonly, by evidence of tubular damage, e.g. proteinuria, glucosuria.

Low-dose methotrexate causes no renal damage but in doses of more than 100 mg/m^2 surface area the drug can be significantly nephrotoxic, even when folinic acid rescue is also given. Toxicity is due to crystallisation in renal tubules of methotrexate and its less soluble 7-hydroxy metabolite. Since methotrexate is primarily excreted by the kidney, a 'vicious circle' is set up and toxicity ensues. With appropriate management renal failure usually resolves within 2–3 weeks, but fatalities can occur, thus underlying the importance of carrying out treatment in centres acquainted with possible hazards (see p. 539). It is absolutely mandatory to monitor both renal function and drug levels during therapy with moderate and high doses of methotrexate (see pp. 539–40). The nephrotoxicity of cis-platinum

Table 19.6 Side-effects of therapy associated with biochemical abnormalities.

Agent	Organ/system involved	Biochemical disturbance
L′-Asparaginase	Liver Pancreas	↓ clotting factor synthesis ↑ amylase ↑ glucose
cis-Platinum	Kidney	↑ creatinine, ↑ urea ↓ GFR ↓ magnesium, ↓ calcium
Cyclophosphamide	Post pituitary/ hypothalmus	ADH ↓ sodium, ↓ osmolarity
Cytosine arabinoside	Liver	↑ ALT, ↑ AST
6MP & 6-thioguanine	Liver	↑ alkaline phosphatase ↑ ALT, ↑ AST
Methotrexate	Liver Kidney	↑ ALT, ↑ AST, ↑ bilirubin ↑ creatinine, ↑ urea ↓ GFR
Nitrosoureas	Kidney	↑ creatinine, ↑ urea
Prednisone	Pancreas	↑ glucose
Radiation therapy (effects enhanced by adriamycin or actinomycin D)	Kidney Liver Pituitary Gonad/pituitary Thyroid	↑ creatinine, ↑ urea ↓ GFR ↑ ALT, ↑ AST, ↑ bilirubin ↓ Growth hormone ↑ Gonadotrophins ↑ TSH
Vincristine Vinblastine	Post pituitary/ hypothalmus	↑ ADH ↓ serum sodium, ↓ osmolality

(Blachley & Hill 1981) is due to deposition of the metal within the kidney where it may persist for a long period. Tubular function is more adversely affected than glomerular function. Plasma creatinine measurements are insufficiently sensitive in monitoring cis-platinum nephrotoxocity and direct measurement of glomerular filtration rate by the ^{51}chromium EDTA method is preferable. Increased renal loss of magnesium, leading to hypomagnesaemia and hypocalcaemia, has been discussed in the previous section. Both radiation therapy and nitrosoureas (BCNU, CCNU) cause renal damage, but the time course is slow. Therefore, urgent serial measurements of renal function are not indicated during treatment but periodic plasma creatinine levels may be indicated in survivors who have received these treatments.

Liver damage

Chronic liver disease is rare in children on or after treatment for malignant disease. Liver damage may occur as a late effect of radiation therapy or of drug treatment, especially 6-mercaptopurine (6-MP) and methotrexate (Perry 1982). Because of the need for repeated blood transfusions, hepatitis due to cytomegalovirus or one of the hepatitis viruses is more common in these children than in the general population. Intermittent-dose methotrexate is probably less likely to give rise to chronic liver disease than when it is given on a continual daily basis; the same probably applies to 6-MP. By decreasing protein synthesis L-asparaginase may lead to impaired synthesis of clotting factors and consequently a bleeding diathesis but the abnormality is rapidly reversible when treatment is discontinued. Since some cytotoxic agents are metabolised or activated in the liver they should be given in modified dosage in the presence of significant liver dysfunction.

Chronic liver disease is rare in children on or after treatment for malignancy. Some of the chronic liver damage previously attributed to methotrexate may in fact be due to a combination of factors, in particular to infections.

Gastrointestinal toxicity

Mucosal cells of the gut are amongst the most rapidly proliferating of all body tissues. As a result, the gastrointestinal tract is particularly susceptible to the effects of chemotherapy and radiation and acute gut toxicity is one of the most important factors limiting treatment dosages. Perhaps because methods of investigation are often invasive, there is little scientific data but there is a growing body of evidence that some degree of biochemically measurable malabsorption does occur in children on prolonged courses of chemotherapy. However, most children on treatment for malignancy continue along their normal height centile, and although some may show evidence of malnutrition, the cause is as likely to be decreased food intake because of chemotherapy-induced nausea of vomiting as malabsorption. Some cytotoxic agents are given by mouth and variable absorption of some, such as methotrexate (Craft *et al.* 1981), may be of significance in the management of the tumour. Hypoallergenic diets have been used, with moderate success, in the management of treatment-induced malabsorption; this suggests that altered intestinal permeability, leading to abnormal absorption of macromolecules, may be one manifestation of treatment-induced gut damage. Standard tests for malabsorption, apart from the xylose absorption test and measurement of iron and folate levels seem to have little place in the routine management of child cancer patients.

Pulmonary toxicity

Lung damage may be the result of surgery, radiotherapy, opportunistic infection or some chemotherapeutic agents, especially bleomycin (Ginsberg & Comis 1982). Blood gas measurements may be required urgently when a child, on treatment for a malignant tumour, presents with an intestitial pneumonitis. Cytomegalovirus, measles and *Pneumocystis carinii* are the most important and dangerous organisms; arterial hypoxaemia is an early manifestation whilst hypercapnia develops later. Serial blood gas measurements are likely to be required by clinicians managing such children. The lung toxicity of bleomycin is dose-dependent and almost invariable when more than 500 units of the drug have been given. Measurement of diffusion capacity is the most sensitive index of lung damage from bleomycin which may be exacerbated if high concentrations of inspired oxygen are given during anaesthesia. Long-term lung damage has occurred following treatment with cyclophosphamide, melphalan, BCNU and busulphan, but the overall incidence is rare in childhood. 'methotrexate lung' was widely reported in the 1970s but it is probable that most of this toxicity was due to opportunistic lung infections secondary to immunosuppression rather than a direct consequence of methotrexate toxicity. The lungs of infants and children have a certain capacity for renewal but older patients may be left with some chronic handicap.

Cardiac toxicity

The anthracyclines (adriamycin and daunorubicin) have dose-dependent cardiotoxicity. However, measurement of intracellular cardiac enzymes (SGOT, SGPT) is of no value in predicting or following cardiomyopathy in children on treatment with these drugs. Two-dimensional echocardiography and radio-isotope scanning are more helpful.

Brain damage

Cranial radiation and drugs (especially methotrexate, vincristine and procarbazine) may have adverse effects on a developing or developed central nervous system. Biochemical monitoring, with the possible exception of beta-2-microglobulin measurements (Mavligit *et al.* 1981), are not helpful in diagnosis and follow up. It is often difficult in patients with brain tumours to distinguish between disease-induced and treatment-induced brain damage.

Pituitary abnormalities

Irradiation of the neurohypophysis may affect the production of several hormones, especially growth hormone (GH). Relatively low doses of radiation may alter the GH response to insulin-induced stress, but physiologically-important dysfunction occurs only if higher doses or dose-rates are used (Shalet *et al.* 1979). Thus, although many children with ALL who have received cranial irradiation (20–25 Gy) as part of their treatment will show an impaired GH response to insulin-induced hypoglycaemia, they continue to grow normally.

Children cured of brain tumours have usually received high radiation doses (35–60 Gy) (Bloom 1982) and, if the pituitary/hypothalamus are within the treatment field, GH deficiency is even more common than in children cured of ALL. Other factors, such as impairment of vertebral growth after spinal irradiation, may contribute to small stature in these patients.

Secretion of other pituitary hormones may be affected by cranial irradiation, but the identification of a clinically important trophic hormone deficiency is extremely unusual; abnormalities in LH and FSH secretion secondary to gonadal damage are discussed in a later section.

Thyroid dysfunction

The thyroid gland inevitably receives radiation during effective treatment for nasopharyngeal or cervical malignancy and for supra-diaphragmatic Hodgkin's disease. Biochemical hypothyroidism occurs in up to 40% of patients with Hodgkin's disease and is easily detected by raised levels of thyroid-stimulating hormone. In patients at risk, routine monitoring of TSH is justified. The management of patients with high TSH levels is controversial; some clinicians argue that thyroid hormone replacement is not indicated unless clinical symptoms of hypothyroidism appear, while others feel that chronic TSH stimulation is likely to increase the incidence of later thyroid malignancy and they therefore recommend replacement therapy.

L-Asparaginase, because of its action in decreasing protein synthesis, may be associated with low thyroxine (T_4) levels, whilst the halide-containing 5-fluorouracil may increase levels of both triiodothyronine and T4 in the serum. However, patients receiving these drugs remain clinically euthyroid, the abnormalities being due to alterations in the characteristics of binding proteins.

Parathyroid function is hardly ever affected by either radiation therapy or chemotherapy.

Gonadal damage

Since curative chemotherapy for patients with ALL and Hodgkin's disease has been available for over 10 years, more data on gonadal function exists for these patients than for those cured of other childhood malignancies. Both endocrine function and fertility are more severely affected in boys than in girls (Chapman 1982; Whitehead *et al.* 1982).

Most boys receiving therapy for ALL or Hodgkin's disease during or after puberty have elevated serum FSH levels, indicating Leydig cell dysfunction, sometimes manifested by gynaecomastia (Shalet *et al.* 1981; Sherins *et al.* 1978). By contrast, pre-pubertal boys less often have raised gonadotrophin levels and almost all of them enter puberty quite normally. Ovarian function is less predictable. In one study (Siris *et al.* 1976) 80% of girls cured of ALL showed normal pubertal progression, but some of those treated during puberty showed evidence of ovarian dysfunction, manifested by either raised or lowered gonadotrophin levels and accompanied in several by menstrual abnormalities.

Damage to fertility is more evident in boys than in girls and is a reflection of histopathologically obvious damage to germinal epithelium. Those who have received alkylating agents or direct gonadal irradiation are very often permanently sterile. On the other hand, recovery of gametogenesis has been described in patients of either sex treated for leukaemia. It is encouraging that there seems to be no increase of congenital abnormalities in babies born to men or women previously cured of childhood cancer.

CONSPECTUS

Recent advances in diagnosis, staging and treatment of childhood malignancies have been heartening. With the most resistant tumours still to be tackled future progress is likely to be slow. The similarity of several of these 'difficult' tumours to adult cancers, e.g. neuroblastoma and oat cell carcinoma of the lung suggests that increasing collaboration with medical oncologists must be fruitful. Besides the introduction of newly discovered single agents and novel drug combinations, new treatment concepts under investigation include more rational 'kinetic' scheduling of drugs and radiation based on a knowledge of the cell-cycle characteristics of both tumour cells and normal proliferating cell populations, and marrow-ablative high dose chemoradiotherapy with marrow transplantation, which is already of established value in the treatment of acute myeloid leukaemia.

Because of the toxicities and sheer unpleasantness of current methods, refinement of therapy is a major current objective in paediatric oncology. Many current clinical trials are, as a principle objective, studying whether the omission of an individual component from combination therapy has deleterious effects on disease-free survival. The occurrence of second tumours in up to 8% of long-term survivors is of particular concern; radiation therapy, aklylating agents and procarbazine are especially culpable and are excluded from treatment programmes wherever possible. The role of 'maintenance' chemotherapy, better called 'continuing' chemotherapy, is under close scrutiny. Such therapy seems to have no value in Hodgkin's disease, acute myelogenous leukaemia, malignant teratomas (in adults) or neuroblastoma. In acute lymphoblastic leukaemia, prolonged 'maintenance' therapy has been associated with a lower relapse rate, but the question of its value after modern intensive induction chemotherapy schedules must be examined anew. More speculatively, there is hope that monoclonal antibodies made against tumour-associated antigens, already of established use in the diagnosis of paediatric tumours (Kemshead *et al.* 1983), can be used as vectors with which to 'target' drugs, radioactivity or toxins (such as ricin or abrin) more selectively to tumour cells than is possible with presently-used non-selective intravenous administration. Hopefully, such approaches to treatment will cause less side-effects and therefore be more tolerable to the patient. At the same time, the efficacy of the anti-tumour effect should be enhanced because of the improved delivery of the therapeutic agent to its 'target'.

REFERENCES

Abt A. B., Kirschner R. H., Belliveau R. E., O'Connell M. J., Slansky B. D., Greene W. H. & Wiernik P. H. (1974) Hepatic pathology associated with Hodgkin's disease. *Cancer* **33**, 1564–71.

Aherne G. W. & Quinton M. T. (1981) Techniques for the measurement of methotrexate biological samples. *Cancer Treat. Rep.* **65** (Suppl. 1), 55–60.

Altman A. J. & Schwartz A. D. (1978) Tumors of germ-cell origin: Embryonal carcinomas, extra-embryonal tumors and teratomas. In Schaffer A. J. & Markowitz M. (eds.) *Malignant Diseases of Infancy, Childhood and Adolescence*, pp. 427–42. W. B. Saunders Co., Philadelphia.

Barber S. G., Smith J. & Hughes R. C. (1978) Melatonin as a tumour marker in a patient with pineal tumour. *Br. Med. J.* **2**, 328.

Blachley J. D. & Hill J. B. (1981) Renal and electrolyte disturbances associated with cisplatin. *Ann. Int. Med.* **95**, 628–32.

Bloom H. J. R. (1982) Intracranial tumour: response and resistance to therapeutic endeavours 1970–1980. *Int. J. Rad. Oncol. Biol. Phys.* **8**, 1083–113.

Bloom S. R., Polak J. M. & Pearse A. G. E. (1973) Vasoactive intestinal peptide and watery-diarrhoea syndrome. *Lancet* **ii**, 14–16.

Bravo E. L., Tarazi R. C., Fouad F. M., Vidt D. G. & Gifford R. W. (1981) Clonidine-suppression test. A useful aid in the diagnosis of pheochromocytoma. *New Engl. J. Med.* **305**, 623–6.

Bravo E. L., Tarazi R. C., Gifford R. W. & Stewart A. H. (1979) Circulating and urinary catecholamines in pheochromocytoma. *New Engl. J. Med.* **301**, 682–6.

Brennan M. F. (1982) Cancer of the endocrine system. In De Vita V. T. Jr, Hellman S. & Rosenberg S. E. (eds.) *Cancer: Principles and Practice of Oncology*, pp. 1019–24. J. B. Lippincott Co., Philadelphia.

BRERETON H. D., SIMON R. & POMEROY T. C. (1975) Pretreatment serum lactate dehydrogenase predicting metastatic spread in Ewing's sarcoma. *Ann. Int. Med.* **83**, 352–4.

BRESLOW N. E., PALMER N. F., HILL L. R., BURING J. & D'ANGIO G. J. (1978) Wilms' tumor: Prognostic factors for patients without metastases at diagnosis. Results of the National Wilms' Tumor Study. *Cancer* **41**, 1577–89.

BROADBENT V. A., BARNES N. D. & WHEELER T. K. (1981) Medulloblastoma in childhood: long-term results of treatment. *Cancer* **48**, 26–36.

CHAPMAN R. M. (1982) Effect of cytotoxic therapy on sexuality and gonadal function. *Semin. Oncol.* **9**, 84–94.

CHESSELLS J. M. (1982) Acute lymphoblastic leukemia. *Semin. Haematol.* **19**, 155–71.

COOPER E. H. & PLESNER T. (1980) Beta-2-microglobulin review: Its relevance in clinical oncology. *Med. Pediat. Oncol.* **8**, 323–34.

CRAFT A. W., RANKIN A. & AHERNE G. W. (1981) Methotrexate absorption in children with acute lymphoblastic leukaemia. *Cancer Treat. Rep.* **65** Suppl. 1, 77–81.

D'ANGIO G. J., EVANS A. E., BRESLOW N., BECKWITH B., BISHOP H., FEIGL P., GOODWIN W., LEAPE L. L., SINKS L. C., SUTOW W., TEFFT M. & WOLFF J. (1976) The Treatment of Wilms' tumor: Results of the National Wilms' Tumor Study. *Cancer* **38**, 633–46.

DEMEOCQ F., MALPUECH G., RAYNAUD E.-J., PLAGNE R. & GAILLARD G. (1981) Elevation of beta$_2$-microglobulin in cerebrospinal fluid after irradiation of the central nervous system. *New Engl. J. Med.* **304**, 1366.

DRAPER G. J., BIRCH J. M., BITHELL J. F., KINNIER WILSON L. M., LECK I., MARSDEN H. B., MORRIS-JONES P., STILLER C. A. & SWINDELL R. (1981) *Childhood cancer in Britain: incidence, survival and mortality.* HMSO, London.

DRASH A., SHERMAN F., HARTMANN W. H. & BLIZZARD R. M. (1970) A syndrome of pseudohermaphroditism, Wilms' tumor, hypertension, and degenerative renal disease. *J. Pediatr.* **76**, 585–93.

EINHORN L. H. & WILLIAMS S. D. (1980) Chemotherapy of disseminated testicular cancer. A random prospective study. *Cancer* **46**, 1339–44.

EVANS A. E., BLORE J., HADLEY R. & TANINDI S. (1971) The Labrosse spot test: a practical aid in the diagnosis and management of children with neuroblastoma. *Pediatrics* **47**, 913–15.

EXELBY P. R., FILLER R. M. & GROSFELD J. L. (1975) Liver tumors in children in the particular reference to hepatoblastoma and hepatocellular carcinoma: American Academy of Pediatrics Surgical Section Survey—1974. *J. Pediat. Surg.* **10**, 329–37.

GANGULY A., GRIBBLE J., TUNE B., KEMPSON R. L. & LUETSCHER J. A. (1973) Renin-secreting Wilms' Tumor with severe hypertension. Report of a case and brief review of renin-secreting tumors. *Ann. Int. Med.* **79**, 835–7.

GEISER C. F. & SHIH V. E. (1980) Cystathioninuria and its origin in children with hepatoblastoma. *J. Pediatr.* **96**, 72–4.

GHOSH L. & MUEHRCKE R. C. (1970) The nephrotic syndrome: a prodrome to lymphoma. *Ann. Int. Med.* **72**, 379–82.

GINSBERG S. J. & COMIS R. L. (1982) The pulmonary toxicity of antineoplastic agents. *Semin. Oncol.* **9**, 52–64.

GITLOW S. E., BERTANI L. M., GREENWOOD S. M., WONG B. L. & DZIEDZIC S. W. (1972) Benign pheochromocytoma associated with elevated excretion of homovanillic acid. *J. Pediat.* **81**, 1112–16.

GLICKSMAN A. S., MAURER H. M. & VIETTI T. J. (1979) Overview of conference on sarcomas of soft tissue and bone in childhood. *Med. Pediatr. Oncol.* **7**, 55–67.

GLUCKMAN P. D., FERGUSON R. S., OSBORNE D. & EVANS M. (1977) Primary hyperparathyroidism in a child. Use of jugular venous catheterization in diagnosis. *Arch. Dis. Child.* **52**, 504–5.

GOLDE E. M. (1979) The Cushing Syndrome: Changing views of diagnosis and treatment. *Ann. Int. Med.* **90**, 829–44.

GOLTZMAN D., POTTS J. T. JR, RIDGWAY E. C. & MALOOF F. (1974) Calcitonin as a tumor marker. *New Engl. J. Med.* **290**, 1035–9.

GROSS-KIESELSTEIN E., NAVON P., BRANSKI A., ABRAHAMOV A. & DOLLBERG L. (1981) Familial erythrophagocytic lymphohistiocytosis in infancy. *Eur. J. Pediatr.* **136**, 223–5.

HANDE K. R., HIXSON C. V. & CHABNER B. A. (1981) Postchemotherapy purine excretion in lymphoma patients receiving allopurinol. *Cancer Res.* **41**, 2273–9.

HANN H.-W. L., EVANS A. E., COHEN I. J. & LEITMEYER J. E. (1981) Biologic differences between neuroblastoma stages IV-S and IV. Measurement of serum ferritin and E-rosette inhibition in 30 children. *New Engl. J. Med.* **305**, 425–9.

HANN H.-W. L., LEVY H. M. & EVANS A. E. (1980) Serum ferritin as a guide to therapy in neuroblastoma. *Cancer Res.* **40**, 1411–13.

HILL B. T. & WHELAN R. D. H. (1981) Assessment of the sensitivities of cultured human neuroblastoma cells to anti-tumour drugs. *Pediatr. Res.* **15**, 1117–22.

HOGAN T. F., CITRIN D., JOHNSON B. M., NAKAMURA S., DAVIS T. E. & BORDEN E. C. (1978) o,p′-DDD (Mitotane) therapy of adrenal cortical carcinoma. *Cancer* **42**, 2177–81.

HOLMES E. C., MORTON D. L. & KETCHAM A. S. (1969) Parathyroid carcinoma: A collective review. *Ann. Surg.* **169**, 631–40.

HOSPITAL IN-PATIENT ENQUIRY (1978) Office of Population Censuses & Surveys: London, HMSO, Series MB4, No. 2.

HOPWOOD J. J. & DORFMAN A. (1978) Glycosaminoglycan synthesis by Wilms' tumor. *Pediatr. Res.* **12**, 52–6.

HOWELL R. R., STEVENSON R. E., BEN-MENACHEM Y., PHYLIKY R. L. & BERRY D. H. (1976) Hepatic adenomata with Type 1 glycogen storage disease. *J. Am. Med. Assoc.* **236**, 1481–4.

JANSEN-GOEMANS A. & ENGELHARDT J. (1977) Intractable diarrhea in a boy with vasoactive intestinal peptide-producing ganglio-neuroblastoma. *Pediatrics* **59**, 710–16.

JAVADPOUR N. (1980) The role of biologic tumor markers in testicular cancer. *Cancer* **45**, 1755–61.

KAPLAN H. (1980) In Kaplan H. (ed.) *Hodgkin's Disease* (2nd Edn) Chap. 4, pp. 116–45. Harvard University Press, Cambridge, Massachusetts.

KEMSHEAD J. T., GOLDMAN A., FRITSCHY J., MALPAS J. S. & PRITCHARD J. (1983) The use of panels of monoclonal antibodies in the differential diagnosis of neuroblastoma and lymphoblastic malignancy. *Lancet* **i**, 12–14.

KNUDSEN A. G. JR (1976) Genetics and the aetiology of childhood cancer. *Pediatr. Res.* **10**, 513–17.

LABROSSE E. H., COMOY E., BOHOUN C., ZUCKER J.-M. & SCHWEISGUTH O. (1976) Catecholamine metabolism in neuroblastoma. *J. Natl. Cancer Inst.* **57**, 633–8.

LAMPERT I. A., CATOVSKY D. & BERGIER N. (1978) Malignant histiocytosis. A clinico-pathological study of 12 cases. *Br. J. Haematol.* **40**, 65–77.

LAUG W. E., SIEGEL S. E., SHAW K. N. F., LANDING B., BAPTISTE J. & GUTERNSTEIN M. (1978) Initial urinary catecholamine metabolite concentrations and prognosis in neuroblastoma. *Pediatrics* **62**, 77–82.

LENNOX E., STILLER C. A., MORRIS JONES P. H. & KINNIER-WILSON L. M. (1979) Nephroblastoma: Treatment during 1970–73 and the effect on survival of inclusion in the first M.R.C. trial. *Br. Med. J.* **2**, 567–9.

LEVIN V. A. & WILSON C. B. (1975) Chemotherapy of brain tumours: Agents in use. *Semin. Oncol.* **2**, 63–7.

LYNCH R. E., KJELLSTRAND C. M. & COCCIA P. F. (1977) Renal and metabolic complications of childhood non-Hodgkin's lymphoma. *Semin. Oncol.* **4**, 325–34.

MANN J. R., LAKIN G. E., LEONARD J. C., RAWLINSON H. A., RICHARDSON S. G. N., CORKERY J. J., CAMERON A. H. & SHAH K. J. (1978) Clinical applications of serum carcinoembryonic antigen and alpha-fetoprotein levels in children with solid tumours. *Arch. Dis. Child.* **53**, 366–74.

MARSDEN H. B. & STEWARD J. K. (1976) Tumours in Children. *Recent Results in Cancer Research* **13**, 2nd Edn. Springer-Verlag, Berlin.

MASON G. A., HART-MERCER J., MILLAR E. J., STRANG L. B. & WYNNE N. A. (1957) Adrenaline-secreting neuroblastoma in an infant. *Lancet* **ii**, 322–5.

MAVLIGIT G. M., STUCKEY S. E., CABANILLAS F. F., KEATING M. J., TOURTELLOTTE W. W., SCHOLD S. C. & FREIREICH E. J. (1980) Diagnosis of leukemia or lymphoma in the CNS by beta microglobulin determination. *New Engl. J. Med.* **303**, 718–22.

MELMON K. L. (1975) The endocrinologic manifestations of the carcinoid tumor. In Williams R. H. (ed.) *Textbook of Endocrinology*, 5th Edn, pp. 1084–104. W. B. Saunders Co., Philadelphia.

MENDELSOHN G., EGGLESTON J. C., OLSON J. L., SAID S. I. & BAYLIN S. B. (1979) Vasoactive intestinal peptide and its relationship to ganglion cell differentiation in neuroblastic tumors. *Lab. Invest.* **41**, 144–9.

MIKE V., MEADOWS A. T. & D'ANGIO G. J. (1982) Incidence of second malignant neoplasms in children: Results of an international study. *Lancet* **ii**, 1326–31.

MITCHELL J. D., BAXTER J., BLAIR-WEST J. R. & MCCREDIE D. A. (1970) Renin levels in nephroblastoma. *Arch. Dis. Child.* **45**, 376–84.

MITCHELL E. P. & SCHEIN P. S. (1982) Gastrointestinal toxicity of chemotherapeutic agents. *Semin. Oncol.* **9**, 52–64.

NADIMINTI Y., WANG J. C., CHOU S.-Y., PINELES E. & TOBIN M. S. (1980) Lactic acidosis associated with Hodgkin's disease. Response to Chemotherapy. *New* Engl. J. Med. **303**, 15–17.

NEUWELT E. A., GLASBERG M., FRENKEL E. & CLARK W. K. (1979) Malignant pineal region tumors. A clinico-pathological study. *J. Neurosurg.* **51**, 597–607.

NIEHAUS C. E., ERSSER R. S. & ATHERDEN S. M. (1979) Routine laboratory investigation of urinary catecholamine metabolite in sick children. *Ann. Clin. Biochem.* **16**, 38–42.

NORGAARD-PEDERSEN B., LINDHOLM J., ALBRECHTSEN R., ARENDS J., DIEMER N. H. & RIISHEDE J. (1978) Alpha-fetoprotein and human chorionic gonadotropin in a patient with a primary intracranial germ cell tumor. *Cancer* **41**, 2315–20.

O'REGAN S., CARSON S., CHESNEY R. W. & DRUMMOND K. N. (1977) Electrolyte and acid–base disturbances in the management of leukemia. *Blood* **49**, 345–53.

OSBAND M. E., LIPTON J. M., LAVIN P., LEVEY R., VAWTER G., GREENBERGER J. S., MCCAFFREY R. P. & PARKMAN R. (1981) Histiocytosis-X: Demonstration of abnormal immunity, T-cell histamine H2-receptor deficiency, and successful treatment with thymic extract. *New Engl. J. Med.* **304**, 146–53.

PALMER M. K. & HANN I. M. (1980) A score at diagnosis for predicting length of remission in childhood acute lymphoblastic leukaemia. *Br. J. Cancer* **42**, 841–9.

PARADINAS F. J., MELIA W. M., WILKINSON M. L., PORTMANN B., JOHNSON P. J., MURRAY-LYON I. M. & WILLIAMS R. (1982) High serum vitamin B12 binding capacity as a marker of the fibrolamellar variant of hepatocellular carcinoma. *Br. Med. J.* **285**, 840–2.

PERRY M. C. (1982) Hepatotoxicity of chemotherapeutic agents. *Semin. Oncol.* **9**, 65–74.

POMBO M., ALVEZ F., VARELA-CIVES R., IGLESIAS-DIZ J. L., MARTINEN F., CASTRO-GAGO M. & PENA J. (1982) Ectopic production of ACTH by Wilms' tumour. *Horm. Res.* **16**, 160–3.

PRITCHARD J. (1979) Histiocytosis X: Natural history and management in childhood. *Clin. Exp. Dermatol.* **4**, 421–33.

PRITCHARD J., DA CUNHA A., CORNBLEET M. A. & CARTER C. J. (1982) Alpha feto-protein (αFP) monitoring of response to adriamycin in hepatoblastoma. *J. Pediatr. Surg.* **17**, 429–30.

PRITCHARD J. & KEMSHEAD J. T. (1983) In Duncan W. (ed.) *Neuroblastoma: recent developments in diagnosis, staging and management*. Symposium, Royal College of Radiologists. Springer-Verlag, Berlin (in press).

REFETOFF S., HARRISON J., KARANFILSKI B. T., KAPLAN E. L., DE GROOT L. J. & BEKERMAN C. (1975) Continuing occurrence of thyroid carcinoma after irradiation to the neck in infancy and childhood. *New Engl. J. Med.* **292**, 171–5.

ROSEN G., CAPARROS B., NIRENBERG A., MARCOVE R. C., HUVOS A. G., KOSLOFF C., LANE J. & MURPHY M. L. (1981) Ewing's sarcoma: Ten-year experience with adjuvant chemotherapy. *Cancer* **47**, 2204–13.

ROUTLEDGE R. C., HANN I. M. & MORRIS-JONES P. H. (1976) Hodgkin's disease complicated by the nephrotic syndrome. *Cancer* **38**, 1735–40.

SATO T. L. (1965) The quantitative determination of 3-methoxy-4-hydroxyphenyl-acetic acid (homovanillic acid) in urine. *J. Lab. Clin. Med.* **66**, 517–25.

SCHEIN P. S. (1972) Chemotherapeutic management of the hormone-secreting endocrine malignancies. *Cancer* **30**, 1616–26.

SCHILSKY R. L. (1982) Renal and metabolic toxicities of cancer chemotherapy. *Sem. Oncol.* **9**, 75–83.

SELDENFELD J. & MARTON L. J. (1979) Biochemical markers of central nervous system tumors measured in cerebrospinal fluid and their potential use in diagnosis and patient management: A review. *J. Nat. Cancer Inst.* **63**, 919–31.

SHALET S. M., HANN I. M., LENDON M., MORRIS-JONES P. H. & BEARDSWELL C. G. (1981) Testicular function after combination chemotherapy in childhood for acute lymphoblastic leukaemia. *Arch. Dis. Child.* **56**, 275–8.

SHALET S. M., PRICE D. A., BEARDSWELL C. G., MORRIS-JONES P. H. & PEARSON D. (1979) Normal growth despite abnormalities of growth hormone secretion in children treated for acute leukaemia. *J. Pediat.* **94**, 719–22.

SIRIS E. S., LEVENTHAL B. G. & VAITUKAITIS J. L. (1976) Effects of leukaemia and chemotherapy on puberty an reproductive function in girls. *New Engl. J. Med.* **294**, 1143–6.

SISSON J. C., FRAGER M. S., VALK T. W., GROSS M. D., SWANSON D. P., WIELAND D. M., TOBES M. C., BEIERWALTES W. H. & THOMPSON N. W. (1981) Scintigraphic localization of pheochromocytoma. *New Engl. J. Med.* **305**, 12–17.

SPARK R. F., CONNOLLY P. B., GLUCKIN D. S., WHITE R., SACKS B. & LANDSBERG L. (1979) ACTH secretion from a functioning pheochromocytoma. *New Engl. J. Med.* **301**, 416–18.

SPIEGEL R. J. & MAGRATH I. T. (1979) Tumor lysis pancreatitis. *Med. Pediatr. Oncol.* **7**, 169–72.

STEWART D. R., MORRIS-JONES P. H. & JOLLEYS P. (1974) Carcinoma of the adrenal gland in children. *J. Pediatr. Surg.* **9**, 59–67.

STRINGEL G., EIN S. H., CREIGHTON R., DANEMAN D., HOWARD N. & FILLER R. M. (1980) Pheochromocytoma in children—an update. *J. Pediatr. Surg.* **15**, 496–500.

STUART M. J., CUASO C., MILLER M. & OSKI F. A. (1975) Syndrome of secretion of antidiuretic hormone following multiple doses of vincristine. *Blood* **45**, 315–20.

TAYLOR W. F., IVINS J. C., DAHLIN D. C., EDMONSON J. H. & PRITCHARD D. J. (1978)

Trends and variability in survival from osteosarcoma. *Mayo Clinic Proc.* **53**, 695–700.

Thorpe W. P., Reilly J. J. & Rosenberg S. A. (1979) Prognostic significance of alkaline phosphatase measurements in patients with osteogenic sarcoma receiving chemotherapy. *Cancer* **43**, 2178.

Tiedemann K., Long R., Pritchard J. & Bloom S. R. (1981a) Plasma vaso-active intestinal polypeptide and other regulatory peptides in children with neurogenic tumours. *Eur. J. Pediatr.* **137**, 147–50.

Tiedemann K., Pritchard J., Long R. & Bloom S. R. (1981b) Intractable diarrhoea in patient with vaso-active intestinal peptide-secreting neuroblastoma. *Eur. J. Pediatr.* **137**, 217–19.

Todani T., Tabuchi K., Watanabe Y. & Tsutsumi A. (1977) True hepatic teratoma with high alpha fetoprotein in serum. *J. Pediatr. Surg.* **12**, 591–2.

Toledano S. (1979) Parathyroid carcinoma in an adolescent. *Med. Pediatr. Oncol.* **7**, 95–102.

Voorhess M. L. (1971) Neuroblastoma with normal urinary catecholamine excretion. *J. Pediatr.* **78**, 680–3.

Voorhess M. L. (1980) Functioning tumors. *Am. J. Dis. Child.* **134**, 14–15.

Whitehead E., Shalet S. M., Morris-Jones P. H., Beardwell C. G. & Deakin D. P. (1982) Gonadal function after combination chemotherapy for Hodgkin's disease in childhood. *Arch. Dis. Child.* **47**, 287–91.

Wiernik P. H. & Serpick A. A. (1979) Clinical significance of serum and urinary muramidase activity in leukemia and other hematologic malignancies. *Am. J. Med.* **46**, 330–43.

Wu J. T., Book L. & Sudar K. (1981) Serum alpha fetoprotein (αFP) levels in normal infants. *Pediatr. Res.* **15**, 50–2.

Wurtman R. J. & Moskowitz M. (1977) The pineal organ. *New Engl. J. Med.* **296**, 1329–33 and 1383–6.

INDEX

Note: Page numbers in **bold** refer to pages on which illustrations/tables appear.